# Strength and Fracture of Engineering Solids

## Second Edition

**DAVID K. FELBECK**
*University of Michigan, Ann Arbor*

**ANTHONY G. ATKINS**
*University of Reading, United Kingdom*

 PRENTICE HALL, INC., Upper Saddle River, NJ 07458

**Librrary of Congress Cataloging-in-Publication Data**

Felbeck, David K.
    Strength and fracture of engineering solids / David K. Felbeck,
Anthony G. Atkins.--2nd ed.
        p.    cm.
    Includes bibliographical references and index.
    ISBN 0–13–856113–3
    1.  Strength of materials.    2. Fracture mechanics.    I. Atkins,
Anthony, 1939–    II. Title.
TA405.F418    1996
620.1'.1—dc20                                                95–34197
                                                                CIP

Acquisitions editor: **BILL STENQUIST**
Editorial/production supervision
    and interior design: **SHARYN VITRANO**
Copy editor: **ANDREA HAMMER**
Cover design: **BRUCE KENSELAAR**
Manufacturing buyer: **JULIA MEEHAN**
Editorial Assistant: **MEG WEIST**
Supplements editor: **BARBARA MURRAY**

 © 1996 by Prentice-Hall, Inc.
Simon & Schuster/A Viacom Company
Upper Saddle River, New Jersey  07458

The author and publisher of this book have used their best efforts in preparing this book.
These efforts include the development, research, and testing of the theories and programs
to determine their effectiveness. The author and publisher make no warranty of any kind,
expressed or implied, with regard to the programs or the documentation contained in this
book. The author and publisher shall not be liable in any event for incidental or conse-
quential damages in connection with, or arsing out of, the furnishing, performance, or use
of these programs.

Printed in the United States of America

10   9   8   7   6   5   4   3   2   1


ISBN 0-13-856113-3


Prentice-Hall International (UK) Limited, London
Prentice-Hall of Australia Pty. Limited, Sydney
Prentice-Hall Canada Inc., Toronto
Prentice-Hall Hispanoamericana, S.A., Mexico
Prentice-Hall of India Private Limited, New Delhi
Prentice-Hall of Japan, Inc., Tokyo
Simon & Schuster Asia Pte. Ltd., Singapore
Editora Prentice-Hall do Brasil, Ltda., Rio de Janeiro

To
Inka and Meg

# Contents

**v**

# *Preface*

Strength and fracture of solids are important topics to most practicing engineers, many technicians, some lawyers, and students preparing for these fields. Theories of strength and practical developments of stronger materials have been maturing and improving rapidly during the past few decades, to the extent that engineers and teachers not specializing in this area are often hard-pressed to keep up with the field. This book is intended to serve these groups as a self-contained presentation of information on the fundamentals of strength, with plentiful data on real materials, presented in the simplest possible way.

This book is written for the second- or third-year college or technical school student who has had a good grounding in chemistry, physics, and a basic course in materials science. We review here material from each of these three subject areas, particularly materials science. Because the topic of strength and fracture is much narrower than materials science, we include much new information beyond the usual materials science coverage. The topics covered here and the method of presentation have been developed through use in third-year courses at the University of Michigan and the University of Reading, through notes, a prior short text, and laboratory experiments.

Three subjects are incorporated here that are new and developing subjects not usually found in fundamental texts: (1) a detailed presentation of the fundamentals and applications of high-performance composite materials (Chapter 17); (2) discussion of brittle fracture from the perspective of both the classic Griffith–Orowan–Irwin analysis and the more general fracture toughness approach of Gurney (Chapter 14); and (3) presentation of the fundamentals of failure analysis (Chapter 19), which is emerging as a separate and growing field of specialization as a result of stricter product and worker safety laws. We have had some experience in these areas and may succeed in providing new perspectives to them.

The book is divided into five general areas: (1) a kind of preamble, or raison d'être, Chapters 1–3; (2) fundamentals of strength of ductile metals, Chapters 4–11; (3) polymers and glasses, Chapters 12–13; (4) fracture of solids, Chapters 14–17; and (5) fatigue and failure analysis, Chapters 18–19. Beginning with Chapter 4, after the introductory material, we have tried to develop the subject from the simple to the complex, with a minimum of reference to subsequent material. We have in most cases selected materials for discussion as examples of certain classes of properties and have made no attempt to provide a comprehensive description of any class of engineering material. We do believe strongly in the value of repetition in learning, so the reader should not be annoyed when a term or symbol is defined a second or third time, perhaps with slightly different words. We believe this practice is better than forcing the reader to search back for the original reference to the term or symbol.

We have incorporated exercises and solutions into the body of the text in order to allow the reader to check on progress and to illustrate particular points or paradoxes. The reader should work out all these exercises as they are presented, for they form an integral part of the presentation.

Many of the problems at the ends of the chapters have been selected from the large collection used in our third-year course. We are indebted to our close colleagues Robert M. Caddell, Julian R. Frederick, and Kenneth C. Ludema for many of these problems and for their other substantial contributions over the years in developing this subject with us. We have been influenced in our approach, consciously and subconsciously, by the work of many other authors, as our list of references shows. We are grateful for their guidance and to the many authors who permitted diagrams and photographs to be reproduced. For numerous original photographs and test results, we are indebted to W. H. Durrant. Thanks are owed to Laurie Power of the Robert Gordon University, and to Bradley Dodd, George Jeronimidis, Richard Chaplin, and other colleagues at the University of Reading for help and advice. Mrs. J. Giacchetto prepared much of the manuscript for publication.

For the second edition, we have first attempted to eliminate, alter, or correct errors, dated statements, and impossible problems (such as the infamous Exercise 12–6) that found their way into the first edition. We have also reduced or deleted a few very nice esoteric subjects that, after teaching from this text for a decade, we were finally convinced should be eliminated as being too advanced. We have completely rewritten and simplified the procedures for solving Mohr's circle problems. We have added a discussion of the field emission scanning electron microscope (FEM), which has come into recent use, and have included some FEM micrographs of possible interest. We have added a section on $K_{Ic}$ testing using the ASTM E399 standard and have included a description of the procedure for evaluating the Weibull modulus. We have added some new problems. We have clarified several figures that were not well done at the outset. We have decided to incorporate a simpler version of specific strength = strength/specific gravity, for we believe it makes more sense. We have incorporated recent advances in the areas of composites and ceramics, as well as developments in high-strength low-alloy steels. Lastly, we have added to Chapters 18 and 19 a number of electron micrographs from our own failure analyses that depict special characteristics of the several fracture modes.

David K. Felbeck
*Ann Arbor, Michigan*

Anthony G. Atkins
*Reading, England*

# 1

# Engineering Design with Materials and the International System of Units

The engineering designer requires the device being created to meet certain design criteria and be manufactured at the lowest possible cost. Therefore, those materials that will best meet these requirements must be selected. That so many different materials are in common use is the result of their diversity of mechanical and physical properties, their differences in cost and geographical distribution, and the propensity of a society to continue to use materials with which it has had experience. This chapter considers in a very general way the procedure followed by the designer in selecting materials for engineering applications.

In its simplest form, the task the designer faces is shown in Fig. 1-1, which represents a portion of an engineering machine or structure. A force $P$ must be transmitted from point $B$ to point $C$, which are located a distance $L$ apart. The part to be designed, shown with dashed lines, must be no higher or thicker than $H$ so that it will not interfere with other parts. The distance between $B$ and $C$ must not increase when under load by more than a length $\delta L_{max}$; *failure* occurs in this case when, through stretching or fracture, $\delta L > \delta L_{max}$. (Other limitations may be imposed as well, such as minimum weight or resistance to a corrosive environment or to repeated loadings, but these are omitted here for simplicity.) The finished product must also be as inexpensive as possible.

The designer's approach to this problem is essentially iterative. First, a material is selected that is low in cost and easily fabricated. A part that will just fit in the space provided can then be made and tested to determine whether it supports the specified force. If it is not strong enough, the designer then selects a stronger material and repeats the process. But testing is very expensive, so the designer prefers to make use of a handbook in which earlier test results are reported and tabulated. Although he or she is interested only in finding information regarding a possible part that will resist the applied force $P$, it is unlikely that test results will be found for a load of magnitude $P$ on the precise material selected, for a part of exactly the required size. (For more complex design problems, the likelihood of finding the exact test results desired becomes virtually zero.)

**Figure 1-1**    The essential design problem.

## STRESS

To overcome this problem of various test specimen sizes, the concept of stress was developed. It is obvious that a larger specimen cross section can support a larger load before fracture, and a little experimentation demonstrates that the maximum load a specimen can support is directly proportional to its cross-sectional area (except for very large or very small areas, which will be discussed in later chapters). Thus, for a *failure load* $P_F$ and initial cross-sectional area $A_0$,

$$P_F = \text{constant} \times A_0 \tag{1-1}$$

The constant is called *failure stress*, given here the symbol $S_F$, and can thus be written

$$S_F = \frac{P_F}{A_0} \tag{1-2}$$

The units of stress are thus force/area, either newtons/square meter ($N/m^2$), also called pascals (Pa), or pounds/square inch ($lb/in.^2$ or psi), or kilopounds/square inch (1000 psi = 1 ksi). (See the discussion of units later in this chapter.)

For the discussion of design procedure in this chapter, the general expression "failure" is used and denoted by the subscript F. However, several explicit terms that will be used in this book should also be defined. Failure load $P_F$ described before may or may not describe the conditions at the instant of fracture. Specifically, the *fracture strength* $S_f = P_f/A_0$, where $P_f$ is the load at fracture. The maximum load may occur at or prior to fracture, and the stress at maximum load is called the *ultimate tensile strength,* $S_u = P_{max}/A_0$. (Note that when $P_f = P_{max}$, $S_f = S_u$.) *Yielding* is the onset of substantial permanent deformation, so yield strength $S_y = P_y/A_0$, where $P_y$ is the load at yielding. Failure by fracture or by excessive elongation may occur at any stage.

The relationships among the various definitions of "failure" may be summarized as follows; failure to perform the desired function occurs at the lowest value of the failure stress $S_F$ that is determined when one of the following occurs:

1. Maximum allowable deflection $\delta L_{max}$ is exceeded. This may occur at either low or high values of $S_F$, depending on the magnitude assigned to $\delta L_{max}$.
2. Discernible permanent (*plastic*) deformation occurs. For many applications, permanent deformation constitutes failure of a part, even though $\delta L < \delta L_{max}$. Then *yield strength* $S_y = P_y/A_0 = S_F$.
3. Fracture occurs. Then $S_F = S_u = P_{max}/A_0$. This can occur either
   (a) at maximum load, in which case $S_F = S_u = S_f$, or
   (b) after maximum load is reached and the load drops before fracture, in which case $S_F = S_u > S_f$.

Now the designer needs to know only the numerical value of $S_F$ for the selected material to determine whether the material is adequate. (A number of factors are involved in selecting $S_F$; these will be discussed in detail as the subjects arise.) If the maximum cross-sectional area of the part in Fig. 1-1 is $H^2$, then the required value of $S_F$ must be at least $P/H^2$ or the part will fail.

## FACTOR OF SAFETY

In order to allow for the possibility of some differences in the strength of the material used, variations in dimensions resulting from manufacturing variables, and service loads that may exceed the design load, a *factor of safety* is employed. Defined as the ratio of the failure stress (the smallest $S_F$ as defined before) to the maximum calculated service stress, the factor of safety takes into account such variables as the precision of the estimate of loads and the calculation of maximum stress and whether human life would be jeopardized by a failure. Thus, the maximum calculated service stress $s_{max}$ is less than the failure stress $S_F$ of the material by a factor equal to the factor of safety (f.s.).

$$s_{max} = \frac{S_F}{\text{f.s.}} \tag{1-3}$$

In Fig. 1-1, the maximum service stress must be $s_{max}$ when the load is $P$, so

$$s_{max} = \frac{P}{A_0} \tag{1-4}$$

Then combining Eqs. (1-3) and (1-4) gives the required area $A_0$:

$$A_0 = \frac{P}{S_F}(\text{f.s.}) \tag{1-5}$$

In Eq. (1-5), $A_0$ is the design size to be calculated, $P$ is the load to be supported by the device, f.s. is the factor of safety, and $S_F$ is the failure stress, a strength-related property of the material to be used. If the $A_0$ required is too large to satisfy the dimensional constraint $H$, then another material must be selected. If the first material selected already has the largest available value of $S_F$, then either the factor of safety must be shaved down (a risky procedure) or the original design must be changed so as to alter the strength and area requirements; otherwise, the device cannot be built.

## STRAIN

The second requirement for the design is that distance $BC$ in Fig. 1-1 not increase by more than a distance $\delta L_{max}$ when load $P$ is applied. In a manner similar to that used in defining stress, the elongation under tensile load is found to be related to the length under stress. The nondimensional term *strain (e)* is thus defined as

$$e = \frac{\delta L}{L_0} \tag{1-6}$$

where $L_0$ is the unloaded length $BC$ in Fig. 1-1.

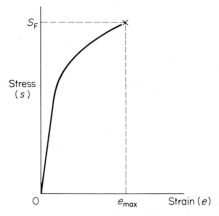

**Figure 1-2** Schematic stress–strain relationship for a moderately ductile solid.

After preliminary selection of the material with failure strength $S_F$ based on $S_y$ or $S_u$, calculations must be made to determine whether the part will extend more than the maximum allowable distance $\delta L_{max}$. The value of strain $e$ for each value of stress $s$ will then be available, for example as in Fig. 1-2, and the value of strain $e_{max}$ corresponding to the failure strength $S_F$ can be established. (Note that this strain is not necessarily the failure strain for the part in Fig. 1-1.) Then, from Eq. (1-6), the $\delta L$ corresponding to $S_F$ can be calculated:

$$\delta L = e_{max} L_0$$

If this $\delta L$ exceeds the allowable $\delta L_{max}$, then the material and/or design must be changed. For example, if the cross-sectional area can be increased without exceeding the limiting dimension $H$ in Fig. 1-1, the stress can be reduced and thus the maximum strain will be reduced.

## MANUFACTURING AND ECONOMIC CONSIDERATIONS

As an integral part of the material selection process, the designer must determine the processes required to manufacture the part. The total cost of the part, including the cost and availability of the raw material or mill product (e.g., rolled plate, extruded beams), and the cost of manufacturing the final part must be estimated. At this stage, the decision-making process becomes quite complicated in considering interactions among material choices, cost of material, and the number of parts to be made by the manufacturing method selected. There are seldom any easy answers, but the extremes of choices in a simple case can illustrate the point.

The design of a single novel automobile carburetor for a research project might specify that the complex shape of this part might best be made by machining it from a small number of solid pieces of free-machining brass, to be brazed together prior to final machining. Such a manufacturing process requires much labor and is therefore expensive. On the other hand, the manufacturer of a million carburetors per year currently makes them of aluminum–zinc alloys cast directly and automatically from molten metal under pressure in a metal mold called a die. The mold can be very expensive, but this cost is spread among all the parts, so the unit cost is much less than for the brass machined part. Improvements in the properties of polymers (plastics) allow the manufacturer an even better choice because polymers melt at a relatively low temperature and thus the dies are cheaper.

## ADDITIONAL MATERIAL PROPERTIES

The general requirements for strength, size, and deformation discussed earlier can result in severe constraints on the range of material selection available to the designer. The situation is made much more difficult by the existence of several different modes of failure of materials, any one of which can lead to the smallest failure stress:

1. Failure from the excessive deformation associated with the onset of permanent (plastic) deformation, called *yielding*
2. Failure from fracture that follows after substantial plastic deformation (*ductile fracture*)
3. Failure from fracture that occurs with little or no plastic deformation (*brittle fracture*) and hence usually without warning
4. Failure from fracture that results from a large number of repeated stress cycles (*fatigue*) at stresses below the stress for yielding
5. Excessive deformation and/or fracture under conditions of elevated temperature (*creep*) at stresses below the stress for yielding
6. Rate-dependent deformation (*viscoelasticity*), as in polymers, where load-elongation behavior depends on speed
7. Crack growth and fracture that occur with time under tensile stresses in certain environments (*hydrogen embrittlement* or *stress-corrosion cracking*)
8. Combinations of the preceding, of which the most conspicuous are fatigue at high temperature and fatigue associated with stress-corrosion environments.

## STRESS CONCENTRATION

The stress considered so far is the average stress over an area, such as in Fig. 1-1. But because a notch, hole, defect, or sudden change in cross section can create a region of locally high stress, the designer must take care to minimize the magnitude of these local stresses. For example, if the part in Fig. 1-1 is a flat plate containing a central hole with a diameter one-tenth the width of the part, it turns out that if no plastic deformation occurs, the maximum stress will be 3.03 times the average stress in the part without the hole; thus, *the elastic stress-concentration factor* is 3.03. The effective stress-concentration factor under fatigue loading conditions will be smaller than the factor under elastic (no plastic deformation), single-tension loading.

Thus, the designer must cope with the additional complexity of calculating the stress concentrations that would result from the use of each design configuration being considered and for the material and manufacturing method used. Because all these variables can influence the stress concentration (e.g., because of differences in gross dimensions, surface finish, inside radii), the maximum stress may occur at different locations in the part for different materials and processes. Before the design can be considered complete, the designer must therefore calculate the maximum stress at every location where a maximum might occur.

## FRACTURE TOUGHNESS

Brittle fracture, where a part fractures with little or no prior plastic deformation, was mentioned earlier as one of the possible failure modes. In many applications, failure in an otherwise strong structure has occurred by brittle fracture as the result of an unanticipated flaw or notch in a material whose resistance under stress to propagation of a flaw (called *fracture toughness)* is too low. The lower the fracture toughness, the smaller the size of a flaw that can lead to failure under ordinary service loads. All practical structures contain flaws or stress-concentrating notches of some size, so fracture toughness is an important criterion for material selection.

Failure of the designer to recognize the importance of fracture toughness of materials and to control the design accordingly has resulted in failures in spacecraft (the *Apollo 13* moon mission experienced fracture of titanium alloy spheres containing liquid oxygen), aircraft (the first commercial jet, *Comet 1*, suffered complete fracture of three aluminum alloy fuselages in service), merchant ships (60 merchant ships with steel hulls have broken completely in two during the past 60 years), several thousand kilometers of oil and gas pipelines, and many oil storage tanks, bridges, and lesser structures.

In the balance of this book, the influence of the mechanical behavior of materials on design and manufacturing will be considered in detail, with the objective of providing the prospective engineering designer with some of the information needed to design safe structures and components that can be reliably manufactured and built.

## THE INTERNATIONAL SYSTEM OF UNITS

The International System of Units, usually called SI (for Système International), is used throughout this book. This system is a modification of the metric system (cgs, MKS), which has been simplified through the elimination of many superfluous units. Of particular interest for the subject area of this book are the following changes: (1) The unit of force, 1 newton (N), is derived as the force required to accelerate a mass of 1 kilogram (kg) 1 meter per second per second (m/s$^2$), where kilogram, meter, and second are base units; (2) a system of prefixes is used to indicate the magnitude of a unit, with each larger prefix increasing by a factor of 1000; and (3) several commonly used combinations of units have been assigned the names of famous people associated with the early use of the unit. The newton has been mentioned; the unit of pressure, or stress, N/m$^2$, is the pascal (Pa).

The six basic units are meter (m) for length, kilogram (kg) for mass, second (s) for time, ampere (A) for electric current, kelvin (K) for thermodynamic temperature, and candela (cd) for luminous intensity. In addition, there are two supplementary units: radian (rad) for plane angle, and steradian (sr) for solid angle. From the first six units all other units are derived.

Some of the derived units pertaining to the materials area are density (kg/m$^3$); energy, work, and quantity of heat (N-m, called the joule, J); force (kg-m/s$^2$, called the newton, N); frequency (cycle/s, called the hertz, Hz); power (J/s, the watt, W); pressure and stress (N/m$^2$, the pascal, Pa); dynamic viscosity (N-s/m$^2$, which is Pa-s); degree Celsius (°C = K − 273.15), which is the common temperature unit used outside thermodynamics and work conducted at very low temperatures.

The prefixes used to indicate multiples of 10$^3$ are given in App. 1. Note that the prefixes for all positive exponents are capitalized, except k, and the prefixes for all negative exponents are lowercase.

Several rules that apply to SI style and usage in written form serve to keep units as simple and clear as possible. Prefixes are used in their simplest forms; for example, use GW (gigawatt), not kMW. Prefixes are used only in the numerator (except for kg, see what follows). For example, use MN/m$^2$ (MPa) for stress instead of N/mm$^2$, and *never* use kg/mm$^2$. The use of centimeter and all other units with prefixes not included in App. 1 is to be avoided, although some residual use of these units will, no doubt, survive.

Because the kilogram is a base unit of SI, its use in the denominator is not a violation and it should be used in place of the gram (g); for example, use MJ/kg, not kJ/g.

Because prefixes are also raised to the power of the unit, mm$^3$ is 10$^{-9}$ m$^3$, not 10$^{-3}$ m$^3$. The convenience of having use of a different prefix for every step by a factor of 10$^3$ is thus lost when exponents greater than 1 are involved. The next unit smaller than m$^2$ is mm$^2$, which is 10$^{-6}$ smaller, and for volume the situation is even worse, with a step of 10$^9$ from m$^3$ to mm$^3$. One solution to this is to use the liter (L), where 1 L is (100 mm)$^3$ = 10$^{-3}$ m$^3$. Then 1 kL = 1 m$^3$ and 1 $\mu$L = 1 mm$^3$. The symbol L for liter has been adopted by the Society of Automotive Engineers (SAE) and avoids the confusion that arises from the lower-case L, which can easily be confused with the number 1. (For area, there should be a separate name for an area of 10$^{-3}$ m$^2$, which if in the shape of a square is 31.62 mm on a side, so we could then apply all the prefixes to this unit in the same way as with the liter.)

Symbols for SI units are usually not capitalized unless the unit is derived from a proper name. One common symbol, m, can mean either "milli" or "meter," except that when two or more units are written together they should be separated by a raised dot or a hyphen. Thus, ms means 10$^{-3}$ second, while m-s means meter-second.

Because of opposite use of decimal points (periods) and commas in different countries, the comma is no longer recommended to be used with numbers under the SI rules. Numbers are simply separated by a space between every group of three on either side of the decimal point, except when a number has only four digits to the left of the decimal point; common practice usually runs these four digits together without a space. The requirement for using spaces instead of commas is probably the one SI rule that is thoroughly ignored throughout the world; in practice, numbers with five or more digits to the left of the decimal point continue to be written with a comma every three spaces to the left of the decimal point. In most engineering applications, this requirement seldom arises, as units greater than 1000 are usually written using the next larger prefix with the unit.

SI symbols are always written in singular form. For example, 100 newtons is 100 N. Also, one space is left between the number and the first symbol: 200 GPa, not 200GPa. Periods should not be used after SI unit symbols except at the end of a sentence.

The SI nomenclature is particularly well suited to use with a hand calculator or computer having exponential notation. With the exception of kg, each prefix in a calculation simply represents its appropriate exponent and is so processed. The exponent in the units of the answer is thus read directly upon completion of the calculation because all conversions of units within the SI are made by multiplying by 1! For example, consider the calculation of the plastic surface energy in Orowan's modification of the Griffith equation as the explanation for the brittle fracture of steel (see Chap. 14). The relation is

$$\sigma \approx \left( \frac{ER}{\pi a} \right)^{1/2}$$

where $\sigma$ is the fracture stress, say, 180 MPa, $E$ is the elastic modulus of steel, 200 GPa, $R$ is the fracture toughness to be calculated, and $a$ is half the crack length in the steel, say, 16/2 mm. The calculation of $R$ is thus

$$R = \frac{\sigma^2 \pi a}{E} = \frac{(180\,\text{MPa})^2(16\,\text{mm}/2)\pi}{200\,\text{GPa}} = 4.1 \ \text{kJ}/\text{m}^2$$

The evaluation involves squaring $(180 \times 10^6)$, then multiplying by $16 \times 10^{-3}$ and $\pi$, then dividing by 2, and finally dividing by $(200 \times 10^9)$. The units of the answer can be determined by writing out in full (if necessary) all units that are abbreviated by names, such as the pascal. The units are thus $(\text{N}^2/\text{m}^4)(\text{m})/(\text{N}/\text{m}^2) = \text{N}/\text{m}$, which is usually written as $\text{N-m}/\text{m}^2 = \text{J}/\text{m}^2$. The exponent of the answer, in this case, +3, dictates the prefix of the answer, k.

Treatment of the kilogram in problems involving mass and force requires a $10^3$ shift because the kilogram is a base unit and $\text{N} = \text{kg-m}/\text{s}^2$. The procedure is to reduce all prefixes for grams by one step of $10^3$, to units of kilograms. For example, the force $F$ required to accelerate a mass of 4 Gg by 7 m/s$^2$ is

$$F = (4\ \text{Gg})(7\ \text{m/s}^2)$$
$$= (4 \times 10^6\ \text{kg})(7\ \text{m/s}^2) = 28 \times 10^6\ \text{kg-m/s}^2$$
$$= 28\ \text{MN}$$

Throughout this book, much graphical data are given in both SI units and the common form of the English system of units. If measurements are taken in SI, there is no reason to convert to or from any other system of units. But because we recognize that we will communicate in two or three systems of units for many years to come, some of the more common conversions are listed in App. 2. All problems in this book are given in SI units, so the only conversion factor that the reader is forced to remember is the number 1.

## REFERENCES

Several references that cover substantial portions of the background and scope of this book are listed here, as well as at the ends of many chapters. The *Metals Handbook* series of the American Society for Metals (ASM) is surely the principal source in the United States for data and practice. New volumes and new editions of the *Metals Handbook* are produced on an ongoing basis, so the reader should also consult those volumes published after the publication of this book.

1. American Society for Metals, *Metals Handbook,* 8th ed. Metals Park, Ohio: American Society for Metals:
   Vol. 1, *Properties and Selection of Metals,* 1961.
   Vol. 2, *Heat Treating, Cleaning, and Finishing,* 1964.
   Vol. 7, *Atlas of Microstructures of Industrial Alloys,* 1972.
   Vol. 9, *Fractography and Atlas of Fractography,* 1974.
   Vol. 10, *Failure Analysis and Prevention,* 1975.

2. American Society for Metals, *Metals Handbook,* 9th ed. Metals Park, Ohio: American Society for Metals:
   Vol. 1, *Properties and Selection: Irons and Steels,* 1978.
   Vol. 2, *Properties and Selection: Nonferrous Alloys and Pure Metals,* 1979.
   Vol. 3, *Properties and Selection: Stainless Steels, Tool Materials and Special Purpose Metals,* 1980.
   Vol. 8, *Mechanical Testing,* 1985.
   Vol. 9, *Metallurgy and Microstructures*, 1985.
   Vol. 11, *Failure Analysis and Prevention,* 1986.
   Vol. 12, *Fractography*, 1987.
   Vol. 13, *Corrosion*, 1987.

3. Alexander, J. M., and R. C. Brewer, *Manufacturing Properties of Materials.* London: Van Nostrand, 1963.

4. Atkins, A. G., and Y.-W. Mai, *Elastic and Plastic Fracture*, 2nd ed. Chichester, U.K.: Ellis Horwood; New York: Simon & Schuster, 1988.

5. Bannantine, Julie A., Jess J. Comer, and James L. Handrock, *Fundamentals of Metal Fatigue Analysis.* Englewood Cliffs, N.J.: Prentice Hall, 1990.

6. Caddell, Robert M., *Deformation and Fracture of Solids.* Englewood Cliffs, N.J.: Prentice Hall, 1980.

7. Cottrell, A. H., *The Mechanical Properties of Matter.* New York: John Wiley, 1964.

8. Dowling, Norman E., *Mechanical Behavior of Materials.* Englewood Cliffs, N. J.: Prentice Hall, 1993.

9. Gordon, J. E., *The New Science of Strong Materials—Or Why You Don't Fall Through the Floor.* London: Penguin, 1968.

10. Gordon, J. E., *Structures—Or Why Things Don't Fall Down.* London: Penguin, 1978.

11. Hertzberg, Richard W., *Deformation and Fracture Mechanics of Engineering Materials.* New York: John Wiley, 1976.

12. Kelly, A., *Strong Solids.* Oxford: Clarendon Press, 1964.

13. Knott, J. F., *Fundamentals of Fracture Mechanics.* London: Butterworths, 1973.

14. LeMay, Iain, *Principles of Mechanical Metallurgy.* New York: Elsevier-North Holland, 1981.

15. McClintock, F. A., and A. S. Argon, eds., *Mechanical Behavior of Materials.* Reading, Mass.: Addison-Wesley, 1966.

16 Rolfe, Stanley T., and John M. Barsom, *Fracture and Fatigue Control in Structures: Applications of Fracture Mechanics.* Englewood Cliffs, N.J.: Prentice Hall, 1977.

17. Rollason, E. C., *Metallurgy for Engineers.* London: Edward Arnold, 1956.

18. Ruoff, Arthur L., *Introduction to Materials Science.* Englewood Cliffs, N.J.: Prentice Hall, 1972.

19. Smith, Charles O., *The Science of Engineering Materials.* Englewood Cliffs, N.J.: Prentice Hall, 1977.

20. VanVlack, Lawrence H., *Elements of Materials Science and Engineering*, 4th ed. Reading, Mass.: Addison-Wesley, 1980.

# 2

# *Traditional Strength Tests and Mechanical Processing*

As preparation for a discussion of the mechanisms that influence the strength of solids, this chapter defines strength and some of the common related mechanical properties. Although metals are often used as examples, the definitions and descriptions of terms apply equally to nonmetallic solids, such as polymers and ceramics. Specific differences in the behavior of polymers and ceramics will be brought out in later chapters. Furthermore, because differences among the standard tests used for measurement of static mechanical properties seldom alter the comparative ranking of different solids, only a few of the many tests are mentioned here. A separate list of references on standard tests is given at the end of this chapter. We show that mechanical working is one often-used method for improving the strength of solids. Some illustrations of the technique, with the increases in strength that may be obtained, are given toward the end of the chapter.

## GENERAL FEATURES OF THE TENSION TEST

When selecting a solid for a structural application, the design engineer's chief concern is that the final part fabricated from this material will be able to support the loads it will experience during its service life. Although these loads may be in the form of compression, tension, torsion, bending, internal or external pressure, or a combination of any of these, the *tension test* (also known as the *tensile test*) usually provides a good measure of the relative ability of materials to resist statically applied stresses at moderate temperatures. Thus, tension-test data are particularly useful to the designer.

The results of tensile tests on low-carbon steel and brass given here show most of the forms of behavior commonly observed, and we will see how the strength-related properties of principal interest have evolved.

A tensile test can be made on a simple cylindrical or flat specimen such as that shown in Fig. 2-1. The portion of the specimen being tested lies within the gage length, which is the distance $L$ between two marks on the specimen. The larger-diameter portions at the ends fit the grips of the *testing machine,* which applies opposing tensile loads $P$. Simultaneous measurement of $P$ and $L$ can be made a number of times during

**Figure 2-1** Simple tensile-test specimen.

a tensile test to produce a plot of load vs. extension, that is, $P$ vs. $\Delta L$ to fracture. Here $\Delta L = L - L_0$, the change in length, where $L$ is the instantaneous length and $L_0$ is the initial length.

   A plot of $P$ vs. $\Delta L$ for a 0.20% carbon steel is shown in Fig. 2-2, together with a plot for a 70% Cu–30% Zn brass. For both metals most of the path $oa$ is a straight line and exhibits *elastic* behavior; Hooke observed that in this region deformation is proportional to load. The curve from $a$ to $b$ shows *yielding* in mild steel, where the behavior departs suddenly from the linearity of the elastic region to give a large deformation with little or no increase in load. Mild steel is *not* typical of most metals in its form of yielding, as seen in comparison with brass, where the horizontal region $ab$ is absent. There a smooth transition occurs between the elastic region, which ends below $a$, and the region of *uniform strain hardening, bu*, where $u$ is the *ultimate* point of maximum load. In steel, $ab$ appears as a discontinuity in the curve up to $u$. Between $b$ and $u$, the specimen elongates and reduces in diameter uniformly, but beyond $u$, *nonuniform strain hardening* occurs. The deformation is concentrated in a neck, ending with fracture at $f$. Note that for most ductile metals, the load at fracture $P_f$ is less than the maximum load $P_u$. Figure 2-3 shows a fractured specimen of mild steel. If the broken parts are fitted back together, it will be seen that the specimen has been *permanently elongated* and *reduced in diameter*. *Plasticity* is the name given to the regions *buf* in Fig. 2-2 that produce this permanent deformation.

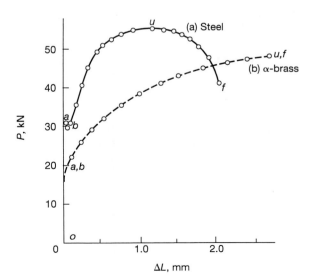

**Figure 2-2** Tensile load $P$ vs. deformation $\Delta L$ for (a) 0.20% carbon steel and (b) 70% Cu–30% Zn brass of initial cross sections 135 mm² and gage lengths 50 mm.

**Figure 2-3**  Fractured tensile specimen of 0.20% carbon steel of initial diameter 13 mm.
(Photograph courtesy of W. H. Durrant.)

## NOMINAL STRESS AND STRAIN

The data given in Fig. 2-2 apply only to a specimen of the dimensions given. What would be the result of doubling $L$ or doubling the cross-sectional area, and so on? Clearly, a multitude of plots for $P$ vs. $\Delta L$ are possible *for the same material,* depending on the size of the test piece.

Because the designer wants to be able to apply the results of such tests to parts ranging widely in size, the test results should be presented in more general form. Furthermore, precise definition of terms that are less ambiguous and more generally useful than "load" and "elongation" make possible a reasonably rigorous description of the conditions for deformation and fracture in solids.

Figure 2-4 shows the test length of the tensile specimens in Fig. 2-1 showing the cross-sectional area $A$. The *nominal* or *engineering stress s* is defined as

$$s \equiv \frac{P}{A_0} \tag{2-1}$$

where the subscript 0 refers to conditions before the load is applied.

The *nominal strain e* is defined as

$$e \equiv \frac{L - L_0}{L_0} \tag{2-2}$$

For any given specimen the values of $A_0$ and $L_0$ are constant, so a plot of $s$ vs. $e$ has exactly the same *shape* as the curve for $P$ vs. $\Delta L$ in Fig. 2-2. However, *all* separate curves for $P$ vs. $\Delta L$ would reduce to *one* plot for $s$ vs. $e$. For example, if the area $A_0$ of a specimen

**Figure 2-4**  The gage section of a circular tensile-test piece.

is doubled, the load $P$ at any time during the test would be doubled, but the nominal stress $s$ will be unchanged.

Some traditional definitions of mechanical properties that have arisen from the tension test are as follows:

*Young's modulus* is the *slope* of the initial reversible straight-line portion *oa* of the plot for *s* vs. *e*.

The *yield strength* is the load at yield divided by the initial area. For annealed mild steel, the yield load is given by $P$ at *a* or *b*. For materials whose $P$–$\Delta L$ plot smoothly curves over, an arbitrary definition of yield load has to be used. One method, when the elastic region is clearly defined, is to use the load at the point of departure from linearity. Another method, when the transition from elasticity to plasticity is imperceptible, is to define an *offset* or *proof yield stress* given by the intersection of an arbitrary line parallel to *oa* with the curve for *s* vs. *e*.

The *tensile strength,* or *ultimate tensile strength,* * is the nominal stress at maximum load, that is,

$$S_u = \frac{P_{max}}{A_0} \tag{2-3}$$

$P_{max}$ is the maximum load during the entire tensile test, irrespective of the strain at which the load reaches a maximum. In Fig. 2-2 for ductile steel, for example, the load reaches a maximum, with $dP/dL = 0$, and then drops to a lower load at fracture. If fracture had occurred between *a* and *u*, the material would be considered to have moderate to poor ductility (see later); if between 0 and *a*, the material would be very brittle (see later). Note that any fracture prior to *u* in Fig. 2-2 would not occur in a zero-slope region (excepting the special case during yielding in the region *ab*). Because the definition of the *ultimate* point of Eq. (2-3) does not account for these important differences in the maximum load, we designate the ultimate that occurs when $dP/dL = 0$ (and $ds/de = 0$) as the *zero-slope ultimate*.

The *percent elongation* is the value of *e* at fracture (converted from a decimal).

The *reduction in area to fracture* (often merely called the *percentage reduction in area* in the tensile test) is given by $(A_0 - A_f)/A_0$, where $A_f$ is the cross-sectional area of the specimen at fracture.

The latter two quantities give some indication of the *ductility* of the solid, that is, of how much deformation can take place before fracture.

For reasons that are explained later, the use of some of the foregoing definitions is to be discouraged as they can lead to confusion. Nevertheless, they are quantities that are often quoted in handbooks of materials properties.

### Example 2-1

A tensile specimen of gage length 50 mm and initial cross-sectional area 10 mm² yields when the load is 4.2 kN, at which instant it has stretched to 50.1 mm. Find the yield stress and Young's modulus.

---

*We use here the capital letter $S$ to indicate that strength is a *property* of the material, in accordance with conventional design notation. The lower case *s* will continue to be used for any value of nominal stress.

**Solution**   $S_y = (4.2 \times 10^3)/(10 \times 10^{-6}) = 420$ MPa
$e_y = (50.1-50)/50 = 0.002$
$E = S_y/e_y = (420 \times 10^6)/(2 \times 10^{-3}) = 210$ GPa

**Example 2-2**

The specimen in Example 2-1 withstood a maximum load of 9 kN when the length was 65 mm and the cross-sectional area 7.7 mm$^2$. It subsequently fractured at a load of 6.5 kN when the length was 90 mm and the cross-sectional area 5 mm$^2$. Find the ultimate tensile strength, the percent elongation at fracture, and the percent reduction in area to fracture.

**Solution**   The ultimate tensile strength $S_u = (9 \times 10^3)/(10 \times 10^{-6}) = 900$ MPa; the percent elongation at fracture is $(90 - 50)/50 = 80\%$; the percent reduction in area to fracture is $(10 - 5)/10 = 50\%$.

## LOADING AND UNLOADING AND WORK HARDENING

The tension tests described so far have taken place under continuous loading, that is, continuous pulling until fracture occurs. What happens if we unload at some stage and reload later? Unloading during the initial elastic loading (*oa* in Fig. 2-2) simply restores the specimen to its initial condition, like unloading a spring. There is no permanent deformation because the yield point was not reached, and there is no evidence at all that the test piece had been pulled. This lack of permanent deformation is the definitive feature of elasticity. However, specimens that are unloaded after the yield stress has been exceeded do display some permanent stretch. If the load and shrinking length are measured during unloading, we obtain curves such as those in Fig. 2-5, where three specimens are represented being loaded (1) from *o* to *g* and back to *h*, (2) *o* to *j* to *k*, and (3) *o* to *m* to *n*. The unloading lines *gh, jk,* and *mn* are all essentially parallel to the initial elastic portion *oa*, even when unloading after passing the zero-slope ultimate load. The extension in the samples just before unloading is given by the values of $\Delta L$ at *q, r,* and *t,* and after unloading *oh, ok,* and *on* represent the permanent (plastic) extension remaining in the bars. The (small) values of *hq, kr,* and *nt* are called the elastic recovery of the bars resulting from unloading. Clearly all these $\Delta L$ values may be converted to permanent strains and elastic recovery strains by dividing by $L_0$. Note that extensions up to *op* in magnitude are possible before yielding takes place, and typically these represent the working deflections (and strains) of elastically deforming bridges, structures, machine parts, and so on.

What happens if a test piece that has been loaded beyond yield and unloaded is subsequently reloaded? Consider a test piece following loading along path *oagh* in Fig. 2-5. A new plot of *P* vs. $\Delta L$ appears as in Fig. 2-6. Here $\Delta L'$ is measured from the new starting length. The slope *oa'* *is* exactly the same as the slope *hg* in Fig. 2-5, and the yield load, $P_{a'}$, is equal to $P_g$ and thus is greater than the initial yield load of the virgin specimen given by $P_a$ in Fig. 2-5. Evidently the process of loading beyond yield and unloading has *strengthened* the material. The process is variously called strain hardening, *work hardening,* or *cold working.* The reasons for the effect are discussed in subsequent chapters, but for the moment, let us simply acknowledge that *mechanical working* can be a means of strengthening a material.

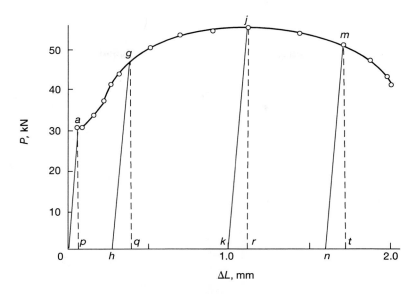

**Figure 2-5**   The effect of loading, unloading, and reloading on $P$–$\Delta L$ tensile plots of a 0.20% carbon steel of initial diameter 13 mm and gage length 50 mm.

The as-received $P$–$\Delta L$ curve is an envelope for all the other curves. In particular, the yield loads of the cold-worked samples are the loads at which the earlier deformation stopped and from where unloading took place. This means that if we know the continuous $P$–$\Delta L$ curve for the annealed material, we have a fairly good idea of the increases in yield load obtainable by cold working. Note that the ductility of the worked material is reduced. Note also that bars cold-worked beyond the zero-slope ultimate (i.e., unloaded from a point such as $m$ in Fig. 2-5) neck immediately upon yielding when reloaded. Thus, reloading a bar cold-worked to $m$ results in simultaneous yield and ultimate, but the ultimate does not exhibit zero slope. When cold-worked to strains less than ultimate, bars of this metal strain-harden after yielding and then display a zero-slope ultimate load.

As explained earlier, it is preferable to normalize loads and extensions into stresses and strains. What does work hardening do to $s$ and $e$? We have said that the continuous $P$–$\Delta L$ curve of the "as-received" material is the same as the equivalent $s$–$e$ curve because $A_0$ and $L_0$ are fixed starting values. However, a piece of material work-hardened less than ultimate and unloaded has a smaller cross-sectional area than the area $A_0$ from which it came. Thus, values of $s$ for cold-worked bars will be *greater* than those based on $A_0$. Again, because there has been prior permanent stretch, the gage length defined by the original marks on the bar will be greater than before, meaning that values of $e$ are less.

Thus, although the $P$–$\Delta L$ curves may be made to coincide by horizontal shifting, the $s$–$e$ curve for the unworked material will be an envelope of the $s$–$e$ curves of the strain-hardened materials *only* if $s$ is defined in terms of the first $A_0$ and $e$ is defined in terms of the first $L_0$.

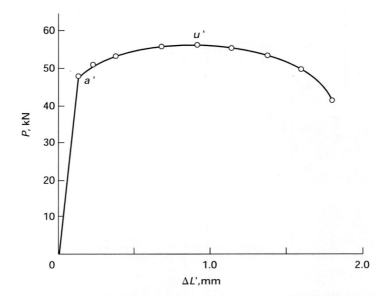

**Figure 2-6**   *P–ΔL'* plot for subsequent testing of specimen prestrained through the loading path *oagh* in Fig. 2-5, where *ΔL'* is measured from the (longer) starting length of the prestrained specimen. The yield strength of the prestrained specimen is increased.

We shall see later that these difficulties can be resolved in terms of so-called *true stress* and *true strain,* but whatever the definition of stress, the fact remains that cold working increases the strength of solids and usually reduces ductility.

**Example 2-3**

A 10-mm diameter bar of an annealed aluminum alloy yields in tension at a load of 6 kN. When the diameter is reduced to 9 mm, the load is 8 kN, at which point the bar is unloaded. Find the nominal yield strength of the metal on first loading and the value upon reloading.

**Solution.**   $S_y$ at first yield is $(6 \times 10^3)/(\pi/4)(10 \times 10^{-3})^2 = 76.4$ MPa; $S_y$ on reloading after cold work is $(8 \times 10^3)/(\pi/4)(9 \times 10^{-3})^2 = 126$ MPa.

## TORSION AND SHEAR

The tension test involves tensile or *normal* stresses which are perpendicular to the plane over which they act; that is, load $P$ is normal to $A_0$ in the definition of $s$.

Consider a round bar that is twisted rather than pulled. A straight line scribed along the length of the bar is distorted, as shown in Fig. 2-7(a), the amount of movement depending on the magnitude of the torque. Radial lines on the end of the bar remain straight, and the twisting deformation process may be viewed as if the bar were made up of a series of thin disks, each one being rotated incrementally relative to its neighbors. Such action is called *shear,* a section of one of the disks being shown in Fig. 2-7(b). Consider a small element. We can define *shear stress* $\tau$ as

**Figure 2-7**   (a) Shear distortion of round bar in torsion; (b) section of "disk" along the length of the bar rotated incrementally relative to its neighbors.

$$\tau \equiv \frac{P}{A} \qquad (2\text{-}4)$$

and the shear strain $\gamma$ as

$$\gamma = \frac{y}{h} \qquad (2\text{-}5)$$

$\tau$ and $\gamma$ are counterparts of $s$ and $e$ (and of $\sigma$ and $\varepsilon$ defined later) when the deformation is by relative sliding rather than pulling apart.

   Yielding can occur in the twisted section if the moment or torque is great enough. However, for small deformations before yield (i.e., at small angles of twist), the behavior is reversibly elastic. Then $\tau$ and $\gamma$ are related by

$$\tau = G\gamma \qquad (2\text{-}6)$$

where $G$ is called the *shear modulus*. $G$ for shear is obviously analogous to $E$ for tension. They are related, and it can be shown that $G \approx E/3$.

   Another illustration of shearing is in the processes of *punching, cropping,* or *blanking,* where a disk, for example, is pushed out from a sheet. Simplistically this is accomplished by shearing along $OA$ and $O'A'$ (Fig. 2-8) until fracture ensues.

**Figure 2-8**   The punching, cropping, or blanking of materials to produce holes or disks. Shear followed by fracture occurs in zones $OA$ and $O'A'$.

**Example 2-4**

A thin-walled tube of mean diameter 12 mm, wall thickness 1 mm, and length 200 mm is twisted by means of a torque wrench attachment, the lever arm of which is effectively 300 mm long. Find the shear stress in the wall of the tube when the arm of the wrench is moved 5 mm if all deformation is elastic ($G$ = 100 GPa).

**Solution**     Viewing the whole length of the tube as shown in Fig. 2-7, we have $y = (5 \times 6)/300$ and $h$ is now the length of the tube, so

$$\gamma = \frac{y}{h} = \frac{5 \times 6}{300 \times 200} = 0.0005$$
$$\tau = G\gamma = (100 \times 10^9)(5 \times 10^{-4}) = 50 \text{ MPa}$$

## COMPRESSION AND COMBINED STRESS STATES

A compression test is the reverse of a tension test; that is, the material is squashed instead of pulled (Fig. 2-9). A plot of compression load vs. decrease in height of the specimen shows the same general features as a $P$–$\Delta L$ plot in tension. There is a reversible elastic region, followed by a plastic region, unloading from which gives a permanently deformed (shorter) cylinder. Upon reloading, the specimen has a greater yield stress than the original starting material. The same difficulty about "starting area" in definitions of $s$ in tension arises with compression, except that because specimens become greater in cross section as the height decreases, $s$ defined in terms of $A_0$ is always *greater* than $s$ defined in terms of the real area. Also, strict application of Eq. (2-2) for strain gives *negative* values in compression because the deformation is in the direction "opposite" to tension.

Although simple in principle, compression tests are complicated by friction between the specimen and the testing machine faces (platens). During deformation, material has to slide in relation to these faces, so the load required to cause a given amount of deformation depends on the friction. Stresses are thus always greater than in the absence of friction, although there are means of compensating for friction.

Material adjacent to the compression faces is restrained by frictional effects, so a test piece often becomes barrel-shaped (Fig. 2-10). The deformation is nonuniform, or nonhomogeneous, with the strain patterns varying through the height of the test piece. Tension tests and frictionless compression tests, on the other hand, produce homogeneous deformation. A useful technique for checking homogeneity of deformation is to scribe grids on the material

**Figure 2-9**     A simple compression-test piece.

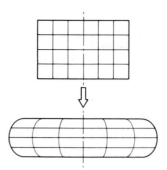

**Figure 2-10** "Barreling" distortion of compression test piece caused by frictional restraint at the platens. Nonhomogeneity of deformation is shown by previously marked rectangular grids becoming skewed.

and compare the patterns before and after loading (Fig. 2-10). Nonhomogeneous deformation is characterized by originally orthogonal (perpendicular) grids becoming skewed.

It is sometimes convenient to regard a skewed pattern such as shown in Fig. 2-10 as being produced by a combination of squashing or pulling (maintaining the right-angled grids homogeneously) and shearing (which skews the grids). In this way, the compression test with friction is an example of *combined stresses*. The frictional stresses are unwanted but must be tolerated. In other combined stress cases, the geometry produces loads acting in different directions. For example, a state of equal biaxial tension exists in the skin of a balloon when blown up and also exists in a metal sheet when bulged into a dome by fluid pressure; see Fig. 2-11(a). Calculations show that a capped thin cylinder under internal pressure has both longitudinal and circumferential stresses in the wall, where the longitudinal stress is half the circumferential stress; see Fig. 2-11(b). Thus, the walls of a beer can have unequal (2:1) biaxial tensions.

It is important, then, to remember that although we emphasize the simple tension test as a means of characterizing materials, there are many situations where materials are subjected to combined stresses. We must therefore have means for predicting the elastic and plastic behavior of solids under combined stresses and means of interpreting data on stress vs. strain obtained from tests. Details are not within the scope of this book, but we will

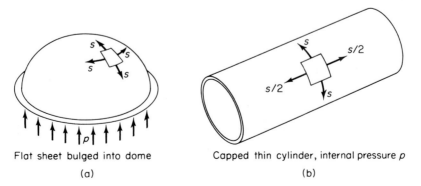

Flat sheet bulged into dome

(a)

Capped thin cylinder, internal pressure *p*

(b)

**Figure 2-11** Combined stress states in (a) a sheet of material bulged into a dome (equal biaxial tensions) and (b) an internally pressurized thin cylinder (unequal biaxial tensions, where the longitudinal stress is one-half the circumferential stress).

touch on  some points (such as the Tresca and Maxwell–von Mises criteria for yielding under combined stresses) in later sections.

One test that occurs under combined stresses is the hardness test, which involves the resistance of a solid to penetration by some sort of indenter. Such tests are described later in this chapter, as they are a means of quickly establishing strength levels without making, for example, specially shaped tensile bars.

## MOHR'S CIRCLE TRANSFORMATIONS

In combined-stress situations, it is often required to establish the magnitude of stresses and strains in directions different from those along which stresses and strains are known by virtue of the external loading. Two reasons for so doing are that brittle fracture is often connected with the maximum tensile stress in a body and that plastic flow is often connected with the maximum shear stress in a body. Given some set of loads on a body, what are the absolutely greatest tensile and shear stresses? On what planes do they occur, and what are the orientations of these planes relative to some known directions in the body? In other words, we must know how to "resolve" stresses and strains in different directions.

### Stress as a Second-Order Tensor

Stress (whether normal or shear) is defined as (force/area). Both force *and* area are vector quantities; that is, they both require magnitude and direction to be fully described. The idea of a force being a vector is, no doubt, familiar to the reader, but perhaps not so for area. Think, however, of the cross-sectional area of a tensile test bar. We usually talk in terms of the size of the normal cross section, for convenience, but we could just as well think of the size of an oblique cross section, the orientation of which would be known; hence "magnitude" and "direction" are required for area. Stress, which is evidently the quotient of two vectors, is *not* a vector itself but rather something called a second-order tensor. The rules for resolving vectors do *not* apply to stress, and we must establish what the correct rules are. We shall learn that a simple graphical technique (Mohr's circle) aids in this process.

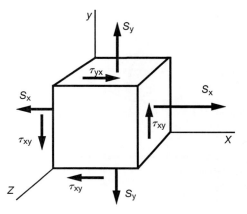

**Figure 2-12**   General two-dimensional stress state, shown in three dimensions, with both normal ($s_x$ and $s_y$) and shear ($\tau_{xy}$ and $\tau_{yx}$) stresses.

Consider a small cubical element of stressed material, such as shown in Fig. 2-12, with tensile stresses $s_x$ on the plane perpendicular to the $x$-direction, defined as the $x$-plane, and tensile stresses $s_y$ on the $y$-plane. Shear stresses are defined by double subscripts: The first subscript defines the plane that the stress acts on, and the second subscript defines its direction. Thus, the shear stress acting on the $y$-plane and with its arrow pointing in the $x$-direction is designated $\tau_{yx}$. The *general* rule (in contrast to the *Mohr* rule, described later) for the sign of a shear stress is as follows: Considering *only* the shear stresses acting on the positive faces of the cube (the planes on the positive or "front" side of the cube, e.g., the planes that are visible in Fig. 2-12), if the direction of the stress is in the positive direction of the second-subscript axis, the shear stress is positive. Thus, the shear stress $\tau_{yx}$ acting on the top of the cube in Fig. 2-12 is positive because it acts in the $+x$-direction; likewise, $\tau_{xy}$ on the right side of the cube in Fig. 2-12 is positive because it points in the $+y$-direction. Because every shear stress must have an equal and opposite stress acting on the "back" plane, the sign of this *pair* of stresses is defined by the general rule just given. In fact, two pair of stresses must always be present in order to provide rotational equilibrium, and these always have the same sign, as defined.

Likewise, the normal stresses should have double subscripts, but because the subscripts are always the same, for example, $s_{xx}$, and because $s$ (and later $\sigma$) is used to represent a normal stress and $\tau$ a shear stress, the meaning remains clear when only a single subscript is used for normal stress.

The existence of the general rule and the different Mohr rule for the sign of a shear stress constitutes one of the chief problems in working with Mohr's circle (see what follows), so it is essential to understand both rules.

### Calculation of $s_\theta$ and $\tau_\theta$

In order to calculate the magnitude and direction of the normal and shear stresses acting on some plane of area $A$ at an angle $\theta$ counterclockwise (positive) from the $y$-direction, as shown in Fig. 2-13, we can resolve the *forces* (not stresses) in the $x$-direction as follows:

$$s_\theta A \sin\theta + \tau_{yx}A \cos\theta = \tau_\theta A \cos\theta + s_x A \sin\theta \qquad (2\text{-}7a)$$

and resolving in the y-direction gives

$$s_\theta A \cos\theta + \tau_\theta A \sin\theta + \tau_{xy}A \sin\theta = s_y A \cos\theta \qquad (2\text{-}7b)$$

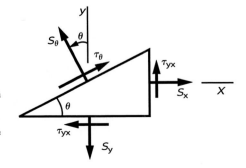

**Figure 2-13**  Stresses $s_\theta$ and $\tau_\theta$ acting on an inclined plane in a general two-dimensional stress system having $s_x$, $s_y$, and $\tau_{xy}$. The general rule for sign of shear stress using double subscripts does not apply to $\tau_\theta$, which is defined here by the diagram.

Manipulation of these equations gives

$$s_\theta = \tfrac{1}{2}(s_y + s_x) + \tfrac{1}{2}(s_y - s_x)\cos\ 2\theta - \tau_{xy}\sin\ 2\theta$$
$$\tau_\theta = \tfrac{1}{2}(s_y - s_x)\sin\ 2\theta + \tau_{xy}\cos 2\theta \tag{2-8}$$

### Example 2-5

A tensile bar has a rectangular cross section 5 mm × 2 mm and is under a vertical load of 3 N. What are the stresses on the plane whose normal is 60° clockwise from the vertical?

**Solution**

$$s_y = \frac{3\,N}{5\,mm \times 2\,mm} = 300\ kPa$$
$$s_x = 0 \quad \tau_{xy} = 0 \quad \theta = -60°(CW)$$

From Eq. (2-8),

$$s_\theta = \tfrac{1}{2}(300) + \tfrac{1}{2}(300)\cos(-120°) = 150 - 75 = 75\ kPa$$
$$\tau_\theta = \tfrac{1}{2}(300)\sin(-120°) = -130\ kPa$$

### Example 2-6

A stress state on two perpendicular planes consists of a normal tension stress on one, a normal compressive stress of equal magnitude on the other, and a pair of complementary shear stresses. On a plane at 60° (counterclockwise) to the plane subjected to the compressive stress, the stress state is 40 MPa tension and 10 MPa shear. What are the values of the unknown orthogonal stresses?

**Solution**    If the magnitude of the unknown tension stress (corresponding to $s_x$ in Fig. 2-13) is $s$, the orthogonal compressive stress (corresponding to $s_y$) is given by $-s$. Then we have in Eq. (2-8)

$$s_\theta = 40 \times 10^6 = \tfrac{1}{2}(-s + s) + \tfrac{1}{2}(-s - s)\cos\ 120° - \tau_{xy}\sin\ 120°$$
$$\tau_\theta = 10 \times 10^6 = \tfrac{1}{2}(-s - s)\sin\ 120° + \tau_{xy}\cos 120°$$

that is,

$$40 \times 10^6 = \left(\frac{1}{2}\right)(-2s)\left(-\frac{1}{2}\right) - \tau_{xy}\left(\frac{\sqrt{3}}{2}\right)$$

and

$$10 \times 10^6 = -s\left(\frac{\sqrt{3}}{2}\right) + \tau_{xy}\left(-\frac{1}{2}\right)$$

From this, $s = 11.3$ MPa and $\tau_{xy} = -39.6$ MPa. Therefore, the unknown stress state is 11.3 MPa tension, 11.3 MPa compression, and complementary shears of magnitude $-39.6$ MPa (with senses opposite those in Fig. 2-13).

## Principal Stresses and Planes

Maximum and minimum values of $s_\theta$ and $\tau_\theta$ may be found by first differentiating Eq. (2-8) with respect to $\theta$ to establish stationary values and then inserting those values of $\theta$ back into the expressions for $s_\theta$ and $\tau_\theta$. Alternatively, we may note that according to Eq. (2-8), $\tau_\theta$ = 0 when $\frac{1}{2}(s_y - s_x) \sin 2\theta = -\tau_{xy} \cos 2\theta$, that is,

$$2\theta = \tan^{-1}\left(\frac{2\tau_{xy}}{s_x - s_y}\right) \tag{2-9}$$

for maximum or minimum $s_\theta$. There will be *two* values of $2\theta$ satisfying this relation, which are 180° apart. In other words, whatever the $(s_x, s_y, \tau_{xy})$ system, there will always be a pair of planes at right angles to each other ($\theta = 90°$) on which there is no shear stress $\tau_\theta$, merely tensile or compressive $s_\theta$. It may be shown (e.g., by differentiation) that the pair of $s_\theta$ are the algebraically greatest and least values of normal stress. These values are called the *principal stresses* of the given $(s_x, s_y, \tau_{xy})$ system, and the planes on which they act are called the *principal planes.*

## Mohr's Circle

Equations (2-8) give the general $(s, \tau)$ combinations on an arbitrary plane oriented at some $\theta$ from the y-plane given an orthogonal set of $(s_x, s_y, \tau_{xy})$. If the orthogonal set of stresses happen to coincide with the principal stress state (i.e., $s_x$ = one principal stress, $s_y$ = the other, and $\tau_{xy} = 0$), Eq. (2-8) will be the special case of giving the general $(s, \tau)$ combinations on an arbitrary plane, oriented at some $\theta$ from a principal plane, in terms of the principal stresses only. If the principal stresses are called $s_1$ and $s_2$ (where $s_1 > s_2$ algebraically), we have

$$s_\theta = \frac{1}{2}(s_1 + s_2) + \frac{1}{2}(s_1 - s_2) \cos 2\theta$$

$$\tau_\theta = \frac{1}{2}(s_1 - s_2) \sin 2\theta \tag{2-10}$$

In a plot of $\tau$ vs. $s$, Eq. (2-10) describes a circle whose center is at $\frac{1}{2}[s_1 + s_2]$, 0 with radius $\frac{1}{2}(s_1 - s_2)$, as shown in Fig. 2-14 (where $s_1$ and $s_2$ are arbitrarily shown as both being positive; they could both be negative or be a mixture, of course, which would shift the circle along the abscissa). The coordinates of point $Q$ on the circle are $s_\theta = OT = [\frac{1}{2}(s_1 + s_2) + \frac{1}{2}(s_1 - s_2) \cos 2\theta]$ along the s-axis and $\tau_\theta = TQ = [\frac{1}{2}(s_1 - s_2) \sin 2\theta]$ along the $\tau$-axis. Thus, *all* stress states in the body, as $\theta$ is varied, lie on the circle. The circle is known as a *Mohr's stress circle* and is equivalent to a graphical tensor transformation.

As angles are in terms of $2\theta$ in the *stress plane,* a rotation of 180° in the *physical plane* of the stressed body (which comes back to the stresses in line before rotation) corresponds with a 360° rotation along the Mohr's circle. To use the circle, an arbitrary sign

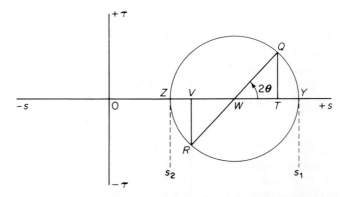

**Figure 2-14**   The basic Mohr's circle representation of stress at a point. The relation between $\tau$ and $s$ given by Eq. (2-10) is a circle with center at $[1/2\,(s_1 + s_2),\,0]$ and radius $1/2\,(s_1 - s_2)$.

convention must be employed for shear stresses. This is the *Mohr rule* for the sign of a shear stress, as mentioned earlier: If the pair of shear stresses $\tau$ *acting on opposite faces of the cube* would produce *clockwise* rotation of the body, it is *positive*. Thus, in Fig. 2-13, $\tau_{xy}$ is negative and $\tau_{yx}$ positive. The shear stress at point $Q$ on the circle (Fig. 2-14) is $TQ$ and is positive; the corresponding complementary shear is $|VR|=|TQ|$, corresponding to the point $R$ diagonally opposite $Q$ (i.e., 180° in the stress plane and 90° in the physical plane), and is negative. Because the shear stresses acting on all four faces of the cube must be of the same magnitude, no subscripts are needed to describe $\tau$. The physical plane displaying the $(OT,\,TQ)$ stress state is located at an angle $\theta$ counterclockwise from the plane in which $s_1$ acts; that displaying $(OV,\,VR)$ is at $(\theta + 90)°$ from the plane of $s_1$; and so on.

In algebraic terms, then, the maximum and minimum principal stresses and the maximum shear stress can be deduced from the Mohr's circle in Fig. 2-14 for given values of $s_x$, $s_y$, and $\tau$. If $s_x = \overline{OV}$, $s_y = \overline{OT}$, and $\tau = \overline{TQ}$ and $\overline{VR}$, then

$$s_1 = \overline{OY} = \overline{OW} + \overline{WY}$$

$$s_2 = \overline{OZ} = \overline{OW} - \overline{WY}$$

$$\overline{OW} = \frac{s_x + s_y}{2}$$

$$\overline{WY} = \text{radius of circle} = \left[\left(\frac{s_y - s_x}{2}\right)^2 + \tau^2\right]^{1/2}$$

$$s_1 = \frac{s_x + s_y}{2} + \left[\left(\frac{s_y - s_x}{2}\right)^2 + \tau^2\right]^{1/2}$$

$$s_2 = \frac{s_x + s_y}{2} - \left[\left(\frac{s_y - s_x}{2}\right)^2 + \tau^2\right]^{1/2}$$

$$(2\text{-}11)$$

The maximum shear stress is the radius of the circle:

$$\tau_{max} = \left[\left(\frac{s_y - s_x}{2}\right)^2 + \tau^2\right]^{1/2} \tag{2-12}$$

## Procedure for Applying Mohr's Circle

Derivation of Mohr's circle for stress has so far been an essentially two-dimensional exercise, with no stresses considered in the $z$-direction. If there is a second set of shear stresses such as $\tau_{zx}$ or $\tau_{zy}$ in addition to $\tau_{xy}$, Mohr's circle *cannot be used*. This problem can be solved by a set of equations called the *stress cubic*, but because this situation is encountered very infrequently, its solution is not covered here.

If the third plane has no shear stresses, then it is a principal stress plane, and any normal stress (including zero stress) in the third direction is a principal stress; such three-dimensional stress systems can be solved using Mohr's circle. Consider a cube with stresses $s_x$, $s_y$, $s_z$, and $\tau_{xy}$, as shown in Fig. 2-15(a). Because $s_z$ acts on a plane without shear stresses, it is a principal stress.

When addressing a Mohr's circle problem, the procedure is as follows:

1. Can it be solved? If one or more planes is a principal stress plane, such as the $z$-plane in Fig. 2-15(a), it can be solved.
2. Viewing the cube from the direction of this principal plane, draw a two-dimensional sketch of the stresses on the other two planes, as in Fig. 2-15(b). Because only one system of shear stresses exists here, $\tau$ is given without subscripts.
3. Construct the Mohr's circle for the two-dimensional system. (Here we have arbitrarily set $s_x > s_y > s_z$.)
   a. On a $\tau$-$s$ coordinate system, label the origin 0 and draw a vertical dashed line at $s = s_x$.
   b. The shear stress $\tau$ on the $x$-plane acts counterclockwise [Fig. 2-15(b)], so it is negative. Draw a horizontal dashed line at $\tau = -\tau$. The intersection of the dashed lines gives the normal and shear-stress coordinates of the $x$-plane; label this point $x$.
   c. Repeat for the normal and shear-stress coordinates for the $y$-plane and label $y$. Note that $x$ and $y$ must be equally spaced above and below the $\tau = 0$ axis, as both have the same magnitude of shear stress $\tau$.
   d. Because $x$ and $y$ lie on a diameter of the Mohr's circle, connect them and construct the circle. Label its center $C$ and its radius $R$.
   e. Calculate $R = \{[(s_x - s_y)/2]^2 + \tau^2\}^{1/2}$
   f. Calculate $\overline{OC} = (s_x + s_y)/2$.
   g. Calculate $s_1 = \overline{OC} + R$; label 1.
   h. Calculate $s_2 = \overline{OC} - R$; label 2.
   i. Calculate the angle from $x$ to 1, $2\theta = \sin^{-1}(\tau/R)$; label $2\theta$.
   j. Draw on the original stressed cube [Fig. 2-15(a)] the principal planes 1 and 2 at $\theta$ from the $x$- and $y$-planes, as shown shaded in Fig. 2-16(a). Draw the principal stresses $s_1$, $s_2$, and $s_3$.

Figure 2-15(a)

Figure 2-15(b)

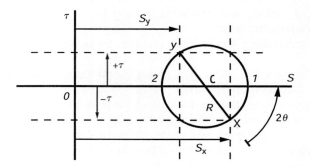

Figure 2-15(c)

**Figure 2-15**   Three-dimensional stress system and resultant Mohr's circle.

4. Because we now know all three principal stresses, $s_1$, $s_2$, and $s_3 = s_z$, and their planes, we can construct the Mohr's circle for the pair of principal stresses $s_1$–$s_3$ and for the pair of principal stresses $s_2$–$s_3$, using exactly the same procedures as given before, except that all shear stresses are zero on the principal planes.

   **a.** From Fig. 2-16(a), draw a two-dimensional sketch with stresses $s_1$ and $s_3$, as viewed in the –2 direction [Fig. 2-16(b)]. Add a label $z$ at $s = s_3 = s_z$, $\tau = 0$, shown in Fig. 2-16(c).

   **b.** Figure 2-16(c) also shows the $x$–$y$ Mohr's circle that was constructed in Fig. 2-15(c), shown here lightly shaded. This is, of course, also the $s_1$–$s_2$ circle. Draw a second Mohr's circle using points 1 and 3 as a diameter. For the stresses chosen in this example, this produces a larger circle than the $s_1$–$s_2$ circle.

   **c.** Repeat for stresses $s_2$ and $s_3$, drawing a third Mohr's circle between points 2 and 3; this is shaded darker in Fig. 2-16(c).

From Fig. 2-16(c), we can learn the following:

- The maximum principal stress on any plane in the cube is $s_1$.
- The minimum principal stress on any plane in the cube is $s_3$. Note, *not* $s_2$ in this problem.

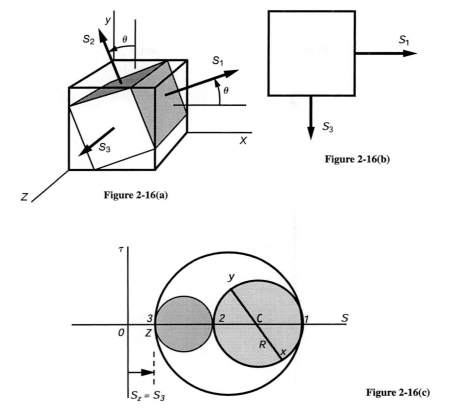

**Figure 2-16(a)**

**Figure 2-16(b)**

**Figure 2-16(c)**

**Figure 2-16** Construction of Mohr's circles for three-dimensional stress system.

- The maximum shear stress occurs in the largest circle, the $s_1$–$s_3$ circle, and $\tau_{max} = (s_1 - s_3)/2$.
- It can also be shown that all stress states on all planes in the cube are represented by the stress states that fall inside and on the largest circle but on or outside the two smaller circles. This information is very important to the design engineer.

**Example 2-7**

Draw the Mohr's circles for the following cases: (a) uniaxial tension prior to necking, (b) direct compression before barreling, (c) simple torsion, and (d) a pressurized beer can. Determine the magnitude of the maximum shear stresses in each case.

**Solution**    The circles are shown in Fig. 2-17(a)–(d), located in the first instance by points $Y$ and $Z$. In cases (a) and (b), the systems are uniaxial, with one principal stress being the applied stress and the other two being zero. Because of the convention that $s_1 > s_2 > s_3$ algebraically, the zero stresses are $(s_2, s_3)$ in Fig. 2-17(a), but $(s_1, s_2)$ in Fig. 2-17(b). In case (d) the stress in the hoop direction is $s_1 = pD/2t$, and in the axial direction, it is $s_2 = pD/4t$, where $p$ is the internal pressure, $D$ is the cylinder mean diameter, and $t$ is the wall thickness. For all practical purposes, $s_3 = p \approx 0$ as $D \gg t$.

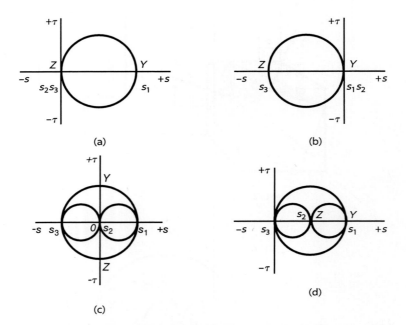

**Figure 2-17**    Solution to Example 2-7, showing Mohr's circles for (a) uniaxial tension prior to necking, (b) homogeneus compression, (c) simple torsion, (d) a pressurized beer can.

The magnitude of the maximum shear stress in cases (a)–(c) is straightforward to calculate and is equal to $YZ/2 = (s_1 - s_3)/2$. In case (d), however, although the maximum shear stress in the plane of the beer can wall is given by $YZ/2 \, (= pD/8t)$, there is a *greater* shear stress $(s_1 - s_3)/2 = s_1/2 = pD/4t$, in the body, which acts in a plane that is normal to a line that lies in the transverse plane of the cylinder and is 45° to the radius.

Note that Fig. 2-17(c) shows that pure shear is equivalent to equal tension and compression at right angles.

### Example 2-8

Find the magnitude of the principal stresses and their orientation for the orthogonal two-dimensional stress system having $s_x = 25$, $s_y = 20$, $\tau_{xy} = 5$ (MPa).

**Solution**    Either the Mohr's circle may be constructed from the coordinate points (25, 5) and (20, –5), or Eq. (2-11) may be solved, giving $s_1 = 28.1$ MPa, $s_2 = 16.9$ MPa, and $s_3 = 0$.

The Mohr's circle construction shows that the radius joining the center of the circle to the point (25, –5) lies at –63.4° (i.e., 63.4° clockwise) from the abscissa. Thus, the $s_1$ plane lies at $63.4/2 = 31.7°$ *counterclockwise* relative to the $x$-plane in which the 25-MPa tension stress acts. Of course, 63.4° is also the solution of Eq. (2-9).

### Example 2-9

Given a stress system (in MPa) where $s_x = 200$, $s_y = -140$, $s_z = 100$, $\tau_{yz} = \tau_{zy} = 80$, calculate the maximum principal stress, the maximum shear stress, and the angles from the $x$-, $y$-, and $z$-axes to the normal to the minimum principal stress plane.

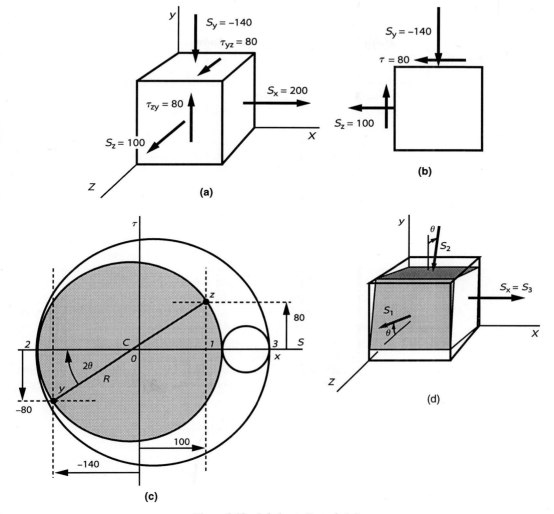

**Figure 2-18**   Solution to Example 2-9.

**Solution**   The stress system described is shown in Fig. 2-18(a). Note that the shear stress is stated with the *general* rule: Because it is positive, $\tau_{yz}$ acts in the +z-direction, and so on. Following the procedure described before for applying Mohr's circle:

1. Yes, this problem can be solved, as the x-plane contains no shear stress and is thus a principal stress plane.
2. Viewed in the −x-direction, the stresses on the y- and z-planes are shown in Fig. 2-18(b).
3. Construct the first Mohr's circle from a diameter with ends at $s_y = -140$, $\tau = -80$, and $s_z = 100$, $\tau = 80$ [Fig. 2-18(c)], and label these points y and z, respectively. Label the origin 0 and the center of the circle C. This Mohr's circle is shown shaded.

• Calculate

$$R = \left[ \left( \frac{s_z - s_y}{2} \right)^2 + \tau^2 \right]^{\frac{1}{2}} = \left[ \left( \frac{100 - (-140)}{2} \right)^2 + (80)^2 \right]^{\frac{1}{2}} = 144.2$$

For precision, retain four significant figures during the calculation and reduce to three significant figures when stating the solution.

• Calculate $\overline{OC} = (s_z + s_y)/2 = (100 - 140)/2 = -20$.
• Calculate $s_1 = \overline{OC} + R = -20 + 144.2 = 124.2$.

Here we run into a minor nomenclature problem, for we have so far stated that $s_1 > s_2 > s_3$. But because we do not yet know the relative magnitudes of all three principal stresses, we can relax this constraint and simply let $s_1$ be the first principal stress calculated, and so on.

On the $\tau$–$s$ plot [Fig. 2-18(c)] label the point at $s_1 = 124.2$, $\tau = 0$ as 1.

• Calculate $s_2 = \overline{OC} - R = -20 - 144.2 = -164.2$. Label the point at $s_2 = -164.2$, $\tau = 0$ as 2.
• Calculate the angle $2\theta$ from $y$ to 2 and label it. $2\theta = \sin^{-1}(\tau / R) = \sin^{-1}(80/144.2) = 33.70°$. This represents a clockwise rotation from $y$ to 2, so the angle $\theta = 16.85°$ (when viewed in the $-x$-direction) will be clockwise from the $y$-direction to the normal to the $s_2$-plane. The $s_2$-plane is shown heavily shaded in Fig. 2-18(d). The $s_1$-plane, shown lightly shaded, is at 90° to both the $s_2$-plane and the $s_x = s_3$-plane.

4. In Fig. 2-18(c), label the third principal stress $s_x = s_3 = 200$, $\tau = 0$ with $x$ and 3. Because all three principal stresses are now known, two more Mohr's circles can be drawn for $s_1$–$s_3$ and for $s_2$–$s_3$ in Fig. 2–18(c).

We are now in a position to answer the questions asked in this problem. The maximum principal stress is $s_x = s_3 = 200$ MPa. The maximum shear stress is the radius of the largest Mohr's circle,

$$\tau_{max} = \frac{s_3 - s_2}{2} = \frac{200 - (-164.2)}{2} = 182 \text{ MPa}$$

From Fig. 2-18(d), the angles from the $x$-, $y$-, and $z$-axes to the normal to the minimum principal stress plane $s_2$ are, respectively, 90°, $-16.8°$, and $-107°$. The last angle, as viewed from the $s_3$-axis, is clockwise from the $z$-axis to the normal to the dark-shaded $s_2$-plane.

## TRUE STRESS AND STRAIN

As mentioned in connection with loading and unloading, the significance of nominal stress as load per original area becomes more and more distorted as the real area over which the load acts changes. For example, in a tension test of ductile metal, the fracture area in the neck $A_f$ may be on the order of half the initial area $A_0$. To provide a more meaningful description of instantaneous load per unit area, the *true stress* $\sigma$ is defined as

$$\sigma \equiv \frac{P}{A} \qquad (2\text{-}13)$$

Thus, $\sigma$ is twice $s$ when $A$ is half $A_0$, and so on.

In the same way, the meaning of nominal strain as deformation per unit length is distorted when length $L$ becomes appreciably larger than $L_0$ during the test. For example, when $e = 1$, the extended length is equal to $2L_0$; that is, we have 100% extension. To what height would we have to compress a cylinder to obtain a compressive strain of 100%? We would have

$$e = -1 = \frac{L_f - L_0}{L_0}$$

Hence, $L_f = 0$, which is impossible.

A better measure of strain is given by referencing changes in length, height, and so on, to the *current* length or height rather than to some original dimension. Then an increment of this *true strain* $\varepsilon$ is given by

$$\Delta\varepsilon = \frac{\Delta L}{L}$$

so                                                                                                      (2-14)

$$\varepsilon = \ln\left(\frac{L}{L_0}\right)$$

Relation (2-14) is valid only when strain is uniform all along $L$ and is thus invalid for tensile strains greater than the necking strain.

**Example 2-10**

Show that doubling the length of a tensile specimen is equivalent, in terms of true strain, to halving the height in compression.

**Solution**   In halving the height, $L = \frac{1}{2} L_0$, so $\varepsilon = \ln(L/L_0) = \ln \frac{1}{2}$. But $\ln \frac{1}{2} = -\ln 2$ and $\ln 2 = \ln(2L_0/L_0)$, which is the true strain when the length is doubled. Thus, with due change of sign for tension and compression, the processes are equivalent. Compare in the text what happens for $e$ under identical conditions.

Another useful form for $\varepsilon$ in the plastic range of strains comes from invoking *volume constancy*. Measurements of diameter and length in the regions *bu* (Fig. 2-2) show that the volume of the gage length is essentially constant, that is,

$$A_0 L_0 = AL$$
$$\frac{\pi d_0^2 L_0}{4} = \frac{\pi d^2 L}{4} \qquad (2\text{-}15)$$

for a round test piece, where $A$, $d$, and $L$ are the values anywhere between $b$ and $u$, either during deformation or after unloading beyond yield. Strictly, there are very small changes in volume, but these are associated with the elastic component of strains and are negligible when compared with plastic strains. Then, from Eq. (2-15),

$$\frac{L}{L_0} = \frac{A_0}{A} = \left(\frac{d_0}{d}\right)^2$$

**Figure 2-19**   Force equilibrium of a thin disk within the neck of a tensils specimen over which the strain may be assumed to be uniform.

thus

$$
\begin{aligned}
\varepsilon &= \ln\left(\frac{L}{L_0}\right) \\
&= \ln\left(\frac{A_0}{A}\right) \\
&= \ln\left(\frac{d_0}{d}\right)^2 = 2\ln\left(\frac{d_0}{d}\right)
\end{aligned}
\right\}
\tag{2-16}
$$

Beyond zero-slope ultimate, $u$ in Fig. 2-2, the deformation, which is concentrated in a neck, is nonuniform both along and across the bar although the volume is still constant. Between $u$ and $f$ in Fig. 2-2, the length over which the uniform deformation occurs is much smaller than $L_0$ and Eq. (2-14) for $\varepsilon$ cannot be used.

However, Eq. (2-16) *can* still be used by imagining a very short cylinder located right in the neck (Fig. 2-19) over which the strain may be assumed to be uniform. If we really knew such a cylindrical length, we could use Eq. (2-14), but it is much easier to measure *diameters* and to calculate $\varepsilon$ from Eq. (2-16). In fact, for the plastic range of deformation, measuring diameter $d$ in the range *buf* (Fig. 2-2) is both easier and more accurate than first measuring lengths from $b$ to $u$ and then having to switch to measuring diameters beyond $u$.

Thus, $\varepsilon = \ln(A_0/A)$ is valid from $b$ to $f$, but $\varepsilon = \ln(L/L_0)$ is valid only from 0 to $u$, the zero-slope ultimate. The elastic strain $oa$ is negligible compared with the plastic strains beyond $a$ in Fig. 2-2, so $\varepsilon = \ln(A_0/A)$ can be used for the full range of plastic strains.

**Example 2-11**

Find the true stresses and strains at ultimate load and fracture for the test piece described in Examples 2-1 and 2-2.

**Solution**

$$\sigma_u = \frac{9 \times 10^3}{7.7 \times 10^{-6}} = 1.17\,\text{GPa}$$

$$\sigma_f = \frac{6.5 \times 10^3}{5 \times 10^{-6}} = 1.3\,\text{GPa}$$

$$\varepsilon_u = \ln(10/7.7) = 0.26 \quad \text{or} \quad \varepsilon_u = \ln(65/50) = 0.26$$

$$\varepsilon_f = \ln(10/5) = 0.69.$$

## RELATIONSHIPS BETWEEN $s$ AND $\sigma$ AND BETWEEN $e$ AND $\varepsilon$

Relationships between $e$ and $\varepsilon$ and between $s$ and $\sigma$ may be obtained as follows. Because

$$e = \frac{L - L_0}{L_0} = \frac{L}{L_0} - 1$$

it follows that

$$1 + e = \frac{L}{L_0}$$

so

$$\ln\left(\frac{L}{L_0}\right) = \ln(1 + e) = \varepsilon \tag{2-17}$$

Also, because $\ln(1 + e) = e - e^2/2 + e^3/3...$, we note that $\varepsilon$ and $e$ are essentially the same at small strains but that up to the zero-slope ultimate, $\varepsilon < e$ as differences become significant at large (plastic) strains.

For a relation between $s$ and $\sigma$, we have

$$s = \frac{P}{A_0} \quad \text{and} \quad \sigma = \frac{P}{A}$$

So we require a relation between $A$ and $A_0$. If elastic volume changes are neglected, it has been shown in Eq. (2-16) that

$$\varepsilon = \ln\left(\frac{A_0}{A}\right)$$

Hence,     (2-18)

$$\exp \varepsilon = \frac{A_0}{A}$$

so

$$s = \frac{P}{A_0} = \frac{P}{A \exp \varepsilon} = \frac{\sigma}{\exp \varepsilon} \tag{2-19}$$

Thus, $\sigma = s \exp \varepsilon$. Again, for small $\varepsilon$, $\sigma \approx s$ because $\exp \varepsilon = 1 + \varepsilon + \varepsilon^2/2!$, but, generally, $\sigma > s$.

### Example 2-12

A round wire of initial diameter 2 mm and length 300 mm yielded at a tensile load of 314 N. Further loading to 400 N increased the length to 320 mm. The maximum load was 450 N when the length was 380 mm. After necking, the wire fractured at a load of 390 N, when the final length and diameter were 430 mm and 1.5 mm, respectively. Compare $s$ and $\sigma$, and $e$ and $\varepsilon$, at various stages in the loading. What is the true stress in the region of the specimen away from the neck at the instance fracture occurs in the neck?

**Solution**   At initial yield $s_y = \sigma_y = 314/(\pi/4)(2 \times 10^{-3})^2 = 100$ MPa; $s_y = \sigma_y$, as there is only an imperceptible change in wire diameter. However, by the time the 400-N load is reached, the diameter is $(300/320)^{1/2}(2) = 1.94$ mm. The true stress is then $400/(\pi/4)(1.94 \times 10^{-3})^2 = 136$ MPa; the corresponding nominal stress is $400/(\pi/4)(2 \times 10^{-3})^2 = 127$ MPa. The associated strains are $\varepsilon = \ln(320/300) = 0.065$ and $e = (320 - 300)/300 = 0.067$. At maximum load, the diameter becomes $(300/380)^{1/2}(2) = 1.78$ mm. The true stress is

$$\sigma_u = \frac{450}{(\pi/4)(1.78 \times 10^{-3})^2} = 181 \text{ MPa}$$

$$s_u = \frac{450}{(\pi/4)(2 \times 10^{-3})^2} = 143 \text{ MPa}$$

Again,

$$\varepsilon_u = \ln\left(\frac{380}{300}\right) = 0.24 \quad \text{and} \quad e_u = \frac{380 - 300}{300} = 0.27$$

At fracture,

$$\sigma_f = \frac{390}{(\pi/4)(1.5 \times 10^{-3})^2} = 221 \text{ MPa}$$

$$s_f = \frac{390}{(\pi/4)(2 \times 10^{-3})^2} = 124 \text{ MPa}$$

Again, the associated strains are $\varepsilon = 2\ln(2/1.5) = 0.58$ and $e = (430 - 300)/300 = 0.43$. This is now less than $\varepsilon$ instead of being greater than $\varepsilon$, as the deformation has become nonuniform. The deformation at the root of the neck, which leads to $\varepsilon$, is much greater than deformations away from the neck. Thus, we cannot use lengths for evaluation of $\varepsilon$ after necking, and so we use diameters. The regions away from the neck remain at the uniform diameter that the whole wire had just prior to necking, that is, 1.78 mm. Thus, the true stress at fracture in the "shoulder" regions away from the neck is $390/(\pi/4)(1.78 \times 10^{-3})^2 = 157$ MPa. We see that these regions elastically unload after necking sets in elsewhere because 157 MPa < 181 MPa, which is the highest true stress experienced by them. The true stress in the neck, of course, continues increasing right up to fracture.

The $P$–$\Delta L$ curve for steel in Fig. 2-2 may be reinterpreted as $s$–$e$ and $\sigma$–$\varepsilon$ curves, as shown in Fig. 2-20.

The numerical differences between the nominal and true stresses and strains for ultimate $u$ and fracture $f$ in Fig. 2-20 can be explained from the definitions of the terms and from the behavior shown in Fig. 2-21. The true stress at ultimate $\sigma_u$ is larger than the nominal stress $s_u$, which is the tensile strength, because the area $A_u$ is smaller than $A_0$. At fracture, $\sigma_f$ is much larger than $s_f$, because the instantaneous area $A_f$ is much smaller than $A_0$.

True strain is the integral of the incremental change in length divided by the instantaneous length [Eq. (2-14)], and nominal strain is the total change in length divided by the initial length [Eq. (2-2)]. Throughout the uniform-strain portion *ou* of the tensile test (with any discontinuous yield behavior in the *ab* region neglected), the instantaneous length $L$ is always greater than the initial length $L_0$; hence, the true strain

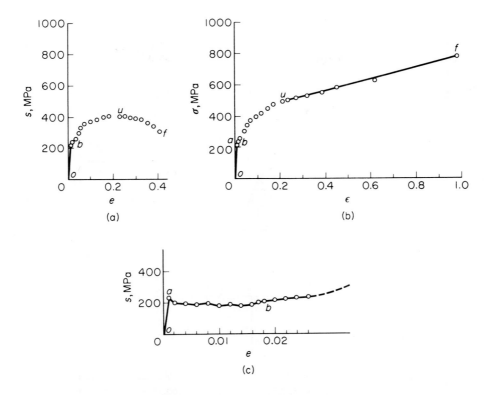

**Figure 2-20** The $P-\Delta L$ data in Fig. 2-2 reinterpreted: (a) as nominal stress *s* vs. nominal strain *e*; (b) as true stress $\sigma$ vs. true strain $\varepsilon$; (c) in a magnified view of the *s–e* curve in the yield region.

is always smaller, in the *ou* region, than the nominal strain. Figure 2-20 shows, for example, that $\varepsilon_u < e_u$.

The most striking difference between the two methods of presentation is that there is no maximum in the $\sigma$–$\varepsilon$ curve. Thus, the curious feature of Fig. 2-2, that there are apparently two compatible *e* values for one *s* value (i.e., $e < e_u$ and $e > e_u$), is eliminated. It is of particular importance that the zero-slope ultimate point ($\sigma_u$, $\varepsilon_u$) is located in an unobvious place somewhere on the $\sigma$–$\varepsilon$ curve.

When loading, unloading, or reloading, $P-\Delta L$ curves are expressed in terms of $\sigma$ and $\varepsilon$, it is found that the concept of "horizontal shifting" works quite well. That is, when a bar is strained to some value $\varepsilon^*$ and then unloaded, the $\sigma$–$\varepsilon$ curve produced by reloading blends into the original $\sigma$–$\varepsilon$ curve if the strain origin is shifted to $\varepsilon^*$ (Fig. 2-22). This suggests that deformation produced in more than one stage is equivalent to the same *total* deformation produced in one step *only* when the deformations are given in terms of $\varepsilon$. That is, deformation of magnitude $\varepsilon_1$, followed by unloading, followed by deformation to $\varepsilon_2$ beyond that, is the same as deformation of magnitude ($\varepsilon_1 + \varepsilon_2$) in one step. This may be

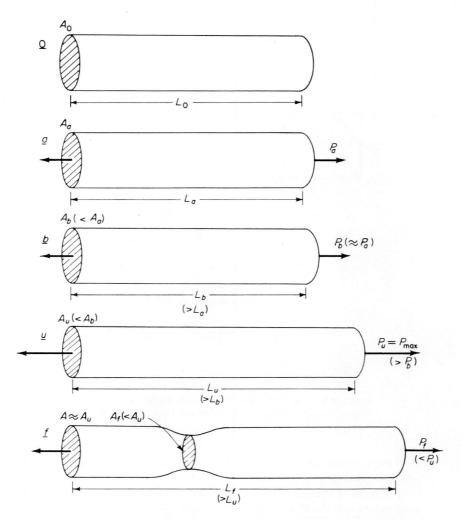

**Figure 2-21**   Five stages of deformation during a tensile test, corresponding to the lettered points in Fig. 2-20.

illustrated as follows: let $\varepsilon_1 = \ln(A_0/A_1)$ and $\varepsilon_2 = \ln(A_1/A_2)$ (remember that $\varepsilon_2$ is measured relative to *its own* starting area). Then

$$\varepsilon_1 + \varepsilon_2 = \ln\left(\frac{A_0}{A_1}\right) + \ln\left(\frac{A_1}{A_2}\right) = \ln\left(\frac{A_0}{A_1}\right)\left(\frac{A_1}{A_2}\right) = \ln\left(\frac{A_0}{A_2}\right)$$

which is the same strain reached by deforming from $A_0$ to $A_2$ in one step.

These arguments do not hold when written in terms of $e$. Thus, true strains are *additive*, but nominal strains are not.

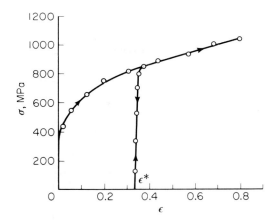

**Figure 2-22** Horizontal shifting of $\sigma$–$\varepsilon$ curves: material prestrained by $\varepsilon^*$ gives a $\sigma$–$\varepsilon$ curve on reloading, which blends in with the curve of the unstrained material if the origin of strain is moved to $\varepsilon^*$.

In the previous example, if we had wished to find the yield strength of the work-hardened bars after the first, second, or total deformation, we would have had to use the $\sigma$–$\varepsilon$ plot and find $\sigma$ at the appropriate $\varepsilon$. Matters would be simplified if there were algebraic relationships between $\sigma$ and $\varepsilon$ in the plastic region. Logarithmic plotting of $\sigma$ vs. $\varepsilon$ for ductile metals shows that over the plastic range *b* to *u*, a reasonable linear relation is followed, that is,

$$\log \sigma = (\text{slope}) \log \varepsilon + \log(\text{constant})$$

(2-20)

or

$$\sigma = K\varepsilon^n$$

where *n* is the slope (called the work-hardening index) and *K* is a constant, which is evaluated by extrapolating the data to $\log \varepsilon = 0$, that is, to $\varepsilon = 1$. Experiments in compression suggest that Eq. (2-20) is valid beyond the necking strain in tension, although data from a tensile test usually show an upward trend after $\varepsilon_u$.

In the elastic region at small strains less than the yield strain, $\sigma = E\varepsilon$. Logarithmic plotting of elastic data gives

$$\log \sigma = \log \varepsilon + \log E \tag{2-21}$$

that is, a straight line of slope (45° on equal scales), with intercept equal to Young's modulus at $\varepsilon = 1$. Hence, plotting both elastic and plastic data gives two linear segments. The transition from one to the other is not necessarily sharp with real-life data, and the yield stress is not necessarily given by the intersection of the lines. It must always be remembered that algebraic $\sigma$–*e* relationships are only approximations of real behavior (Fig. 2-23).

When reloading data are plotted on logarithmic paper, where the plastic strains are measured relative to the bar's new starting size, the applicability of Eq. (2-20) comes into question, as the prestrained $\sigma$–$\varepsilon$ data tend to follow shallow concave-up curves (Fig. 2-23). Even so, some workers draw straight lines through the data points and quote different pairs of (*K*, *n*) values for each prestrain value. However, in the spirit of horizontal shifting, it is

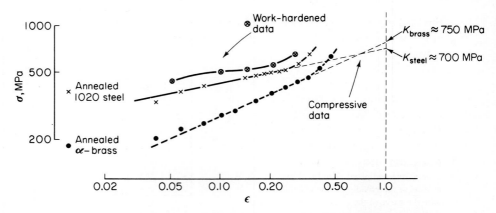

**Figure 2-23**   Logarithmic plotting of $\sigma$–$\varepsilon$ data showing reasonable agreement with a relationship of the form $\sigma = K\varepsilon^n$ in the plastic range.

found that all data fit on the one envelope curve if the prior strain from cold work $\varepsilon_{cw}$ is added to the true strains measured in the reloading test before the data are plotted.

Thus, the full version of Eq. (2-20), allowing for any prestrain $\varepsilon_{cw}$, is

$$\sigma = K(\varepsilon_{cw} + \varepsilon)^n \qquad (2\text{-}22)$$

where $\varepsilon$ is the strain measured beyond the condition of prior cold work. Now $K$ and $n$ take the same values for *all* degrees of prestrain.

If we think of initial yielding of a cold-worked bar as being expressed by $\varepsilon = 0$ in Eq. (2-22), we have, for the yield strength $\sigma_y$ of a material prestrained an amount $\varepsilon_{cw}$,

$$\sigma_y = K\varepsilon_{cw}^n \qquad (2\text{-}23)$$

Thus, solids with large $n$ undergo greater changes in strength when cold-worked. Representative values of $K$ and $n$ for many metals are given in App. 3; Chap. 12 discusses the effects of strain rate and temperature on these parameters.


## TENSILE INSTABILITY, POLYMER "DRAWING," AND LÜDERS BANDS

During the early portion of the strain-hardening region of the stress–strain curve, the load increases because the rate of increase of true stress resulting from strain hardening is greater than the rate of decrease of area. With further strain, the rate of increase of stress [i.e., the slope $ds/de$ in Fig. 2-20(a)] gradually decreases. The zero-slope ultimate condition represents a turning point where the increase in load resulting from strain hardening is exactly equal to the decrease in load associated with the area reduction. Beyond ultimate the decrease in load from decrease in area exceeds the increase in load from increase in true stress. The reduction in area resulting from the first increment of strain beyond ultimate thus lowers the load-carrying capacity of the region where this strain occurs below that of the other regions along the length of the specimen. Because the load required to cause additional strain is lower in this region than elsewhere, further strain beyond ultimate is confined to this region of instability. Characteristic necking results.

For a dead-load tensile test, in which masses are hung on a tensile bar, the reduction of load beyond ultimate is never observed because the specimen fractures when the total mass reaches a value corresponding to $s_u$. In the usual laboratory hydraulic or constant-speed electric tensile-testing machine, however, the machine is designed to pull with force $P$ just enough to maintain the existing strain and cause a small increment of additional strain and is thereby capable of picking up the decreasing load beyond the zero-slope ultimate condition.

Figure 2-21 shows schematically the test section of the specimen at each point labeled on the stress–strain curve in Fig. 2-20. Beyond $u$, associated with the drop in load $P$, is the local reduction in the cross-sectional area already mentioned. The last part in Fig. 2-21 shows the localization of strain and resultant necking as it appears just prior to fracture at $f$.

Because the calculation of true stress accommodates the effect of reduction in area, the $\sigma$–$\varepsilon$ curve in Fig. 2-20(b) shows a continuous increase of true stress (although at a reduced rate) with strain from $b$ to $u$ to $f$. The true-stress–true-strain curve beyond $u$ indicates the average conditions at the minimum diameter of the neck, where most of the deformation takes place. Because the load beyond ultimate is less than the ultimate load, no substantial deformation occurs in regions away from the neck, which in fact elastically unload beyond $u$.

The load maximum occurs at the onset of necking, and the incremental change in load is zero, that is, $\Delta P = 0$. Because $P = \sigma A$ at all points on the $\sigma$–$\varepsilon$ curve,

$$\Delta P = \sigma(\Delta A) + A(\Delta \sigma)$$

Hence, at the point of tensile instability (zero-slope ultimate),

$$\sigma_u(\Delta A_u) + A_u(\Delta \sigma) = 0$$

that is,

$$\frac{d\sigma}{\sigma_u} = \frac{-dA}{A_u} \tag{2-24}$$

If we again invoke plastic volume constancy, we can show that $-dA/A = d\varepsilon$. This follows from the volume being equal to $AL$ for all $A$ and $L$, whence

$$\Delta(\text{volume}) = A \, \Delta L + L \, \Delta A$$

and if $\Delta(\text{volume}) = 0$,

$$\frac{+dL}{L} = \frac{-dA}{A} = +d\varepsilon$$

Thus, in the equation describing the changes at necking, we have

$$\frac{d\sigma}{\sigma_u} = d\varepsilon \quad \text{or} \quad \left(\frac{d\sigma}{d\varepsilon}\right)_u = \sigma_u \tag{2-25}$$

Instability points are thus given when the *slope* of the $\sigma$–$\varepsilon$ curve is numerically equal to the *magnitude* of the true stress at that point. Hence, point $u$ in Fig. 2-20(a) is the same as point $u$ in Fig. 2-20(b).

**Example 2-13**

Show that if $\sigma = K(\varepsilon_{cw} + \varepsilon)^n$, the strain at the zero-slope ultimate is given by $\varepsilon_u = n - \varepsilon_{cw}$. Hence, determine the length at neck initiation of a tensile bar initially prestrained by $\varepsilon_{cw} = 0.05$, that is, 50 mm long after prestraining ($n = 0.2$).

**Solution**    From Eq. (2-25), $d\sigma/d\varepsilon = \sigma_u$, so

$$\frac{d[K(\varepsilon_{cw} + \varepsilon)^n]}{d\varepsilon} = K(\varepsilon_{cw} + \varepsilon_u)^n$$

that is,

$$nK(\varepsilon_{cw} + \varepsilon_u)^{n-1} = K(\varepsilon_{cw} + \varepsilon_u)^n \quad \text{or} \quad \varepsilon_{cw} + \varepsilon_u = n \quad \text{or} \quad \varepsilon_u = n - \varepsilon_{cw}$$

Note that for the case of $\varepsilon_{cw} = 0$, $\varepsilon_u = n$. These relationships hold only for the zero-slope ultimate. If the bar is prestrained by $\varepsilon_{cw} = 0.05$, $\varepsilon_u = 0.2 - 0.05 = 0.15$. Hence, $\ln(L_u/50) = 0.15$ or $L_u = 58$ mm. See Example 2-18.

**Example 2-14**

A tensile specimen, initially of 11.3 mm diameter, has a diameter at the zero-slope ultimate of 10 mm and a diameter at fracture of 8.5 mm. If the maximum load during the test is 75 kN, find the load at fracture. Assume that $\sigma = K\varepsilon^n$.

**Solution**    The true strain at ultimate is $2 \ln(11.3/10) = 0.24$. Therefore, $n = 0.24$, as there is no prestrain. The true stress at ultimate is $(75 \times 10^3)/(\pi/4)(10 \times 10^{-3})^2 = 955$ MPa. If $\sigma = K\varepsilon^n$, $K = (955 \times 10^6)/(0.24)^{0.24} = 1.35$ GPa. The true strain at fracture is $2 \ln(11.3/8.5) = 0.57$. The true stress at fracture is $(1.35 \times 10^9)(0.57)^{0.24} = 1.18$ GPa. The load at fracture is $(1.18 \times 10^9)(\pi/4)(8.5 \times 10^{-3})^2 = 67$ kN.

In the tensile test of most ductile metals, the necked region continues to thin down until fracture takes place. With some polymers and superplastic alloys in tension, a neck forms, but it grows stably, material being fed into the necked region from the thicker adjacent areas. This behavior is called *drawing* in the polymer literature and is discussed in Chap. 12, where it is contrasted with the unstable propagation of necks, which is the subject of this section. Also in Chap. 12 the similarity between regions of discontinuous yielding [such as $ab$ in Fig. 2-20(a) and (c)] and the growth of stable necks is described. Discontinuous yielding is often accompanied by a series of striations called *Lüders bands*, which ripple along the gage section in the manner of stable necks; however, Lüders bands occur at much smaller strain levels than the strains encountered in stable-neck propagation in polymers. It must be noted that not all polymers display stable necks, and likewise only some metals display discontinuous yielding behavior.

**Example 2-15**

A bar of initial area 25 mm$^2$ is reduced to an area of 22 mm$^2$ and unloaded. At what area and load will it neck if reloaded? Assume $\sigma = 1500\, \varepsilon^{0.4}$ MPa.

**Solution**    $\varepsilon_{cw} = \ln(25/22) = 0.13$, and $\varepsilon_u = (n-\varepsilon_{cw}) = 0.4 - 0.13 = 0.27$. The area at ultimate is obtained from $\ln(22/A_u) = 0.27$, so $A_u = 17$ mm$^2$. The true stress at ultimate is $1500(0.4)^{0.4}$, where the true strain at ultimate is $0.4 = 0.27 + 0.13$, which becomes 1.04 GPa. Therefore, the necking load at ultimate is $(1.04 \times 10^9)(17 \times 10^{-6}) = 18$ kN.

## HARDNESS

Although the tensile test is usually a good measure of the ability of a material to resist plastic deformation and failure, it is expensive and destroys the specimen. Hardness tests, of which several kinds are in use, can overcome both of these disadvantages.

Hardness is a measure of the resistance to penetration under load by some form of indenter. In the *Brinell* hardness test, for example, hardness is determined by the diameter of the permanent deformation left by a 10-mm-diameter hard spherical indenter: The larger the diameter, the lower the Brinell hardness number (HB), which is calculated by dividing the load by the spherical area of the indentation. The *Vickers* hardness number is measured in terms of indentation by a diamond pyramid. Neither of these is a "direct-reading" test, and the indentation (less than a few millimeters in size) has to be measured with an eyepiece and the hardness number looked up in a table.

*Rockwell* hardness machines give hardness numbers directly on a dial or digital display on the machine, as they rely on the in situ depth of indentation. They are thus quicker to use. Various different indenters and loads are used for various ranges of hardness: Thus, Rockwell C (HRC) is used for the harder steels [the indenter being a diamond-tipped cone (brale) under a 150-kg$_f$ load], and Rockwell B (HRB) (where the indenter is a hardened steel ball of $^1/_{16}$ in. diameter under a 100-kg$_f$ load) for softer steels. Because indentations recover more in their depth than in their cross-sectional area, Rockwell tests tend to give higher hardnesses for materials with larger Young's modulus. Nevertheless, there are approximate relationships among the various scales, and tables of equivalent numbers are available (App. 4). The Rockwell tests produce much smaller indentations than the Brinell test.

Although these and other hardness standards may differ in the means of producing and measuring plastic formation, they all depend on the resistance of the metal to plastic deformation over a wide range of strain. The strain field under an indentation is in general complicated and varies from point to point, but it may be expected that there is a "representative" value of the yield stress, characteristic of some average strain produced by the indentation.

**Figure 2-24**   The origin of plastic constraint in a hardness test: (a) a hypothetical unconstrained indentation test on a free column of material where the applied stress (i.e., hardness) would be about the same as uniaxial yield stress; (b) the actual constrained conditions of a practical hardness test where the applied stress is about 2.5 to 3 times the uniaxial yield stress.

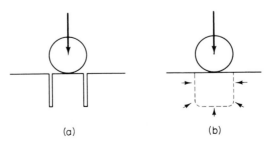

(a)                                                  (b)

Because a hardness measurement is a "constrained" compression test (Fig. 2-24), Brinell hardness and Vickers hardness numbers turn out to be between 2.5 and 3 times this representative yield stress and not merely the same as the yield stress; for these tests, the units of hardness developed historically as $kg_f/mm^2$. Consequently, measurements of hardness in the laboratory are good indicators of changes in strength produced by cold working as described in this chapter or changes in strength produced by alloying or heat treatment described in later chapters. Hardness measurements are also useful for the determination of force required to produce any given observed indentation that resulted from contact between two metal parts. This technique is discussed in Chap. 19.

For many *steels* (but not other materials), the representative yield stress of material deformed by a Brinell indenter is approximately numerically equal to $S_u$, so we have

$$S_u \approx \frac{HB}{2.85} \ (kg_f/mm^2)$$

$$\approx 3.45 HB \ (MPa)$$

(2-26)

where HB is the Brinell hardness and 2.85 is the constraint factor relating hardness to uniaxial tensile or compressive yield strength. Figure 2-25 shows this relation for steels. For

**Figure 2-25**   Brinell hardness number vs. tensile strength for steels of a wide range of carbon content and heat treatment.

other materials with different work-hardening indices, this $S_u$/HB relation is not followed, nor does it hold for materials that fracture in tension with little plastic deformation (compare data points for some of the martensitic steels in the figure and see Chap. 10).

The effects of creep and time-dependent mechanical behavior on hardness are described in Chap. 12.

**Example 2-16**

The Brinell hardness of a certain steel is 207. The work-hardening index of the same steel is 0.15. Determine the value of $K$ in $\sigma = K\varepsilon^n$ for the steel.

**Solution**  $S_u = 3.45\text{HB} = 714$ MPa. The true strain at ultimate is $n = 0.15$. From Eq. (2-19), $\sigma_u = S_u \exp n = 714 \exp 0.15 = 830$ MPa. But $\sigma_u = Kn^n$, and so $K = 830/0.15^{0.15} = 1.1$ GPa.

## DYNAMIC AND NOTCHED-IMPACT TESTING, CREEP TESTING, AND FATIGUE TESTING

The tension, compression, and hardness tests described in this chapter are quasi-static, that is, they are slow and take place under low rates of straining or stressing. There is also a need for mechanical-property information under *dynamic* or *high-strain rate* conditions, where, in broad terms, data are obtained from "fast" tensile tests. Such experiments are described in Chap. 12, where we also discuss *creep* testing, which concerns the behavior of materials under very *low* rates of strain. Again, materials can respond differently if subjected to many repeated applications of the load instead of only one or a small number of applications; such fatigue behavior is the subject of Chap. 18.

In *notched-impact testing*, the energy of a pendulum (swinging under standardized conditions) that is absorbed in fracturing a notched test piece is measured. In a *Charpy test*, the specimen is broken in three-point bending by being struck behind the V-notch; in an *Izod test*, the specimen is broken as a notched cantilever with the notch facing the direction of the blow from the pendulum. As shown in Chap. 14, the energies thereby determined are a rough measure of the area under the $\sigma$–$\varepsilon$ curve as distorted by the complex nature of the deformation at the notch. It is often said that Charpy and Izod data reflect the *toughness* (or resistance to cracking) of materials under dynamic conditions. However, the results of the test are determined more by the notch than by the rate of straining, and although notched-impact data give some qualitative indication of crack resistance, it is preferable to use the quantitative concepts of *fracture toughness* and *fracture mechanics* discussed in Chap. 14 to assess the possibility of cracking leading to failure in a given situation.

## MECHANICAL PROCESSING

We have seen that cold working increases the strength of ductile materials. However, simple stretching as a means of strengthening is limited by necking, which means that cold work no greater than $\varepsilon_{cw} = n$ may be imparted. Thus, in a medium-carbon steel for which $\sigma = 1500\varepsilon^{0.15}$ MPa, the maximum increased yield strength is

$$\sigma = 1500(0.15)^{0.15} \approx 1130 \text{ MPa} = 1.13 \text{ GPa}$$

In compression, however, tensile instabilities are absent, so it seems that greater amounts of cold work could be imparted. This is so, and the widely used process of squeezing material through *rollers* may be thought of as continuous compression (Fig. 2-26). In this way, true strains as high as unity can be enforced and appreciable improvements in strength thus gained, along with desirable changes in shape.

In other types of mechanical processing, also shown in Fig. 2-26, material is pushed or pulled through a converging channel (die) to effect a reduction in cross section. Pulling through a die, which is called *drawing*, is the means by which wire is made. The difference between simple tension and wire drawing is the presence of sideways compressive stresses in the die. They aid the plastic flow, and the state of combined stress within the die requires a *smaller* external stress than the equivalent reduction in cross-sectional area would require in simple tension. True strains greater than $n$ can be readily imparted in wire drawing, the limit of which occurs when the drawing stress becomes equal to the strength of the drawn material (see Example 2-20). Pushing through a die, which is called *extrusion,* is the same as the action in squeezing a toothpaste tube. There is no obvious limiting strain in this process, but friction in particular effectively limits the available reductions. Complicated sections may be made from round or rectangular starting materials if the rollers or dies are profiled (compare the action of tubes for cake frosting or icing decoration.)

**Figure 2-26**   Various types of mechanical processing, such as rolling, drawing, and extrusion, showing also the production of profiled sections.

In mechanical processing, it is common to use the percentage reduction in area $r$ as a measure of strain, given by $r = (A_0 - A)/A_0$. Clearly, this expression has the same basis as the percentage reduction of area to fracture in the tensile test, but here $A$ represents the new cross-sectional area produced from the initial area $A_0$ by processes of rolling, drawing, extrusion, and so on. Because $1 - r = A/A_0$, the relationship between $\varepsilon$ and $r$ is

$$\varepsilon = \ln\left(\frac{1}{1-r}\right) \tag{2-27}$$

Note that percentage reductions, unlike true strains, are *not* additive.

### Example 2-17

A strip of initial thickness 10 mm is rolled down in successive stages to 9 and 5 mm. Find the true strains imparted and demonstrate that percentage reductions in area are not additive. Assume that there is no sideways spread in rolling.

**Solution**    The cross-sectional area of the strip is $tw$, where $t$ is the thickness and $w$ is the width. Thus, from Eq. (2-16), $\varepsilon = \ln(A_0/A) = \ln(t_0 w_0/tw)$. But if there is no sideways spread in the rolling process, $w$ is constant, and so $\varepsilon = \ln(t_0/t)$ alone. Hence, for rolling from 10 mm to 9 mm (abbreviated 10 mm $\to$ 9 mm) and using subscripts before and after $\varepsilon$ to indicate the thickness before and after rolling, for 10 mm $\to$ 9 mm, the strain is $_{10}\varepsilon_9 = \ln \frac{10}{9} = 0.105$ and for 9 mm $\to$ 5 mm, the strain is $_9\varepsilon_5 = \ln \frac{9}{5} = 0.588$. Had the reduction been performed in a single rolling pass from 10 mm to 5 mm, the strain would be $_{10}\varepsilon_5 = \ln \frac{10}{5} = 0.693$. Note that $10 \to 5 = 10 \to 9 + 9 \to 5$.

In terms of reductions of area, again because $w$ is constant, $r = (t_0 - t)/t_0$. Therefore (using the same subscript notation as for $\varepsilon$), $_{10}r_9 = (10 - 9)/10 = 0.1 = 10\%$; $_9r_5 = (9 - 5)/9 = 0.44 = 44\%$; $_{10}r_5 = (10 - 5)/10 = 0.5 = 50\%$. We see that $10\% + 44\% \neq 50\%$. The true strains above could, of course, have been evaluated alternatively using Eq. (2-27), that is, $_{10}\varepsilon_9 = \ln[1/(1 - 0.1)]$, and so on.

### Example 2-18

If the material in Example 2-17 obeys $\sigma = 2000 \ \varepsilon^{0.2}$ MPa, determine the yield strengths that the rolled material would have after the two stages of rolling and sketch the tensile nominal-stress–nominal-strain curves that each would display.

**Solution**    After a 10 mm $\to$ 9 mm rolling pass, $\sigma_y = 2000(0.105)^{0.2} = 1.27$ GPa. $S_y$ will be the same referenced to the 9-mm thickness of the rolled sheet from which tensile test pieces would be made. Again, after the 9 mm $\to$ 5 mm rolling pass, $\sigma_y = 2000(0.105 + 0.588)^{0.2} = 1.86$ GPa and $S_y = 1.86$ GPa referenced to a 5-mm-thick test piece. As far as $\sigma$–$\varepsilon$ curves are concerned, the two samples of cold-rolled sheet would behave as shown in Fig. 2-22, with the prestrained curves blending into the $\sigma$–$\varepsilon$ of the unrolled sheet, the $\varepsilon^*$ values being 0.105 and 0.693 (i.e., 0.105 + 0.588). However, the details of the response of test pieces made from the two samples of cold-rolled sheet would be quite different. According to Example 2-8, the true strain at ultimate of a sample prestrained by $\varepsilon_{cw}$ is $\varepsilon_u = n - \varepsilon_{cw}$. The prestrains in the two sheets here are 0.105 and 0.693, and $n = 0.2$. This suggests that $\varepsilon_n = 0.2 - 0.105 = 0.095$ and $\varepsilon_u = 0.2 - 0.693$, a negative value. This brings out a very important point, which is that samples for which $\varepsilon_{cw} < n$, when tested in tension, display uniform plastic flow over the strain interval from initial yield to $n - \varepsilon_{cw}$, at which point the specimen necks. However, specimens already cold-worked greater than the strain at zero-slope ultimate ($\varepsilon_{cw} > n$) show *no uniform*

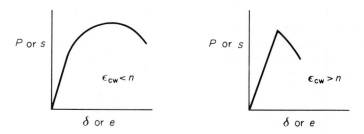

**Figure 2-27**    Solution of Example 2-18, showing how annealed materials prestrained less than $n$ (the strain at the zero-slope ultimate) display uniform plastic flow prior to necking in subsequent tensile tests, but how materials cold-worked to a strain greater than $n$ neck immediately on yielding.

*reduction* in cross section after initial yield and instead *neck immediately*; yield and ultimate coincide. Thus, load–extension (or $s$–$e$) curves for the two samples of cold-rolled sheet are as shown in Fig. 2-27. Compare the difference with the $\sigma$–$\varepsilon$ curves (Fig. 2-22). Clearly, when $\varepsilon_{cw} > n$, the method of prestraining cannot be direct tension.

## WORKING LOADS

Approximate values for the loads required in mechanical processing may be obtained by considering work. Consider first the work done in a simple tension test, which we use as a reference process. The force on the specimen at some instant during deformation is given by $\sigma A$, where $\sigma$ and $A$ are the current values of true stress and cross-sectional area, respectively; in mechanical processing, we often call the current value of true stress during deformation the *flow stress*. The increment· of work done $\Delta W$ in extending the specimen a further $\Delta L$ is given by force × distance moved, so

$$\Delta W = \sigma A\, \Delta L$$

The work done per unit volume is

$$\frac{\Delta W}{V} = \sigma \frac{\Delta L}{L} = \sigma\, \Delta\varepsilon$$

where the volume of the test piece is $V = AL$. If we assume constancy of volume, the total work done per unit volume in extending the test piece from an initial length $L_0$ to a final length $L^*$ [i.e., from $\varepsilon = 0$ to $\varepsilon^* = \ln(L^*/L_0)$] is

$$\frac{W}{V} = \int_0^{\varepsilon^*} \sigma\, \Delta\varepsilon \tag{2-28}$$

which is the area under the $\sigma$–$\varepsilon$ curve up to $\varepsilon^*$, as shown in Fig. 2-28.

If we have a highly work-hardened material, with a flat $\sigma$–$\varepsilon$ curve or if we assume some average yield stress $\bar{\sigma}$, we may write

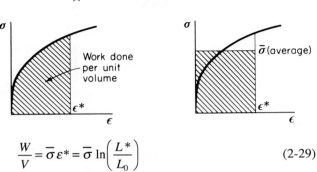

**Figure 2-28**  Work done in a simple tension test. The area under the $\sigma$–$\varepsilon$ curve out to some strain $\varepsilon^*$ is the work done per unit volume of gage section in attaining that strain $\varepsilon^*$. The area in a work-hardening case may be viewed as an equivalent rectangle over the same strain base but with average (constant) flow stress $\overline{\sigma}$.

$$\frac{W}{V} = \overline{\sigma}\,\varepsilon^* = \overline{\sigma}\,\ln\!\left(\frac{L^*}{L_0}\right) \tag{2-29}$$

with units of J/m$^3$, for example. In the case of a material that work-hardens according to $\sigma = K\varepsilon^n$,

$$\overline{\sigma} = \frac{1}{\varepsilon^*}\int_0^{\varepsilon^*} K\varepsilon^n\,d\varepsilon = \frac{K(\varepsilon^*)^n}{n+1} \tag{2-30}$$

Equation (2-29) gives the work required for *homogeneous deformation* (discussed in the section on compression and combined stress states earlier in this chapter; see Fig. 2-10). If we are prepared to neglect the distortion effects of friction in the die and so-called redundant deformation (which concerns the work of "skewing the grids" and which, although inherent in a given process, is not necessary as far as the external shape change is concerned), we can use this work expression to estimate extrusion and drawing loads.

Consider extrusion, where a "slug" originally of length $L_0$ and cross-sectional area $A_0$ becomes a length $L$ of extrudate of area $A$ (Fig. 2-29). The work done by the extrusion ram force $P$ in pushing out the slug is force × distance moved = $PL_0$. From Eq. (2-29),

$$PL_0 = V\overline{\sigma}\ln\left(\frac{L}{L_0}\right)$$

So

$$P = \left(\frac{V}{L_0}\right)\overline{\sigma}\ln\left(\frac{L}{L_0}\right)$$

or

$$\begin{aligned}
\sigma_{\text{extrusion}} = \frac{P}{A_0} &= \overline{\sigma}\,\ln\left(\frac{L}{L_0}\right) \\
&= \overline{\sigma}\,\ln\left(\frac{A_0}{A}\right) \\
&= \overline{\sigma}\,\ln\left(\frac{1}{1-r}\right)
\end{aligned} \tag{2-31}$$

**Figure 2-29**   The extrusion of a billet where a length $L_0$ of area $A_0$ becomes a length $L$ of extrudate with area $A$.

Thus, the work done per unit volume, the area under the $\sigma$–$\varepsilon$ curve between the strains in question, and the magnitude of the extrusion stress are all identical.

The same arguments apply for working stresses in wire drawing and other processes, where the assumption of homogeneous deformation is made. Where friction and redundant deformation are significant, and thus homogeneous work underestimates the working loads, an efficiency factor $\eta$ can be introduced, so

$$\sigma_{\text{extrusion}} \text{ or } \quad \sigma_{\text{drawing}}, \text{ and so on, is given by} \left(\frac{1}{\eta}\right)\overline{\sigma}\varepsilon * \qquad (2\text{-}32)$$

The effect of strain rate in altering $\overline{\sigma}$ in working processes is discussed in Chap. 12.

### Exercise 2-19

Find the load necessary to extrude a round bar 100 mm in diameter made of an aluminum alloy of average yield strength 300 MPa through a hole of 20 mm diameter.

**Solution**   The extrusion pressure, given by Eq. (2-31), is $(300 \times 10^6)2 \ln(100/20) = 966$ MPa. The cross-sectional area at the ram is

$$\frac{\pi}{4}(100 \times 10^{-3})^2 = 7.85 \times 10^{-3}\ \text{m}^2$$

So the extrusion load is $(966 \times 10^6)\ (7.85 \times 10^{-3}) = 7.6$ MN.

### Exercise 2-20

Determine the maximum attainable reduction in wire drawing.

**Solution**   The drawn wire will yield and flow will take place downstream of the die when the drawing stress is as great as the yield stress of the drawn material. From Eq. (2-31), the drawing stress is $\overline{\sigma}\varepsilon$, where $\overline{\sigma}$ is the average flow stress within the die. If the material has a constant yield stress $\overline{\sigma}$, the yield strength of the drawn (and, indeed, the undrawn) material will be the same, and so under these circumstances the maximum strain attainable in wire drawing will be set by $\overline{\sigma} = \overline{\sigma}\varepsilon$ or $\varepsilon = 1$ or $r = 63\%$. If the material work-hardens through the die according to $\sigma = K\varepsilon^n$, then from Eq. (2-30), $\overline{\sigma} = K\varepsilon^n/(n + 1)$, but the yield strength of the wire as it emerges from the die will be $\sigma_y = K\varepsilon^n$. The limit of drawing is then set by $K\varepsilon^n = [K\varepsilon^n/(n + 1)]\varepsilon$ or $\varepsilon = 1 + n$, which is greater than before; for example, with $n = 0.1$, $r = 70\%$, and with $n = 0.5$, $r = 78\%$. Note that because $\varepsilon$ (equal to 1 or to $1 + n$) is greater than $n$, yielding and necking coincide at the exit to the die (see Example 2-18). In practice, because of friction and redundant deformation, practical limiting reductions in wire drawing are smaller than these values.

## ANISOTROPY

The discussions of strength so far have not suggested that the properties of materials can vary with direction. That is, we assume that materials are *isotropic* and that if test pieces are taken at various directions from a large piece of material, they will all give essentially the same $\sigma$–$\varepsilon$ curve. Almost no materials are truly uniform and isotropic. An obvious example is wood, which has different properties along and perpendicular to the grain. Rope has little strength across its width. Figure 2-30(a) shows the results of tension tests of specimens cut, at different angles, from a sheet of a cold-rolled aluminum alloy. Figure 2-30(b) gives similar data for Ti-6Al-4V sheet material. It is seen that there are considerable variations in strength with orientation.

These effects are called *anisotropy*. The magnitude of the differences in strength and ductility depends on the degree of cold work, and anisotropy is commonly encountered in metals and polymers that have been rolled, drawn, extruded, forged, and so on. Directional variations may be present in solids for other reasons, the bases of which are covered in later chapters. For example, single-phase crystalline materials have some degree of variation in grain size associated with point-to-point variations in cooling rates. Two-phase

(a)

**Figure 2-30**   Anisotropy in (a) commercial-purity aluminum (99.0% Al) in the "half-hard" condition (i.e., cold rolled to a reduction of about 21% in area after the last anneal), and (b) Ti-6Al-4V alloy sheet (8 mm wide by 500 $\mu$m thick).

(b)

**Figure 2-30**   (cont.)

**Figure 2-31**   Elliptical cross section of neck in tensile bar of 1100 aluminum. (Photograph courtesy of W. H. Durrant.)

materials show in addition compositional variations that result from finite cooling rates during solidification. Some structural reasons for anisotropic effects are mentioned in Chap. 5, and fiber-reinforced composites, which are inherently anisotropic, are discussed in Chap. 17. In general terms, amorphous materials in bulk can be more nearly isotropic than crystalline materials, except for fabrication effects. It is important to be aware that *preferred orientation* or *texture effects* (other names associated with anisotropy) can persist to some extent even after so-called thorough homogenizing or annealing heat treatments, as mentioned in later chapters such as Chap. 6.

An important indicator of anisotropy in metals is the $r$ factor (not to be confused with cold work, which uses the same symbol), which can be deduced from measurements of the strains in a tension test. As a specimen extends in one direction, it shrinks in the other directions; from considerations of constant plastic volume, we have

$$\varepsilon_L + \varepsilon_w + \varepsilon_t = 0 \tag{2-33}$$

where $\varepsilon_w$ and $\varepsilon_t$ are (negative) width and thickness strains. In an isotropic material, with no preferential orientations, $\varepsilon_w = \varepsilon_t$. In a round bar, for example, the gage section remains circular and reduces uniformly. In an anisotropic material, the gage section may become elliptical, with some directions shrinking preferentially; again, ball hardness indentations are often noncircular in anisotropic solids. This type of effect is shown in a tensile test specimen in Fig. 2-31. Then, and in sheet specimens for which width and thickness have obvious meanings, $\varepsilon_w \neq \varepsilon_t$. The degree of anisotropy may be quantified in terms of the ratio $r = \varepsilon_w/\varepsilon_t$. This can be obtained by mounting strain sensors on tensile specimens in the width and thickness directions. Sometimes, with very thin sheet material, $\varepsilon_t$ is too small to measure accurately. In that case, $\varepsilon_t$ can be inferred from $\varepsilon_w$ and $\varepsilon_L$ by means of Eq. (2-33).

Anisotropy can be beneficial in some situations if the major component of load is always in one direction. Furthermore, in sheet metal working certain types of anisotropy help to prevent premature failure of the metal during forming processes.

**Example 2-21**

A strip of low-carbon steel, 1 mm thick, 10 mm wide, and 75 mm long, is observed to become 9.5 mm wide when it is stretched to 80 mm in length. Find the $r$ factor for this steel.

**Solution**    $\varepsilon_w = \ln(9.5/10) = -0.051$. $\varepsilon_L = \ln(80/75) = 0.065$.
$\varepsilon_t = -(-0.051 + 0.064) = -0.014$. Therefore,

$$r = \frac{-0.051}{-0.014} = 3.6$$

# REFERENCES

1. American Society for Metals, *Metals Handbook*, 8th ed., vol. 1, Properties and Selection of Metals. Metals Park, Ohio: American Society for Metals, 1961.

2. American Society for Testing and Materials, *Index to Standards*. Philadelphia: American Society for Testing and Materials (use latest edition). The index will give the location in the *Book of ASTM Standards* for any ASTM test.

3. American Welding Society, *Welding Handbook,* 5th ed., sec. 1. New York: American Welding Society, 1963.

4. Crane, F. A. A., *Mechanical Working of Metals.* London: Macmillan, 1964.

5. Dieter, George E., Jr., *Mechanical Metallurgy.* New York: McGraw-Hill, 1961.

6. Fenner, A. J., *Mechanical Testing of Materials.* London: George Newnes, 1965.

7. Gordon, J. E., *The New Science of Strong Materials—Or Why You Don't Fall Through the Floor.* London: Penguin Books, 1968.

8. Gordon, J. E., *Structures—Or Why Things Don't Fall Down.* London: Penguin Books, 1978.

9. Marin, Joseph, *Mechanical Behavior of Engineering Materials.* Englewood Cliffs, N.J.: Prentice Hall, 1962.

10. McClintock, Frank A., and Ali S. Argon, eds., *Mechanical Behavior of Materials.* Reading, Mass.: Addison-Wesley, 1966.

11. O'Neill, H., *Hardness Measurement of Metals and Alloys,* 2nd ed. London: Chapman and Hall, 1967.

12. Polakowski, N. H., and E. J. Ripling, *Strength and Structure of Engineering Materials.* Englewood Cliffs, N.J.: Prentice Hall, 1966.

13. Tabor, D., *The Hardness of Metals.* Oxford: Clarendon Press, 1951.

## PROBLEMS

**2-1.** A tensile test on 1020 steel gives the following results:

|   | Load, kN | Diameter, mm | Length, mm |
|---|---|---|---|
| 0 | 0 | 12.8 | 50.800 |
| 1 | 22.2 | — | 50.848 |
| 2 | 28.5 | — | Yielding begins |
| 3 | 51.2 | 12.2 | 56.1 |
| 4 | 51.2 | 10.4 | 67.3 |
| 5 | 43.6 | — | 69.8 (fracture) |

No data were taken at maximum load.
(a) Calculate the elastic modulus.
(b) Calculate the maximum load during the test.
(c) Calculate the maximum nominal strain.
(d) Calculate the strength of this steel.

**2-2.** A tensile test of 0.40% carbon steel produced the following data:

|   | Load, kN | Diameter, mm | Length, mm |
|---|---|---|---|
| Initial | 0 | 12.8 | 35.6 |
| Maximum load | 88.1 | 11.9 | 40.6 |
| Fracture | 75.0 | 9.22 | 46.7 |

(a) Calculate the strength of this specimen.
(b) Calculate the maximum nominal strain during the test.

**2-3.** The following data were taken during a tensile test of a 0.20% carbon steel.

| Reading | 0 | 1 | 2 | 3 | Fracture |
|---|---|---|---|---|---|
| Load, kN | 0 | 40.0 | 49.8 | 50.8 | 46.3 |
| Length, mm | 50.0 | 52.8 | 56.9 | 65.3 | 69.1 |
| Diameter, mm | 12.8 | 12.6 | 12.1 | 11.0 | 9.65 |

None of these readings was taken at maximum load.
**(a)** Calculate the maximum nominal strain during the test.
**(b)** Calculate the nominal stress at reading 2.
**(c)** Calculate the maximum true stress during the test.
**(d)** Calculate the strength of this specimen. (Show clearly the data you have used.)

**2-4.** A tensile test of 70-30 brass produced the following data (*none* of these readings was taken at maximum load):

| | Load, kN | Diameter, mm | Length, mm |
|---|---|---|---|
| Initial | 0 | 12.8 | 50.8 |
| Reading 1 | 53.8 | 11.4 | 63.5 |
| Reading 2 | 55.1 | 10.7 | 68.6 |
| Fracture | — | 8.26 | 74.9 |

**(a)** Calculate the strength of this specimen.
**(b)** Calculate the maximum reduction in area during the test.
**(c)** Calculate the maximum nominal strain during the test.
**(d)** Calculate the diameter of the specimen at the instant of maximum load.

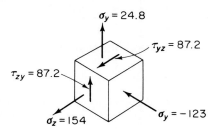

**Figure 2-32**

**2-5.** Calculate the magnitude of the maximum shear stress in the stressed element in Fig. 2-32. All stresses are in MPa.

**2-6.** Calculate the maximum normal stress and the maximum shear stress in the cube shown in Fig. 2-33.

**Figure 2-33**

**2-7.** Calculate the maximum normal stress and the maximum shear stress in the cube shown in Fig. 2-34.

**Figure 2-34**

**2-8.** **(a)** Calculate the magnitude of the maximum tensile stress in the cube shown in Fig. 2-35.
  **(b)** Calculate the angles (degrees) between the normal to the plane of maximum stress and the $x$-, $y$-, and $z$-axes.
  **(c)** Calculate the maximum shear stress in the cube.

**Figure 2-35**

**2-9.** An element of a solid exposed to a stress system is shown in Fig. 2-36. Normal stress fracture will occur when the maximum normal stress is 323 MPa; shear fracture occurs when the maximum shear stress is 192 MPa.
  **(a)** If $A$ is gradually increased from zero, calculate $A$ (in MPa) when fracture occurs.
  **(b)** By which mode does fracture occur, normal stress or shear stress?

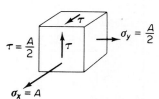

**Figure 2-36**

**2-10.** The graph in Fig. 2-37 shows results of a tensile test of 1020 steel. Specific values at several points are given in the table.

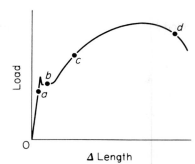

**Figure 2-37**                                Δ Length

(a) Calculate the constants $K$ and $n$ for the plastic strain-hardening region.
(b) Calculate the modulus of elasticity for this steel.
(c) Estimate the yield strength.
(d) Calculate the maximum load during the test.

|          | Load, kN | Gage length, mm | Diameter, mm |
|----------|----------|-----------------|--------------|
| Initial  | 0        | 50.800          | 12.800       |
| a        | 23.6     | 50.848          | 12.796       |
| b        | 27.1     | 51.900          | —            |
| c        | 46.4     | 55.6            | 12.4         |
| d        | 52.5     | 62.5            | 10.9         |

**2-11.** You are given one half of a fractured brass tensile test piece (Fig. 2-38) and told that the initial diameter was 12 mm and the maximum load during the test was 36 kN. Calculate the numerical value of $K$ in the relation $\sigma = K\varepsilon^n$.

10.5 mm                          9 mm

**Figure 2-38**

**2-12.** Annealed steel wire and cold-worked aluminum wire kink when bent; cold-worked steel wire and annealed aluminum wire do not kink but bend "smoothly." Explain this seeming paradox.

**2-13.** A laboratory experiment involved a tensile specimen that had been pulled beyond necking, unloaded, and machined down to a diameter slightly larger than the neck. In the subsequent test, why did the specimen not break where the neck had been?

**2-14.** Plot separately the nominal stress–strain and true stress–strain curves to the same scale (but without numbers) for the situations indicated. Assume that elastic deformation is negligible compared to plastic deformation. Make your graphs for parts (a)–(c) consistent with each other with respect to relative magnitude. Label each curve *a*, *b*, *c*, or *d*. Plot *s–e* and $\sigma$–$\varepsilon$ curves for:

   **(a)** A specimen of ductile metal that necks; indicate the ultimate point on both curves with *u*.
   **(b)** A specimen machined from a bar of the same kind of metal as used in part (a). The bar was subjected earlier to a stress about halfway between yield and ultimate.
   **(c)** A specimen machined from a bar of the same kind of metal as used in part (a). The bar had earlier been cold-rolled to a longitudinal true strain well beyond the ultimate true strain but less than the fracture strain.
   **(d)** A ductile material that does not strain-harden.

**2-15.** A tensile test of a metal was run in the laboratory, but only the maximum load was measured during the test (it was 59 kN). All other data were taken before and after the test as follows:
Initial diameter, 13.0 mm
Minimum diameter after test (at point of fracture), 10.0 mm
Maximum diameter after test (away from fracture), 11.6 mm
   **(a)** Find the strain-hardening exponent *n* in the equation $\sigma = K\varepsilon^n$ for this metal.
   **(b)** Find the load on the specimen at the instant just prior to fracture.

**2-16.** The following data were obtained in a tensile test of naval brass:

|            | 0    | 1    | Fracture |
| ---------- | ---- | ---- | -------- |
| Load, kN   | 0    | 50.7 | 48.0     |
| Diameter, mm | 12.8 | 11.9 | 8.5    |

Condition 1 as measured was *not* the maximum load condition. Calculate the strength of this brass.

**2-17.** For the plastic strain-hardening graph for metals A and B shown in Fig. 2-39,
   **(a)** Find the value of *K* for metal A.
   **(b)** Find the value of *n* for metal B.
   **(c)** Determine which is stronger, A or B.

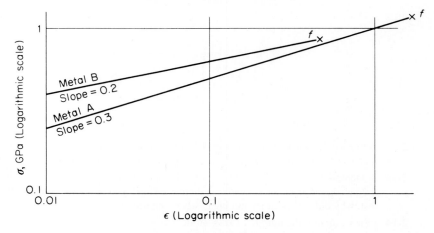

**Figure 2-39**

**2-18.** For the log $\sigma$ vs. log $\varepsilon$ plots shown in Fig. 2-40, which specimen has
  (**a**)  The largest $K$ ?
  (**b**)  The largest true strain at ultimate?
  (**c**)  The largest strength?
  (**d**)  The largest elastic modulus?

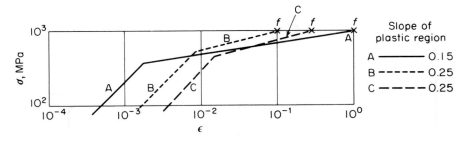

**Figure 2-40**

**2-19.** Why are strength and hardness usually related? For what general category of metals are strength and hardness less likely to be related and why?

**2-20.** If a steel bar undergoing a Rockwell B test has a burr on it that contacts the anvil of the testing machine, will the hardness reading obtained be too high, unchanged, or too low?

**2-21.** When the HRB hardness at the bottom of several Brinell indentations was measured, these data resulted:

| Material | Annealed | 500-kg$_f$ Load | 1000 kg$_f$ Load |
|---|---|---|---|
| 1020 steel | 46 | 72 | 77 |
| Brass | −35 | 66 | 84 |

  (**a**)  Estimate the HRB that you would expect to measure at the bottom of a 1500-kg$_f$ Brinell indentation for 1020 steel and for brass.
  (**b**)  Which material has the higher value of $n$?
  (**c**)  Which material has the higher value of $K$? Describe exactly the line of reasoning you used to reach the answer.

**2-22.** Two separate Brinell hardness tests using a ball 10 mm in diameter are made at different locations on the same piece of 70-30 brass, (**a**) at 500 kg$_f$ (4.90 kN) and (**b**) at 1500 kg$_f$ (14.7 kN). Which test gives the higher hardness, (a) or (b)? Explain why this happens.

**2-23.** An annealed metal follows $\sigma = (700 \text{ MPa})\varepsilon^{0.3}$. To what final thickness must a piece of this metal, initially 15 mm thick, be cold-rolled to increase its hardness to 160 HB? (Assume $S_u = 3.45 \times$ HB MPa.)

**2-24.** A low-carbon steel that follows $\sigma = (630 \text{ MPa})\varepsilon^{0.18}$ is annealed and then cold-worked 55%.
  (**a**)  Find the Brinell hardness before the cold working and after the cold working if $S_u = 3.45 \times$ HB.

(b) The reduction of area to fracture from the annealed condition is 70%. If you give the 55% cold-worked specimen to a stranger who does not know its history, what value will be found for reduction of area (cold work) to fracture?

**2-25.** An annealed 70–30 brass bar yields the following when rolled:

| Condition | Thickness, mm | HRB |
|---|---|---|
| A (initial) | 12.7 | −12 |
| B | 11.5 | 57 |
| C | 7.5 | 83 |

(a) Calculate the longitudinal (i.e., rolling direction) true strain from A to B.
(b) Calculate the longitudinal true strain from B to C.
(c) Contrast the true strain increments in parts (a) and (b) with the changes in hardness and explain why the changes are not proportional.

**2-26.** Data from the cold rolling of annealed brass, initially 12.7 mm thick, are as follows:

| Thickness, mm | 12.7 | 11.5 | 10.0 | 9.0 | 7.9 | 6.3 |
|---|---|---|---|---|---|---|
| HRB | 8 | 51 | 75 | 82 | 88 | 95 |

If it is known from tensile data that this brass follows the relationship $\sigma = (700)\varepsilon^{0.5}$ MPa, find the HRB hardness of the broken tensile bar at a point midway between the fracture surface and the shoulder at the end of the test length.

**2-27.** Three hardness readings during the cold rolling of 70-30 brass are as follows: initially 13 mm thick, 13 HRB; 12 mm thick, 40 HRB; 6.4 mm thick, 85 HRB. Calculate $n$ in the relation HRB = constant $\times \varepsilon^n$, where $\varepsilon$ is the strain in the rolling direction.

**2-28.** A square bar of brass that fractures at a true tensile strain of 0.6 is annealed, then cold-rolled so that the reduction in thickness is 35%, and given to you. If you were unaware that the bar had already been cold-rolled, what additional percentage of reduction in thickness would you obtain if you cold-rolled it until fracture occurred, assuming that the true tensile strain at fracture applies to rolling ?

**2-29.** A fabricator wishes to use some aluminum sheet in a rolling operation. He conducts a tensile test on a sample initially 6 mm × 6 mm in cross section and determines that this material follows $\sigma = (210)\varepsilon^{0.21}$ MPa and has a final area at fracture of 19 mm². The cold-rolled sheet must have a yield strength of 170 MPa and be 10 mm thick. Find the *required initial thickness* of the sheet before rolling.

**2-30.** You are given several identical metal bars of 25-mm-square cross section and about 300 mm in length. You conduct a tensile test on one bar and observe that the maximum load during the test is 200 kN and that the cross-sectional area is 500 mm² when maximum load is reached. The minimum cross-sectional area at fracture is 280 mm². You wish to produce a bar of 12 mm × 25 mm cross section that will support a maximum load of 140 kN; such a bar could be made of one of the 25-mm-square bars by cold rolling to get the strength and then machining to get the correct dimensions. If the final bar is to have the highest possible ductility (i.e., do not cold-roll any more than necessary), to what thickness (before final machining) should the 25-mm-square bar be rolled?

**2-31.** Sketch a typical work-hardening stress–strain curve and superimpose the "drawing stress–drawing strain" relationship *on the same axes,* using Eq. (2-30) for the stress in wire drawing. Hence show how a maximum attainable reduction comes about in drawing and evaluate it for $\sigma = K\varepsilon^n$. What factors does the simple theory above neglect? What effect do they have on the value of maximum reduction? What effect do they have on the mechanical properties of the final product?

**2-32.** Find the extrusion stress necessary to extrude iron, already cold-worked by 10%, through a 30% reduction in area. Assume that $\sigma = (560)(\varepsilon_{cw} + \varepsilon)^{0.32}$ MPa.

**2-33.** Experiments were performed to determine the "best efficiency" in drawing through some fixed-angle dies. Best efficiency means the highest ratio of the theoretical drawing stress for homogeneous deformation to the *actual* drawing stress. The results for copper wire are shown in Fig. 2-41 for reductions between $\varepsilon = 0.45$ and $\varepsilon = 0.70$. Copper wire is to be drawn in the factory from a diameter of 6.5 mm to 2.5 mm. It has been decided that the drawing stress should be no more than 75% of the wire flow stress in order to prevent fracture. If the copper being drawn has an almost constant flow stress (i.e., displays only minor work-hardening), how many drawing passes will be required to end up with the correct size wire?

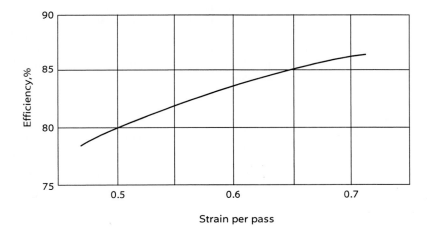

**Figure 2-41**

**2-34.** What is the $r$ value of an isotropic plate?

**2-35.** Dimensions of the gage section of a piece of cold-rolled aluminum strip were measured before and after a tensile test which was halted before necking occurred. The results were as follows:

|         | Length, mm | Width, mm | Thickness, mm |
|---------|------------|-----------|---------------|
| Initial | 159        | 27.2      | 5.36          |
| Final   | 200        | 24.4      | 4.71          |

Calculate the *r*-value for this aluminum.

**2-36.** Two flat strip tensile specimens, having gage section dimensions of $L_0$ = 50 mm, $w_0$ = 19 mm, and $t_0$ = 1.2 mm, are made from a sheet of metal 1 × 2 m whose thickness is 1.2 mm. One specimen is made parallel to the 2-m length, while the second is at right angles to the first. The following tensile results were obtained:

|  | Specimen 1 | | | Specimen 2 | | |
|---|---|---|---|---|---|---|
|  | Load, $N$ | $L$, mm | $w$, mm | Load, $N$ | $L$, mm | $w$, mm |
| (i) Yield | 720 | 50.1 | 19.0 | 720 | 50.1 | 19.0 |
| (ii) | 1100 | 58.0 | 17.7 | 1000 | 58.0 | 17.4 |
| (iii) Max | 1400 | 64.0 | 17.0 | 1280 | 63.0 | 16.5 |

As a means of studying anisotropy:
(a) Determine the $r$ values for these specimens.
(b) What is the thickness of specimen 2 at the maximum load?
(c) What is the true stress at maximum load for specimen 1?
(d) Compare the values of the tensile strengths of these specimens.

**2-37.** An internally pressurized aluminum sphere, of radius 73 mm and initially 1.65 mm thick, fractures at a pressure of 5 MPa. The thickness at fracture is 1.07 mm. Calculate the equivalent through-thickness compressive stress at fracture. If the sphere is isotropic, calculate the tensile circumferential strain at fracture.

**2-38.** The following data were taken on a specimen of 0.20% carbon steel (1020 steel):

| Load, kN | Length, mm | Dia., mm | Load, kN | Length, mm | Dia., mm |
|---|---|---|---|---|---|
| 0 | 50.8 | 12.8 | 48.9 | 55.6 | 12.3 |
| 32.9 | 52.1 | 12.7 | 51.9 (max.) | 64.0 | 11.4 |
| 37.4 | 52.8 | 12.6 | 48.9 | 68.6 | 10.1 |
| 40.0 | 53.1 | 12.6 | 41.8 (fract.) | 71.1 | 9.07 |
| 44.5 | 53.8 | 12.5 |  |  |  |

(a) Calculate the strength of this steel.
(b) Calculate the maximum nominal strain during the test.
(c) Calculate the maximum true stress during the test.

**2-39.** A tensile test on a specimen of annealed metal X results in the following data:

| Condition | Load, kN | Minimum diameter, mm |
|---|---|---|
| Initial | 0 | 12.83 |
| Yield | 48.9 | 12.80 |
| Maximum load | 75.6 | 9.45 |
| Fracture | — | 7.98 |

(a) Find the maximum nominal stress during the test.
(b) Find the maximum true stress during the test.
(c) You are given a 25-mm-thick plate of annealed metal X. What is the minimum thickness to which this plate can be cold-rolled without further heat treatment?

**2-40.** A standard tensile test is run on a piece of steel, and the following data are obtained:

| Condition | Load, kN | Minimum diameter, mm |
|---|---|---|
| Initial | 0 | 12.8 |
| Maximum load | 103 | 12.2 |
| Fracture | — | 8.94 |

(a) Calculate the exponent $n$ in $\sigma = K\varepsilon^n$.
(b) Calculate $K$ in the same equation.
(c) Find the true strain at fracture.
(d) Calculate the true stress at fracture.
(e) This was an actual test. The load at fracture was measured as 81.8 kN. What value does this give for true stress at fracture? Explain any discrepancy with the answer in part (d).

**2-41.** A low-carbon steel is found to follow $\sigma = (717 \text{ MPa})\varepsilon^{0.234}$. If its initial diameter is 12.8 mm and its diameter at fracture is 9.27 mm, calculate the maximum tensile load (kN) that is reached during a tensile test of this steel.

**2-42.** An element of material is subjected to the following stresses: $\sigma_x = 60$ MPa, $\sigma_y = 140$ MPa, $\sigma_z = 220$ MPa, $\tau_{xz} = 50$ MPa. Calculate the maximum shear stress in the element.

**2-43.** For a cube stressed by $\sigma_1 = 110$ MPa, $\sigma_2 = -70$ MPa, $\sigma_3 = -150$ MPa, what value of compressive stress must $\sigma_3$ reach in order for $\tau_{max}$ to be equal to 150 MPa?

**2-44.** A specimen of annealed copper of initial hardness HRE 30 is subjected to an axial true strain of 0.27. Estimate its Rockwell E hardness after this strain, show the source of all your numbers, and state any assumptions you have made.

**2-45.** A tensile test specimen provides the following data when tested:

| Condition | Load, kN | Min. dia., mm |
|---|---|---|
| Initial | 0 | 12.8 |
| Maximum load | 62.7 | 10.3 |
| Fracture | — | 8.42 |

Calculate the maximum true stress during this test.

**2-46.** Calculate the magnitude of the maximum normal stress at a point in a solid that is subjected to the following stresses:

$$\sigma_x = 327 \text{ MPa}, \ \sigma_y = 463 \text{ MPa}, \ \sigma_z = -115 \text{ MPa}, \ \tau_{yz} = 134 \text{ MPa}.$$

**2-47.** A tensile test of a 70–30 brass produces the following data:

| Diameter, mm | Load, kN |
|---|---|
| 12.8 | 0 |
| 11.9 | 28.0 |
| 11.1 | 37.4 |

Calculate the strain-hardening exponent for this alloy.

**2-48.** The outer surface of a thin, hollow cylinder that is internally pressurized and also subjected to a twisting moment is found to experience the following stresses (all in MPa): $\sigma_x = 89.6$, $\sigma_y = 44.8$, $\tau_{xy} = 29.4$. Calculate the maximum shear stress at this point in the cylinder.

**2-49.** A simple torsion test on a metal X causes failure by slip when the maximum shear stress is 150 MPa. A uniaxial tension test on the same metal X causes cleavage fracture when the maximum normal stress is 250 MPa. If a tensile test is performed on metal X in a chamber under uniform compressive hydrostatic pressure of 80.0 MPa, then calculate the following:
(a) The failure mode: slip or cleavage fracture.
(b) The tensile stress applied by the tensile testing machine at the instant of failure.

**2-50.** A nonstandard tensile test on metal Z gives the following data:

| Condition | Min. diameter, mm | Load, kN |
|---|---|---|
| Initial | 14.6 | 0 |
| 1 | 14.1 | 35.5 |
| Fracture | 12.7 | 59.0 |

Calculate:
(a) The constants in the strain-hardening equation, $\sigma = K\varepsilon^n$.
(b) The strength of the specimen.

**2-51.** A point in a solid is subjected to the following stresses:

$$\sigma_x = 280 \text{ MPa}, \ \sigma_y = -142 \text{ MPa}, \ \tau_{xz} = 130 \text{ MPa}$$

(a) Calculate the maximum principal stress at this point.
(b) Calculate the angle from the normal to the x-plane to the normal to the plane of maximum principal stress.
(c) On a cube with x-, y-, z-planes as its faces, draw with some precision the plane of maximum principal stress and label the angle calculated in part (b).

**2-52.** A cylindrical specimen of metal X is loaded in a tensile testing machine until it fractures. The results of this test are as follows:

| Condition | Load, kN | Min. diameter., mm |
|---|---|---|
| Initial | 0 | 11.6 |
| Max. load | 85.7 | 11.1 |
| Fracture | — | 8.16 |

Calculate the true stress at fracture.

**2-53.** A piece of chalk that is built in at one end, as shown in Fig. 2-42, fractures when it is subjected to a torque $M_t$ of 1.67 N-m and an axial force $F$ of 1.43 kN. The maximum torsional shear stress $\tau$ for a cylinder of diameter $d$ under torsional torque $M_t$ is

$$\tau = \frac{16M_t}{\pi d^3}$$

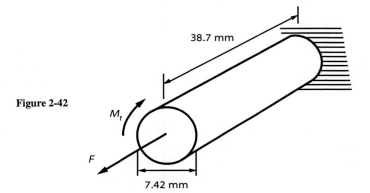

**Figure 2-42**

Calculate the maximum tensile stress in the chalk at the instant of fracture.

# 3

# *Classes of Properties of Engineering Materials*

The variety of materials used for engineering purposes is the result of differences in mechanical and physical properties, cost, availability, and prior experience by the designer or user. Eskimos build igloos of ice because it is handy (cheap), an adequate insulator, strong enough in compression, and easily fabricated, and because they know how to build an ice igloo. Pueblo Indians used adobe (clay) for exactly the same reasons. Each designer has obviously made a wise application of the materials available, although the choices are quite limited.

In contrast, many more factors influence the choice of material for a component of a modern engineering structure, to the extent that often the better choice between two materials can be made only after trials of both. For example, the choice among gray cast iron, aluminum–silicon alloy, and magnesium alloy for automobile engine blocks remains unresolved to the present day. Attempts have been made in the past to produce large engine blocks [with a displacement on the order 5 L (305 in.$^3$)] of aluminum alloy, but most aluminum and magnesium alloy engine blocks are found in smaller engines. The better thermal conductivity and lighter weight of the aluminum and magnesium alloys must be evaluated against the lower cost and greater experience with gray cast iron.

In this chapter, we examine the mechanical properties exhibited by the different classes of metals and polymers of engineering interest. The properties of particular interest are given in App. 5. Structural materials such as concrete, brick, asphalt, soil, and wood (and ice and adobe, as well) are not included here, as all except wood lack sufficient tensile strength for general mechanical use, and wood, while a magnificently balanced structural material, is relatively immune to changes in its properties by humans. The mechanical properties of greatest importance are described in the next three paragraphs.

*Stiffness*, which for a fixed-size piece of material is another word for Young's (elastic) modulus, is the ratio of stress to strain in the linear elastic portion of the stress–strain graph, the lower portion of line *oa* in Fig. 2-2. Stiffness controls the deflection under load and is thus important in retaining close tolerances of parts at stresses lower than yield strength. In parts exposed to bending, failure by buckling or excessive deflection can occur at low stresses if the elastic modulus is too low.

The *strength* of any engineering material is the maximum nominal stress in a tensile test [Eq. (2-3)]. For a solid to be able to support the forces that it will experience during service, it must at least have sufficient strength.

*Toughness* is the ability of a material to resist cracking and will be described more explicitly in Chap. 14. Green wood is tough, and window glass is not tough (brittle); a wide range of materials have toughness between these extremes. One rough indicator of toughness is the nominal strain to fracture $\varepsilon_f$. Another approximate measure of toughness is $S_u/S_y$, because large values of this ratio indicate large plastic deformation after yield and hence high toughness. [Be aware, however, of the "scale effect" in fracture (Chap. 14) where very large structures made of materials that are ductile in small (laboratory-size) test pieces, may behave in a brittle fashion.]

The properties of metals and alloys depend in part on the properties of the pure elements from which they are made. Appendix 6 lists some approximate properties of pure metals that are useful for reference.

## DUCTILE STEELS

Ductile steel, usually steel that is not heat-treated, is used more than any other single class of metal in the world today because of its high stiffness, high toughness, low cost, and adequate strength. The microstructural characteristics of ductile steel that lead to its good ductility will be developed in Chaps. 4–9.

Steel, both ductile and high-strength, has the highest elastic modulus (about 200 GPa) of any ordinary structural material except the high-performance composites (Chap. 17). The strength of ductile steels usually exceeds about 350 MPa, with an upper limit on the order of 700–800 MPa. This range of strength for the cheapest steels corresponds approximately to the strength range of the highest-strength heat-treated aluminum alloys and greatly exceeds the maximum strength of wood (which is on the order of 100 MPa) and polymers (which ranges from 5 to 190 MPa).

Ductile steel is one of the toughest metals, giving it ease of fabrication by rolling, for example, and high reliability in service because of its relative insensitivity to notches and cracks. The nominal strain to fracture $\varepsilon_f$ is on the order of 0.2–0.5 for ductile steels.

The widespread distribution and availability of iron in the earth's crust is a happy accident that has profoundly influenced technological development.

## CAST IRONS

Cast irons were discovered by ancient people as a consequence of the high carbon content of the iron and the resulting lower melting temperature (see Chaps. 8 and 15). For thousands of years, gray cast iron has found applications in civilization even though it is very brittle. The nominal fracture strain of most gray and white irons is virtually zero. Nevertheless, this castable metal has found many applications because it is relatively easy to fabricate using casting methods, where molten metal is poured into a mold of the desired shape, where it freezes. While steel can also be cast, it has higher melting temperatures. The elastic modulus of cast iron is about half that of steel and pure iron (about 100 GPa), it has moderate strength, and it can be made very hard and wear-resistant (white iron). Furthermore, modern technology has discovered how to produce cast iron of greatly improved toughness and thus has widened the range of its applications.

Castings are machined afterward to produce very flat surfaces or precise holes. Most cast irons are easier to machine than ductile steels simply because of their low ductility and tensile strength. In many applications, stiffness and low cost of fabrication are more important than high strength and ductility, so cast iron is used.

Strengths of gray cast irons range from 140 to 320 MPa, depending on chemical composition (see Chap. 15), and are thus close to the range of strength of ductile steel. Higher-ductility cast irons, called *nodular, ductile,* or *malleable,* can be made very ductile ($\varepsilon_f = 0.20$, $S_u = 380$ MPa) or very strong ($\varepsilon_f = 0.04$, $S_u = 650$ MPa), but they cost more.

The hardness of cast iron ranges from about 150 HB to 260 HB for gray irons, 110 HB to 240 HB for ductile irons, and up to 670 HB for specially treated alloy white iron.

## ALUMINUM ALLOYS

The mass density of pure aluminum is 2700 kg/m$^3$, compared with 7870 kg/m$^3$ for pure iron. This nearly threefold weight advantage of aluminum over iron alloys accounts for many of its applications. Pure aluminum is much more resistant to atmospheric corrosion than steel or pure iron, a quality that leads to additional applications. But with an elastic modulus of 70 GPa, aluminum has about 35% of the stiffness of steel.

Aluminum alloys usually include a few percent magnesium, silicon, or copper to increase strength and hardness through mechanisms described in Chaps. 8 and 9. These alloys can be cast or wrought. Pure aluminum melts at 660°C, and these alloys melt at even lower temperatures, so high-production parts are frequently cast under pressure in molds made of steel; this is called *die casting.* Wrought aluminum alloys are hot- or cold-rolled or extruded as parts of constant cross section such as bars, sheets, ell or tee sections, and tubes.

The complexity of cast aluminum parts does not permit the same control over microstructure that is possible for wrought shapes, so cast parts generally have lower strengths. Characteristic tensile strengths and fracture strains are given in App. 5; the maximum strength of cast aluminum alloys is on the order of 300 MPa.

Wrought aluminum alloys have maximum strengths of 650 MPa and generally much better ductility than cast alloys. The opportunity for cold-forming these alloys also provides an additional strengthening mechanism (see Chap. 6). The maximum *specific strength*, or strength/specific gravity, of wrought aluminum alloys is 250 MPa, or about twice that of ductile steel. The higher specific strength thus accounts for the wide application of aluminum alloys to subsonic aircraft skin, where good ductility is essential for forming the product. Even if ductile steel possessed the same specific strength as wrought aluminum alloy, the better corrosion resistance of aluminum would probably dictate its use in many applications despite the higher elastic modulus of steel.

## POLYMERS

Recent decades have seen an enormous expansion in the engineering applications of synthetic polymeric materials. A detailed discussion of the properties of polymers is presented in Chap. 13. Polymers exhibit mechanical behavior different from that of the metals already described in this chapter. The most obvious difference is that most polymers continue to deform for a long time after a stress is applied; that is, their behavior is *time-dependent*. This characteristic of polymers is called *viscoelasticity*. Continued deformation over a long time thus limits stresses to values much lower than the stresses allowable for a short-term loading. Such behavior requires the designer of a part made of a polymer to consider the stress–time history of the part in addition to the factors already described for metals.

In conjunction with their time-dependent behavior, most polymers exhibit a strong sensitivity of mechanical properties to temperature. Many polymers show marked changes in properties in the range between –50 and 150°C. For example, with modest increases in temperature in this range the deformation rate of polymers can increase markedly to the point where the material is a viscous liquid. With reduction of temperature, many rubbery polymers become brittle and unusable at –50°C. Such behavior severely limits the upper and lower temperatures for polymers, and often this is the chief reason they cannot be used in many applications.

When a polymer changes during cooling from a flexible, rubbery material to a brittle, glassy state, this transition is called the *glass transition*. The highest-strength polymers are generally brittle at 20°C, a reflection of their having high *glass-transition temperatures*, well above 20°C. Even low-strength polymers become brittle at sufficiently low temperatures. Thus, the glass-transition temperature $T_g$ is an important property of each polymer. Applications requiring rubbery behavior, such as flexible tubing, demand polymers with a $T_g$ well below the operating temperature; anyone who has tried to uncoil a plastic garden hose outdoors in winter knows how stiff and useless the hose is at temperatures approaching $T_g$.

## SPECIAL APPLICATIONS

The categories of materials described earlier include most of the common structural materials. In addition, special requirements dictate the use of materials having special mechanical properties. Some of these materials are described in this section. In achieving some special property, we usually pay a higher price for the material and often must also accept inferiority in some of its other properties.

**Low-alloy steels.** These steels usually contain 0.30–0.50% carbon (see Chap. 10), have small (0–5%) amounts of other elements in iron, and can be heat-treated to have strengths up to about 2 GPa. However, their toughness is lower than that of plain-carbon steels, and nominal fracture strain can be as low as 0.05. It is possible to manufacture low-alloy steels of still higher strength, but the resulting loss in ductility causes them to be of only limited value. The maximum specific strength of low-alloy steel of about 260 MPa leads to its use for many structural parts in aircraft, as it has slightly higher specific strength than wrought aluminum alloys. The better fatigue properties (long-term strength under varying cyclic stresses) of low-alloy steels give them a further advantage over aluminum alloys in many applications. Low-alloy steels are widely used in vehicles and machinery wherever a combination of high strength and good hardness is desired and where moderate ductility is tolerable.

**Tool steels.**     These steels have high hardness and wear resistance, often at elevated temperatures. These properties are achieved through high-carbon, high-alloy content (manganese, silicon, nickel, chromium, vanadium, tungsten, molybdenum, and cobalt) and heat treatment. Hardness up to 68 Rockwell C and good stability and hardness at temperatures to 600°C are attainable. Tool steels are also used in a limited number of aerospace structural applications where their high hardness, strength, and temperature resistance provide the only material available, but low toughness is usually the penalty for their use.

**Sintered carbides.**     These carbides are usually mixtures of very hard intermetallic compounds formed by being pressed together at high temperature, usually with cobalt as a binder. Typical compounds in use are carbides of tungsten, titanium, and tantalum. The hardness is above 70 HRC at room temperature, and hardness up to 67 HRC may be retained at 760°C. Carbides thus provide hardness at high temperature that cannot be attained by any steel, as well as extraordinary wear resistance. Carbide-tipped tools generally have longer lives than tool-steel tools in the same application, but of course carbide tools cost more. The good corrosion resistance of the sintered carbides in many environments provides an added advantage for many applications, such as for mining machinery and oil well drilling bits.

**High-temperature materials.**     These materials are intended for service at temperatures above 650°C. The most common are alloys of iron with chromium, nickel, and cobalt, often called *superalloys*. The alloy content is generally high, up to 30% chromium, for example, in contrast to that of low-alloy steels, which generally contain no more than 5% of alloying elements. Typical good high-temperature materials can support a tensile stress of 140 MPa at 650°C, 50 MPa at 750°C, and 25 MPa at 900°C for acceptable rates of *creep deformation*, deformation that continues with time. (High-temperature creep is similar to viscoelastic deformation of polymers; see Chap. 12.) High-temperature materials usually also have good corrosion resistance at high temperature. However, in many cases the corrosion rate rather than the creep rate can limit the operating temperature of a material. Applications of these materials are in the high-temperature components of engines, turbines, furnaces, and manufacturing equipment.

**Ceramics.**     These materials are compounds of metals and nonmetals, often of very complex structure, which are used for a wide range of structural and nonstructural applications. Brick and concrete are structural ceramics that are widely used because of their low cost and good compressive strength. Glass is used because it is corrosion-resistant, transparent, easily formed to complex shapes, and cheap. For applications at very high temperatures (to 2500°C), refractory materials such as MgO are used. Ceramics generally have very low ductility and associated low tensile strength, but some "high performance" ceramics such as stabilized zirconia have improved toughness.

**Stainless steels.**    These steels are designed to resist corrosion. Since corrosion resistance and high-temperature resistance are found in similar alloys, many materials serve both purposes. Stainless steels are iron alloys with 12–30% chromium, low carbon (usually less than 0.30%), and often substantial amounts of nickel (3–22%). Strengths range up to 1.4 GPa for heat-treated or severely cold-worked grades, with nominal fracture strains of 0.10–0.30 and acceptable levels of toughness. In spite of their much higher cost over iron and ductile steel, stainless steels have numerous applications such as in chemical processing (often combined with high temperatures), in food-handling equipment, and where good resistance to environmental corrosion is required.

**Copper.**    Copper is used in nearly pure form (at least 99.9% Cu) because it is an excellent electrical conductor and is characterized by excellent corrosion resistance and ease of mechanical forming.  Gold and silver are excellent conductors but are too expensive for mass use. Alloys of copper retain much of the corrosion resistance of pure copper and exhibit good strength and ease of fabrication. Although beryllium–copper has an acceptable specific strength of 170 MPa, most copper alloys, usually called brasses or bronzes, fall below 100 MPa. Thus, the high cost of copper alloys and their low specific strength restrict their use in structural applications to corrosive environments (especially fresh and salt water) in which no other material can serve as well.

**Titanium alloys.**    These alloys provide a strong, lightweight metal capable of withstanding operating temperatures considerably higher than aluminum alloys can. Pure titanium has a mass density of 4510 kg/m$^3$ (between those of iron and aluminum), a melting temperature of 1668°C (130°C above that of iron), and an elastic modulus of 116 GPa. Alloys of titanium involve aluminum and vanadium up to about 10% in many alloys, plus smaller quantities of up to about eight other elements. The highest-strength alloys have maximum specific strengths of 350 MPa, the highest of any material discussed so far in this chapter. Titanium alloys retain useful ductility to very low temperatures, say –200°C. Furthermore, commercially pure titanium has excellent corrosion resistance in many environments, while titanium alloys have somewhat poorer but still adequate corrosion resistance. Production and fabrication costs for titanium are high, so its use is restricted to those applications for which it is uniquely suited.

Retention of useful strength (1.7 GPa maximum at 430°C for one alloy) up to 500°C allows titanium alloys to be used in supersonic aircraft, where structural components may be subjected to high continuous operating temperatures. For example, the temperature of the air at the leading edge of the wing of an aircraft flying at three times the speed of sound at an altitude of 20 km is about 340°C. As will be discussed in Chap. 9, aluminum alloys rapidly lose strength above 200°C, with the best aluminum alloys useful at reduced stresses up to a maximum of about 300°C. For this reason, aircraft made of aluminum alloys are limited to a little more than twice the speed of sound in air, for which the leading-edge air temperature is 120°C at 20 km altitude. The first commercial supersonic aircraft, the *Concorde*, is limited to such speeds because it uses long-established aluminum alloys. Titanium alloys provide the highest specific strength of any commercially available metal suitable for high-speed aircraft.

**Magnesium alloys.**     These alloys have the lowest density of any structural metal; pure magnesium has a mass density of 1740 kg/m$^3$, and the common alloying elements of magnesium, such as aluminum, zinc, manganese, and thorium, raise the density to a maximum of about 1840 kg/m$^3$. Thus, magnesium is used in many applications where low mass is important, such as aerospace structures, truck and trailer body components, small portable power tools, and ladders. Magnesium alloys can be cast and can be wrought by processes such as extrusion and rolling, in many ways similar to aluminum alloys. In certain applications magnesium alloys offer better performance than aluminum alloys. On the basis of a structure having the same mass, the lower mass density of magnesium permits, for example, the use of sheet of about 50% greater thickness and a consequent reduction of maximum stress in bending by a factor of about 2.3. In a design in which bending stresses are critical it is possible to fabricate from magnesium alloys a structure that has a lower mass than the same structure made of aluminum alloys. By the same reasoning, even with the lower elastic modulus of magnesium taken into account (see App. 5), deflections from bending stresses in sheet material are lower in magnesium by a factor of about 2.2 for a given mass.

In simple tension and compression, magnesium and aluminum alloys provide nearly equal load-carrying capacity for the same mass, so the lower cost of aluminum (per kilogram) dictates its widespread use in such applications. Magnesium alloys are more resistant to ordinary atmospheric corrosion than steel but less so than aluminum alloys. Magnesium alloys require surface protection when used in saltwater environments.

**Composite materials.**     Composites are synthetic combinations of materials on a very fine scale (1—100 $\mu$m) that provide properties unattainable in any one component material alone. Very fine filaments usually have much greater strength than the same material in bulk (see Chap. 17, in which composites are discussed in detail), so surrounding an array of strong filaments with a low-density matrix material can produce a composite material of high specific strength. By proper selection, special desired properties such as corrosion resistance or electrical resistance can be obtained.

Typical properties of two composite materials are given in App. 5. Moldable glass fiber–resin is a commercial mixture of very fine glass fibers, either in short lengths or as continuous filaments, which can be produced in complex shapes by high-production equipment. Although still expensive compared with ductile steel, this class of materials provides specific strength up to twice that of ductile steel. Low ductility and elastic modulus restrict uses of this material and require careful design. Applications include molded boats, automobile panels, and equipment housings.

Because of the extraordinary strength and stiffness of the graphite filaments, carbon fiber composite exhibits by far the highest specific strength of any commercial material, combined with the stiffness of steel. Low ductility and high cost will continue to limit its applications. Graphite epoxy has been used for components of orbiting space vehicles, helicopter rotors, aircraft frames, bicycles, and fishing rods.

This chapter has examined the wide range of properties that govern the selection of different materials for different applications. In many cases materials are selected because they have good nonmechanical properties, such as conductivity or corrosion resistance. Usually such physical properties cannot be changed for a given material, so the designer must change materials in order to change physical properties. However, most mechanical properties can be changed. The balance of this book is directed toward the fundamentals that control mechanical properties. Once the fundamentals are understood, methods for improving mechanical properties can be determined. We start with the basic structure of simple solids and then advance to more complex materials.

## R E F E R E N C E S

1. American Society for Metals, *Metals Handbook*, 8th ed., vol. 1, *Properties and Selection of Metals*. Metals Park, Ohio: American Society for Metals, 1961.

2. American Society for Metals, *Metals Handbook*, 9th ed., vol. 1, *Properties and Selection: Irons and Steels*. Metals Park, Ohio: American Society for Metals, 1978.

3. American Society for Metals, *Metals Handbook*, 9th ed., vol. 2, *Properties and Selection: Nonferrous Alloys and Pure Metals*. Metals Park, Ohio: American Society for Metals, 1979.

4. American Society for Metals, *Metals Handbook*, 9th ed., vol. 3, *Properties and Selection: Stainless Steels, Tool Materials and Special-Purpose Metals*. Metals Park, Ohio: American Society for Metals, 1980.

5. Brick, R. M., R. B. Gordon, and A. Phillips, *Structure and Properties of Alloys*, 3rd ed. New York: McGraw-Hill, 1975.

6. DeGarmo, E. Paul, *Materials and Processes in Manufacturing*, 4th ed. New York: Macmillan, 1974.

7. Ruoff, Arthur L., *Introduction to Materials Science*. Englewood Cliffs, N.J.: Prentice Hall, 1972.

8. Smith, Charles O., *The Science of Engineering Materials*, 2nd ed. Englewood Cliffs, N.J.: Prentice Hall, 1977.

9. Suh, N. P. and A. P. L. Turner, *Elements of the Mechanical Behavior of Solids*. New York: McGraw-Hill, 1975.

# 4

# Structure of Solids

Chemical bonds hold matter together. We may distinguish between those forces that hold atoms together within a molecule and those that attach molecules to one another. The strong primary valency forces associated with covalent, ionic, and metallic bonds are in the former category, while van der Waals forces are weak secondary valency forces. The state of the material (solid, liquid, or gas) depends on whether the internal energy of the material (which is increased by increasing the temperature) is capable of overcoming its internal cohesion, which is a manifestation of chemical bonding. The commonly experienced strength properties of solids, discussed throughout this book, are also manifestations of solid-state bonding forces that resist deformation.

## CRYSTALLINE AND AMORPHOUS SOLIDS

At absolute zero all atoms and molecules are at rest, and everything must be a hard solid. Increase of temperature produces random vibrations of atoms and molecules, which is another way of describing heat. Eventually the internal cohesion of a solid (where the atoms and molecules have little freedom) is lost and the molecules acquire the sort of "shuffling freedom" that characterizes a liquid. Further increase of temperature eventually leads to boiling, where the molecules are released from any kind of mutual cohesion. It is possible under some conditions for the liquid stage to be missed in this sequence (i.e., sublimation), but that need not concern us here for common engineering solids. Cooling reverses the process, gases condensing to liquids and liquids freezing to solids.

In the solid state, atoms and molecules are arranged in either of two ways:

1.  In a *crystalline* form, where the atoms or molecules are packed together in a highly regular, definite geometric form, or
2.  In an *amorphous* manner, with no specific form.

A crystal is analogous to bricks laid regularly in a wall, and the same number of bricks heaped in a pile is the equivalent of an amorphous solid. Clearly, amorphous solids are porous and less dense than the corresponding crystalline solid.

All materials have a tendency to become crystalline on freezing from a melted state, but not all achieve it. Metals have highly regular arrangements of molecules, with systematic three-dimensional geometric packing. Other substances may be crystalline to a greater or lesser extent and some engineering solids have no clear packing of atoms or molecules. It all depends on the magnitude of the attractive forces between the atoms and molecules in the solid state and on how successful these cohesive forces are in drawing them together, that is, whether the "bricks" are drawn together into a regular pattern or whether they end up in a higgledy-piggledy fashion. Some materials can exist readily in either form; others exist with mixtures of both forms in the same solid; and in others the crystalline or the amorphous state predominates. For example, amorphous metals *can* be produced by tremendously fast cooling, but for most practical purposes metals are crystalline.

## GLASSES

The condition of the liquid just before freezing and the rate at which it is cooled play important roles in determining the form of the solid. If a liquid just before freezing is very viscous (where the molecules "interfere" with one another) or if the cooling rate at freezing is very rapid, the molecules will not be able to choose the closely packed sites of a crystalline arrangement but will rather end up in an irregular, amorphous array. The general term for such an amorphous solid is a *glass*. Sugar is normally crystalline, but if cooled very quickly it becomes toffee, which is a glass. Sometimes a bit of crystallization does take place; thus fudge is partially crystallized toffee. Sand (silica, $SiO_2$) melts as a viscous liquid and on cooling gives one of the types of glassy, amorphous solids used for windows, beer bottles, and the like.[*] Hence the word "glass" is commonly used, but the term really has the wider, precise scientific meaning just given. Silica also exists in a crystalline form, which is the naturally occurring mineral quartz; this had plenty of time to arrange its atoms into crystals as it was formed during slow cooling in the earth's crust.

**Example 4-1**

Give a possible reason why windows in medieval churches have cracked and fallen out in recent times.

**Solution**   The transition from the glassy state to the crystalline state, which is called *devitrification*, can occur for some materials over very long periods of time at ambient temperature. Since solids prefer to be in the crystalline form, church window glass has devitrified over the

---

[*]As the melting point of pure silica is about 1600°C, practical glasses have lime and soda added to allow easier fusing at lower temperatures.

centuries. The crystalline form is more dense than the amorphous form, so the glass has shrunk, leading to cracks and misfits. (See Example 16-3 for a general discussion of this matter.)

**Example 4-2**

Expensive glass articles are often called "crystal glass," but glass is amorphous. Explain.

**Solution**   Scientifically speaking, this is an incorrect use of the word. It arises because the type of glass designated "crystal" and used for crystal chandeliers, decanters, and so on, is much better than common glass and in the early days was considered to "sparkle" and reflect as well as rock crystal (another name for quartz). But crystal glass is still amorphous.

## VISCOSITY

The actual behavior of a thick, sluggish liquid just before freezing depends on the form of the molecules and the way they associate above the melting point. Although we use the words "melting" and "freezing," with the thickest liquids it may be impossible to know precisely what we mean by these terms. Viscosity is one measure of the amount by which the liquid molecules, instead of remaining independent, form torpidly mobile networks or chains. The coefficient of viscosity ($\eta$) relates the applied stress to the strain rate (see Chap. 12). Changes in $\eta$ with temperature may be used to distinguish between, for example, the clear-cut transformation from ice to water at one end of the spectrum of melting and freezing behavior and the behavior of treacly molasses or pitch at the other.

The distinction between fluids and solids is arbitrarily drawn at a viscosity of 100 TPa-s. At this value, a cube with 25-mm sides could support a person's mass for a year without sinking more than a couple of millimeters. Insofar as the mechanical properties of crystalline solids are time-dependent at temperatures just below their melting points, the choice of 100 TPa-s attempts to indicate how "thick" a treacly liquid must be to have properties more-or-less similar to those of crystalline solids.

## GLASS-TRANSITION TEMPERATURE

Figure 4-1 shows how $\eta$ changes typically with temperature for the two extreme conditions of solidification down to crystalline and to amorphous solids (and, of course, in reverse for melting). Approximate values for the viscosity of gases at boiling and condensing points and of liquids at melting and freezing points are 10 $\mu$Pa-s and 1000 $\mu$Pa-s, respectively. If crystallization occurs, the viscosity changes immensely at the melting and freezing point, up to perhaps 100 PPa-s (recall the SI prefixes from App. 1: P = peta = $10^{15}$), and goes on rising as the crystalline solid cools down. If, on the other hand, the liquid fails to crystallize, although $\eta$ increases as the temperature drops, the increase is continuous and gradual, with no sharp rise in $\eta$ at a melting or freezing point. When such a "supercooled liquid" attains the critical 100 TPa-s, the structure is such that we have effectively an amorphous solid, thereafter called a *glass*. The temperature at which 100 TPa-s is reached is called the *glass-transition temperature* $T_g$. For many materials it is about one-third of the boiling temperature. The glass-transition temperature also reflects changes in

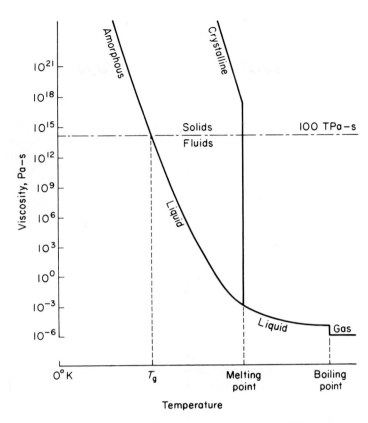

**Figure 4-1**   Changes in viscosity with absolute temperature for crystalline and amorphous solids.

other properties, such as the so-called secondary thermodynamic quantities (e.g., the coefficient of expansion); elastic moduli also change at $T_g$, glasses stiffening as the temperature drops (see Chap. 13). Different materials behave in different ways in the transition regions where $\eta$ ranges from 1 mPa-s to 100 TPa-s; some get stiff gradually, whereas others are more "liquidy" for a longer temperature (or time) span. Others display rubbery behavior at temperatures just above $T_g$, with enormous reversible mechanical extensions. These differences are related to the different molecular structure and arrangements in the cooling range of temperatures down to $T_g$.

## TYPES OF CRYSTAL STRUCTURES

Metal alloys freeze over a *range* of temperature (between the liquidus and solidus on the phase diagram, Chap. 8), during which the liquid is transformed through a slushy, partly liquid, partly solid state to a complete solid, and the almost vertical line for pure elements marked "crystalline" in Fig. 4-1 has a clear slope over this range of temperature. Unlike the curved line marked "amorphous" in Fig. 4-1, however, the alloy line still has obvious kinks and almost discontinuous changes in viscosity with temperature.

Metals and simple inorganic substances readily form *crystals*. Rocks, gemstones, and the like, have had eons to cool in their formation and so are crystalline. Indeed, the first postulates about crystal structures were based on models of spheres (atoms) packed in different ways to give the external shapes of naturally occurring inorganic crystals such as quartz, calcite, and salt. There are basically six patterns of crystal structure, of which only two are important in traditional engineering solids. These patterns are the shapes taken by the building bricks of the crystal. In Fig. 4-2 the spheres at the corners of the bricks represent atoms reduced from their actual size; in a pure element, the atoms are all the same, but in a metal alloy or crystalline chemical compound they are of more than one type. The edges of the bricks represent the directions of the bonding forces, and the whole network of edges in three dimensions is called a *space lattice*.

The types of space lattice are

1. *Cubic*, where all edges are mutually at right angles and the edges are of equal lengths (many metals fit this category)
2. *Tetragonal*, where all edges are mutually at right angles and two of the edges are equal in length, but the third is different
3. *Orthorhombic*, where all edges are mutually at right angles and all the edges have different lengths (a common brick is of this sort)
4. *Monoclinic*, where only two of the edges are mutually perpendicular and the other edge is at another angle, and where all edges have unequal lengths
5. *Triclinic*, where none of the edges is mutually perpendicular and where none of the edges is equal to another (a special case of this, where the edges do have equal lengths, is called *rhombohedral*)
6. *Hexagonal*, where the base of the building brick is a regular hexagon.

A given lump of a solid may be a *single crystal* (perhaps a few millimeters in extent), but common engineering solids are *polycrystalline*, having many small, individual crystals (or grains)—some only a micrometer in extent—which have grown from the melt in different directions, leading to *grain boundaries* between the individual crystals. These topics are dealt with in succeeding chapters.

## TYPES OF POLYMER STRUCTURES

In contrast to crystals, easy glass formers are substances with molecules of complex shapes (which find it difficult to pack together and which interfere with one another in the liquid state) and/or substances whose molecules link together strongly in the liquid to form long chains or networks. Thus, many oxides, silicates, borates, and phosphates readily form glasses. Again, many molecules have the propensity to form long chains where thousands of repeat units link up. The repeat unit (or *monomer*) can be a single atom, such as sulfur or selenium, but commonly it is understood that the monomer is a molecule. Furthermore,

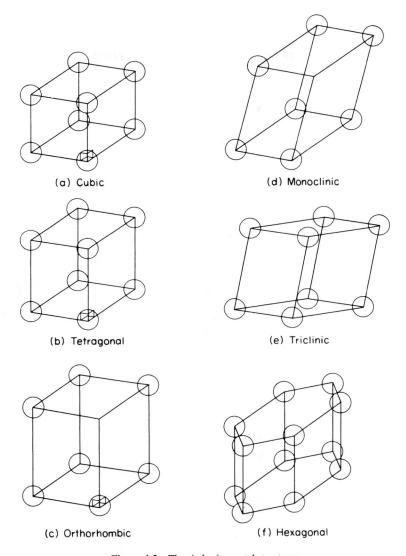

(a) Cubic　　　　　　　　　　　　　(d) Monoclinic

(b) Tetragonal　　　　　　　　　　　(e) Triclinic

(c) Orthorhombic　　　　　　　　　　(f) Hexagonal

**Figure 4-2**　The six basic crystal structures.

these molecules can be inorganic, such as chrysolite ($3MgO \cdot 2SiO_2 \cdot 2H_2O$), in asbestos and silicone oil [$Si(CH_3)_3O$], in silicones; or organic, such as ethylene in polyethylene, iso-prene in natural rubber, urethane in polyurethane, amides in nylon, saccharides in cellulose (wood, for example) and formaldehydes in bakelite. It is the latter carbon-based organic materials that are familiarly called *polymers* or *plastics,* though strictly "polymer" means "many mers," whatever their nature. Most animal and vegetable living matter is in the form of organic polymers. Some of the mers bring atoms other than carbon into the main

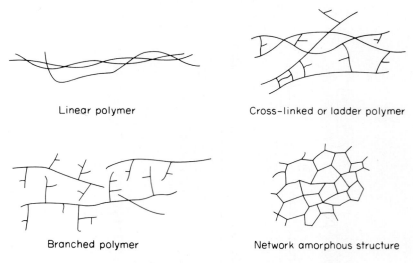

Figure 4-3    Various types of amorphous polymer structures.

chain; for example, there is oxygen in cellulose and nitrogen in nylon. Again, structural units with excess valency or spare bonds in the main chain allow side chains to grow out like branches on a tree. These may link up with side groups on other chains, giving a *cross-linked* polymer; if this process occurs many times, *network* polymers are produced. Figure 4-3 illustrates some of these amorphous structures. Polymers are often visualized as tangled masses of spaghetti or string; they can therefore be quite simple or extremely complicated in their makeup and behavior, as discussed in Chap. 13. A characteristic of some polymers is a range of temperature over which the mechanical response is rubberlike, with tremendous reversible strains. When the is rubbery range encompasses room temperature, the polymers are called *elastomers.*

## COMPARISON OF TYPES OF MATERIALS

The schematic in Fig. 4-4 attempts to interrelate most engineering solids. It should be viewed in conjunction with the viscosity–temperature plot in Fig. 4-1. Three distinct types of materials are identified:

1. *Crystals* (with clear-cut melting/freezing points)
2. *Glasses* (with no sharp transition to the solid state)
3. *Polymers* (also with no sharp transition to the solid state but with the feature of rubbery mechanical behavior).

It is seen that (2) and (3) are subdivisions of amorphous solids that do not have specific melting/freezing points. Each type of material is represented by a cylindrical column

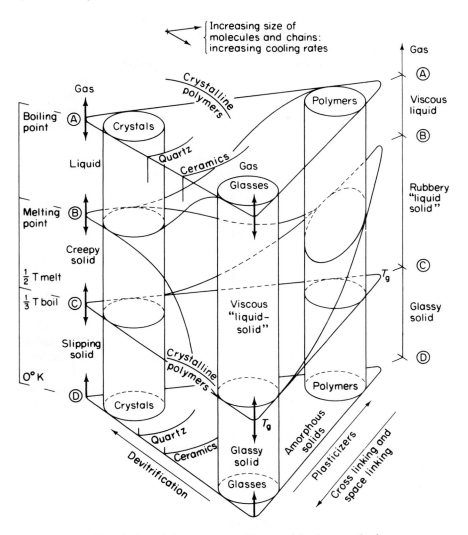

**Figure 4-4** The interrelation among crystalline materials, glasses, and polymers.

that forms the edges of a triangular prism. Temperature is the vertical scaling, 0°K being represented by the base plane, and temperatures being to an arbitrary scale where boiling and melting temperatures are normalized to the same level.

Let us trace what happens when materials are cooled from the gaseous state, that is, follow down through the temperature levels A, B, C, and D. Vapors condense to liquids at the boiling point. Then, for crystalline materials, liquids freeze solid at the melting point, with marked changes in viscosity; for metal alloys, freezing may occur over a small range of temperature around point B. Between levels B and C such a solid is a "creepy" material whose mechanical properties are time-dependent (Chap. 12). Time dependency progressively diminishes below about one-half of the absolute melting temperature (below level C) until the solid is not markedly time-dependent (Chaps. 2 and 5). Glasses, on the other hand, remain liquidy as they cool and get thicker and thicker, until at about one-third of the absolute boiling temperature (which corresponds, more or less, with $T_g$) they reach the borderline viscosity of 100 TPa-s. Below the temperature at which the viscosity attains this level, they are considered glassy solids whose mechanical behavior is not markedly affected by time and may be related merely to stress, like slipping crystals (Chap. 5). Polymers too are liquidy above $T_g$, but they also can exhibit a range of rubbery behavior between the viscous liquid and glassy regimes.

Many solids in practice show characteristics between the foregoing distinct groupings and could be located vertically along the sides of the schematic prism in Fig. 4-4. For example, many ceramics consist of small crystals of metallic silicates and oxides kept together by a thin layer of a glassy material (Chap. 16). Similarly, there are crystalline polymers that have regions within the bulk of the tangled mass of polymer chains in which the chains are oriented in a regular, repeatable manner. Some short-chain polymers with simple mers can crystallize almost completely on cooling (e.g., polyethylene), but small, oriented regions called *crystallites*, if they occur in commercial polymers, exist in isolated regions, with the same molecular chain often extending through both crystalline and amorphous regions. Again, between the polymer column and the column representing glasses in Fig. 4-4 an immense range of properties can be produced by *cross linking* and/or *space linking*, tending to eliminate the rubbery regime from the polymer starting position; the vulcanization of natural rubber (where sulfur is added to cross-link the long chains of natural rubber molecules) is that very thing. Indeed, it is convenient sometimes to think of silica glass as a densely cross-linked polymer. These ideas are further developed, and other methods of altering the properties of polymers are described (e.g., by *plasticization*) in Chap. 13.

The overall pattern of events depicted in the three-dimensional schematic drawing in Fig. 4-4 can be markedly dependent on cooling rate. Increased rates of cooling, where less time is allowed for molecular sorting out, tend to move things away from the crystalline behavior edge of the prism.

# REFERENCES

1. Cottrell, A. H., *The Mechanical Properties of Matter.* New York: John Wiley, 1964.
2. Gordon, J. E., *The New Science of Strong Materials—Or Why You Don't Fall Through the Floor.* London: Penguin Books, 1968.
3. Harris, B., and A. R. Bunsell, *Structure and Properties of Engineering Materials.* London: Longmans, 1977.
4. Gordon, J. E., *Structures and Materials.* New York: Scientific American, 1988.

# 5

# Crystal Imperfections and Slip

Strength is one of the most important properties of a solid, and an examination of the atomic-scale behavior of solid materials is essential to an understanding of the reasons for strength. This chapter provides (1) an introduction to the rudiments of crystallography, as many solids of engineering importance are crystalline in nature, (2) the classical development of the theoretical shear stress for slip in a perfect crystal, and (3) a description of the nature of imperfections in real crystals and how they affect strength.

## EQUILIBRIUM SPACING OF ATOMS AND CRYSTAL STRUCTURE

Atoms can be regarded as relatively hard spheres because the outer electron shell (or the electron cloud in metals) acts as a strong and very short-range repelling force to other atoms. Whereas the positively charged nucleus, or center, of an atom is in force equilibrium with its own surrounding electrons, the balance of forces is altered by the negative charge of the electrons of a nearby atom. The classical representation of these forces, usually neglecting the small repelling force between the two positive nuclei, is shown in Fig. 5-1. The spacing $a_0$ where the net force between atoms is zero represents the equilibrium spacing, that is, the spacing of the atoms in the absence of other forces.

The idea of two isolated atoms as hard spheres resting a distance $a_0$ apart is valid only at absolute-zero temperature (0°K). At any other temperature, atoms are in motion and thus do not have the same equilibrium spacing $a_0$ as their average position.

Let us first consider a large number of like atoms (i.e., atoms of the same element) together at 0°K and then later gradually increase the temperature. At 0°K we expect the atoms to assume some equilibrium spacing because they are completely at rest. The simplest and commonest model is what we get when we carefully fill a box with ping-pong balls, a layer at a time. The resulting spacing and positions of the ping-pong balls are analogous to one kind of *crystal structure*. For real atoms, the rules that hold for equilibrium spacing of only two atoms become much more complicated when each atom is close to many other atoms; the equilibrium spacing is different, but the principle is the same. Thus,

**Figure 5-1**    Schematic representation of forces between two atoms. Equilibrium spacing is $a_0$.

at 0°K a large number of like atoms assemble together in some equilibrium geometric pattern that is the characteristic crystal structure for that element. This three-dimensional structure, with imaginary lines connecting the atom centers, is called a *crystal lattice*. Different elements are found to prefer different crystal structures; after a brief excursion into the subjects of melting and vaporization, we will consider the three most common crystal structures in metals.

## SOLIDS, LIQUIDS, AND VAPORS

Everyone knows that solids retain their shape, liquids fill a container up to a certain level, and vapors (gases) expand to fill a container no matter how big it is. The enormous mechanical differences among these three usual forms, or *phases,* tend to cloud the simple changes that occur with increasing temperature. But what about solids that ooze, or liquids that do not flow, or the processes of evaporation (liquid to vapor) and sublimation (solid to vapor)? Our simple atomic model, if realistic, should help to explain some of these phenomena.

We return now to the solid crystal at 0°K, with the hard-sphere atoms at rest and in a fixed crystal structure. If heated, the individual atoms will vibrate at their positions, maintaining nearly the same average spacing as before. The average spacing increases a little with temperature, a process called *thermal expansion.* This vibration is the essence of increased temperature, so we are actually using a kind of circular argument in saying increased temperature leads to increased atomic motion. Nevertheless, as temperature (vibration) increases further, atoms eventually vibrate so much that some actually switch positions at random, a process called *diffusion*; at still higher temperatures some atoms may vibrate out of their positions and not be replaced, leaving holes or *vacancies*.

The *melting temperature* represents a transition in atomic structure, on heating, from an ordered crystalline structure *(solid)* to a substantially disordered, or amorphous, structure *(liquid)* of enormously lower *viscosity*. Viscosity, which is the ratio of shear stress to the resulting shear strain rate, has already been discussed in Chap. 4 and will be considerted further in Chaps. 12 and 13. The transition from solid to liquid when a pure element is heated occurs quite suddenly, and the melting point is uniquely defined; in the case of . alloys (mixtures of elements, described later), melting occurs over a range of temperatures. As heat flows to the solid, individual atoms of the solid acquire much more energy and thus vibrate so much that they break away from their crystal lattice sites. The *heat of fusion* of a material is simply the total energy necessary to allow all atoms of a given mass to acquire enough energy to break away from the ordered crystalline structure. The atoms are still close enough together in the new liquid state to experience the bonding forces in Fig. 5-1, but their mobility is so great that they have no fixed equilibrium positions with respect to each other.

*Vaporization* occurs when an atom receives enough energy so that it can break loose from the local forces in Fig. 5-1. Vaporization can occur either from the liquid state, in which case the energy required is the *heat of vaporization;* or from the solid state, a process called *sublimation,* which requires the sum of the energies required for melting and vaporization. Atoms or molecules (groups of atoms) of a vapor, or gas, are usually widely spaced, move at high velocities, and experience occasional collisions with each other.

Thus, in a rather oversimplified way, the essential changes from solid to liquid to vapor can be understood in terms of whether the balance between bonding forces and atomic vibration constrains the atoms to a closely spaced regular structure (solid), a closely spaced near-random structure (liquid), or a widely spaced random structure (vapor). As we shall see, although the transformation from solid to liquid is very sharp and occurs at a fixed temperature for pure elements, the atoms in solids just below their melting temperature have sufficiently high mobility to take on some of the characteristics of liquids.

One more point should be noted before we return to the crystal structure of solids. It is possible to consider the phase changes described previously in terms of *surface free energy,* which is the energy required to create a free surface such as exists at the interface between a solid and its vapor. An atom on the surface of a solid is only partly surrounded by adjacent atoms, in contrast to an atom in the interior of the solid. Clearly, there are fewer bonds associated with surface atoms, and energy is required to break free from some of the bonds of an interior atom when the atom becomes [by diffusion, say, or fracture (see Chap. 14)] an atom on the surface of the crystal. An atom that has no fixed bonds, such as in the liquid state, has thus broken even more bonds and requires still higher energy. And an atom in the vapor state has virtually no bonds, so it requires the highest energy to attain that state. Thus, for an interior atom in a solid to sublime requires first that it diffuse to the free surface through accidentally acquiring energy from the vibrations of its neighbors. Then it must be given a relatively large "kick" to impart the energy required for complete separation from the solid crystal and into the vapor state.

*Vapor pressure* is the equilibrium pressure of the vapor atoms in the vicinity of the free surface (between either solid or liquid and the vapor) at a given temperature, where as many atoms drop back into the solid crystal (or liquid) as are ejected. It is fortunate that the vapor pressure for solids is usually extremely small, or the earth's atmosphere would be filled with atoms of metals, say, and no solid would last very long. For example, molybdenum at 2000 K, which is 69% of its melting temperature, has a vapor pressure of 105 $\mu$Pa. This is a rather low pressure (atmospheric pressure being about 101 kPa): At 2000 K, 96 g (1 mol) of Mo would just fill, with vapor at a pressure of 105 $\mu$Pa, a cubic box roughly 500 m on each side. As the melting temperature is approached, vapor pressure increases rapidly; the vapor pressure of Mo at 2566 K (88.5% of melting temperature) is 133 mPa, about 1300 times larger than at 2000 K.

## CRYSTAL STRUCTURE OF METALS

Of the metallic elements, 44 solidify in one or more of three simple crystal structures. Of these three structures, two are versions of the cubic system [Fig. 4-2(a)] and the other is a version of the hexagonal system [Fig. 4-2(f)]. The common metals that are exceptions are manganese (complex cubic), mercury [rhombohedral, Fig. 4-2(e) with all edges of equal length], and tin [a body-centered version of the tetragonal system, Fig. 4-2(b)]. References to thorough discussions of all crystal systems are given at the end of this chapter. Appendix 6 lists the crystal structures of many metals of engineering interest.

**Body-centered cubic.**     The geometry of the *body-centered cubic* (bcc) crystal lattice consists of a simple cube with atom centers at each corner, plus an atom positioned at the cube center, as shown in Fig. 5-2. Such a characterization of the crystal is useful for remembering the spatial relationships of the various atoms, but the lines in Fig. 5-2 are solely for this purpose and do not necessarily represent the most important atomic bonds. For example, the distance between corner atoms and the center atom is less than the dis-

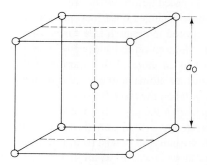

**Figure 5-2**   Body-centered cubic unit cell, with small circles representing the centers of atoms. Dashed lines are only to show the position of the center atom.

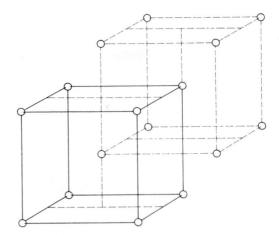

**Figure 5-3** A bcc crystal regarded as interlocking simple cubes, one set shown with
solid lines and one set with dashed lines.
The atom at the center of the solid set is thus
the corner atom of the dashed set, and vice
versa.

tance $a_0$ between corner atoms. The cube formed by the solid lines is called the *unit cell*,
and the complete crystal is formed of many repetitions of this unit cell. Appendix 6 gives
the size of the *lattice constant* $a_0$ for common bcc metals.

There is no difference at all between the corner and center atoms in Fig. 5-2,
although a superficial examination of the unit cell might suggest otherwise. We could just
as well have drawn a unit cell using the center atom as one of its corners. In fact, we could
even think of the bcc structure as a set of interlocking simple cubes (Fig. 5-3), but this is a
more complicated model to treat mathematically and has, therefore, been replaced by the
simpler bcc model.

For clarity, we have so far shown the bcc crystal lattice by locating only the center
points of each atom. However, since the crystal is really (in our thinking, anyway) made up
of hard spheres that are touching, the bcc unit cell really appears as in Fig. 5-4. Only one-
eighth of each corner atom is located within each unit cell cube, while the entire body atom
is located within the unit cell. Thus, the requirements that the atoms be touching and be situated in the bcc geometry lead to a fixed relationship between the lattice constant $a_0$ and the
size of the atoms.

**Example 5-1**

Calculate the ratio $a_0/r$ for a bcc crystal where $r$ is the radius of the atoms.

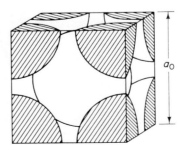

**Figure 5-4** The bcc unit cell, with atoms
shown as full-sized spheres.

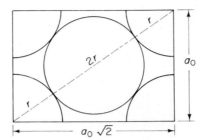

**Figure 5-5**   Solution to Example 5-1.

**Solution**   Cut the unit cell along a diagonal plane through the center atom (Fig. 5-5). The line of atoms touching is thus from one corner atom through the body atom to the opposite corner atom, shown dashed in Fig. 5-5. This diagonal is $a_0\sqrt{3}$, equal to four atomic radii. Thus, $4r = a_0\sqrt{3}$, and so $a_0/r = 4/\sqrt{3} = 2.31$.

**Face-centered cubic.**   The other common cubic crystal structure in metals is *face-centered cubic* (fcc), shown in Fig. 5-6. Here, in addition to atoms at each corner of the cube, atoms are located at the center of every face of the imaginary cube surface. The differences between bcc and fcc are much greater than might appear from a comparison of Figs. 5-2 and 5-6. Figure 5-7 shows the atoms, full-sized, in a unit fcc cell.

**Example 5-2**

How many atoms are in the bcc unit cell? How many in the fcc unit cell?

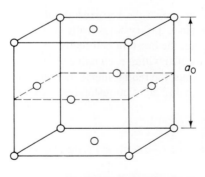

**Figure 5-6**   Face-centered cubic crystal unit cell, with small circles representing centers of atoms. Dashed lines are only to show the positions of the face atoms.

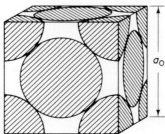

**Figure 5-7**   The fcc unit cell, with atoms shown as full-sized spheres.

**Solution**   From Fig. 5-4, for bcc there is 1 center atom and 8 corner parts, each $1/8$ atom, and so the total is $1 + 8(1/8) = 2$ atoms. From Fig. 5-7, for fcc there are $8(1/8)$ corner atoms and $6(1/2)$ face atoms, and so the total is $8(1/8) + 6(1/2) = 4$ atoms. We thus conclude that for about the same density of atoms, the fcc cube has roughly twice the volume.

**Example 5-3**

Calculate $a_0/r$ for fcc.

**Solution**   From Fig. 5-7, the atoms touch along the face diagonal, which is thus $4r$ long. Thus, $a_0\sqrt{2} = 4r$, and so $a_0/r = 4/\sqrt{2} = 2.83$.

**Hexagonal close-packed.**   The third common crystal structure of metals is *hexagonal close-packed* (hcp), shown in Fig. 5-8. The atoms of the top and bottom planes in this sketch are centered on the outlines of hexagons and at the hexagon centers. Six spheres perfectly surround and touch a seventh identical sphere if all are on a plane; these close-packed planes are called *basal planes*. Between the top and bottom planes lies an identical plane of atoms (also a basal plane) shifted so that the spherical atoms rest in the valleys between the spheres in the top and bottom planes. Figure 5-9 shows a side view of the hcp structure. The ratio of the unit-cell height $c$ to the hexagonal side dimension $a$ would be $c/a = 1.633$ if the spacing of the planes depended solely upon the size of the atoms as hard spheres. In hcp metals $c/a$ varies from the theoretical value of 1.633 by as much as 15%, so we may conclude that factors not incorporated into our model influence the spacing of the basal planes.

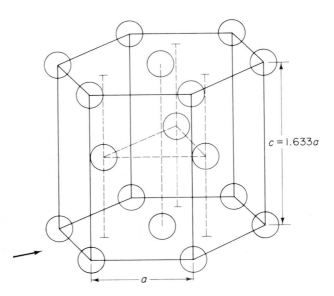

$c = 1.633a$

**Figure 5-8**   The hexagonal close-packed unit cell, with small circles representing centers of atoms. Solid lines outline the unit cell, and dashed lines show the relative positions of atoms. The arrow indicates the viewing direction in Fig. 5-9.

$a$

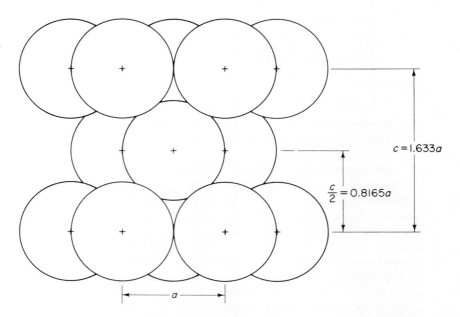

**Figure 5-9**   Hexagonal close-packed crystal, as viewed from the direction of the arrow in Fig. 5-8.

## MILLER INDICES

The strength of crystalline materials (such as metals) can usually be related to events that take place in individual crystals. To help in describing the geometry of events on a crystalline scale, a spatial nomenclature called *Miller indices,* akin to vector notation, has been developed. The principles of vectors are incorporated in the use of Miller indices, but there are differences in application, such as the method of defining planes, and the symbols used are different. Miller notation is helpful in understanding strength-related processes such as slip.

      **Directions in cubic crystals.**     A direction vector in a cubic crystal *always starts at the origin* of the three axes lying parallel to the three edges of the unit cubic cell. The coordinates of the point in space where the vector ends are given in *square brackets* [*xyz*], where *x*, *y*, and *z* are the number of unit-cell dimensions ($a_0$) along each respective axis. If the coordinate is negative, a bar is placed above it.

      An example for the face-centered cubic (fcc) system is the [110] direction. In Fig. 5-10, a single cell of an fcc metal is shown, with small circles to represent each atom. The [110] direction is thus one unit-cell dimension ($a_0$) in the *x*-direction, one $a_0$ in the *y*-direction, and zero in the *z*-direction. The vector always starts at the origin and here ends at the point [$x = 1a_0$, $y = 1a_0$, $z = 0$]. If we wish to talk about the same direction on the top face of the unit cell, we can also call it the [110] because once the vector is

**Figure 5-10**  Directions in fcc.

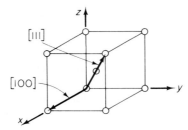

**Figure 5-11**   Directions in bcc.

defined, it can be moved anywhere in space as long as it remains parallel to its original direction.

For clarity the unit cell dimension is sometimes given as well, for example, $a_0[110]$. This procedure then allows us to vary the length of the direction vector. For example, the vector from the center of the atom at the origin in Fig. 5-10 to the center of the atom at the center of the bottom face is thus $a_0/2[110]$; this is often stated as $1/2$ [110]. The family of directions crystallographically equivalent to [110] is designated as <110>. This includes [101], [01$\bar{1}$], [$\bar{1}$10], [$\bar{1}$ $\bar{1}$ 0], and so on.

For the body-centered cubic (bcc) system the same rules apply. Figure 5-11 shows the [111] direction and the [100] direction.

**Directions in hexagonal close-packed crystals.**    The hexagonal close-packed (hcp) system is geometrically more complicated and has four indices, one of which is redundant. Here the first three coordinates $(a_1, a_2, a_3)$ are at 120° to each other, and the fourth coordinate $(c)$ is perpendicular to the plane of the three coordinates. Figure 5-12 shows the hcp system, with the intermediate (identical) layer of atoms positioned by dashed lines. Note that $a_3$ is positive toward the back. In order that directions be perpendicular to planes with the same indices (which is probably not apparent at this stage), the directional indices must be selected so that the $a_3$ index is equal to the negative of the sum of the $a_1$ and $a_2$ indices. For example, the [11$\bar{2}$0] direction is shown: one unit-cell (here atom-to-atom) dimension in the $+ a_1$ direction, one unit-cell dimension in the $+ a_2$-direction, two unit-cell dimensions in the $- a_3$ direction, and zero in the $c$-direction. The vector connecting the origin atom with the first atom in the $- a_3$ direction is thus $1/3$ [11$\bar{2}$0]. Selection of indices to describe a desired direction is not easy because of the constraining rule that $a_1 + a_2 = - a_3$.

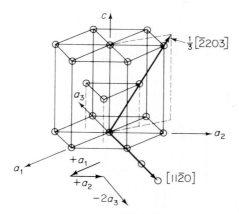

**Figure 5-12**   Directions in hcp.

When the $c$-direction component is not zero, the directions are even more complicated; for example, the $1/3$ $[\bar{2}203]$ is shown in Fig. 5-12. (See Example 5-4.) This direction
passes through the atom in the intermediate plane.

**Example 5-4**

Draw enough unit cells of the hcp system to show that the tip of the vector $1/3$ $[\bar{2}203]$ lies at
the point shown in Fig. 5-12.

**Solution**   The construction is given in Fig. 5-13. Start from the origin in the bottom left cell,
move $-2a_1$, then $+2a_2$, then $0\ a_3$, and then $+3c$. Connect the origin to the terminal point for

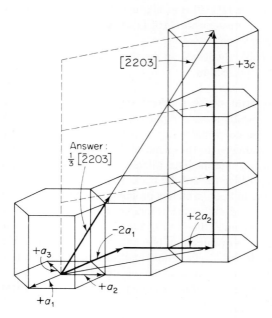

**Figure 5-13**   Solution to Example 5-4.

vector $[\bar{2}20\bar{3}]$. This vector lies in the plane described by dashed lines parallel to the basal direction $[\bar{2}200]$. Thus, $^1/_3 [\bar{2}20\bar{3}]$ ends one unit cell up from the starting point.

**Planes in cubic crystals.**   Planes are described by the reciprocal of their intersections with the perpendicular coordinate axes $x, y, z$. To be described, a plane cannot pass through the origin. (Although this obscure procedure may seem to be a mad plot against the reader's sanity, this nomenclature is useful and mathematically the easiest to handle once it is mastered.) As will be evident later, to be crystallographically important, planes must pass through significant numbers of atoms; fortunately, this restriction greatly limits the number of important planes.

Planes are indicated by Miller indices in curved brackets (parentheses). Figure 5-14 shows the atoms in the (110) plane, shown shaded, in an fcc crystal, with most of the face atoms omitted here for clarity. The indices are developed as follows: $x$-intercept = $1a_0$; $y$-intercept = $1a_0$; $z$-intercept = $\infty$. Thus, the plane is ($^1/_1$ $^1/_1$ $^1/_\infty$) or (110). Indices are normally reduced to the smallest possible set of integers. The family of planes equivalent to (110) is denoted {110}. One memory aid for distinguishing the indices for directions and planes is to recall that directions are straight lines and Miller indices for directions are enclosed in square brackets, which are made up of straight lines; curved and curly brackets then apply to planes.

The same procedure applies to the bcc system. Figure 5-15 shows four unit cells of bcc, which are needed in order to describe the intercepts for the (112) plane. Only one body atom is shown, in the cell that contains the origin. The atoms shown lie in this plane.

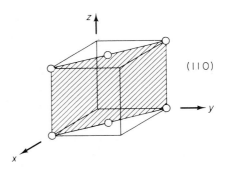

**Figure 5-14**   The (110) plane in fcc.

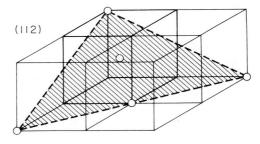

**Figure 5-15**   The (112) plane in bcc.

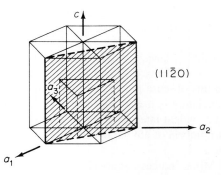

$(11\bar{2}0)$

**Figure 5-16**   The $(11\bar{2}0)$ plane in hcp.

Note that the intercepts are $x = 2a_0$, $y = 2a_0$, $z = 1a_0$, giving $(^1/_2\ ^1/_2\ ^1/_1)$ or $(112)$ to the smallest integer.

**Planes in hexagonal close-packed crystals.**   The procedure for hcp planes is the same as for the cubic system except that the redundant $a_3$ must be accommodated. Because numerical values of indices for a plane are the same as for the direction normal to the plane (which is why reciprocals are used to define a plane), the rule that $a_1 + a_2 = -a_3$ must also hold for planes. Figure 5-16 thus shows the $(11\bar{2}0)$, for intercepts of $a_1 = 1$, $a_2 = 1$, $a_3 = -1/2$, $c = \infty$, to give $\left(\frac{1}{1}\frac{1}{1}\frac{1}{-1/2}\frac{1}{\infty}\right)$ or $(11\bar{2}0)$. Note that the $a_3$ index is fixed as soon as $a_1$ and $a_2$ are set.

**Manipulations and tricks with Miller indices.**   For a cubic system of lattice constant $a_0$, the spacing $d_{(hkl)}$ between the set of planes defined by *(hkl)* can be found from

$$d_{(hkl)} = \frac{a_0}{\sqrt{h^2 + k^2 + l^2}}$$

It is essential that the set of planes *(hkl)* includes *all* atoms in the crystal.

**Example 5-5**

Calculate the spacing of the $(110)$ in bcc.

**Solution**   $d_{(110)} = a_0/(1^2 + 1^2 + 0^2)^{1/2} = a_0/\sqrt{2}$. The set of planes of this spacing includes all the atoms in the crystal.

An example where the planes as defined by the lowest common integers do not include all atoms is the $(110)$ in fcc. Figure 5-17 shows how this family of planes excludes, for example, the atoms marked with $x$ that lie between the two planes of open circles. The above equation gives $d_{(110)} = a_0/\sqrt{2}$ for the spacing of planes, but this includes only half the atoms. To avoid this problem, the integer rule must be violated and the planes containing the complete set of atoms must be designated $(220)$. This gives the correct spacing as $a_0/2\sqrt{2}$.

A useful result of Miller indices is that the normal has the same indices as its plane. (Miller indices were chosen with this in mind.) Consider the problem shown in Fig. 5-

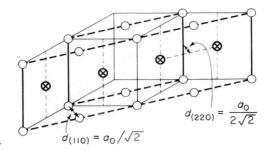

**Figure 5-17**  Spacing of (110) in fcc.

18(a), where a bcc slip plane, the (110), contains a slip direction $[\bar{1}11]$. It is desired to find the direction lying in (110) that is perpendicular to $[\bar{1}11]$. This is solved as a geometry problem in Fig. 5-18(b), and the direction is found to be $[\bar{1}1\bar{2}]$. One must demonstrate that $[\bar{1}1\bar{2}]$ is perpendicular to $[\bar{1}11]$. Even for a simple case this is not easy. For more complex cases the geometry rapidly becomes very messy.

Using Miller indices, consider the same problem, restated in Fig. 5-19, without regard to the geometry of the cubic unit cell. (The reader who has not studied vector

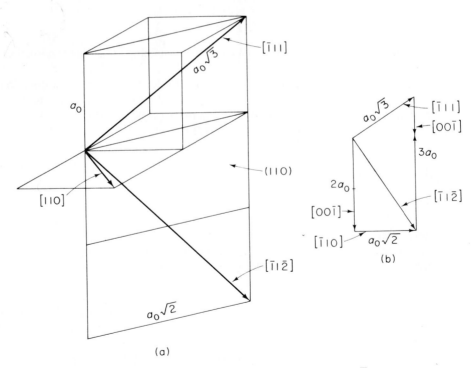

(a)

(b)

**Figure 5-18**  To find the direction in (110) that is perpendicular to $[\bar{1}11]$ in bcc.

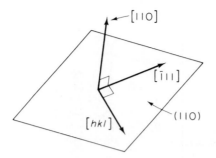

**Figure 5-19** The same problem as in Fig. 5-18, solved using Miller indices.

manipulation can skip the next two paragraphs.) The slip direction $[\bar{1}11]$ lies in (110). The problem is to find *(hkl)*, lying in (110), normal to $[\bar{1}11]$. First construct the normal [110] to the plane (110). Then, $[hkl] = [\bar{1}11] \times [110]$; for $x, y, z$ as unit vectors of length $a_0$ in the three directions,

$$[hkl] = \begin{vmatrix} x & y & z \\ \bar{1} & 1 & 1 \\ 1 & 1 & 0 \end{vmatrix}$$
$$= x(0-1) - y(0-1) + z(-1-1)$$
$$= [\bar{1}1\bar{2}]$$

This quick solution also tells us that the plane containing $[\bar{1}11]$ that is normal to the (110) is the $[\bar{1}1\bar{2}]$.

Miller indices also provide a quick test to determine whether a direction lies in a plane, because the dot product of this direction and the direction normal to the plane must then be zero. Does a direction [ghj] lie in a plane *(def)*? Yes, if *gd + he + jf = 0*. Thus, in Fig. 5-18 we see that $[\bar{1}1\bar{2}]$ does lie in the (110) since $-1 + 1 + 0 = 0$.

## SPACING OF ATOMS

William H. Bragg and his son William L. Bragg first demonstrated quantitatively with X-rays that the atoms in a solid metal crystal exist in a regular, repeated pattern with remarkably constant spacing between atoms. Using X-rays of wavelength $\lambda$ striking a metal crystal at an angle $\theta$ with a set of planes of atoms spaced $d$ apart, they showed by simple geometry that the X-rays were diffracted at the angle $\theta$ to the planes if

$$\lambda = 2d \sin \theta \tag{5-1}$$

Diffraction occurs, in fact, if the left side of the equation is equal to any integral number of wavelengths, although the higher orders are not usually important. The significance here of the Braggs' work is that it shows the high precision and regularity of atomic spacing within a metal, thus confirming the hypothesis that atoms are found in a regular three-dimensional array, with each atom at a fixed mean position relative to each other atom, in a characteristic crystal structure.

**Example 5-6**

A single crystal of nickel (face-centered cubic, atomic radius = 124.3 pm) gives first-order diffraction with X-rays from a copper target (wavelength = 154 pm) at an angle $\theta$ = 26.0° (0.454 rad). Find which set of planes is producing this diffraction.

**Solution**   $n\lambda = 2d \sin \theta$ and $d = n\lambda/(2 \sin\theta) = (1 \times 154$ pm$)/(2 \sin 26°) = 176$ pm. For Ni, $a_0 \sqrt{2} = 4r$ and $a_0 (4 \times 124.3$ pm$)/\sqrt{2} = 352$ pm. Thus, $d/a_0 = 176/352 = 0.5$, and the diffraction is from {200} planes. If this problem is still not clear, a review of simple crystal structures and Miller indices discussed earlier in this chapter may be in order.

If the crystal structure includes a relatively small number of instances where the atomic spacing is not uniform, diffraction will still occur. Thus, X-ray diffraction cannot readily discriminate between crystals that are perfect and those that have a few instances of nonuniform spacing and are thus slightly imperfect. It will be shown later in this chapter that imperfections do exist. If the number of imperfections grows to a relatively high value in a crystal, the X-ray diffraction will be blurred by the irregularities in crystal spacing and orientation.

## SLIP IN PERFECT CRYSTALS

Assume that a metal crystal is perfect. The yield strength of this crystal must be related to the way in which atoms move relative to each other to cause strain and to the magnitude of forces required to produce this motion. Observation of a large number of deformed crystals reveals that the most common motion is the sliding of one plane of atoms relative to an adjacent plane of atoms. This is called *slip*.

The essential aspects of the process of slip can be described with a two-dimensional diagram. Although some observed behavior is attributable to the three-dimensional geometry of a real crystal, Fig. 5-20 shows how slip might occur in a perfect crystal. Figure 5-20(a) shows the initial position of 16 atoms lying in a plane and having a uniform crystal structure. The diagonal line will be used to keep track of the relative positions of the atoms.

A simple tensile or compressive stress acting on the crystal in any direction except vertical or horizontal will have some shear-stress component $\tau$ acting on the horizontal and vertical faces of the crystal; for simplicity, the vertical shear stresses are not shown in Fig. 5-20(b). When $\tau$ becomes large enough, the interatomic bonds are overcome on some plane of atoms, shown in Fig. 5-20(b) as the middle plane, and sliding begins.

If $\tau$ is further increased, the upper eight atoms can move to the right to a new equilibrium position, shown in Fig. 5-20(c). The magnitude of this relative motion is the amount of slip occurring during the complete process. Because the atoms are in new equilibrium positions, no stress is required to hold them there; thus, the deformation, or slip, that has taken place represents permanent *plastic* deformation that remains after the stress is removed. The crystal in Fig. 5-20(c) is just as perfect as in Fig. 5-20(a), so it would not be possible to say that plastic deformation had occurred if one did not know the history of the crystal.

**Figure 5-20** Schematic diagrams of slip in a perfect crystal: (a) initial position; (b) an intermediate position, with shear stress $\tau$; and (c) final position, no stress.

## MAXIMUM THEORETICAL STRESS FOR SLIP

When the relative sliding of one layer of atoms over another reaches the midpoint between the two equilibrium positions, the shear stress is zero. By symmetry it can be seen that the upper two layers of atoms in Fig. 5-21, which represents such a condition, could fall with equal ease to the right or to the left. Thus, the shear stress required to move the upper group of atoms to the right from the initial equilibrium position must reach a maximum before the middle position shown in Fig. 5-21. The variation of shear stress with sliding distance $x$ from the initial equilibrium position can be roughly approximated by a sine function:

$$\tau = \tau_{max} \sin\left(\frac{2\pi x}{b}\right) \tag{5-2}$$

where $\tau_{max}$ is the maximum shear stress, $b$ is the distance between equilibrium positions, and $x$ is the relative slip between the two planes at the instant the applied shear stress is $\tau$. Figure 5-21 thus shows the case where $x = b/2$.

If strength is related to the shear stress required to cause slip to occur, then the highest value of shear stress $\tau_{max}$ should give an indication of the strength of a perfect crystal. If

**Figure 5-21** Unstable midpoint of slip process; $\tau = 0$ when $x = b/2$.

$\tau_{max}$ is not reached, extensive slip will not occur. A rough approximate calculation of $\tau_{max}$ should, therefore, provide an order-of-magnitude estimate of the strength to be expected. This is done in the following paragraphs.

**Example 5-7**

At what value of $x$ will $\tau = \tau_{max}$ in a perfect crystal?

**Solution**    Since $\tau = \tau_{max} \sin(2\pi x/b)$ from Eq. (5-2), $\tau = \tau_{max}$ when $\sin(2\pi x/b) = 1$; thus, $2\pi x/b = \pi/2$, and $x = b/4$. Other information suggests that the maximum stress occurs at a smaller value of $x$ because the stress does not follow an exact sine relationship, but this does not alter the order of magnitude of $\tau_{max}$.

The slope of the $\tau$–$x$ curve can be calculated for small values of $x$ and equated to the measured slope of the elastic stress–strain curve. This step can be taken because elastic behavior occurs at relatively small, reversible displacements of atoms from the equilibrium position, a condition represented by the $\tau$–$x$ curve at $x \ll b$. From Eq. (5-2),

$$\frac{d\tau}{dx} = \tau_{max} \frac{2\pi}{b} \cos\left(\frac{2\pi x}{b}\right)$$

For $x \ll b$, $\cos(2\pi x/b) \approx 1$:

$$\frac{d\tau}{dx} = \tau_{max} \frac{2\pi}{b}$$

Elastic deformation resulting from shear stress can be represented by Fig. 5-22, where the dashed lines show the position of the upper two atoms after the shear stress $\tau$ is applied. The resultant angular shear strain $\gamma$ on the entire crystal can be measured, and $\gamma$ is found to be directly related to the shear stress $\tau$:

$$\frac{\tau}{\gamma} = G \tag{5-3}$$

$G$, the *shear modulus of elasticity,* is a constant for each elastic solid.

In Fig. 5-22, by geometry, for $x \ll a$, $\gamma = x/a$. Combining with Eq. (5-3) gives $x/a = \tau/G$. In the straight-line elastic range, the slope is $d\tau/dx = \tau/x = G/a$. Assuming that $a \approx b$, as can be seen in Fig. 5-22, and combining with Eq. (5-2), we obtain

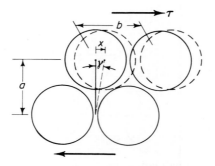

**Figure 5-22**   Elastic displacement under low stress $\tau$.

$$\tau_{max} \frac{2\pi}{b} = \frac{G}{a}$$

$$\tau_{max} \approx \frac{G}{2\pi} \tag{5-4}$$

In Chap. 14 relating to fracture, a similar type of calculation for *pulling* atoms apart (as opposed to their *sliding* over one another as in this case) gives the theoretical tensile fracture strength required for a perfect crystal as $E/2\pi$ (where $E$ is the tensile modulus of elasticity).

## MEASURED STRESSES FOR SLIP

Theoretical values [from Eq. (5-4)] and average measured values of shear stress for slip for three metals of fairly high purity are given in the accompanying table.

|  | Theoretical $\tau_{max}$, MPa | Measured $\tau_{slip}$, MPa |
|---|---|---|
| Zone-refined iron | 14,000 | 35 |
| 99.996% aluminum | 3500 | 7 |
| Electrolytic copper | 7000 | 17 |

The theoretical strengths are thus 400 or 500 times greater than the measured strengths of high-purity metals. Similar results are obtained for a wide range of metals.

Even with the inaccuracies in the assumptions involved in the derivation of Eq. (5-4), it is unlikely that a discrepancy on the order of 500 can be explained away. Therefore, the basic premise that crystals are perfect and that slip occurs as described in Fig. 5-20 must be invalid. If slip occurs in some other way not involving simultaneous movement between two planes of atoms, then the crystal cannot remain perfect. This conclusion has far-reaching implications for the nature of strengthening mechanisms, as will be shown in the next section.

One exception to the relatively low strength of real metals has been observed. In some metal crystals in the shape of a very fine filament or whisker, shear stresses on the order of 10 GPa have been measured. Because the theoretical maximum stress is of this same magnitude, we may conclude that these fine crystals either are perfect or contain imperfections that do not substantially influence strength. Applications of these very-high-strength filaments are discussed in Chap. 17.

## DIRECT OBSERVATION OF IMPERFECTIONS

The tension test reveals a wide range of behavior among different metals and alloys. For example, at the same temperature, tensile strengths of different metals and alloys can differ by a factor of about 100, and there is a wider range for true strain at fracture, which can vary by a factor of about 1000. There is no consistent correlation between these two mechanical properties, although high fracture strain is more often associated with low strength than with high strength. Furthermore, a wide range of mechanical properties may be produced in a single alloy of constant chemical composition through heat treatment and mechanical working. The fundamental reasons behind these variations in properties must be understood if the selection, heat treatment, and properties of metals and alloys are not to appear as some mysterious and inconsistent sort of alchemy.

Because the assumption of perfectly regular crystal structure in metals leads to calculated stresses for slip that are several orders of magnitude larger than the measured values, the search for an understanding of strengthening mechanisms leads to a study of the nature of possible imperfections. These imperfections must provide a mechanism for permitting slip to occur at relatively low stresses. We now turn to the current evidence for plausible imperfections that may exist in crystals, with the objective of finding general rules governing strength.

It is possible to observe the shape and distribution of imperfections in many solids. The scale of observation necessary to view imperfections directly must be much finer than the wavelength of light, which is on the order of 500 nm, because the interatomic spacing of most metals falls in the range 0.2–0.4 nm. For the observation of imperfections on this size scale, an electron beam accelerated through, say, 100 kV is well suited, as it has a wavelength of 3.7 pm and can penetrate a specimen of metal 10–100 nm thick. An electron microscope arrangement for observing atomic-scale defects in crystals is shown

**Figure 5-23**   Schematic arrangement for thin-foil electron microscopy.

**Figure 5-24** Dislocations in an electron micrograph of a thin foil of 18% chromium–8% nickel stainless steel. Magnification is given by the 1-$\mu$m scale marking, which represents the size a 1-$\mu$m dimension would have if enlarged the same amount as the specimen. (Photograph courtesy of P. D. Goodell.)

in Fig. 5-23. An electron beam traveling through a thin crystal specimen may find some defects oriented so as to cause local Bragg diffraction; when this condition occurs, the defects are effectively opaque to these electrons. The electrons that do not pass near any defect are not diffracted, and after enlargement by the lens system darken all of the photographic film except in the region of the defect.

It is important to note that the study of thin foils by the method in Fig. 5-23 is not the same as a shadow picture. The conditions for Bragg diffraction are quite critical, and very few of the defects present in a crystal are visible at any one time. Tilting the specimen to change its angle to the electron beam can cause the first defects to disappear and other defects to appear.

Figure 5-24 is a typical electron micrograph of a thin specimen of stainless steel. The dark lines are imperfections in the crystal; these imperfections appear as light lines in a negative exposed to the electron beam, and the reversal occurs in the print made from the negative. A series of such pictures made of a single thin specimen, but in different regions and with different orientation to the electron beam, reveals similar lines throughout the specimen. If the specimen is stressed so that slip occurs, the pattern of lines changes. Thus, imperfections must be associated with slip on two counts: (1) The observed imperfection configuration is altered with slip, and (2) the strength of a real metal is much lower than the calculated strength of a perfect crystal.

**Example 5-8**

What two characteristics must imperfections have in order to explain the behavior of real metals?

**Solution**  Imperfections must (1) provide a mechanism by which slip can occur, and (2) be capable of reducing the strength by several orders of magnitude below the strength of a perfect crystal. In addition, any imperfection being considered must be stable; that is, it must not disappear when the part is stressed. Many kinds of imperfections can be imagined that do not have both of these features, and so the number of kinds of imperfections of interest here is reduced substantially.

Other experimental methods have been applied successfully to the observation of imperfection patterns. These include etching (accelerated corrosion) of a polished surface to reveal pits where imperfections intersect the surface; accumulation of opaque foreign atoms in the locally distorted region surrounding an imperfection in a transparent crystalline solid; and formation of an image by emission of ions from a very small spherical surface, which reveals the local surface configuration of atoms. Although these methods reveal some additional information, they are not further discussed here. The important result of the several independent methods of observing imperfections is that they agree on the existence of the imperfections described in the next section.

## EDGE DISLOCATION

Figure 5-25 shows a plane of atoms of a simple cubic crystal arranged in a perfect matrix. (The cubic matrix is used here instead of the 60° matrix in Fig. 5-20 because the cubic structure is simpler to view. No metals have this simple cubic structure, but the argument that follows applies equally well to the more complex crystal structures.) Note in Fig. 5-25 that all atoms, here represented by small circles, follow a regular, repeated pattern, the necessary requirement for a perfect crystal. The lines between circles represent the principal bonds between each atom and its neighbors. In three dimensions, lines would also be drawn normal to the plane of this figure to connect these atoms with atoms lying in the adjacent parallel planes.

**Example 5-9**

Does the next parallel plane of atoms lying in back of those in Fig. 5-25 have atoms (a) directly in back of the atoms in the diagram, (b) directly in back of the squares in the diagram, or (c) elsewhere?

**Solution**　(a) Directly in back of the atoms in the diagram. If the next atoms were behind the squares, we would have a body-centered cubic structure. For the simple cubic case, then, a line of atoms in the two-dimensional diagram represents a whole plane of atoms perpendicular to the plane of the diagram and lying in back of (and in front of) the line. The equivalent three-dimensional diagram is shown in Fig. 5-26.

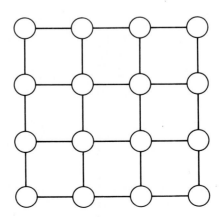

**Figure 5-25**　Plane of a simple cubic crystal without imperfections.

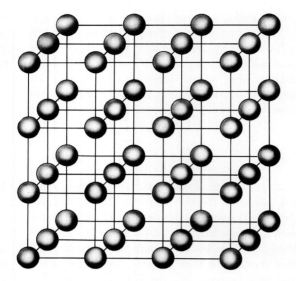

**Figure 5-26**   Perfect simple cubic crystal, in three-dimensional diagram.

A plane of atoms containing an imperfection, called a *dislocation*, is shown in Fig. 5-27. This figure differs from Fig. 5-25 in that two extra atoms are present. These represent an extra plane of atoms that ends within the crystal. Figure 5-28 shows this configuration in three dimensions. In Fig. 5-28 a line along the lower edge of the plane that ends within the crystal defines the region of atomic discontinuity; such a line describes an *edge dislocation*. The interatomic spacing cannot be exactly the same both above and below the dislocation line because in Fig. 5-28 there are four spacings across the top and three spacings across the bottom. (Real dislocations distort a crystal over a much wider region than this simple diagram shows.) Thus, an electron beam may be diffracted in the region of the dislocation line without being diffracted in the bulk of the crystal, and so the dislocation is visible on the photographic film because some electrons did not travel in a straight path through the aperture (see Fig. 5-23).

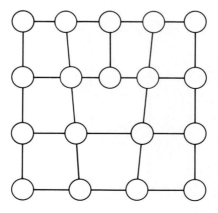

**Figure 5-27**   Plane of a simple cubic crystal that contains an edge dislocation.

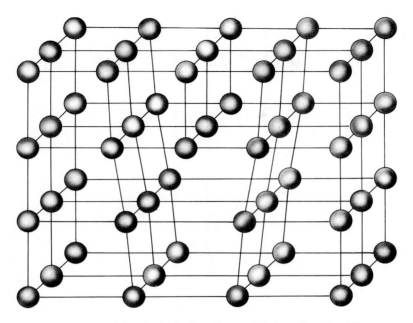

**Figure 5-28**    Edge dislocation in simple cubic crystal, in three-dimensional diagram.

The edge dislocation was first hypothesized in 1934 by E. Orowan, M. Polanyi, and G. I. Taylor, each working independently, to explain the observed low strength of real crystals. This occurred long before the development of the electron microscope; because other techniques for dislocation observation had not yet been developed, many years were to pass before the first direct observation of dislocations and confirmation of their existence.

## SLIP BY DISLOCATION MOTION

If the edge dislocation is to qualify as an imperfection of interest here, it must provide a mechanism for slip that differs from the way slip occurs in a perfect crystal. The edge dislocation is the line where the extra plane of atoms ends. If this extra half-plane is able to shift slightly so as to form interatomic bonds with a single half-plane, rather than being situated between two planes, then the dislocation line will be moved laterally to a new extra half-plane. Figure 5-29(a) shows one plane of a stable edge dislocation, with atom A initially representing the bottom row of atoms in the extra plane. We can consider the dislocation as a line perpendicular to the plane of the diagram that lies immediately underneath the plane of atoms represented by the single atom A. When a shear stress $\tau$ is applied [Fig. 5-29(b)], the upper set of atoms is displaced to the right initially through elastic strain. At first, the bond between atoms B and C remains, even though A has moved closer to C than in Fig. 5-29(a). With a further increase of $\tau$, atom A moves closer than atom B to atom C, and the C–B bond is broken and replaced by the C–A bond

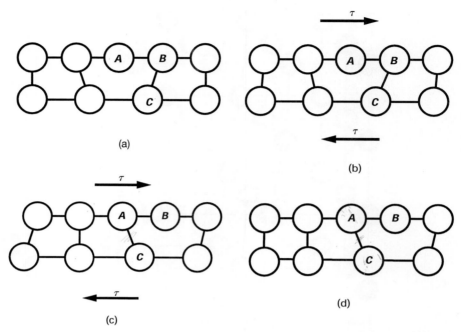

**Figure 5-29**   Motion of edge dislocation.

[Fig. 5-29(c)]. The dislocation has now moved one interatomic spacing to the right and lies under the plane of atoms represented by atom B. Finally, if at this stage the shear stress $\tau$ is removed, those atoms remote from the dislocation, such as at the left edge of this set, resume their equilibrium positions [Fig. 5-29(d)]. Note that with the shear stress removed, the dislocation is stable at its new location B; it does not return to its original position under the plane of atoms represented by A. This process has, therefore, produced *plastic deformation*, although the magnitude is much too small to be measured. For the full edge dislocation *line* to move to the right during this process, all the bonds between the out-of-plane atoms in back of B and C must likewise be broken and replaced by bonds between the atoms in back of A and C.

The total motion of the five upper atoms in Fig. 5-29 relative to the four lower atoms is very small during this process. If slip is to occur, that is, if all the atoms above the *slip plane* (a horizontal plane through the middle of Fig. 5-29) are to move one interatomic spacing to the right, the dislocation must move further than one interatomic spacing. The sequence in Fig. 5-30 shows how slip can result from dislocation motion. Figure 5-30(a) is a perfect crystal into which a dislocation can be forced by an indentation at the upper left edge (not shown). This is not the most common way that dislocations are produced, but it serves as a simple illustration. An applied shear stress then causes the dislocation to move by shifting of interatomic bonds [Fig. 5-30(b) and (c)] until the extra plane emerges from the crystal and appears as a step on the right edge [Fig. 5-30(d)]. Note that while the dislocation moves the full width of the crystal, each of the upper atoms moves only one inter-

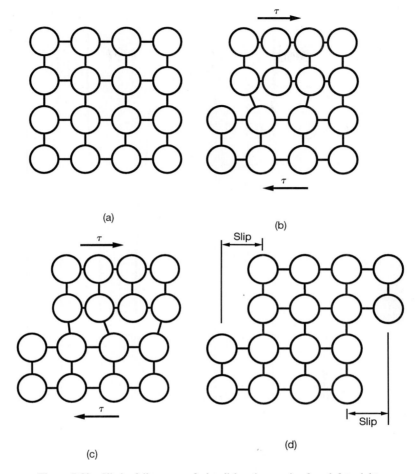

**Figure 5-30** Slip by full traverse of edge dislocation moving from left to right.

atomic space to the right. In this case slip can be measured at the left or right edge and is equal to one interatomic spacing in the direction parallel to the direction in which the dislocation moves. To produce a slip step large enough to be observed in an optical microscope, many dislocations must move along or near the same slip plane.

**Example 5-10**

Slip in nickel commonly occurs in a <110> direction, as is generally true of fcc metals. If the atomic radius of Ni is 124.3 pm, how many edge dislocations must emerge from a crystal of Ni to produce a step that is just visible in an optical microscope, say, a step 1 $\mu$m high?

**Solution**   Interatomic spacing in <110> in fcc is $2r = 248.6$ pm. (See App. 6 for properties of pure metals.) 1 $\mu$m/248.6 pm = 4020 dislocations.

(a)

(b)

(c)

**Figure 5-31**    Three ways to represent the same edge dislocation.

A simple symbol for the edge dislocation can be introduced here to replace the cumbersome diagrams of individual atoms. Figure 5-31(a) shows an edge dislocation of the form already discussed, but it shows solid lines connecting atoms only in the extra plane. The dashed line represents the slip plane, where the extra plane of atoms terminates. In Fig. 5-31(b) the individual atoms have been replaced by lines representing planes perpendicular to the diagram. The symbol used in Fig. 5-31(c) shows the location of the same edge dislocation by its position, the direction of the extra plane by the single vertical leg of the symbol, and the horizontal slip direction of the dislocation by the horizontal line of the symbol. The edge-dislocation symbol can be used in any orientation in a plane diagram.

## STRAIN FROM DISLOCATION MOTION

It is possible to calculate the shear strain that results from the motion of one or more edge dislocations moving on parallel planes in a single crystal. Figure 5-32(a) shows the shear strain $\gamma$ resulting when one dislocation moves all the way across the crystal. If $\gamma$ is always measured by the angular rotation of a line from the atom in the lower left corner to the one in the upper left corner, then no matter where horizontal slip occurs, the shear strain $\gamma$ will be the same for the same total slip. For Fig. 5-32(a), $\gamma = b/h$, where $b$ *is* the slip caused by the one dislocation and $h$ is the height of the crystal. If $n$ dislocations traverse the crystal, the total slip will be $nb$, and if $nb \ll h$, $\gamma = nb/h$. These $n$ dislocations need not occur on the same plane, as each parallel slip displacement $b$ will move the atom in the upper left corner a distance $b$ to the right. Figure 5-32(b) shows the result of complete traverse by three dislocations; here $\gamma = 3b/h$.

In real crystals the number of dislocations involved in a simple slip process is very large, as the calculation for Example 5-10 demonstrates. Thus, the average motion on a gross scale resulting from many small dislocation motions is measured as shear strain. Consider Fig. 5-33, which shows the influence on the atom in the upper left corner (any point on the upper surface will serve, in fact) of the motion of an edge dislocation from position $A$ to position $B$, where the distance moved is half the width, $w/2$. In the vicinity of

(a)  (b)

**Figure 5-32** Shear strain associated with traverse of a crystal by a dislocation: (a) one dislocation; (b) three parallel dislocations.

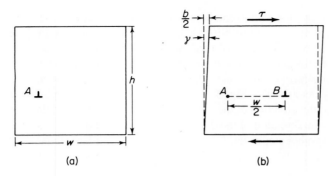

**Figure 5-33**   Limited motion of a dislocation from position $A$ to position $B$.

the dislocation, individual atom motion ranges from zero to $b$, but at a large distance from
the slip plane, the average displacement during this process must be $b/2$. Thus, for this
case, $\gamma = b/2h$, or for the general case of $n$ parallel edge dislocations of magnitude $b$ moving an average glide distance $g$ in a crystal of height $h$ and width $w$, the shear strain $\gamma$ is

$$\gamma = \left(\frac{nb}{h}\right)\left(\frac{g}{w}\right) \tag{5-5}$$

Such highly simplified parallel dislocation motion seldom occurs in real metals, so Eq. (5-5)
must usually be further modified to accommodate complexities of geometry and dislocation
configuration.

### Example 5-11

A crystal face 15 mm high and 20 mm wide intersects $10^8$ parallel edge dislocations on parallel slip planes. The interatomic distance in the direction of dislocation motion is 0.30 nm in
this crystal. How far must each dislocation move, on the average, to give a shear strain of
0.001?

**Solution**    Equation (5-5) gives $\gamma = nbg/hw$; $g = \gamma hw/nb$, $g = (0.001 \times 15 \text{ mm} \times 20 \text{ mm})/(10^8$
$\times 0.30 \text{ nm}) = 10 \ \mu\text{m}$. In the real case it is more likely that new dislocations will be generated
and some of the existing dislocations will not move, but the effect is the same.

## STRESS FOR DISLOCATION MOTION

The second requirement for a dislocation is that, in addition to providing a means for producing slip and shear strain, it must lower the strength of a perfect crystal by several orders
of magnitude.

The maximum shear stress required for slip results from a summation of the forces
required at any one time to break all atomic bonds involved in the slip process. Consider,
for example, a perfect crystal cube with 30-mm edges and a slip plane parallel to one of
its faces. Since the spacing of metal atoms in the slip direction is on the order of 0.30
nm, the length of each edge contains about $10^8$ atoms, and a slip plane thus has $10^{16}$
atoms. If slip occurs by dislocation motion, then only a small number of atomic bonds

across the slip plane are broken at any one time. It is possible to estimate the maximum number of atoms that can be involved in a dislocation at any one instant in order to account for the observed strength-reduction factor of about 500 below the theoretical shear stress. For the 30-mm single crystal cube having $10^{16}$ atoms on its shear plane, the number of atomic bonds being broken at any one time must be less than $10^{16}/500 = 20 \times 10^{12}$. If we consider an edge dislocation running the full depth of the crystal, the effective width of this dislocation will be less than $(20 \times 10^{12})/10^8 = 200,000$ atoms wide. For interatomic spacing of approximately 0.30 nm, a single dislocation will thus be less than $200,000 \times 0.30$ nm $= 60$ $\mu$m wide. Actual observed dislocations are much narrower than 60 $\mu$m, so this requirement is easily met even for a large number of dislocations that move at the same time.

Thus, we see that the essential reason for the low stress for slip is that the dislocation is narrow compared to the width of the slip plane, so only a relatively small number of atomic bonds across the slip plane are being broken at any one time. A number of analogies have been noted that exhibit this effect. For example, an earthworm moves by first extending its head (front end) so as to create a short tensile extension, or dislocation, just behind its head. This dislocation then travels backward the length of its body and eventually causes the tail to move forward the same distance that the head moved at the outset, with a net forward motion of the whole worm. Another example is the local vertical displacement, or hump, of a carpet that can be easily moved from one edge of the carpet to the other so as to shift the whole carpet slightly. [Note that the net motion of the worm is opposite the direction of motion of the extensional dislocation, but the net motion of the carpet is in the same direction as the motion of the compressive (actually in the third dimension) dislocation.] These two analogies exhibit one feature in common with dislocations in crystals: The forces resisting motion are being overcome over only a relatively small fraction of their total area at any one time, so the maximum force, or stress, is much less than would be required in the absence of a dislocation.

The conclusion reached through the foregoing arguments is that dislocations can provide a mechanism for slip at stresses much lower than the theoretical stress required for a perfect crystal and that the observed widths of dislocations are consistent with the magnitude of strength reduction required. The essential mechanism for reduction of strength by a dislocation is in the breaking of a few interatomic bonds at one time rather than the breaking of all bonds at once, as would be required in the perfect crystal.

## DISLOCATION DENSITY IN REAL SOLIDS

A common way to measure dislocation density is to cut a crystal and count the number of dislocations that intersect the cut face. The *areal density* $\rho$ is then the number of dislocations divided by the area. Using the dimensions as defined in Fig. 5-33(a) and assuming that $n$ is the total number of dislocations present, we have

$$\rho = \frac{n}{hw} \tag{5-6}$$

Measurements of real metals give dislocation densities of $10^{10}$–$10^{11}$/m$^2$ for annealed metals and on the order of $10^{16}$/m$^2$ for severely cold-worked metals. The causes of increase in dislocation density and the resulting changes in strength are discussed in Chap. 6.

## STRENGTH AND DISLOCATION MOTION

For those materials in which slip is significant, strength is the stress required to produce a certain strain. A commonly used value for strength is nominal stress at maximum load. A possibly more important measure of strength is the yield stress, the stress required to produce the onset of plastic deformation; lower stresses produce only reversible elastic deformation. Sometimes the departure from elastic to irreversible plastic deformation is difficult to identify, in which case the strain needed for the determination of yield stress is assigned an arbitrary value such as 0.002 (see Chap. 2). Whichever value of strain is used for specifying strength—the strain required for ultimate or the strain required for initial plastic deformation—comparison of strengths for two different solids should be made at the same strain. The following discussion will show how such a comparison leads to a correlation between strength and dislocation glide distance.

The general relationship between normal and shear strain follows a Mohr's circle form of representation (see Chap. 2), so the maximum shear strain can be used as a general measure of strain. Equation (5-5) can be simplified by Eq. (5-6) to give

$$\gamma = \rho b g \qquad (5-7)$$

where $\gamma$ is the plastic shear strain, $\rho$ is the density of *moving* dislocations, $b$ is the local slip resulting from movement of the parallel edge dislocations, and $g$ is their average glide distance. Equation (5-7) is not of general quantitative application because of the specialized geometric restrictions in its derivation, but qualitatively it shows that the plastic shear strain is directly related to the density of active dislocations and their average glide distance. The value of $b$ is normally a constant for a given solid and is called the *Burgers vector* of a dislocation. The Burgers vector is thus the slip that would result from motion of a dislocation across the full width of a crystal, as shown in Fig. 5-32(a). Numerically, minimum and maximum values of $b$ differ by a factor of about 2 for all common metals. For applications covered in this book, $b$ is equal to twice the atomic radius.

Geometric complexities of dislocations also make the behavior of real solids more difficult to predict than the previous case suggests. If slip in a metal crystal involved only slip of parallel dislocations completely traversing the crystal, all specimens of this metal would have the same strength. Tests of single crystals substantially fulfilling these requirements do, in fact, show nearly constant low shear stress for slip over moderate ranges of strain. But, in fact, dislocations usually move on more than one plane, do not remain in straight lines, and interact strongly with each other; so the average dislocation glide distance can be greatly altered by the structure of the solid. Furthermore, the geometry of real dislocations is seldom as simple as for the special case of the edge dislocation. If an edge-dislocation line turns 90° in the slip plane within a crystal, the discontinuity that results will have the form of the axis of a spiral staircase, where it is impossible to match up planes; this is called a *screw dislocation* and is described in the next section. Most of the strength-related properties of edge dislocations are valid for screw dislocations and for dislocations that are part edge and part screw.

Dislocations often occur as loops that grow in size rather than as straight lines. (Loops are made possible by the presence of a screw dislocation.) The shear strain that results from the motion of straight, parallel dislocations [Eq. (5-5)] may be further modified by the ratio of the length $t$ of the loop perpendicular to the glide distance $g$ in a crystal of depth $d$:

$$\gamma = \left(\frac{nb}{h}\right)\left(\frac{g}{w}\right)\left(\frac{t}{d}\right)$$

This situation is shown in Fig. 5-34. As the dislocation has moved a distance $g$ and thus has swept out the shaded area $gt$, this equation is geometrically equivalent to the strain from a dislocation loop that has grown from zero area to the present area $gt$. (Since it is geometrically impossible for a dislocation to end within a crystal, only a dislocation loop can in fact exist entirely within a crystal.) If the swept area $gt$ of one dislocation is replaced by $A_{\text{swept}}$ and the denominator $hwd$ is replaced by the crystal volume $V$, the shear strain becomes

$$\gamma = \frac{nbA_{\text{swept}}}{V} \tag{5-8}$$

This equation is valid for the real case of a very large number of nearly circular dislocation loops growing from point sources in crystals.

To obtain an estimate of the number of dislocations involved in any slip process of a real polycrystalline metal, some modification of the assumption of parallel slip planes must be made to suit the individual case.

Equation (5-7) links strain with active dislocation density and glide distance. To show how strength, or stress at a given strain, is related to dislocation behavior, compare two crystals identical in active dislocation density but with crystal A having a greater average dislocation glide length than crystal B. Assume further that the second dislocation in

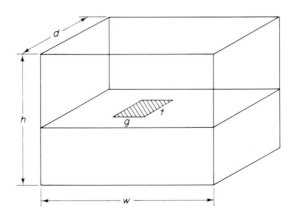

**Figure 5-34** Rectangular dislocation loop of swept area *gt* in a crystal of volume *hwd*. (Real loops are more nearly circular.)

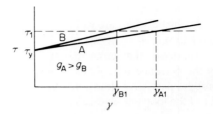

**Figure 5-35** Shear stress vs. shear strain, where crystal A has greater average dislocation glide length than crystal B.

each crystal moves at a stress slightly higher than the stress needed for the first dislocation, and so on. At each stress level a certain number of dislocations will have moved. Thus, if some minimum stress $\tau_y$ is required for motion of the first dislocation in each crystal, a plot of $\tau$ vs. $\gamma$ for crystals A and B might look like Fig. 5-35. Consider some stress $\tau_1 > \tau_y$. If the same number of dislocations in crystals A and B have moved, then from Eq. (5-7), $\gamma_{A1} > \gamma_{B1}$ because $g_A > g_B$. For a comparison of strengths at the same strain, Fig. 5-35 shows that for any fixed strain, say $\gamma_{B1}$, $\tau_B > \tau_A$. This leads to the conclusion that strength can be increased by a reduction in average dislocation glide distance.

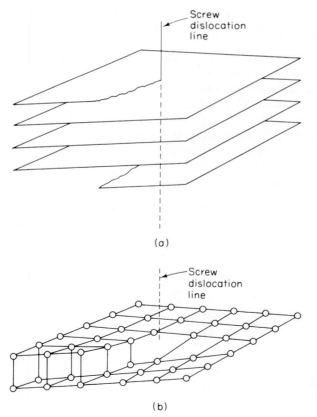

**Figure 5-36** A screw dislocation in a simple cubic crystal: (a) showing only the atomic planes, and (b) showing some of the atoms in the top two planes.

## SCREW DISLOCATION

The edge dislocation must be, by definition, a straight line along the edge of an extra plane of atoms which comprises an atomic-scale discontinuity. In a similar way we can describe a *screw dislocation* because it is also a straight line along an atomic discontinuity, but the discontinuity is the center of a helix, or spiral, of atoms that are not properly matched. Figure 5-36 shows the essential geometry. Note that if we follow a single plane of atoms in a path completely around the screw dislocation line, we will then be on a higher or lower plane than at the start. The closest physical reality to this geometry is found in an automobile parking structure consisting of a continuous, flat spiral ramp of the general shape shown in Fig. 5-36(a). Just as there can be no atoms at the exact center of a screw dislocation, a spiral ramp must have a hole or step along its axis.

The manner in which the screw dislocation permits slip at low average stress is shown in Fig. 5-37. The presence of the dislocation permits the bond between atoms A and B to be broken; when this happens all along the dislocation line, the dislocation moves into

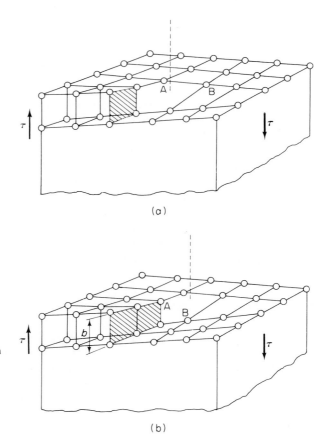

**Figure 5-37** Motion of a screw dislocation under shear stress. (a) Initial position, with dislocation indicated by dashed line. (b) When right side moves down and left side moves up under stress, dislocation has moved one interatomic spacing. Note the new positions of atoms A and B.

the position shown in Fig. 5-37(b). The Burgers vector *b* for this screw dislocation is thus the slip that results from the motion of the dislocation, which can be seen to be the height of the shaded step in Fig. 5-37. The Burgers vector *b* is shown here with arrowheads on both ends, as we have not assigned any sign convention to slip. For more complicated situations it is necessary to have a system of sign conventions, including *b*.

The progress of a screw dislocation across a crystal is shown in Fig. 5-38. Like the edge dislocation, slip is always in the direction of the shear stress, but the motion of the screw dislocation is *perpendicular* to the shear stress and slip. This characteristic makes the geometry of the screw dislocation more difficult to follow.

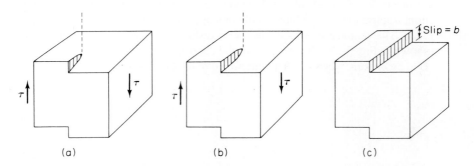

**Figure 5-38**    Slip caused by motion of a screw dislocation. (a) Dislocation line shown dashed, near front edge. (b) Dislocation has moved to center of crystal. (c) Dislocation has emerged from the far side of the crystal, and the entire crystal has undergone slip.

Screw dislocations have some properties that differ from those of edge dislocations. The most obvious is that the slip direction *b* is *parallel* to the screw dislocation line but *perpendicular* to the edge dislocation line [see Fig. 5-30(d)]. The slip plane must contain both the dislocation line and the Burgers vector, so an edge dislocation is always restricted to a single plane defined by the dislocation line and a perpendicular line containing the Burgers vector. But any plane that cuts a screw dislocation also contains the Burgers vector, so a semi-infinity of planes can be described that contain this single line. However, crystallographic considerations limit the number of possible slip planes for a given screw dislocation.

## MIXED DISLOCATIONS AND FORCES

Real dislocations are usually curved; thus, they are a mixture of edge and screw components and vary from point to point depending on the orientation of the dislocation line segment with respect to the Burgers vector. Figure 5-39(a) shows such a dislocation, a curved line from *S* to *E* that lies in a vertical slip plane. Point *S* shows the location of the screw portion of the dislocation, where *b* is parallel to the dislocation line. The vertically shaded step on top of the crystal represents an external slip step that has already occurred. The darker shaded region represents the internal area over which the dislocation *SME* has already swept. The

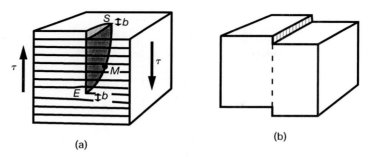

(a)                                                    (b)

**Figure 5-39**   (a) Dislocation line *SME* varies from screw (*S*), where it is parallel to *b*, to mixed (*M*), where it is neither parallel nor perpendicular to *b*, to edge (*E*), where the dislocation line is perpendicular to *b*. (b) A smaller view of this crystal after the dislocation has run fully through it.

horizontal lines along the front face indicate atomic planes and illustrate how the dislocation emerges as an edge dislocation at point *E*, where *b* is perpendicular to the dislocation line. As the stress $\tau$ is applied, the screw dislocation portion *S* moves toward the back of the crystal, the edge dislocation portion *E* moves downward parallel to $\tau$, and point *M*, which is a portion of the dislocation that is part edge and part screw, exhibits a combination of the two motions. Thus, the dislocation arc expands and eventually causes slip in the entire crystal [Fig. 5-39(b)]. At this final stage it is impossible to tell what kind of dislocation caused the slip; in fact, if the steps on the top and bottom of the crystal are cut off, it is impossible to even know that slip has occurred, because the crystal is perfect in Fig. 5-39(b).

Screw dislocations of the same sign (e.g., if both are right-handed, as in Fig. 5-36) repel each other when they are parallel and on the same slip plane, with the repulsive force varying with $1/x$, where *x* is the distance between them. Screw dislocations of opposite sign attract each other, and if allowed to come fully together, will annihilate each other. Both statements also hold for interactions between edge dislocations. But a screw dislocation and an edge dislocation parallel to it have no interaction, as their Burgers vectors are mutually perpendicular. Two mixed dislocations interact to the extent that their screw components interact and their edge components interact.

In general, dislocations tend to repel each other, and so in addition to permitting slip at low stresses, they are self-inhibiting. We will see in Chap. 6 how this behavior is in itself an important strengthening mechanism.

## CRYSTALLOGRAPHIC CONSTRAINTS
## TO DISLOCATION MOTION IN METALS

It is observed that dislocations usually move on those planes that are the most widely spaced apart: the {111} in fcc, the {110} in bcc, and the {0001} in hcp. The most widely spaced planes must thus also be the most densely packed planes, because the average number of atoms in any volume must be constant. The {111} in fcc and the {0001} in hcp are of maximum packing density, called *close-packed,* and both have an identical hexagonal

**Figure 5-40**   Atoms on the (111) in fcc. If the atoms were drawn full sized, they would be touching along <110>.

matrix. (The reason that fcc and hcp are so very different lies in the stacking sequence perpendicular to the hexagonal planes, in that the third plane in fcc does not lie above either the first or second planes, while the third plane in hcp lies above the first plane and thus begins the stacking sequence again.) Figure 5-40 shows the (111) in fcc, Fig. 5-41(a) shows how the atoms in the (111) slip plane are ordered in fcc, and Fig. 5-41(b) shows the hexagonal matrix of the basal planes of hcp.

The planes in bcc of maximum packing density are the {110}, but because the atoms in these planes are not in the close-packed hexagonal configuration of the {111} in fcc and {0001} in hcp, the {110} planes in bcc do not achieve the maximum theoretical packing density. Figure 5-41(c)–(e) shows these differences. No other planes in bcc are even as densely packed at the {110}. Thus, at least in part because of the absence of a plane of maximum theoretical density, slip in bcc iron, for example, occurs on {110}, {112}, and {123}.

**Example 5-12**

Calculate the ratio of spacing $d$ of planes to lattice constant $a_0$ for each of the planes in Fig. 5-41.

**Solution**

(a) For (111) in fcc: Use $d_{(hkl)} = a_0 / (h^2 + k^2 + l^2)^{1/2}$ or

$$\frac{d}{a_0} = \frac{1}{(h^2 + k^2 + l^2)^{1/2}} = \frac{1}{(1+1+1)^{1/2}}$$

$$= \frac{1}{\sqrt{3}} = 0.577$$

(b) For (0001) in hcp: Obviously, the spacing of planes is the same (theoretically) as that of (111) in fcc, except that for hcp $a$ is twice the atomic radius $r$. Thus, for hcp $d/a$ is equal to $d/a_0$ for fcc, corrected for the ratio $a_0/a$. In fcc, $a_0 \sqrt{2} = 4r$, so $a_0 = 4r/\sqrt{2}$. Thus,

$$\frac{a_0}{a} = \frac{4r/\sqrt{2}}{2r} = \sqrt{2}$$

and so

$$\frac{d}{a} = \frac{d\sqrt{2}}{a_0} = \sqrt{\frac{2}{3}} = 0.8165$$

(Note that this spacing is exactly half the theoretical $c/a$ spacing of 1.633 given earlier, where $c$ is the spacing between alternate planes.)

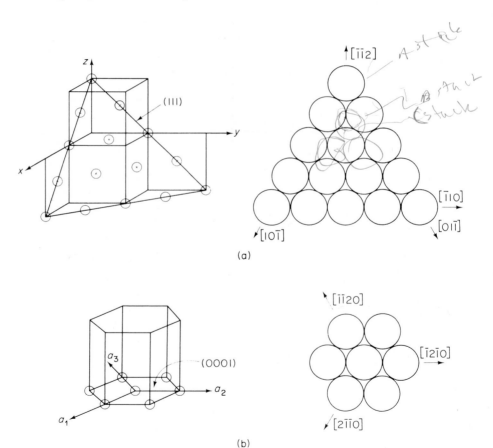

**Figure 5-41** The common slip planes in metals: (a) (111) in fcc; (b) (0001) in hcp, (c) (101) in bcc; (d) (112) in bcc; (e) (213) in bcc.

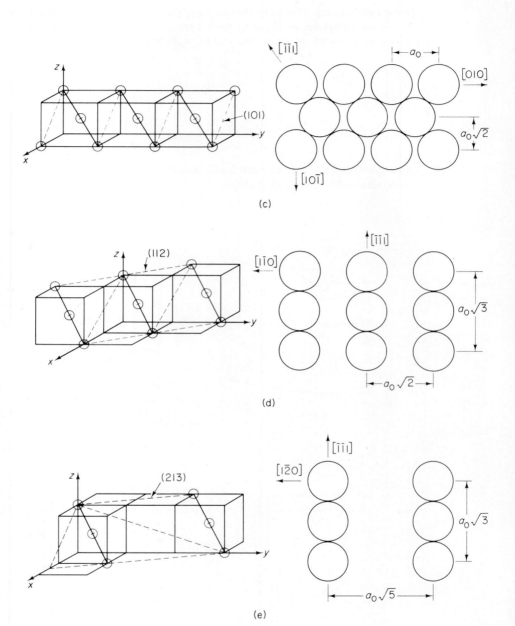

**Figure 5-41**   (cont.)

(c) For (101) in bcc:

$$\frac{d}{a_0} = \frac{1}{(1+0+1)^{1/2}} = \frac{1}{\sqrt{2}} = 0.707$$

(d) For (112) in bcc:

$$\frac{d}{a_0} = \frac{1}{(1+1+4)^{1/2}} = \frac{1}{\sqrt{6}} = 0.408$$

(e) For (213) in bcc:

$$\frac{d}{a_0} = \frac{1}{(4+1+9)^{1/2}} = \frac{1}{\sqrt{14}} = 0.267$$

It is surprising that slip occurs between such closely spaced planes as the (213) compared with the (101), which are 2.65 times as widely spaced. The reader may correctly conclude that our hard-sphere model of atoms is too simple to explain this observation.

Metals slip in the direction of closest atom-to-atom distance, which is the direction along which they touch in Fig. 5-41. Since this slip occurs by dislocation motion, it has the magnitude of the Burgers vector $b$, which is thus the distance from the center of one atom to the center of an adjacent atom that it touches. Thus, $b = 2r$, as has already been noted. The detailed mechanism by which the dislocation moves is more complex than we have described here because the dislocation often breaks up into *partial* or incomplete components. But since the net effect on slip is the same, we will consider throughout this text that $b = 2r$, without regard to the exact mechanism.

The slip directions are thus <110> for fcc, <11$\bar{2}$0> for hcp, and <111> for bcc. A *slip system* consists of a slip plane and direction. Some of the possible slip directions are shown in Fig. 5-41; thus, (111) [$\bar{1}$10] in Fig. 5-41(a) for fcc describes a slip system. Since <111> is the only possible slip direction in bcc, the three different slip planes for bcc must contain a <111>. In iron (bcc), slip often leaves wavy traces, resulting from the shifting of screw dislocations from one of the three classes of slip planes to another and back while maintaining the same Burgers vector. For example, the cases shown in Fig. 5-41(c)–(e) all have $b = (a_0/2) [\bar{1}\,\bar{1}\,1]$, which can include any of the three possible classes of slip systems having the same Burgers vector.

Slip in fcc metals usually occurs much more easily than in hcp metals. Examples of fcc metals are Al, Ni, Cu, Au, Ag, and Pb; hcp metals are Mg, Ti, and Zn. This behavior can be explained in part by the uniform distribution in space of possible slip systems in fcc, in contrast with the limitation of easy slip systems in hcp to the basal plane, (0001). In fcc each {111} has 3 possible <110> lying on it; there are 4 possible {111}, so fcc has 12 possible slip systems. (Note that directions and planes that are the full negatives of other directions and planes are not really different. The ($\bar{1}\,\bar{1}\,\bar{1}$) is parallel to the (111), and the [1$\bar{1}$0] is the same slip direction as the [$\bar{1}$10] but opposite in sign.) In contrast, hcp has only three common slip systems; however, when constrained by the applied shear stress or by the presence of alloying elements, hcp metals slip on planes other than the basal plane, but the Burgers vector does not change.

Slip direction, which is always parallel to the Burgers vector, is usually stated as, for example in Fig. 5-41(a), $[\bar{1}10]$. Strictly speaking, if $b = 2r$, this value is twice too large because $[\bar{1}10]$ will reach from one corner of the unit cell across the face to the other corner. Thus, we say that $b = (a_0/2)\,[\bar{1}10]$ to be precise; or often just $b = {}^1/_2\,[\bar{1}10]$ because it is clear that the basic unit-cell dimension is $a_0$; or, even more simply, $b = [\bar{1}10]$, which gives direction only, the $a_0/2$ being understood.

## RESOLVED SHEAR STRESS

A *polycrystalline* specimen of metal contains many crystals of more-or-less random orientation. When subjected to stress, the specimen first experiences slip in the crystal most favorably oriented to allow dislocation motion. Thus, the critical condition is the magnitude of the net shear stress acting on a slip plane and parallel to a slip direction.

Because tensile forces are much simpler to apply than shear forces, consider the geometry of a tensile test, shown in Fig. 5-42. The area of the slip plane is $A/\cos\phi$, where $A$ is the area perpendicular to the tensile force $P$ and $\phi$ is the angle between $P$ and the slip plane normal. Thus, the average applied axial stress is $P/(A/\cos\phi)$. The component of this stress parallel to the slip direction $b$ is the resolved shear stress $\tau$ and is the axial stress times $\cos\lambda$:

$$\tau = \frac{P\cos\phi\cos\lambda}{A}$$

Because $P/A$ is the average applied tensile stress $\sigma$,

$$\tau = \sigma\cos\phi\cos\lambda$$

**Figure 5-42**   Calculation of resolved shear stress parallel to slip direction $b$, from tensile force $P$.

The maximum value of cos $\phi$ cos $\lambda$ is reached when $\phi$ and $\lambda$ are both 45°, and cos $\phi$ cos $\lambda$ = 0.5.

Schmid's law, the result of measurements on real crystals, states that slip occurs in a crystal when $\tau$ reaches the critical resolved shear stress $\tau_c$ for that material. E. Schmid measured $\tau_c$ for high-purity magnesium to be 181 kPa, independent of the crystal orientation with respect to the tensile stress $\sigma$. We would indeed expect that, for a given material, dislocations would begin to move at a fixed level of resolved shear stress.

## CROSS SLIP

It was noted earlier that a screw dislocation lies parallel to its Burgers vector and is thus not necessarily restricted to a single slip plane. We can now see that the possible slip systems in real crystals limit the number of planes on which the screw dislocation can move. Figure 5-43 shows two planes, (111) and (11$\bar{1}$), in fcc. Dislocation $AB$ lies parallel to its Burgers vector $b = [\bar{1}10]$ and is thus a screw dislocation. Assume that the resolved shear stress is sufficient to move this dislocation on the (111) to position $CD$, at which time two possible events can cause cross slip: (1) The resolved shear stress changes so that it is greater for this dislocation on the (11$\bar{1}$) than on the (111); or more commonly, (2) the dislocation is blocked from further motion on the (111), such as by fine particles (Chap. 9) or by other dislocations, and the resolved shear stress on the (11$\bar{1}$) is sufficient for motion on the (11$\bar{1}$). In either case, the screw dislocation then leaves the (111) and move onto the (11$\bar{1}$), a process called cross slip.

Cross slip is very important in ductile metals as it permits more dislocation motion, and hence more plastic deformation, than would otherwise be possible. Cross slip is a

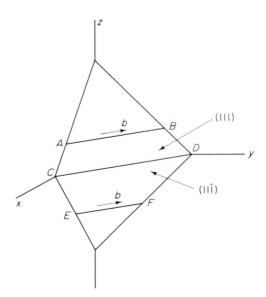

**Figure 5-43** Cross slip of a screw dislocation segment in an fcc crystal. Dislocation $AB$ on (111) moves to $CD$ and then moves on (11$\bar{1}$) to $EF$.

mechanism for bypassing dislocation barriers. *Double cross slip* often occurs when a blocked screw dislocation cross slips to position such as *EF* in Fig. 5-43 and then returns to a new (111) that is not blocked. The net result is to avoid the barriers on the original (111) that first stopped the dislocation at *CD*.

The wavy slip associated with bcc metals is a form of cross slip, except that it appears to be more the result of the absence of a close-packed slip plane (such as {111} in fcc) rather than the result of a blocked slip plane. However, the result is the same as in fcc in that a blocked screw dislocation can easily move to another plane. In Fig. 5-41(c)–(e), note how easily a screw dislocation with $b = [\bar{1}\,\bar{1}\,1]$ can shift from (101) to (213) to (112), in contrast to the relatively large direction change required in Fig. 5-43 for the $[\bar{1}10]$ screw dislocation to shift from (111) to (11$\bar{1}$).

# REFERENCES

1. Birchenall, C. Ernest, *Physical Metallurgy.* New York: McGraw-Hill, 1959. Chapter 2 describes crystal systems, X-ray techniques, and the categories of possible dislocations.

2. Cottrell, A. H., *Dislocations and Plastic Flow in Crystals.* Oxford: Clarendon Press, 1953. One of the two classic early texts on dislocation theory. For advanced-level reading.

3. Friedel, Jacques, *Dislocations.* Reading, Mass.: Addison-Wesley, 1964. A comprehensive dislocation text on an advanced level.

4. Hull, Derek, *Introduction to Dislocations.* Oxford: Pergamon Press, 1965. Elementary theory, complete enough for most needs, with numerous illustrations of real materials.

5. LeMay, Iain, *Principles of Mechanical Metallurgy.* New York: Elsevier North-Holland, 1981. Chapter 4 discusses in detail the properties of dislocations and their interactions.

6. McClintock, Frank A., and Ali S. Argon, eds., *Mechanical Behavior of Materials.* Reading, Mass.: Addison-Wesley, 1966. Chapter 4, "Dislocation Mechanics," by Ali S. Argon, gives a thorough and well-illustrated introduction to the important properties of dislocations.

7. McLean, D., *Mechanical Properties of Metals.* New York: John Wiley, 1962. Chapter 2 provides a rapid summary of dislocation theory.

8. Orowan, E., "Fracture and Strength of Solids," *Reports on Progress in Physics*, 12 (1949), pp. 185–232. A classic early review of the origins of strength.

9. Polakowski, N. H., and E. J. Ripling, *Strength and Structure of Engineering Materials.* Englewood Cliffs, N.J.: Prentice Hall, 1966. Chapter 4 introduces the types of forces between atoms and describes the crystallography of metals and some nonmetals. Chapter 7 describes the role of dislocations in plastic deformation.

10. Read, W. T., Jr., *Dislocations in Crystals.* New York: McGraw-Hill, 1953. The other classic early text (with Cottrell's) on dislocation theory; also on an advanced level.

11. Reed-Hill, Robert E., *Physical Metallurgy Principles.* Princeton, N.J.: Van Nostrand, 1964. Chapter I provides a review of crystal structures and Miller indices.

12. Weertman, Johannes, and Julia R. Weertman, *Elementary Dislocation Theory.* New York: Macmillan, 1964. A brief, complete, beginning-level text on the theory.

## PROBLEMS

**5-1.** Suppose a single crystal of iron (body-centered cubic, atomic radius = 0.124 nm) of volume $10^{-6}$ m$^3$ contains $10^9$ atoms imperfectly situated in the crystal. Calculate the percentage of atoms that are still perfectly situated.

**5-2.** Derive the equation for the maximum theoretical shear stress for slip in a perfect crystal and write out the assumptions made.

**5-3.** What other mode of elastic deformation, in addition to the shear strain $\gamma$ in Fig. 5-22, can occur in a real crystal subjected to a tensile stress?

**5-4.** One of the most uncertain assumptions in the derivation of Eq. (5-4) is the sinusoidal variation of $\tau$ in Eq. (5-2). What evidence suggests that this assumption is reasonable?

**5-5.** Develop the geometry for the Bragg equation [Eq. (5-1)], relating wavelength $\lambda$ of X-rays, the angle $\theta$ made with atomic planes, and the spacing $d$ of the atomic planes.

**5-6.** Show how an edge dislocation can produce slip as it moves across an otherwise perfect crystal; use several sketches and indicate the dislocation by the symbol ⊥ but do not show individual atoms.

**5-7.** Draw a two-dimensional sketch of a dislocation in a 60° lattice (such as bubbles form).

**5-8.** Why cannot an edge dislocation end *within* a crystal? Draw such a situation for a simple three-dimensional cubic lattice and show what happens.

**5-9.** The strength of a specimen of iron that is 0.5 $\mu$m in diameter may be on the order of 10 GPa, yet the strength of a bar of the same metal that is 10 mm in diameter is on the order of 10 MPa. Briefly explain why this difference in strength occurs, in terms of the atomic mechanisms that most influence strength.

**5-10.** A single crystal of height $y$ and width $x$ contains a total of $n$ edge dislocations of like and opposite signs lying in parallel planes, as shown in Fig. 5-44. Find an expression for the shear strain $\gamma$ associated with the motion of the dislocations over an average glide distance $L$. State any additional assumptions that you make.

**Figure 5-44**

**5-11.** Calculate the ratio of the Burgers vector to the lattice constant for an edge dislocation in a
  **(a)** Simple cubic lattice.
  **(b)** Body-centered cubic lattice.
  **(c)** Face-centered cubic lattice.

**5-12.** What must the angle between the Burgers vector of an edge dislocation and the dislocation line be? Do both the dislocation line and the Burgers vector lie in the slip plane?

**5-13.** A single crystal of fcc metal of $a_0 = 0.361$ nm has the dimensions given in Fig. 5-45. All dislocations are edge (as shown), lie parallel to the $x$-direction, and move parallel to the $xy$-plane. The $y$-
direction of slip corresponds to [110], the usual slip direction in fcc. If the average dislocation
glide distance is 1 $\mu$m, how many active dislocations must there be to cause a strain $\gamma$ of 0.15?

**Figure 5-45**

**5-14.** Using a three-dimensional matrix of simple cubes to represent unit cells, for bcc iron,
   **(a)** Draw the (312), containing at least five atoms; show all atoms that are on this plane.
   **(b)** Draw a Burgers vector for slip that lies on the (312), being precise as to length.
   **(c)** Label the Burgers vector with precise Miller indices for magnitude and direction.

**5-15.** A single crystal of aluminum is shown in Fig. 5-46. The plane outlined with a dashed line is
the (111), and the slip direction [$\bar{1}$10] is shown. One of a series of parallel dislocations lying
on the (111) is shown; during slip it moves in the [$\bar{1}$10] for one-fourth of the total possible distance across the crystal.
   **(a)** Calculate the magnitude of the shear strain if $2.67 \times 10^5$ dislocations so move on other parallel (111) planes.
   **(b)** Show this shear strain by drawing a sketch of the crystal after the dislocations have moved.

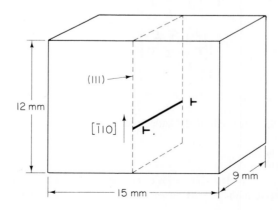

**Figure 5-46**

**5-16.** Using a simple three-dimensional cubic matrix, for an fcc crystal,
   **(a)** Draw a ($1\bar{1}1$) that intersects at least five atoms. Show all atoms on this ($1\bar{1}1$) with small circles.

**(b)** Draw a Burgers vector for one of the possible slip directions that lies on this ($1\bar{1}1$); be sure its length is correct.

**(c)** Label this Burgers vector with its Miller indices, being precise as to direction and magnitude.

**5-17.** A single crystal of hcp titanium of initial dimensions 14 mm × 14 mm × 14 mm is deformed by a shear stress so as to produce a shear strain $\gamma = 0.083$. If $1.74 \times 10^{10}$ dislocation loops of average diameter $D$ are counted on all the parallel slip planes (which are parallel to one of the faces of the 14-mm cube) and if the original dislocations were infinitesimal loops, calculate the value of $D$.

**5-18.** In a crystal of copper 17.3 mm on each side, calculate the number of dislocations required to cause a shear strain of 0.129 if the average glide distance is 16.4 $\mu$m. State any assumptions that you make.

**5-19.** The copper single crystal shown in Fig. 5-47 contains parallel dislocations on parallel slip planes, some of which are shown. If each dislocation glides an average of 2.69 $\mu$m, calculate the total dislocation line length if the shear strain is 0.127.

**Figure 5-47**

**5-20.** A single crystal of lead has a shear stress applied to its (111) in the [$\bar{1}10$] as shown in Fig. 5-48. If the average dislocation is a circular loop that is initially of diameter 24.7 $\mu$m and grows to diameter 67.1 $\mu$m, calculate the number of such dislocations that must be present to cause a plastic shear strain of 0.0736.

**Figure 5-48**

**5-21.** Calculate the maximum theoretical tensile strength of a perfect filament of pure magnesium.

**5-22.** A single crystal of iron slips on the {112}<111> system.

    **(a)** Draw any {112}, label it, and then draw a Burgers vector in this plane and label it with the correct Miller indices.

    **(b)** Calculate the scalar magnitude of the Burgers vector using the lattice constant $a_0$ from App. 6.

**5-23.** How large (in mm) will the slip step be in a single magnesium crystal after $3 \times 10^6$ dislocations move on the same slip plane and emerge at an edge of the crystal?

**5-24.** Calculate the maximum theoretical tensile load that a whisker of aluminum of 0.234 $\mu$m diameter can support. How many 1500-kg automobiles could be supported by a 0.1-$m^2$ cross section of these whiskers in tension? How many such automobiles could your pencil support in tension if it were made of aluminum whiskers?

**5-25.** Using a three-dimensional simple cube representing one unit cell, draw and clearly label the following: [210], (101), and (010).

**5-26.** Draw the (121) in bcc with the Burgers vector $(a_0/2)$ [$\bar{1}1\bar{1}$] lying in the (121).

**5-27.** Suppose that slip occurs in a nickel crystal as shown in Fig. 5-49. If $10^8$ parallel dislocation segments lie on parallel (111) slip planes, oriented in the same way as the dislocation of length $c$ in Fig. 5-49, and the applied shear stress $\tau$ is parallel to the Burgers vector of the dislocations, calculate the shear strain in the crystal if all the dislocation segments of length $c = 14$ $\mu$m bow out to semicircular shape (shown as a dashed line) and then move no more.

**Figure 5-49**

**5-28.** A single rectangular crystal of aluminum is oriented with its edges parallel to the coordinates $x$, $y$, $z$, with the $x$-edge 12.2 mm long, the $y$-edge 15.7 mm long, and the $z$-edge 8.5 mm long. The $z$-plane is (111), and the $+y$-direction is [$\bar{1}10$]. The crystal contains $10^8$ parallel edge dislocations that lie parallel to the $x$-axis on (111) planes. If a shear stress $-\tau_{zy}$ produces a shear strain $\gamma_{zy} = -0.0500$, calculate the average distance that each dislocation has moved.

**5-29.** On an $x$-, $y$-, $z$-coordinate system draw a cube that starts at the origin. Then draw three more identical cubes adjacent to it: one on the $x$-side of the first cube, one on the $y$-side, and one on the $z$-side. Assume that these cubes represent four unit cells of bcc iron.

    **(a)** Draw the [$1\bar{1}1$].

    **(b)** Draw a {110} that includes this [$1\bar{1}1$] and write its Miller indices.

    **(c)** Calculate the ratio of spacing of {110} to the spacing of {112} in bcc.

# 6

# *Pure Elements*

This chapter and the five that follow apply to the study of practical metals the relationship between strength and dislocation motion that was developed in Chap. 5. The wide range of mechanical properties of metals and their alloys can be attributed largely to the range of conditions required for slip. In many commercial metals, the strength achieved is the result of several different mechanisms acting simultaneously to limit dislocation motion. These mechanisms can best be developed and described by considering real metals, beginning with the simplest structures. This chapter, therefore, examines the kinds of strengthening mechanisms possible in solids containing only one kind of atom, that is, in pure elements.

## SINGLE CRYSTAL OF LOW DISLOCATION DENSITY

A pure element in the form of a single crystal containing a few dislocations experiences local slip of a few interatomic spacings at a relatively low stress. In a given crystal, a minimum stress apparently exists for dislocation motion, called the *friction stress* or *Peierls–Nabarro stress*. This is the minimum stress required to move an otherwise unimpeded dislocation, and its magnitude is ordinarily much smaller than that of the yield stress because yielding requires the movement of a large number of dislocations. A sufficient number of dislocations usually do not already exist before yielding, so they must be created and set in motion. As will be shown in the discussion that follows, this requires a substantial stress, and then the new dislocations interact and impede their own motion. Thus, because the strain from the unimpeded motion of one dislocation is infinitesimal and the strain for yielding is finite and measurable, the stress needed for yielding, sufficient to generate and move a large number of dislocations in a real crystal with many barriers to motion, is much greater than the Peierls–Nabarro stress.

It is possible to create new dislocations by several mechanisms. For example, they can be created by a force applied to a free surface, as was shown in Fig. 5-30. A similar form of this mechanism can exist at a grain boundary when a slip step from one grain acts against another grain to generate new dislocations. Cross slip within a crystal can lead to

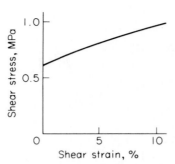

**Figure 6-1** Stress–strain curve for a 99.96% pure magnesium single crystal loaded in tension at 203 K. [From H. Conrad and W. D. Robertson, "Effect of Temperature on the Flow Stress and Strain-Hardening Coefficient of Magnesium Single Crystals," *Trans. AIME,* 209 (1957), p. 503. © by AIME and reprinted with its permission.]

the generation of a large number of dislocation loops through a mechanism called the *Frank–Read source.* The important point here is that large numbers of new dislocations can be created. If these dislocations do not interfere with each other, then slip can continue indefinitely at this low stress. Although such a set of conditions is seldom achieved for a real crystal, careful experiments on high-purity single crystals have approached these conditions and have resulted in the expected low stress for slip. For example, Fig. 6-1 shows the early part of a stress–strain curve for high-purity magnesium. Because slip in magnesium, which is hexagonal close-packed, tends to be limited to the parallel basal (0001) planes, dislocations experience less interaction with each other than in metals that have several intersecting slip systems. Thus, the stress required for increasing amounts of strain in a single crystal of magnesium does not increase very rapidly until the strain is large. From similar experiments with a wide range of materials, it can be concluded that very pure elements in the form of single crystals of low (but not zero) dislocation density and with little dislocation interaction have very low strength. In the previous chapter, strength was shown to be inversely related to the average dislocation glide distance. This agrees with experimental observations of pure single crystals in which the average dislocation glide distance is relatively large and the strength therefore low.

### Example 6-1

What would happen to the stress required for slip in a single crystal of magnesium if a slip system not parallel to the basal plane also came into action?

**Solution**   The stress required for slip should increase. Data on magnesium show that when slip occurs on another system, such as on $\{10\bar{1}0\}$ planes in $<1\bar{2}10>$ directions, the stress needed for slip at room temperature is more than 10 times that required for basal slip.

## INTERACTION OF DISLOCATIONS

In most crystals, only limited slip can occur before dislocations begin to interfere with each other. The influence of dislocation interaction on the stress needed to move dislocations can be understood from a simple energy analysis. In the region of a dislocation line, the atoms are severely displaced from their equilibrium positions (see, for example, Fig. 5-28). The energy of displacement of the atoms is recoverable elastic energy, which is the same kind of energy stored by the displacement of atoms in a stressed spring. When the atoms return to their equilibrium positions, they do work on the surrounding atoms; how-ever, the atoms in a dislocation cannot return to their equilibrium positions except by mov-ing the dislocation away, and this simply shifts the site of atom displacement to a new group of atoms, resulting in almost no energy change. When a stress is applied, this shift of dislo-cation position results in slip. Because the stored energy in the dislocation does not depend on its general location within the crystal, there has been no change in energy of the disloca-tion (if its shape and length remain unchanged during slip), so the stress required is low.

To show what can happen when one dislocation approaches another, the spring anal-ogy is particularly useful. Figure 6-2 shows a straight-line, load-elongation relationship for a simple elastic spring in tension. For a given displacement of magnitude $a$, the energy required is represented by the area under the curve, triangle $OAa$. Suppose, however, that the spring is already stressed so that its initial elongation is $a$. What is the energy required for an additional elongation of $a$? Figure 6-2 shows that the additional energy required to deform the spring from elongation $a$ to $b$ (where $b = 2a$) must be the area under the curve from A to B, or $ABba$. This area is substantially larger (in this case 3 times larger) than the area under the curve from $O$ to $A$. Thus, deformation of an already stressed elastic system requires more energy than the same deformation in an unstressed system.

Consider what happens in the simple case of an edge dislocation approaching another edge dislocation on the same slip plane and with the same sign (i.e., with the extra planes of atoms of both dislocations on the same side of the slip plane). This situation is shown in Fig. 6-3, where $A$ is a stationary dislocation that dislocation $B$ is approaching. Four atoms adjacent to the slip plane are shown, already displaced from their equilibrium positions by dislocation A before dislocation B approached. Note that even at some dis-tance from A, the atoms above the slip plane (which is indicated by a horizontal dashed line) are compressed from their equilibrium positions to accommodate the extra plane of atoms above the slip plane and that these atoms have a spacing $x$. The atoms below the slip plane are correspondingly stretched apart, to a spacing $y$. Thus, if $b$ is the Burgers vector of the dislocation, $x < b < y$. As dislocation B approaches, its effect is to further compress the

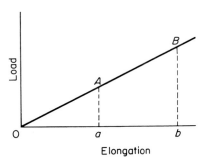

**Figure 6-2** Load-elongation of an elastic spring.

**Figure 6-3** Interaction between edge dislo-cations of like sign.

atoms above the slip plane and to stretch the atoms below the slip plane further apart. Then the spring analogy discussed previously applies and the additional energy required for the displacement of these four atoms by dislocation $B$ must be greater than the energy required for the initial displacement of these same atoms by dislocation $A$.

The source for this required extra energy must be the applied shear force moving through some distance. If the atoms shown in Fig. 6-3 are halfway between $A$ and $B$, it can be seen by symmetry that if $A$ approached first from the right and then stopped, and then $B$ approached from the left, the energy required to move $B$ into its present position must be greater than the energy that was required to move $A$ into position. But the shear strain associated with the motion of $A$ and $B$ is related only to the distance each has moved [see Eq. (5-5)], so the distance that the applied shear force moves is the same for both cases. Therefore, if the energy (work done) is to be greater for case $B$, the shear force needed to move $B$ into position after $A$ is in position must be greater than the shear force required initially to move $A$ into position. In other words, as dislocation $B$ approaches dislocation $A$, the shear stress must increase.

The shear stress required to move a dislocation does not always increase as it approaches other dislocations. In a few special cases, the shear stress can decrease. An obvious example is to change the sign of dislocation $B$ in Fig. 6-3 so that the extra plane of atoms is below the slip plane. Then interatomic spacing $x$ would be increased and $y$ decreased by dislocation $B$, resulting in a less severely strained lattice and less stored elastic energy.

The stress field surrounding an edge dislocation can be approximated from elasticity theory. For an edge dislocation lying along the $z$-axis, perpendicular to the $xy$-plane, with the extra plane of atoms in the $-y$-direction (see Fig. 6-4), the shear stress $\tau_{xy}$ at any point is

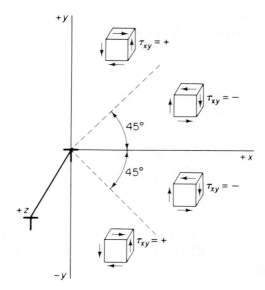

Figure 6-4   Shear stresses resulting from an edge dislocation parallel to the $z$-axis.

$$\tau_{xy} = \frac{-Gb}{2\pi(1-v)}\left[\frac{x(x^2-y^2)}{(x^2+y^2)^2}\right] \tag{6-1}$$

where $G$ is the shear modulus of elasticity (see App. 6 for values for metals), $b$ is the Burgers vector, $v$ is Poisson's ratio (usually about 0.3), and $x$ and $y$ are the coordinates of the point in question. Note that $\tau_{xy}$ does not vary with $z$. The directions of the shear stresses in four regions are shown in Fig. 6-4; the shear stresses shown acting on the cubes also serve to define the directions of positive and negative shear stresses.

If the extra plane of the edge dislocation in Fig. 6-4 is in the $+y$-direction, then the sign of Eq. (6-1) will be reversed. Figure 6-5(a) shows such a dislocation $A$ located at the origin. To determine the direction of stress, and hence the force, on any dislocation $B$ located at $x$, $y$, Eq. (6-1) is used with a sign change. If $B$ is located as in Fig. 6-5(a), where $x > y$, then the sign of $\tau_{xy}$ from Eq. (6-1) is positive. A positive $\tau_{xy}$ will cause a force to the right on dislocation $B$; this will cause $B$ to separate from $A$ and is thus repulsive. Figure 6-5(b) shows the relative magnitude and direction of the force between $A$ and $B$ for any position $x$, $y$ of $B$. For every case, the force on $B$ is always in the $\pm x$-direction.

The most usual interaction between two parallel edge dislocations of like sign occurs when they lie on the same plane, $y = 0$. This situation cannot be shown in Fig. 6-5(b) because $x$ is plotted in terms of $y$, but from Eq. (6-1), for the dislocation in Fig. 6-4, with $y = 0$, we have

$$\tau_{xy} = \frac{-Gb}{2\pi(1-v)}\frac{1}{x} \tag{6-2}$$

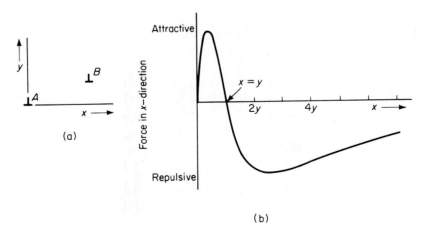

(a)

(b)

**Figure 6-5**   Interaction forces between parallel edge dislocations of like sign. (a) Relative positions of the dislocations, where $A$ is at the origin and the position of $B$ is $x$, $y$. (b) Variation in force between $A$ and $B$. [From A. H. Cottrell, *Dislocations and Plastic Flow in Crystals.* (Oxford: Clarendon Press, 1953), p. 48. Reprinted with permission of the publisher.]

Thus, edge dislocations of like sign on the same plane repel each other; when the signs are opposite, they attract.

The stresses surrounding a screw dislocation are radially symmetrical and are thus most simply described in cylindrical coordinates. Figure 6-6 shows a screw dislocation having the shape of a left-handed helix lying along the $z$-axis. The shear stress $\tau_{\theta z}$ shown varies only with the radius $r$:

$$\tau_{\theta z} = \frac{-Gb}{2\pi}\, \frac{1}{r} \qquad (6\text{-}3)$$

where $G$ is the shear modulus of elasticity and $b$ is the Burgers vector. Note that this equation differs from Eq. (6-2) only by omission of the factor $(1-v)$.

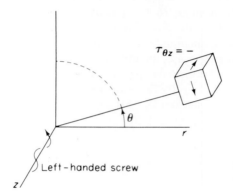

**Figure 6-6**  Shear stresses resulting from a left-handed screw dislocation parallel to the $z$-axis.

Any two parallel screw dislocations of the same sign (both right-handed or both left-handed) repel each other. If one is a right-handed screw and one left-handed, they will attract.

Edge dislocations do not cause a $\tau_{\theta z}$ shear stress, and screw dislocations do not cause a $\tau_{xy}$ shear stress. Thus, an edge dislocation and a parallel screw dislocation do not interact. The forces for the general case between nonparallel dislocations are very complex and vary from point to point along each dislocation.

**Example 6-2**

If dislocations exert forces on each other, why do they not move in the absence of external stresses into positions where the forces are minimized?

**Solution**   Although a complete answer to this question is not known, loss of energy (associated with the Peierls–Nabarro force already discussed) during dislocation motion apparently prevents free motion of dislocations. The nature of this lost energy may result in part from a "twanging" of atoms as the dislocation passes by; such vibration of atoms would be dissipated as heat, a common observation during plastic deformation. When the interaction forces between dislocations are great enough, the dislocations move by themselves to lower their interaction forces, but dislocation motion stops before the forces vanish.

## STRAIN STRENGTHENING (WORK HARDENING)

From interaction forces between dislocations it is an easy step to strain strengthening. The preceding section showed how the stress required for dislocation motion can be increased in the presence of another dislocation. A simple instance of strain strengthening is shown in Fig. 6-7, where dislocation $A$ originating from a dislocation source at the left has been stopped by a barrier of some kind, such as a grain boundary (see the later discussion on this point), and dislocations $B$ and $C$ originating from the same source and moving under applied shear stress $\tau$ are prevented from moving as far as $A$ by the repulsive force of $A$. (Dislocation $C$ is repelled by both $A$ and $B$, so the separation of $B$ and $C$ is greater than the separation of $A$ and $B$.) Therefore, the stress required to move dislocation $C$ a given distance from the source is greater than the stress that was required to move $B$ the same distance from the source. Thus, $C$ moves a shorter distance under the same stress. Under a given stress the dislocations find equilibrium positions and move no further.

**Figure 6-7**   Pile-up of edge dislocations.

If additional strain is required, the stress must be raised. Then one or more of at least four possible events can occur: (1) the spacing between $A$, $B$, and $C$ decreases, (2) the source generates new dislocations; (3) another set of dislocations elsewhere begins to move; and (4) dislocation $A$ breaks through or moves around its barrier and allows further strain. The first three of these events promote further strain strengthening; the fourth tends to diminish the rate of strain strengthening. Because billions of dislocations are ordinarily present in a standard test specimen, it is unrealistic to speak of just one dislocation system acting at any instant. Many thousands of such systems must be acting, but as their dislocations either pile up or are hindered in some other way (other mechanisms have been hypothesized), the stress required for additional motion increases in the same way as described before. The simple argument for a single dislocation system should, therefore, remain valid for the real case. In this way it is possible to understand the dislocation behavior that causes strain strengthening or, as it is often called, *strain hardening* or *work hardening*. A thin-foil electron micrograph of piled-up dislocations in Type 304 stainless steel is shown in Fig. 6-8. Note the tendency toward increased spacing of dislocations with distance to the left of the barrier.

In contrast to the easy slip of a magnesium single crystal (Fig. 6-1), most metals experience a marked increase in stress needed for slip as strain increases. The lower curve in Fig. 6-9 is the stress–strain curve for a copper single crystal, which is face-centered cubic. This curve is much steeper than the one for magnesium because many equivalent slip systems in the copper crystal begin to act, intersect each other, and provide many barriers to dislocation motion not initially present in the copper specimen.

2 μm

**Figure 6-8**  Thin-foil transmission electron micrograph of dislocations piled up against a grain boundary (visible at upper right), in moderately cold-worked Type 304 stainless steel. The short segments of dislocations run through the thickness from top to bottom of this very thin specimen. (Micrograph courtesy of Stephen J. Krause.)

### Example 6-3

(a) Draw two intersecting {111} slip planes and show their <110> slip directions for an fcc structure. Then sketch a third {111} <110> system that intersects both of the others.

(b) Is it possible to have a single fcc crystal so oriented that two or more intersecting slip systems can act at the same applied stress?

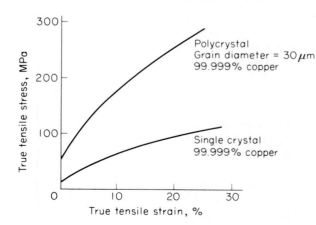

**Figure 6-9**  Representative stress–strain curves for single crystal and polycrystalline copper. (Reprinted with permission from *Acta Metallurgica,* Vol. 1, R. P. Carreker, Jr., and W. R. Hibbard, Jr., "Tensile Deformation of High-Purity Copper as a Function of Temperature, Strain Rate, and Grain Size." Copyright 1953, Pergamon Press, Ltd.)

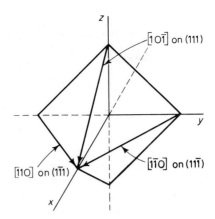

**Figure 6-10**   Three intersecting slip systems in fcc.

**Solution**   (a) Figure 6-10 shows three different {111} planes and three intersecting <110> directions lying in these planes. The edges of the triangular portions of the planes are shown as the <110> directions for clarity. Note that each direction lies in two {111} planes; for example, the [10$\bar{1}$] lies in both (111) and (1$\bar{1}$1). This even further complicates the possible slip that can occur. There are 12 equivalent slip systems in fcc: three <110> directions on four different {111} planes.

(b) By symmetry, a tensile stress applied in the x-direction would exert exactly the same shear stress on all three slip systems shown in Fig. 6-10. Thus slip on intersecting systems, called *multiple glide,* can occur at a very early stage.

The upper curve in Fig. 6-9 shows the behavior of polycrystalline copper. The number of crystals capable of easy glide (i.e., slip at the low stress levels of single-crystal copper) is small and represents a very small fraction of the total number of crystals (grains) present in the specimen. Thus, an easy glide portion of the curve is not seen because easy glide of the most favorably oriented individual crystals ends in this case after an extremely small cumulative strain. The diminishing strain strengthening in Fig. 6-9 (i.e., decreasing $d\sigma/d\varepsilon$) apparently results when dislocations are able to break through or go around barriers, as has already been discussed. The higher the stress, the more likely dislocations are to move through or around barriers, so it might be expected that the rate of strain strengthening diminishes with increasing strain.

## VACANCIES

The simplest possible kind of defect is a vacancy, where a single atom is missing from an otherwise perfect crystal structure. A simple model of a vacancy in a simple cubic lattice is shown in Fig. 6-11, where the center atom is missing. The region of displacement of the surrounding atoms from their equilibrium positions extends much farther in a real case than is shown here.

The smooth shifting of atomic bonds that occurs when a dislocation moves in a perfect lattice is disrupted in the region at the center of the vacancy and for many at inter-

**Figure 6-11**  Vacancy in simple cubic lattice.

atomic distances away from the center. The argument given earlier for additional energy to move a dislocation into the strain field of another dislocation also holds in the region of a vacancy, as many of the atoms adjacent to the vacancy have already been displaced from their equilibrium positions, a condition that sets up a strain field. However, in contrast to interaction between parallel dislocations, only a short length of the dislocation is directly affected by a single vacancy. But because a dislocation cannot be broken and thus terminated within a crystal, holding up one segment of a dislocation tends to hold up the whole dislocation. In practice, many vacancies are probably present to influence each dislocation at a number of points.

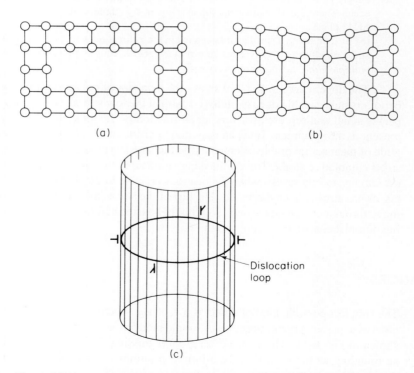

**Figure 6-12**    (a) Circular disk of vacancies prior to collapse. (b) Formation of circular edge–dislocation loop through collapse of vacancy disk. (c) The dislocation loop is constrained to move only on the shaded slip "surface," which is unlikely.

Observation of thin metal foils containing vacancies reveals that in some cases the vacancies come together to form flat disks of vacancies, one of which is shown in Fig. 6-12(a). Figure 6-12(b) shows how these disks can collapse to form a very special kind of edge-dislocation loop. This dislocation is special because the loop does not lie in the slip plane but is perpendicular to it. For this loop to move, slip has to occur on a cylindrical or prismatic surface, so it is called a *prismatic dislocation loop.* Figure 6-12(c) shows the loop in Fig. 6-12(b) in three dimensions, with the slip "surface" shaded. Such motion is unlikely to occur, so these loops remain fixed in position and act as barriers to ordinary dislocation motion in the same way as for other very small point defects such as individual vacancies.

Although vacancy strengthening is very effective when 0.1–1% of the atom sites are empty, the normal concentration of vacancies at room temperature is several orders of magnitude smaller. Thus, vacancy strengthening is not a practical means for strengthening metals, except for those of very high purity, and is not considered further here.

## GRAIN BOUNDARIES

The boundary between two crystals is a very thin region of imperfection where the crystal orientation must shift from one crystal lattice to the other. In the region where this change in crystal orientation occurs, some atoms are out of equilibrium position because they are attracted by atoms from both adjacent crystals. This region is the *grain boundary* (a grain is one crystal of a polycrystalline specimen), and it can be seen on a polished and etched surface because the higher energy of the grain-boundary atoms leads to etching at a higher rate at the boundary than elsewhere in the grains. The resulting etched grooves in the polished surface can be seen in an optical microscope. Figure 6-13 shows these grooves as lines on a polished and etched specimen of iron containing very little carbon.

When some part of a dislocation approaches a grain boundary, more energy, or higher stress, is required to move the dislocation through the boundary. Even if the dislo-

**Figure 6-13**   Optical micrograph of grain boundaries in Armco iron of 0.02% C.

cation moves through the grain boundary, the slip plane in the next crystal will be at some angle to the original slip plane. The dislocation must then become longer in order for part of the dislocation to follow the new slip plane, and part the old. Figure 6-14 illustrates schematically a very simple case of a dislocation crossing grain boundaries. A single crystal, as in Fig. 6-14(a), contains a dislocation source S which generates one dislocation loop that moves on the slip plane, shown shaded, until it is close to the outer edges of the crystal. Figure 6-14(b) shows three grains, A, B, and C, of the same total size as the single crystal in Fig. 6-14(a) and having a source S in grain A. When the dislocation loop reaches the boundaries between A and B and between A and C, it is forced to follow the new slip planes, shown shaded. In general, the boundary between B and C is not a common slip plane, so the loop is pinned at its intersection with the three grains, point a. The final dislocation length is then greater than for the single-crystal case, and both cases have roughly the same area swept by the dislocation. An increase of the dislocation length increases stress, by the following argument: It has already been shown that the atoms in a dislocation have greater elastic energy than if they were located in a perfect crystal lattice. Because the total energy is proportional to the number of atoms in the region of the dislocation, a longer dislocation has greater total elastic energy because it involves more atoms. For this reason, lengthening a dislocation to allow it to lie on two intersecting slip planes requires greater total energy. This extra energy can come only from an increase in stress during deformation, as was concluded earlier in this chapter. Thus, the polycrystalline specimen in general has higher strength than the single-crystal specimen.

Grain boundaries therefore hinder dislocation motion. Because the boundary of a given grain fully envelops dislocations in that grain, the dislocations cannot move around the grain boundary in the same way that they can sometimes bypass point defects and other

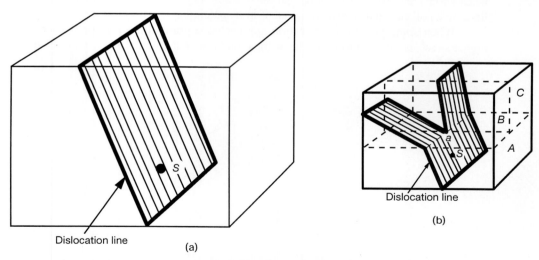

Dislocation line

(a)

Dislocation line

(b)

**Figure 6-14**    (a) Single crystal with a dislocation source at S. (b) Three crystals, A, B, and C, with a source S in grain A.

**Figure 6-15** Variation of yield stress with grain diameter for copper [From D. McLean, *Mechanical Properties of Metals* (New York: John Wiley & Sons, 1962), p. 76. Reprinted with permission of the publisher.]

dislocations. If grain boundaries are spaced closer together (i.e., for smaller grain diameters), the average dislocation glide distance is reduced and the stress must be increased for further dislocation motion. [Recall Eq. (5-5): $\gamma = nbg/hw$.] Figure 6-15 shows the marked increase of stress for initial slip (yield stress) for copper as the grain size (diameter) decreases.

**Example 6-4**

Develop the equation for the grain size relationship in Fig. 6-15.

**Solution** The constants in an equation of the form $y = mx + b$ are $b = y$ at $x = 0$ and $m =$ slope. From Fig. 6-15, $b = 20$ MPa and roughly $m = (80$ MPa $- 20$ MPa$) / (0.3 \ \mu m^{-1/2} - 0) = 200$ MPa-$\mu m^{1/2}$. Thus, $\sigma_y = 20$ MPa $+ (200$ MPa-$\mu m^{1/2})d^{-1/2}$, where $d$ is the grain diameter in micrometers. This is usually known as the *Petch equation* and written in the form

$$\sigma_y = \sigma_i + k_y d^{-1/2} \tag{6-4}$$

where $\sigma_y$ is the yield stress and $\sigma_i$ and $k_y$ are constants developed previously. One would thus expect a single crystal of this copper to have a yield strength of about $\sigma_i = 20$ MPa. Figure 6-9 shows that the stress in a single crystal of copper is close to 20 MPa at $\varepsilon = 0.5\%$.)

Strain strengthening (Fig. 6-9) and reduction of grain size (Fig. 6-15) are the two most powerful and practical means for increasing the strength of pure or nearly pure metals.

## SUBBOUNDARIES AND RECRYSTALLIZATION

A ductile metal that has been plastically strained has higher dislocation density and, hence, higher stored elastic energy than before straining. From thermodynamic reasoning and by observation, it is known that the specimen tends toward a minimum stored elastic energy under favorable conditions. One such condition is high temperature. As temperature is raised, the increased kinetic energy of individual atoms promotes dislocation motion at lower and lower stresses until a temperature range is reached where dislocations can move because of the stress fields of other dislocations. Thus, dislocations that attract move together, and dislocations that have repulsive forces move apart. A simple example is shown in Fig. 6-16, where Fig. 6-16(a) shows the strained structure of dislocations on three parallel glide planes and Fig. 6-16(b) shows a feasible lower energy alignment of these dislocations after some time at elevated temperature. Figure 6-5 has already shown that dislocations of like sign tend to attract and thus to stack above each other when $x < y$. Edge dislocations must also space themselves uniformly above and below each other in order to minimize the total elastic energy. Such motion perpendicular to the slip plane is called *dislocation climb*. For climb, atoms must be added to or removed from the edge of the extra plane, a process requiring substantial diffusion, which therefore can occur only at elevated temperatures. (See Chap. 12 for a discussion of high-temperature creep.) Vacancies created during the cold-working operation can contribute to the process of dislocation climb by migrating to the vicinity of edge dislocations. The presence of a vacancy allows individual atoms to move more freely and thus aids diffusion and climb. The configuration of dislocations shown in Fig. 6-16(b) has a lower elastic energy than a random distribution of dislocations. The vertical dotted lines represent two-dimensional surfaces called *low-angle grain boundaries* because the angle between adjacent subgrains is very small. Because of the shape of the subgrains that result, the change from Fig. 6-16(a) to Fig. 6-16(b) is called *polygonization*.

### Example 6-5

Calculate the angle between adjacent subgrains in Fig. 6-16(b) for a crystal of Burgers vector $b$ and vertical spacing $d$ as shown.

**Solution**    Figure 6-17 shows five dislocations in a subboundary. The left and right subgrains are at an angle $\theta$ because of the presence of the extra half-planes of atoms that end at the dislocations, acting as a wedge between the subgrains. If the distance $d$ separating the dislocations is large compared with the Burgers vector $b$ of each dislocation, then $\theta/2 = \tan^{-1}(b/2d)$; thus, $\theta = b/d$.

(a)                                                        (b)

**Figure 6-16**    (a) Simple model of a strained (bent) crystal. (b) Formation of low-angle grain boundaries, shown as vertical dashed lines.

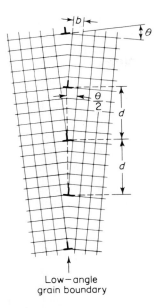

**Figure 6-17**   Calculation of angle between subgrains.

Low−angle
grain boundary

Dislocations are impeded by subboundaries in the same way as they would be by any set of dislocations. Because the dislocation density, and hence average spacing of dislocations, does not change during polygonization, the hardness and strength should not change very much. Figure 6-18 shows the room-temperature hardness and electrical resistivity that result after a number of specimens of identically cold-worked nickel are heated to different temperatures. In the region labeled "Recovery," no change in an initially polished and etched surface is visible in an optical microscope. Yet the resistivity drops in this range. This apparently results from the formation of low-angle grain boundaries and the absorption of vacancies by dislocations. This is the range in which polygonization occurs. The grain boundaries remain fixed in this temperature range; that is, recrystallization has not yet started.

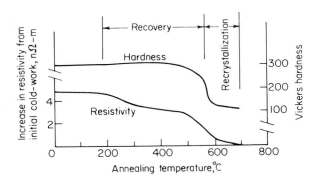

**Figure 6-18**   Room-temperature hardness and change in resistivity vs. annealing temperature for cold-worked 99.6% nickel. [From L. M. Clarebrough, M. E. Hargreaves, and G. W. West, "The Release of Energy During Annealing of Deformed Metals," *Proc. Royal Soc.* (London), 232A (1955), p. 252. Reprinted with permission of The Royal Society.]

If other identically prestrained specimens are heated to higher temperatures, dislocations can glide completely out of each grain so as to allow the formation of completely new grains of very low dislocation density. This is called *recrystallization* and can be observed in an optical microscope that operates at elevated temperatures. The dislocation density and elastic energy of the new crystals are much less than for crystals containing subboundaries. The strength and hardness of the recrystallized metal are therefore markedly reduced, as shown in the region in Fig. 6-18 labeled "Recrystallization."

**Example 6-6**

Compare qualitatively the dislocation density in a pure metal after (a) slow cooling from the melt, (b) cold working, (c) recovery, and (d) recrystallization.

**Solution**   Dislocation density is lowest after (a) slow cooling, has virtually the same high value after (b) cold working and (c) recovery, but is again much less after (d) recrystallization.

If the purity of the nickel is increased, in this case from 99.6 to 99.85% nickel, recrystallization will occur at a lower temperature, probably because fewer impurity atoms are present to impede dislocation motion and diffusion. A mechanism by which solute atoms can interfere with dislocation motion is described in Chap. 7 and shown in Fig. 7-6.

The process of recovery and recrystallization is used to eliminate the effect of strain hardening and to control the grain size. The latter is possible because if the initial strain is increased, more elastic energy will be available to force the rate of nucleation of new grains to exceed the rate of growth of each grain, so a larger number of smaller grains will be formed. Except by melting, it is not possible to alter the grain size of pure single-phase solid elements without first introducing some plastic strain.

# REFERENCES

1. American Society for Metals, *Strengthening Mechanisms in Solids*. Metals Park, Ohio: American Society for Metals, 1962. Chapter 3, "Strain Hardening," by Jack Washburn, discusses strain-hardening mechanisms in detail.

2. Caddell, R. M., *Deformation and Fracture of Solids*. Englewood Cliffs, N.J.: Prentice Hall, 1980. Chapter 7 provides a thorough basic description of dislocation theory.

3. Guy, Albert G., *Elements of Physical Metallurgy*, 2nd ed. Reading, Mass.: Addison-Wesley, 1959. Chapter 12 gives step-by-step descriptions of recovery and recrystallization.

4. McLean, D., *Mechanical Properties of Metals*. New York: John Wiley, 1962. See Chap. 5, "Strain Hardening in Pure Metals."

5. Reed-Hill, Robert E., *Physical Metallurgy Principles*. Princeton, N.J.: Van Nostrand, 1964. Chapter 5 describes grain boundaries.

6. Weertman, Johannes, and Julia R. Weertman, *Elementary Dislocation Theory*. New York: Macmillan, 1964. Chapters 2 and 3 analyze the stresses and forces associated with dislocations; Chap. 5 discusses the dislocation pile-up.

7. Zackay, Victor F., ed., *High-Strength Materials*. New York: John Wiley, 1965. Chapter 12, "Slip Mechanism in Single Crystals of Hexagonal Close-Packed Phases," by J. E. Dorn and J. B. Mitchell, covers modes of slip and strengthening mechanisms in hcp metals.

## PROBLEMS

**6-1.** Slip in bcc crystals is observed to occur in <111>, the close-packed direction. What is the most widely spaced set of planes that contain this direction? How many slip systems involving this set of planes and the <111> directions are possible?

**6-2.** (a) What is the effect of cold rolling a bar of pure copper on its hardness? Explain why this occurs.
(b) What is the effect of cold rolling a bar of pure copper on its load-carrying capacity? Explain why this occurs.

**6-3.** You are given an annealed, coarse-grained nickel bar. Describe in detail at least two different ways that you can strengthen it without changing its chemical composition.

**6-4.** If two parallel edge dislocations of opposite sign are on the same slip plane, they will come together (if they are not impeded) and annihilate each other. What will happen if they are of opposite sign, separated by 50 nm on the slip plane, but one dislocation is displaced to another parallel slip plane 30 nm away? If a small shear stress that tends to force the dislocations together is then applied, what happens?

**6-5.** Describe briefly two different atomic mechanisms for strain hardening.

**6-6.** If you heat a piece of cold-worked copper to a sufficiently high temperature (short of melting) and water-quench it, it becomes softer. Why?

**6-7.** Sketch the approximate curves for hardness vs. annealing temperature (omit numbers) for 20% cold-worked copper and 50% cold-worked copper. Label each curve.
(a) Explain the difference in recrystallization temperature.
(b) Which has the smaller grain size after recrystallization at the lowest possible temperature? Why?
(c) Which has the greater room-temperature strength after cold working but before annealing? Explain why, describing the basic mechanism(s) briefly.
(d) Which has the greater room-temperature strength after annealing at the lowest possible temperature? Explain why, describing the basic mechanism(s) briefly.

**6-8.** What is the essential difference between recovery and recrystallization in a cold-worked metal?

**6-9.** A long bar of pure metal that follows the relationship $\sigma = (500 \text{ MPa})\varepsilon^{0.2}$ is initially 50 mm thick and 100 mm wide. It is subjected to the following rolling and heating sequence: 40% reduction of area, full annealing (recrystallization), 40% reduction of area, full annealing, and then 35% reduction of area. (Each reduction is calculated on the basis of the area just before each rolling operation is started.) Assuming no change in width during this sequence, find the maximum tensile load the bar can support after this sequence.

**6-10.** A 30% cold-worked brass (82 HRB) is put into a furnace for 0.5 h at 350°C and water-quenched; its final hardness is 82 HRB. Another 30% cold-worked brass is held in a furnace for 0.5 h at 370°C; after water quenching, its hardness is 68 HRB.
Explain why the hardness changed at 370°C but did not change at 350°C and describe the fundamental mechanism(s) involved.

**6-11.** An annealed 70-30 brass bar is initially 12.70 mm thick, and its hardness is –9 HRB. It is then rolled to the thicknesses indicated and the hardness measured.

|                          | Initial | 1     | 2    |
| ------------------------ | ------- | ----- | ---- |
| Thickness, mm            | 12.70   | 11.76 | 9.88 |
| Rockwell B hardness, HRB | –9      | 51    | 74   |

(a) Calculate the true longitudinal strains from initial to condition 1, and from condition 1 to condition 2.

(b) Explain, from a fundamental point of view, why the changes in hardness do not correspond to the changes in true strain.

**6-12.** Two separated Brinell hardness tests using a ball 10 mm in diameter are made on the same piece of 70-30 brass:

(a) At 500 $kg_f$ (4.90 kN).

(b) At 1500 $kg_f$ (14.7 kN).

Which test gives the higher hardness, (a) or (b)? Explain why this happens.

**6-13.** The line force acting in a direction perpendicular to a dislocation line can be expressed as

$$F = \tau b$$

where $F$ is the line force (in N/m), $\tau$ is the shear stress acting on the plane of dislocation motion, and $b$ is the Burgers vector. Calculate the magnitude and direction of the force acting on an edge dislocation in a single crystal of pure vanadium as the result of a second edge dislocation of the same sign lying parallel to and in the same slip plane as the first dislocation and separated by a distance of 35.7 nm.

**6-14.** Calculate the maximum shear stress at a point 27.3 $\mu$m away from the axis of a screw dislocation in iron.

**6-15.** Two parallel screw dislocations of the same sign lie along parallel <110> directions in a single crystal of aluminum, separated by a distance of 89.4 $\mu$m. If both dislocations are 175 $\mu$m long, calculate the direction and magnitude of the force that each exerts on the other.

**6-16.** Three parallel edge dislocations, all of length 471 $\mu$m, lie on a single slip plane in an iron crystal, as shown in Fig. 6-19. Dislocations $A$ and $C$ are of the same sign, and dislocation $B$ is of the opposite sign. Find the direction and magnitude (in N) of the net force acting on dislocation $B$.

Figure 6-19

# 7

# *Single Phases*

Solids seldom consist of atoms of only one element, so the influence of foreign atoms is therefore an important factor in the strength of most real solids. This chapter describes those strengthening mechanisms that can be attributed to the presence of foreign atoms distributed more-or-less uniformly on an atomic scale; that is, it describes the relative strengths of mixtures of elements in contrast to the strengths of the separate elements. Such mixtures of metals are called *alloys*. These effects on strength can act in addition to the strengthening mechanisms for pure elements that were described in the previous chapter.

## PHASES

When applied to solids, the term *phase* means any collection of atoms that is homogeneous on a fine scale. When the optical microscope was still the most powerful magnifying device available, any observably homogeneous portion of the structure as seen in the microscope could be conveniently termed a phase; multiple phases are described in Chaps. 8 and 9. However, with the advent of electron and ion microscopes, which can reveal details on the order of atomic dimensions, the earlier definition of a phase loses much of its meaning. For example, sometimes unlike atoms segregate themselves into a configuration that is not homogeneous on the scale of a few atoms but is homogeneous on a slightly larger scale and behaves physically and mechanically like a solid of uniform properties; such a solid can here be properly termed a *phase*. The distinction between one- and two-phase solids might therefore be established as homogeneity on a scale slightly larger than atomic. A hypothetical test to determine whether a solid is a single phase could therefore be to select samples containing 100 to 1000 atoms from any two regions in the solid. If the crystal structure and percentages of atoms of each element present are the same in both samples, then the solid is a single phase. This test can, of course, be applied to a pure element, which is single phase because it has no second element present. A solid chemical compound consisting of molecules with a fixed ratio of two or more elements is also a phase. Other kinds of phases are described in the following discussion.

## INTERSTITIAL SOLID SOLUTIONS

A solid solution is a solid phase that may have a wide range of ratio of foreign (solute) atoms to solvent atoms. If the solute atoms are substantially smaller than the solvent atoms, the solute atoms may find relatively large holes between the sites of the solvent atoms. Figure 7-1 shows three solute atoms in such voids; this is called an *interstitial solid solution.* It should be made clear that a two-dimensional figure does not show how much room is really present. For example, in a simple cubic lattice such as is shown in Fig. 7-1, if the solvent atoms are touching each other, the maximum size of an interstitial atom that can be placed between the atoms in their (100) plane without distorting the lattice is substantially smaller than the maximum size of an atom that can be placed out of the plane of Fig. 7-1 in the voids at the centers of the cubic cells.

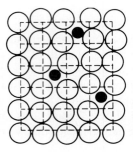

**Figure 7-1** Schematic sketch of three interstitial atoms in solid solution in a simple cubic lattice.

### Example 7-1

Calculate the maximum ratio of interstitial atomic radius to solvent atomic radius for the simple cubic structure for (a) the two-dimensional case, and (b) the three-dimensional case.

**Solution**  Let $R$ = solvent radius and $r$ = solute radius (see Fig. 7-2). (a) $2R\sqrt{2} = 2R + 2r$; thus, $r/R = \sqrt{2} - 1$. (b) $2R\sqrt{3} = 2R + 2r$; thus, $r/R = \sqrt{3} - 1$. This represents an increase of $(0.732 - 0.414)/0.414 = 77\%$ in the size of an atom that can fit into the space lattice. Similar calculations can be made for the other crystal structures.

Experimental evidence indicates that interstitial solid solutions do not form unless the ratio of the radius of solute atoms to the radius of solvent atoms is less than 0.59.

(a)

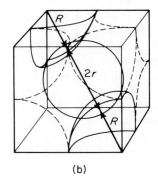

(b)

**Figure 7-2**  Calculation of atomic radii for Example 7-1.

This upper limit would involve large distortions in fcc and bcc crystals. Elements with very small atomic radii are favored as interstitial solutes, so the most important interstitial solutes in metals are carbon, boron, nitrogen, and oxygen. For example, iron at elevated temperatures can contain up to about 2% interstitial carbon, which more than covers the range of alloys known as steels. Interstitial hydrogen is usually undesirable in most metals because it can cause cracking and embrittlement (see Chap. 18). For reasons not fully understood, these elements form interstitial solid solutions more readily with the transition metals (those with incomplete inner electron shells) than with nontransition metals.

When two elements are in contact, there is a tendency for their atoms to mix, which follows from the thermodynamic argument that systems tend toward maximum disorder. Counteracting this tendency toward mixing is a tendency toward complete separation because energy is required to distort the lattice *elastically* when the atoms are mixed, as shown in Fig. 7-1. A balance between these two effects occurs. Usually the maximum solute/solvent ratio is higher at higher temperatures; if an insufficient supply of solute atoms is available, then the ratio will be less than the maximum. For interstitial solid solutions the maximum ratio of number or mass of solute atoms to total number or mass of atoms, the *solubility,* is seldom more than a few percent.

Interstitial atoms increase strength in the same way that vacancies do (see Chap. 6), by elastically distorting the regular lattice structure of atoms and impeding dislocation motion. Even interstitial atoms smaller than interstitial holes cause some distortion by altering the motion of nearby solvent atoms. Although the extent of distortion shown in Fig. 7-1 is very limited, the effect of interstitials in real solids extends much farther. In a crystal containing 1% interstitial atoms, these atoms are distributed throughout the crystal so as to be less than five interatomic spacings apart. Thus, the strain produced by interstitial solute atoms is general throughout the matrix, not limited to small isolated regions surrounding each solute atom. In this way, the solute atoms slightly increase the average crystal lattice spacing of the solvent atoms. The extent of strengthening of a pure metal by an interstitial element is shown in Fig. 7-3.

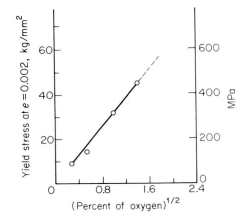

**Figure 7-3** Strengthening of pure zirconium by interstitial oxygen. [From D. G. Westlake, "The Combined Effects of Oxygen and Hydrogen on the Mechanical Properties of Zirconium," *Trans. AIME,* 233 (February 1965), p. 368. © 1965 by AIME and reprinted with its permission.]

**Example 7-2**

A calculation of the solubility of interstitial carbon in bcc iron at room temperature gives a maximum value of $2.3 \times 10^{-7}$ mass percent C. Approximately how many unit-cell spacings $(a_0)$ of Fe separate each C atom in this solid solution?

**Solution**   Let $N$ represent the number of atoms. Then,

$$\frac{N_{Fe} \times 55.85}{N_C \times 12} = \frac{1}{2.3 \times 10^{-9}}$$

Thus, $N_{Fe}/N_C = 9.34 \times 10^7$. Fe is bcc, so each unit cell contains two Fe atoms, giving ($9.34 \times 10^7$)/2 = $4.67 \times 10^7$ unit cells of Fe per C atom. Assume for simplicity that each C atom lies at the center of a cube containing $4.67 \times 10^7$ unit cells; then the closest distance between C atoms will be the same as the width of the large cube, which is $(4.67 \times 10^7)^{1/3}$ = 360 unit-cell widths.

## SUBSTITUTIONAL SOLID SOLUTIONS

Instead of placing small solute atoms in the interstices between solvent atoms, as described previously, the solvent atoms themselves may be replaced by other, different atoms, giving a *substitutional solid solution.* While interstitial solid solutions are limited to small amounts of solute, substitutional solid solutions can have a wide range of solubility. The extent to which unlike atoms form a substitutional solid solution is determined in part by their similarities in atomic radius (which should not differ by more than 15%), crystal structure, and electron structure. Several pairs of elements are *completely soluble* in each other at all temperatures. Among the more common metals, the copper–nickel and cobalt–nickel systems are fcc examples, and the molybdenum–tungsten system is a bcc example. In addition, many other combinations of metals form substitutional solid solutions ranging up to 5 or 10% in temperature ranges of practical importance; thus, these alloys exhibit *limited solid solubility.* Substitutional solid solutions are much more common than interstitial solid solutions in metals.

(a)

(b)

**Figure 7-4**   Substitutional solid solutions, viewed schematically, with (a) solute radius larger than solvent radius, and (b) solute radius smaller than solvent radius.

Figure 7-4 depicts schematically the distortion caused by solute atoms larger and smaller than the solvent atoms. The appearance of the matrix in both cases is similar to that of the distorted matrix in Fig. 7-1, and for the same amount of distortion the effect on dislocation motion is similar.

The effect of a substitutional solute should therefore be to strengthen the solvent because the distorted lattice interferes with dislocation motion. More solute should distort the lattice more and thus cause further strengthening. Figure 7-5 shows that the strength of pure copper is increased by nickel in a substitutional solid solution. This effect is not due to nickel being stronger than copper because addition of copper to pure nickel (on the right side of Fig. 7-5) also increases strength. At some point between these extremes of pure elements, at about 70% nickel, the strengthening effect from lattice distortion reaches a maximum. Similar results are found for solid solutions with a limited range of solubility, although a maximum such as that observed in copper–nickel usually does not occur in the solubility range.

**Example 7-3**

Copper forms a substitutional solid solution with zinc up to about 38% Zn. Which alloy should be stronger: (a) 95% Cu–5% Zn, or (b) 90% Cu–10% Zn?

**Solution**   90% Cu–10% Zn should be stronger because the increase in the amount of Zn in this range should further distort the lattice. Typical measured values are 260 MPa for Cu–10% Zn and 240 MPa for Cu–5% Zn.

The practical utilization of solid-solution strengthening is that it provides a large number of alloys having a wide range of properties. If high strength were the only criterion, then the alloy and grain size giving the highest strength would be the only ones used. However, differences in the cost of metals encourage the use of alloys richer in the lower-cost metal; differences in corrosion resistance promote use of the more resistant metal; and variations in toughness tend to be inversely related to strength, so an alloy of less than maximum strength is often used in order to meet the ductility requirements of service or of the fabrication process (see Chap. 14). The single-phase copper–zinc alloys (brasses), ranging from 0 to 38% Zn, provide examples of these factors.

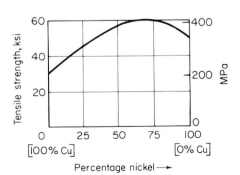

**Figure 7-5**   Effect of composition of substitutional solid solution on strength of Cu–Ni alloys. (Adapted from ASM data.)

## SUPERSATURATION

The degree of solubility may change with temperature, so the limits to solubility may change (usually decreasing) as the temperature decreases. At a temperature approaching the melting temperature, diffusion in a solid can be very rapid. Alloys at high temperatures thus reach stable equilibrium concentrations much more rapidly than at absolute temperatures that are low compared with the melting temperature. If a solid-solution alloy in equilibrium at high temperature is suddenly cooled to a low temperature where the maximum equilibrium percent solute is less than that of the alloy mixture, the solid solution is *supersaturated.* More solute atoms are in solution at this lower temperature than would be present after slow cooling. (Note that for as long as the atoms remain uniformly distributed, such a solid solution is still a single phase.)

All supersaturated solid solutions will theoretically reach equilibrium concentrations sooner or later. But if the time required at low temperature is on the order of hundreds or thousands of years, such a solid solution may be considered stable for engineering design purposes. The strengthening mechanism associated with a supersaturated solid solution is the same as for a solid solution in equilibrium, except that the presence of additional solute atoms and the resulting severe distortion cause even greater strengthening in the supersaturated solid solution. Thus, if the amount of supersaturation is diminished by a reduction in the fraction of solute atoms in solution, the strength of the solid solution will be reduced. In the next two chapters, other changes that can occur in such solutions will be discussed; these changes can result in the introduction of additional strengthening mechanisms not considered here.

## SOLUTES AND DISLOCATIONS

The atoms in the vicinity of a dislocation are situated at sites substantially displaced from their equilibrium sites (see Chap. 6). In an edge dislocation such as shown in Fig. 7-6, the atoms below the slip plane are in tension and those above are in compression. Figure 7-6 shows sites in an edge dislocation where interstitial atoms and large and small substitutional atoms could reduce the lattice strain energy. That is, the strain energy is less in Fig. 7-6 than in the case where the solute atoms are located far away from dislocations. Thus, when these solute atoms "fall" into sites near dislocations, there is less probability that they will leave, and if enough solute atoms are located along the line of the dislocation, it is said to be *pinned.* Because oxygen and nitrogen atoms are common pinning interstitials, this concentration of interstitials is called an *atmosphere.* In order to move away from the solute atoms, the dislocation must return the elastic energy that was dissipated when the solute atoms reached the dislocation. At normal temperatures this extra energy can be provided only by increasing the stress required to free the dislocation from its interstitials, thereby increasing the yield strength. Once free of its interstitials, the dislocation can continue to move at a lower stress. This accounts for the drop in stress often observed in annealed mild steel following initial yielding [see Fig. 2-20(c)].

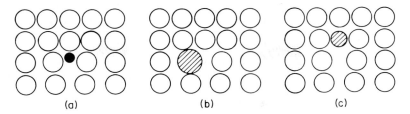

**Figure 7-6**  Energetically favorable lattice positions in an edge dislocation for (a) interstitial atom, (b) larger substitutional atom, and (c) smaller substitutional atom.

Strengthening by dislocation pinning forms an important part of full solid-solution strengthening, and this effect acts in addition to the lattice-distortion effect described earlier. Solute atoms can reach dislocations in two obvious ways: by diffusion of the solutes and by dislocation motion intersecting the solutes.

Diffusion of solute atoms is promoted by higher temperatures; and interstitial solute atoms, being smaller, diffuse more rapidly than substitutional solute atoms. For example, if a mild-steel tensile specimen is pulled slowly at 200–250°C, the carbon (and probably nitrogen) interstitial atoms can diffuse fast enough to catch the dislocations after they have broken away from their atmospheres of interstitials. In the strain-hardening region of the stress–strain curve, below ultimate, after a small amount of strain has occurred, enough interstitials will have caught and pinned dislocations to cause an increase in stress similar to initial yielding. This is followed by a drop in yield stress when the dislocations again break away from their interstitial atmospheres. This process is repeated many times to produce a sawtooth-shaped stress–strain curve of successive increases and sudden drops in stress; the phenomenon is called the *Portevin–Le Châtelier effect.* A return of the drop in stress at yielding can also occur in mild steel at room temperature following a small plastic deformation, but weeks are required for the diffusion of interstitials to dislocations at room temperature. This process, whether occurring at room temperature in a week or at 150–250°C in a few minutes, is called *strain aging.*

The other means for bringing solute atoms and dislocations together is through dislocation motion that intersects the solute atoms. Since dislocations are generated and move at room temperature when plastic strain occurs, the increase in strength through the intersection of dislocations and solutes represents a secondary kind of strain hardening. Figure 7-7 shows the effect of a small amount of substitutional copper on the strain-hardening behavior of aluminum. (The slight difference in grain size of the two specimens has only a very small influence on strain hardening.) The upward shift of the stress–strain curve cannot be attributed entirely to the pinning of dislocations by solutes; the basic solid-solution strengthening mechanism from lattice distortion is also in operation. The higher strength of the solid solution forces the generation of new dislocations, which then interact with each other to increase the rate of strain hardening and to delay the onset of mechanisms by which dislocations can bypass barriers to their motion. The amount of strengthening due to each mechanism is not known.

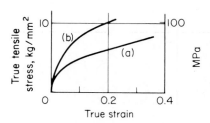

**Figure 7-7**  Stress–strain curves at 26°C for (a) 99.98% Al, 0.33-mm grain size, and (b) Al–0.23 atomic % Cu, 0.30-mm grain size. [From D. McLean, *Mechanical Properties of Metals* (New York: John Wiley & Sons, 1962), p. 173. Reprinted with permission of the publisher.]

## ORDER–DISORDER EFFECTS

Substitutional atoms usually exhibit either a preference for atoms of the same element as adjacent neighbors—a type of behavior called *clustering*—or a preference for atoms of an unlike element as adjacent neighbors called *short-range order.* In either case, such ordering results in localized inhomogeneities in the atomic ratio when interatomic attractions and repulsions cause preferential positioning of atoms into ordered groups. The extreme example of clustering is thus complete separation of the two dissimilar elements. It is clear that 1 kg of Ni and 1 kg of Cu will have lower energy when they are completely separated than when they are melted together to form a 2-kg mixture and then cooled, because the resultant mixture of different atoms will result in lattice strains no matter what positions the atoms select after cooling.

Although the precise reasons for these dissimilar forces between atoms are not yet fully understood, it is clear that the equilibrium-ordered configuration has lower energy than the random solid solution. Figure 7-8(a) shows a solid solution of equal numbers of atoms of two elements that have a preference for unlike atomic bonds. After passage of a dislocation and consequent slip of one interatomic spacing, the number of unlike bonds is reduced on the average at least to half and usually to less than half because of the initial ordering of the structure. Thus, the strain energy of the crystal in Fig. 7-8(b), surface effects being ignored in this small sample of atoms, is greater than in Fig. 7-8(a), which is nearer the equilibrium configuration. The source of this energy input must be the increase in stress required to pass the dislocation through the region, which is thus an increase in strength. If a second dislocation immediately followed the first, the stress required for its motion might be smaller, because the ratio of unlike atomic bonds would probably be increasing. But usually some time passes before another dislocation traverses this same slip plane, and by then the atoms have rearranged themselves through motion over very short distances into the lower-energy configuration, a new short-range

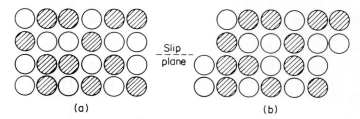

(a)                                              (b)

**Figure 7-8**  Order–disorder effect in a crystal composed of equal numbers of two kinds of atoms. (a) Ordered structure having five unlike bonds and one like bond across slip plane. (b) After slip of one interatomic space, structure has one unlike bond and four like bonds across slip plane.

order similar to that in Fig. 7-8(a). The process of forcing disorder by dislocation motion must then be repeated; therefore, the stress for slip, and consequently the strength, remains high.

### Example 7-4

If a preference for unlike atomic bonds, such as shown in Fig. 7-8, results in strengthening, would a preference for like atomic bonds cause weakening?

**Solution**    No. If more like bonds existed in equilibrium before slip, then dislocation motion must decrease the number of like bonds and thus increase the local energy. The point of the argument is that dislocation motion disrupts the preferred local order of either type and thus requires a higher stress.

## REFERENCES

1. American Society for Metals, *Metals Handbook,* 9th ed., vol. 1, *Properties and Selection: Irons and Steels.* Metals Park, Ohio: American Society for Metals, 1978.

2. American Society for Metals, *Metals Handbook,* 9th ed., vol. 2, *Properties and Selection: Nonferrous Alloys and Pure Metals.* Metals Park, Ohio: American Society for Metals, 1979.

3. American Society for Metals, *Metals Handbook,* 9th ed., vol. 3, *Properties and Selection: Stainless Steels, Tool Materials and Special-Purpose Metals.* Metals Park, Ohio: American Society for Metals, 1980.

4. American Society for Metals, *Strengthening Mechanisms in Solids.* Metals Park, Ohio: American Society for Metals, 1962. Chapter 2, "Solid Solution Strengthening," by P. A. Flinn, includes extensive theoretical treatment of this subject.

5. Brick, Robert M., Robert B. Gordon, and Arthur Phillips, *Structure and Properties of Alloys,* 3rd ed. New York: McGraw-Hill, 1965. Chapter 3 discusses diffusion and properties of solid solutions.

6. LeMay, Iain, *Principles of Mechanical Metallurgy.* New York: Elsevier North-Holland, 1981. Chapter 6 provides thorough coverage of strengthening mechanisms.

7. McLean, D., *Mechanical Properties of Metals.* New York: John Wiley, 1962. Chapter 6 gives a thorough analysis of the effects of solutes.

8. Reed-Hill, Robert E., *Physical Metallurgy Principles.* Princeton, N.J.: Van Nostrand, 1964. Chapter 8 discusses the interactions between dislocations and solute atoms.

9. Zackay, Victor F., ed., *High-Strength Materials.* New York: John Wiley, 1965. Chapter 9, "On Lattice Effects and the Strength of Alloys," by T. L. Johnston, A. S. Tetelman, and A. J. McEvily, Jr., emphasizes the importance in some materials of inherent lattice resistance to dislocation motion.

## PROBLEMS

**7-1.** The largest interstitial holes in fcc crystals occur at $1/2,0,0$ and $1/2,1/2,1/2$ positions in the lattice, which are crystallographically equivalent. If solvent and solute atoms are assumed to be hard spheres, calculate the maximum possible ratio of solute-atom radius to solvent-atom radius without lattice distortion.

**7-2.** The largest interstitial holes in bcc crystals occur at $1/2, 1/4, 0$ positions. If solvent and solute atoms are assumed to be hard spheres, calculate the maximum possible ratio of solute-atom radius to solvent-atom radius without lattice distortion. Why is the $1/2, 1/2, 0$ position not the largest hole, because by inspection of a cube face it can be seen that there is more space between the corner atoms in the unit cell at $1/2, 1/2, 0$ than there is at $1/2, 1/4, 0$?

**7-3.** Using the results for Probs. 7-1 and 7-2, show why the bcc form of iron (atomic radius = 0.124 nm) might have lower solubility for interstitial carbon (atomic radius = 0.071 nm) than does the fcc form of iron (atomic radius = 0.128 nm).

**7-4.** What is a phase? Give three examples of solid phases.

**7-5.** Is it possible to produce a supersaturated Cu–Ni solid solution? Explain.

**7-6.** The *Bauschinger effect* is a reduction of the yield stress when the direction of applied stress is reversed following initial deformation beyond yield. If solute atoms are viewed as local barriers to dislocation motion, how can this model be used to explain the Bauschinger effect?

**7-7.** In bcc iron, interstitial carbon can occupy a crystallographically different lattice site when the crystal is under tension or compression than when it is unstressed (this is called the *Snoek effect*). Because this shift in interstitial position requires energy, describe a mechanism by which the Snoek effect could contribute to solid solution strengthening.

**7-8.** Calculate the diameter of the maximum interstitial hole in aluminum.

# 8

# Phase Diagrams

The presence of even a small amount of a second phase in the microstructure of a metal can markedly change its mechanical properties; in particular, strength usually increases. Control of the amount and distribution of the second phase provides substantial control over strength. This chapter presents a brief review of equilibrium-phase diagrams in order to show the origins of the structure and strength of many commercial alloys and to develop the reasoning behind heat-treating operations.

## COPPER–NICKEL, A SIMPLE EQUILIBRIUM PHASE DIAGRAM

Pure copper melts at 1084.5°C, and pure nickel melts at 1455°C. If 3 kg of nickel and 7 kg of copper are heated to 1500°C, all will be liquid. If the liquid nickel and copper are stirred together, they will form a liquid solution because their atoms will be mixed together on a very fine scale. At what temperature will this mixture become solid on cooling, and in what form will the atoms be combined in the solid? These questions can be answered through use of the *equilibrium-phase diagram* (often called simply the phase diagram) for copper and nickel (Fig. 8-1).

This phase diagram tells what phases are present and their relative amounts, for any mixture of copper and nickel, at any temperature reached through a process of very slow cooling from the melt or very slow heating of a solid. To answer our initial question, a mixture of 3 kg of nickel and 7 kg of copper is 30% Ni. In Fig. 8-1, 30% Ni at 1500°C is point *A*, lying in a region labeled *L* (liquid), so the phase is a liquid solution and remains so on cooling until it reaches about 1240°C, point *B*.

On cooling just below 1240°C the solution enters the two-phase region, and small amounts of a solid will precipitate, or freeze. (This solid is here designated as *α*; Greek letters are conventionally used to designate solid phases in equilibrium diagrams.) The two-phase region represents combinations of composition and temperature that are impossible; that is, neither the liquid nor solid phase can have a composition of 30% Ni–70% Cu within this region. Rather, the phases present at, say, 1239°C are a solid of about 42% Ni

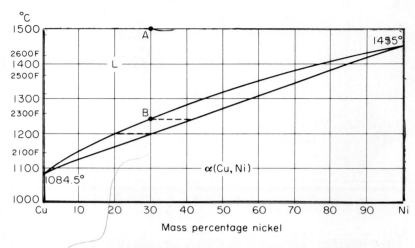

**Figure 8-1**   Copper–nickel equilibrium phase diagram. [From "Phase Diagrams," *Metals Handbook: Metallography, Structures and Phase Diagrams,* 8th ed., vol. 8, T. Lyman, ed. (Metals Park, Ohio: American Society for Metals, 1973), p. 294.]

and 58% Cu and a liquid of slightly less than 30% Ni and slightly more than 70% Cu. Nothing exists having a composition between those of these two phases. Of course, the overall masses of Ni and Cu, taken together in the two phases, remain as 3 kg of Ni and 7 kg of Cu.

Upon further slow cooling, more solid precipitates, but its nickel content gradually diminishes until at about 1200°C the solid is 30% Ni. At the same time, the content of all the solid that has precipitated earlier also changes to equal the composition of the solid currently precipitating. Thus, very slow cooling is essential for true equilibrium.

At the same time, the composition of the remaining liquid at each stage changes from its initial 30% Ni at 1240°C to about 20% Ni at 1200°C. It is obvious that this must occur because at all times the average composition of all the solid plus all the liquid must remain 30% Ni–70% Cu.

Below 1200°C, a single-phase solid solution of copper and nickel exists, with no further structural changes on cooling. The existence of a single-phase solid solution, such as this one, over the complete composition range is relatively rare.

## SILVER–COPPER, A MORE COMPLEX SYSTEM

At a given temperature, if the amount of second element present in a two-element solid exceeds the maximum that can normally be in solid solution with the first element (i.e., there is limited solid solubility), the excess of second element will usually be found in combination with the first element in the form of a separate and different phase. That this separate phase is different can be established by a marked difference in atomic ratio and usually a difference in the crystal structure of the second phase.

The second phase forms because the energy of a single-phase solid solution of the same overall composition is greater than the total energy of the two separate phases. Each of these phases is then saturated with its respective solute element. Strain energy is required to displace solvent atoms from their equilibrium positions enough to permit solute atoms to be held in the matrix. As more solute atoms are added, the strain energy usually increases rapidly. Because the atomic-level phenomena that influence this strain energy are still not fully known, the energies cannot be calculated exactly. The atomic concentrations required for saturation are thus only observed, not calculated. With the assumption that a system tends to change toward a minimum-energy condition, observation of real binary (two-element) combinations that are cooled very slowly usually reveals the saturation concentrations.

Thus, the equilibrium-phase diagram is obtained from experiments. Data are taken with a range of concentrations of the elements involved. The discussion that follows is limited to binary systems; because several ternary (three-element) systems are commercially important, the detailed descriptions of their use given in the references at the end of this chapter should be consulted. As an example of a binary system, Fig. 8-2 is the phase diagram for the silver–copper system. This diagram indicates what phases are present under equilibrium conditions for each temperature and each possible composition, which is the percentage by mass (as given here) or by ratio of the number of atoms of each element present. Each solid phase is designated by a Greek letter, in this case $\alpha$ and $\beta$.

**Figure 8-2** Silver–copper equilibrium phase diagram. [From "Phase Diagrams," *Metals Handbook: Metallography, Structures and Phase Diagrams,* 8th ed., vol. 8, T. Lyman, ed. (Metals Park, Ohio: American Society for Metals, 1973), p. 253.]

At sufficiently high temperatures, all mixtures of silver and copper are liquids; the elastic strain energy associated with a distorted solid-crystal structure is therefore absent. Solidification of a pure element—of pure copper at temperatures below 1084.5°C, for example—may be looked upon as a transformation to a lower energy configuration; otherwise, the copper would remain liquid. Solid copper at 1050°C thus has a lower energy than liquid copper at 1050°C, so if liquid copper is cooled quickly to 1050°C, it will become supercooled and soon transform to the solid form, giving off heat in the process. The freezing temperature of 1084.5°C does not mean that liquid copper cannot be cooled below that temperature and remain liquid but rather that under equilibrium conditions (very slow cooling) pure cooper freezes at 1084.5°C. The rest of the equilibrium-phase diagram must be interpreted in the same way.

**Example 8-1**

Name the phases present in a silver–copper alloy under the following conditions: (a) 30% Cu at 900°C; (b) 98% Cu at 800°C; (c) 3% Cu at 750°C; (d) 70% Cu at 850°C; (e) 50% Cu at 300°C.

**Solution**    The answer to each question is shown by the symbols in the area in Fig. 8-2 where the selected composition and temperature lie. (a) Liquid solution; (b) $\alpha$; (c) $\beta$; (d) $\alpha$ + liquid; (e) $\alpha + \beta$.

The unshaded areas of the phase diagram in Fig. 8-2 indicate regions where a single phase has the lowest energy and thus exists alone under equilibrium conditions. The shaded areas represent temperature–composition combinations where no minimum-energy single phase can exist in equilibrium. If a single phase exists at a temperature and composition that fall in a shaded area, the energy of this single phase will be greater than the energy of some combination of two phases. Thus, the shaded areas may be considered forbidden regions where no *stable* phase can exist. A mixture of silver and copper with an average composition that falls within a shaded area when it is cooled from liquid thus achieves the lowest possible energy by forming two phases from the adjacent unshaded areas. The relative amounts of these two phases are exactly those that give the overall composition of the original liquid solution.

As an example, consider slow-cooling a 30% Ag–70% Cu (by mass) liquid solution from 1100°C. This composition is shown as a vertical dashed line in Fig. 8-2. At 900°C, point A, the phase diagram indicates that this composition is forbidden, that under these conditions a solid solution rich in Cu (arbitrarily called $\alpha$) of about 8% Ag (point B) and a liquid solution of about 41% Ag (point C) together provide the lowest possible energy. This forbidden region is labeled "$\alpha + L$" because these are the two phases present, $\alpha$ and liquid, and each has a composition that falls on its boundary with the two-phase region.

The masses of $\alpha$ of 8% Ag and liquid of 41% Ag (we assume for now that these are exact values) must be such as to give an *average* overall composition of 30% Ag. If the average overall composition of the mix at 900°C were 8% Ag (point B in Fig. 8-2), there would be all $\alpha$ and no liquid. At the other extreme, if the average composition were 41% Ag (point C), there would be all liquid and no $\alpha$ present at 900°C. The fraction of liquid present varies linearly from 0 to 1.00 of the total mix as the average composition is changed from 8% Ag to 41% Ag. Since the average composition for this problem is 30% Ag, the proportion that is liquid is the fraction *BA/BC,* or

$$\frac{0.30-0.08}{0.41-0.08}=0.667$$

The proportion of $\alpha$ present is $1 - 0.667 = 0.333$. To check this result, assume an initial mix of 50 kg containing 15 kg of Ag and 35 kg of Cu, which is thus 30% Ag. At 900°C, Cu will be found in the 0.333 fraction of the mix that is $\alpha$ (92% Cu) and in the 0.667 fraction of the mix that is liquid (59% Cu). The total mass of Cu will be distributed as follows:

| | | |
|---|---|---|
| Cu in $\alpha$: | $(0.92)(0.333)(50 \text{ kg}) =$ | 15.3 kg |
| Cu in liquid: | $(0.59)(0.667)(50 \text{ kg}) =$ | 19.7 kg |
| Total Cu | | = 35.0 kg |

This calculation fully accounts for all the copper. A similar calculation can be made for the silver in the two phases.

　　The fraction of one phase of a two-phase mixture is simply the fraction of the distance across the two-phase region, moving *toward* the single-phase region of the phase in question. Thus, for the above problem as illustrated in Fig. 8-2, the fraction of liquid for a 30% Ag mixture at 900°C has been shown to be $\overline{BA}/\overline{BC}$ or $(30 - 8)/(41 - 8) = 0.667$. Likewise, the fraction of $\alpha$ is $\overline{CA}/\overline{CB}$ or $(41 - 30)/(41 - 8) = 0.333$; in this case, the fraction of $\alpha$ varies from 0 to 1.00 *from right to left* across the $(\alpha + L)$ region, toward the $\alpha$ region. The same procedure applies to calculation of the masses of liquid and solid $\beta$ (silver-rich) for compositions greater than 71.9% Ag.

**Example 8-2**

　　A mixture of 30 kg of silver and 70 kg of copper is slowly cooled from liquid to 800°C (see details in Fig. 8-2). Find (a) the total mass of $\alpha$, and (b) the mass of silver in $\alpha$.

　　**Solution**　From Fig. 8-2, the average composition is 30 kg/(30 + 70) kg = 0.3, or 30% Ag, the same as in the previous discussion.
　　(a) The mass of $\alpha$ at 800°C is $[(67 - 30)/(67 - 8)](100 \text{ kg}) = 62.7$ kg.
　　(b) The mass of silver in 62.7 kg of $\alpha$, composition 8% Ag, is $0.08 \times 62.7 \text{ kg} = 5.0$ kg.

## EUTECTIC TRANSFORMATION (SILVER–COPPER)

Although the calculations in the previous section always give the correct masses of phases in equilibrium, they tell nothing about the size, shape, or distribution of the phases. Because the configuration of the phases can have a large influence on strength (see Chap. 9), some special kinds of transformations that affect phase configuration must be studied. The Ag–Cu diagram in Fig. 8-2 is characteristic of a large number of binary systems in that it exhibits a *eutectic transformation* where a single-phase liquid of 28.1% Cu and 71.9% Ag transforms at a constant temperature (780°C) on cooling to two solid phases $\alpha$ and $\beta$ in a unique proportion. Because the entire transformation occurs at constant temperature— the *eutectic temperature*—the two solid phases that are formed are usually mixed together on a microscopic scale in a characteristic geometry that can be identified under the optical microscope and is called the *eutectic microstructure.* (Note that a microstructure may contain more than one phase.) A photomicrograph of the characteristic as-cast Ag–Cu eutectic microstructure is shown in Fig. 8-3(a). For contrast, Fig. 8-3(b) shows Ag–Cu of eutectic composition that has been cooled under controlled solidification conditions. (The effect of shape and spacing of phases is discussed in Chap. 9.) Before studying the special mechanical properties that result from the eutectic transformation, methods must be developed for calculating masses of eutectic microstructure present in compositions deviating somewhat from the eutectic.

The eutectic is not a phase: It is a microstructure made up of two phases. Therefore, the amount of eutectic present cannot be read directly from the equilibrium diagram. But since the eutectic always forms from the complete transformation of a single-phase liquid of eutectic composition at the eutectic temperature, the amount of eutectic must be the same as the amount of liquid from which the eutectic was formed. The eutectic forms *only* from this liquid (71.9% Ag at 780°C, in the Ag–Cu case), so the mass of this liquid fully accounts for the mass of eutectic formed. The horizontal constant-temperature line at the eutectic temperature divides liquid–solid and solid–solid regions, so it is not possible to calculate the amounts of phases present at this exact temperature. For example, for a eutectic composition (71.9% Ag), the equilibrium phases present at exactly the eutectic temperature (780°C) can be all liquid, liquid + $(\alpha + \beta)$, or all $(\alpha + \beta)$; how much liquid remains at exactly 780°C depends on the cooling conditions and cannot be determined from the phase diagram. For this reason, calculation of amounts of phases must be made at temperatures slightly above or below the eutectic temperature. Instead of writing 780⁺°C or 780⁻°C, a convenient convention is to add 1°C to or subtract 1°C from the eutectic temperature, a shift so small as to cause negligible changes in the composition of the phases. (For example, the saturation composition of $\beta$ at 779°C can be considered to be 91.2% Ag, the same as $\beta$ at 780°C.) No ambiguity results from this adjustment because for equilibrium conditions a mixture of 28.1% Cu and 71.9% Ag at 781°C is all liquid, and at 779°C it is all solid $(\alpha + \beta)$.

The next step, determination of the amount of eutectic for, say, 50 kg of a 70% Cu–30% Ag mixture, involves the calculation of masses of phases by the method already described. At 781°C all the liquid present is of eutectic composition (71.9% Ag), so the mass of liquid at 781°C is equal to the mass of eutectic solid at 779°C.

**Example 8-3**

Calculate the mass of liquid in 50 kg of a 70% Cu–30% Ag mixture at 781°C.

**Solution**   The composition of liquid is 71.9% Ag, and that of $\alpha$ is 7.9% Ag. Thus, the fraction liquid at 781°C (from left to right) is

$$\frac{30 - 7.9}{71.9 - 7.9} = 0.345$$

The mass of liquid is thus (0.345)(50 kg) = 17.3 kg.

From the solution to Example 8-3 we see that the mass of solid eutectic at 779°C is also 17.3 kg, the same as the mass of liquid at 781°C. During the eutectic transformation,

(a)

**Figure 8-3**   Silver–copper eutectic (71.9% Ag) in two forms, where the dark phase is $\alpha$ and the light phase is $\beta$ (see Fig. 8-2): (a) as-cast, and (b) cooled under controlled conditions to produce directional solidification. (Photographs courtesy of Robert R. Jones.)

(b)

the amount and composition of *free $\alpha$ (noneutectic)* remains unchanged. At 781°C, 50 – 17.3 = 32.7 kg of free $\alpha$ is present with the liquid; at 779°C, the same 32.7 kg of free $\alpha$ is still present with the eutectic. So $\alpha$ is now found in two places in the microstructure: as free $\alpha$ and as part of the eutectic, mixed intimately with $\beta$ in the special eutectic configuration. If the average composition were, say, 85% Ag, free $\beta$ instead of free $\alpha$ would be present with the eutectic.

   Further slow cooling of the 50 kg of 30% Ag results in no phase changes that have much influence on mechanical properties, but in some special heat treatments these changes can be important. Note that in this temperature range the solubility of both $\alpha$ and $\beta$ decreases on cooling. The eutectic at 779°C contains $[(71.9 - 7.9)/(91.2 - 7.9)] \times 17.3$ kg = 13.3 kg of $\beta$ of 91.2% Ag and 17.3 kg – 13.3 kg = 4.0 kg of $\alpha$ of 7.9% Ag. At 400°C the eutectic contains $\alpha$ of about 2% Ag and $\beta$ of about 99% Ag. These changes in composition as the eutectic cools do not alter the total mass (17.3 kg) or average composition (71.9% Ag) of the eutectic. The composition changes of the two phases occur by diffusion of Cu to $\alpha$ and of Ag to $\beta$ across the boundaries between the $\alpha$ and $\beta$ in the eutectic. The intimate mixture of the two phases provides very large surface areas of phase boundaries to accommodate this diffusion. Because the mass of metal atoms that diffuses is a small fraction of the total mass of eutectic, the shape of the eutectic microstructure changes very little in cooling from 779°C to 400°C. The mass proportions of $\alpha$ and $\beta$ in the eutectic can also change slightly by diffusion during cooling. However, many phase diagrams have nearly symmetrical $\alpha$ and $\beta$ *saturation lines* [separating the $\alpha$ region and the $\beta$ region from the $(\alpha + \beta)$ region], so the change in proportions of the two solid phases is small.

**Example 8-4**

Find the masses of $\alpha$ and $\beta$ in 17.3 kg of Ag–Cu eutectic at 400°C.

**Solution**    Calculating from right to left in Fig. 8-2 at 400°C, for 71.9% Ag (the eutectic composition), we find that the mass of eutectic $\alpha$ is

$$\left(\frac{99 - 71.9}{99 - 2}\right)(17.3 \text{ kg}) = 4.8 \text{ kg}$$

The mass of eutectic $\beta$ is 17.3 kg – 4.8 kg = 12.5 kg. Notice that the relative masses of $\alpha$ and $\beta$ at 400°C are slightly different from the amounts calculated at 779°C, already calculated earlier as 4.0 kg and 13.3 kg.

At 779°C the 30% Ag mixture also contains 32.7 kg of free $\alpha$. During further cooling some diffusion will occur between the free $\alpha$ and the eutectic. Because these components are not mixed on such a fine scale as the $\alpha$ and $\beta$ in the eutectic, the phase-boundary area is much smaller, and thus diffusion between the free phase and the eutectic is assumed to be negligible. This assumption then allows a conceptual separation of the eutectic liquid and the noneutectic at 781°C, so they can be treated as separate problems. This argument justifies the statement in the previous paragraph that the total mass of eutectic remains constant during cooling. The mixture of 70% Cu–30% Ag can now be broken up and thought of as two separate components at 781°C: 17.3 kg of eutectic liquid of average composition 71.9% Ag and 32.7 kg of free $\alpha$ of 7.9% Ag. The eutectic component of the problem was covered above. The noneutectic component must maintain the same average composition as the $\alpha$ at 781°C. The phase diagram shows that a mixture of average composition 7.9% Ag cannot be a single phase at 400°C and that both $\alpha$ and $\beta$ are present.

**Example 8-5**

Find the mass of $\beta$ present after 32.7 kg of free $\alpha$ of 7.9% Ag at 781°C is cooled to 400°C.

**Solution**    Calculating at 400°C, 7.9% Ag, from left to right in the two-phase region, we find that the mass of $\beta$ is

$$\left(\frac{7.9 - 2}{99 - 2}\right)(32.7 \text{ kg}) = 2.0 \text{ kg}$$

The mass of $\beta$ calculated in Example 8-5 forms gradually as the $\alpha$ is slowly cooled from 781°C to 400°C. This $\beta$ is observed to form at the $\alpha$ grain boundaries because grain boundaries are regions of higher energy where nucleation and growth of the second phase can provide larger possible energy reduction.

Figure 8-4 shows a slowly cooled 90% Ag–10% Cu (which was used for many years for silver coins in the United States until the high price of silver forced a change to a copper–nickel alloy), which results in a mixture of noneutectic $\beta$ and eutectic. The small

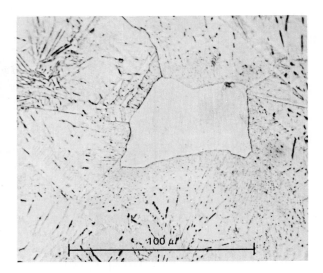

**Figure 8-4**    Slowly cooled 90% Ag–10% Cu (coin silver) showing a light patch of silver-rich $\beta$ surrounded by eutectic.

amount of grain boundary $\alpha$ that should form on slow cooling, in the manner described in the previous paragraphs, may account for the black line that can be seen along part of the $\beta$ grain boundary.

The logic followed in the procedure previously described for calculating the relative amounts of eutectic, noneutectic, free, and grain-boundary phases can be expressed graphically. Figure 8-5 shows the method of solving the problem involving 30% Ag–70% Cu cooled under equilibrium conditions from liquid. Start with 50 kg of liquid at point $A$ in Fig. 8-5. At point $B$, just above the eutectic temperature, the liquid (17.3 kg) and the free $\alpha$ (32.7 kg) can be separated for calculation purposes by arrows from $B$ to $C$ and $D$. The liquid of eutectic composition at $C$ first transforms to eutectic solid ($\alpha + \beta$) as it passes through the eutectic temperature and then cools to 400°C. The amounts of eutectic $\alpha$ and $\beta$, 4.8 kg and 12.5 kg, respectively, at 400°C are shown by arrows from $F$ to the saturation lines for $\alpha$ and $\beta$ at the edges of the two-phase ($\alpha + \beta$) region. Likewise, the noneutectic (free) $\alpha$ at point $D$ cools and precipitates out grain boundary $\beta$. At point $G$, 400°C, the noneutectic can be split into 30.7 kg of noneutectic (free) $\alpha$ and 2.0 kg of noneutectic (grain boundary) $\beta$. Note that after the separation at point $B$ into eutectic and noneutectic components, the average composition value of 30% Ag–70% Cu is not used again in any calculation because the compositions of the eutectic and noneutectic components control the relative amounts of $\alpha$ and $\beta$ present.

## EUTECTOID TRANSFORMATION (IRON–CARBON)

A *eutectoid transformation* is the cooling transformation at a constant temperature from a single-phase *solid* to two new solid phases that are mixed together in a characteristic manner. Thus, all that was said in the previous section about a eutectic transformation from liquid to solid applies to a eutectoid transformation, except that the initial phase in the eutectoid transformation is a solid, and thus this is a *solid-state transformation*. The eutec-

**Figure 8-5** Procedure for calculating amounts of phases present during equilibrium cooling. E, eutectic; NE, noneutectic. [Adapted from "Phase Diagrams," *Metals Handbook: Metallography, Structures and Phase Diagrams,* 8th ed., vol. 8, T. Lyman, ed. (Metals Park, Ohio: American Society for Metals, 1973), p. 253.]

toid transformation occurs for a unique composition in a given system and at a unique equilibrium temperature. The methods for calculating amounts and compositions of phases in the eutectoid are the same as for the eutectic.

The most important of all eutectoid systems is the iron–carbon system, shown in Fig. 8-6. Each forbidden two-phase region is labeled with the two phases present. It is still necessary to observe experimentally the configuration in which these phases are mixed together, which has earlier been described as the *microstructure.* The upper left region of the diagram, at 1495°C, contains a *peritectic transformation* (solid + liquid transforms on cooling to a different solid) which will not be considered further in this discussion because, in most cases, transformations on further cooling obliterate any influence of the peritectic on mechanical properties at room temperature. The Fe–C system also has a eutectic transformation for 4.30% C at 1148°C, which is of importance in cast irons (Chap. 15).

Pure iron below 912°C is a bcc structure called *ferrite ($\alpha$)*, which can hold only very small amounts of carbon in interstitial solid solution; in calculations the carbon content at 20°C is taken as zero. At temperatures above 912°C the equilibrium structure of interstitial carbon in iron is an fcc structure called *austenite ($\gamma$)*, which can have up to 2.11% carbon in solid solution (at 1148°C). Iron and carbon can also form a compound, $Fe_3C$, when enough carbon is present. $Fe_3C$ has acquired several names: *iron carbide, carbide,* and

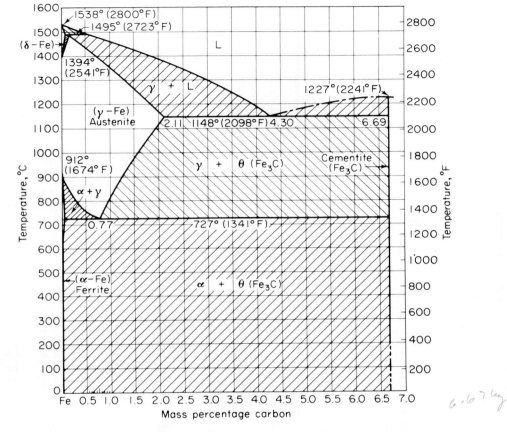

**Figure 8-6**   Iron–carbon equilibrium phase diagram in the absence of free graphite.
[From "Phase Diagrams," *Metals Handbook: Metallography, Structures and Phase Diagrams*, 8th ed., vol. 8, T. Lyman, ed. (Metals Park, Ohio: American Society for Metals, 1973), p. 275.]

*cementite* are the most common. It is usually assigned the symbol $\theta$, the customary designation for intermetallic compounds in phase diagrams. $Fe_3C$ is orthorhombic in crystal structure and is composed of 6.69% C by mass.

**Example 8-6**

Show that $Fe_3C$ has 6.69% C by mass.

**Solution**   The atomic ratio in $Fe_3C$ must be 3 Fe:1 C. Using atomic masses of each element gives

$$\text{Mass fraction C} = \frac{1 \times 12.011}{(1 \times 12.011) + (3 \times 55.85)} = 0.06689 = 6.69\%$$

This is the desired result.

**Figure 8-7**   Optical micrograph of pearlite.

The *eutectoid composition* of iron and carbon is 0.77% C. When an iron–carbon solution of this composition is cooled slowly, at 727°C the austenite transforms at constant temperature to ferrite and iron carbide, which grow simultaneously during the transformation in the form of a layered structure called *pearlite* (because it looks, under a microscope, somewhat like mother-of-pearl). Pearlite is composed of alternating layers of ferrite and $Fe_3C$. A photomicrograph of pearlite is shown in Fig. 8-7. Details of the structure can be seen better in the electron micrograph in Fig. 8-8, which shows the form of the etched surface of pearlite. The width of these layers can be controlled by

**Figure 8-8**   Electron micrograph of replica of polished and etched surface of annealed carbon steel with less than 0.77% carbon. Upper right portion is pearlite; note the irregularities in the formation of the layers of carbide and ferrite. The bottom and upper left regions are ferrite.

altering the cooling rate; faster cooling, such as by cooling a small specimen of steel in air or even oil quenching, produces finer spacing of the layers of ferrite and iron carbide in pearlite. This narrowing of the layers with faster cooling occurs because the transformation in this *quasi-equilibrium* transformation takes place at a temperature significantly lower than 727°C, and at lower temperatures the diffusion required to effect the transformation is more limited, so the transformation begins at more locations throughout the austenite and thus the pearlite is finer. Thus, it is not a true equilibrium process, but because the products of the transformation are the same, it is called a quasi-equilibrium process.

**Example 8-7**

Calculate the mass fraction of iron carbide in pearlite at room temperature.

**Solution**   From Fig. 8-6 the percentage of carbon in ferrite at room temperature is nil, so, moving across the $(\alpha + \theta)$ region from left to right, we find that the mass fraction of $\theta$ is

$$\frac{0.77 - 0}{6.69 - 0} = 0.115 = 11.5\%$$

For *noneutectoid* compositions, the procedure used to calculate amounts of different microstructures is exactly the same as for noneutectic systems. Figure 8-9 gives more detail in the low-carbon region of the Fe–C diagram than can be seen in Fig. 8-6. From Fig. 8-9, slow cooling of 15 kg of, for example, 0.40% C gives the following structures:

At 820°C:   100% austenite, or 15 kg of austenite.

At 728°C:   Fraction $\gamma = \dfrac{0.40 - 0.02}{0.77 - 0.02} = 0.507 = 50.7\%$

$0.507 \times 15$ kg $= 7.60$ kg $\gamma$ and thus 7.40 kg $\alpha$

At 726°C:   Separate into two isolated systems:

(a) $\gamma \Rightarrow$ pearlite (abbreviated P), giving 7.60 kg P($\alpha + \theta$).
(b) $\alpha \Rightarrow \alpha$ (no change), 7.40 kg $\alpha$. This is called *proeutectoid* or *free ferrite* to distinguish it from the ferrite in the pearlite.

At 20°C:

(a) Mass of pearlite is unchanged, 7.60 kg, but composition of ferrite changes from 0.02% C to virtually zero C, while the fraction of $Fe_3C$ increases very slightly.
(b) $\alpha \Rightarrow \alpha + \theta$. The final ferrite has virtually zero C. The very small amount of $Fe_3C$ forms at the grain boundaries and in most cases can be neglected. Mass of grain-boundary $Fe_3C$ is

$$\frac{0.02 - 0}{6.69 - 0} \times 7.40\,\text{kg} = 0.022\,\text{kg}$$

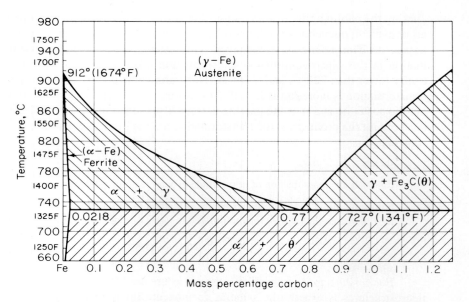

**Figure 8-9**   Iron–carbon equilibrium phase diagram, to 1.2% C. [From "Phase Diagrams," *Metals Handbook: Metallography, Structures and Phase Diagrams,* 8th ed., vol. 8, T. Lyman, ed. (Metals Park, Ohio: American Society for Metals, 1973), p. 275.]

In calculating rough mass fractions for the 0.40% C–Fe, little error is introduced by neglecting the 0.02% C composition of ferrite at 726°C and roughly approximating the eutectoid composition as ~0.8% C. Thus, the fractions of pearlite and free ferrite in this problem would be approximately 0.5 and would remain thus throughout the range from 726°C to 20°C. This simplification can be used for approximate calculation of mass fractions for carbon contents above about 0.15% without significant error. However, for calculating compositions, the values shown in the equilibrium diagram must be used without alteration.

### Example 8-8

Calculate the fraction of pearlite at 20°C in 0.60% C–Fe that has been slowly cooled *(annealed)* in a furnace from austenite to room temperature.

**Solution**   From Fig. 8-9 and the foregoing simplification, the fraction of austenite at 728°C is $(0.60 - 0)/(0.80 - 0) = 0.75$. All austenite at 728°C goes to coarse pearlite, so the fraction of pearlite at 726°C is 0.75. This fraction is unchanged by further cooling, so the fraction of pearlite at 20°C is 0.75.

Thus, equilibrium-cooled steels have pearlite fractions approximately as follows: 0.20% C steel, 25% pearlite; 0.40% C steel, 50% pearlite; 0.60% steel, 75% pearlite; eutectoid steel, 100% pearlite.

Compositions with a carbon content greater than 0.77% produce pearlite plus iron carbide. For example, consider 15 kg of 1.00% C cooled from austenite:

At 860°C:    100% austenite, 15 kg.

At 728°C:    Fraction $\theta = (1.00 - 0.77)/(6.69 - 0.77) = 0.0389 = 3.9\%$
The other 96.1% is $\gamma$.
Mass $\theta = 0.0389 \times 15 = 0.58$ kg. This is found at the austenite grain boundaries.

At 726°C:    Separate into two isolated systems.

**(a)** $\gamma \Rightarrow$ pearlite $(\alpha + \theta)$, $15 - 0.58 = 14.4$ kg.
**(b)** $\theta \Rightarrow \theta$ (no change), which now is found at the boundaries of the prior austenite and thus comprises a three-dimensional network within the pearlite.

At 20°C:    No further significant change occurs below 726°C under equilibrium cooling conditions.

## SUMMARY

Equilibrium-phase diagrams reveal the composition and amount of each phase present under equilibrium conditions. These are the phases toward which a nonequilibrium system will tend to move if given sufficient time for the necessary diffusion of atoms. The form and distribution of the phases (the *microstructure*—see Chap. 9) are not given by the equilibrium-phase diagram and therefore must be found by experiment and observation. In many binary metal systems the equilibrium phases can also be produced by different nonequilibrium processes so as to have very different shapes and distributions. The consequent differences in strength and hardness are discussed in Chap. 9.

## REFERENCES

1. American Society for Metals, *Metals Handbook,* 9th ed. vol. 8, *Metallography, Structures, and Phase Diagrams.* Metals Park, Ohio: American Society for Metals, 1973. This volume contains an extensive and readily available collection of binary and ternary phase diagrams.
2. Birchenall, C. Ernest, *Physical Metallurgy.* New York: McGraw-Hill, 1959. Pages 87–92 give an introduction to ternary systems.
3. Brophy, Jere H., Robert M. Rose, and John Wulff, *Thermodynamics of Structure.* New York: John Wiley, 1964. Chapter 3 discusses the conditions for the formation of surfaces between phases, that is, phase boundaries. Chapter 6 describes nucleation and growth of second phases.
4. Reed-Hill, Robert E., *Physical Metallurgy Principles.* Princeton, N.J.: Van Nostrand, 1964. Chapter 13 thoroughly discusses all important types of phase diagrams.
5. Rhines, F. N., *Phase Diagrams in Metallurgy.* New York: McGraw-Hill, 1956. A test completely devoted to the subject of phase diagrams.

## PROBLEMS

**8-1.** For the binary system in Fig. 8-10, calculate the mass of $\beta$ in the eutectic at 50°C if 3 kg of element A is mixed with 7 kg of element B and slowly cooled from 400°C.

**8-2.** Three 100-kg bars of element A and one 100-kg bar of element B are put in a crucible, heated to 400°C, and then slowly cooled to 100°C. Using the equilibrium diagram in Fig. 8-10, find, at 100°C:

(a) The mass of $\beta$ solid solution in the eutectic.

(b) The mass of element B not in the eutectic.

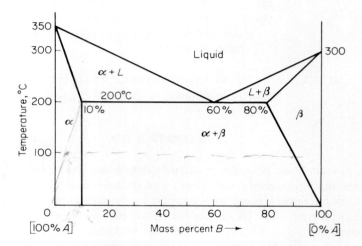

**Figure 8-10**   Equilibrium phase diagram for fictitious elements A and B (for use in problems).

**8-3.** A mixture of 3 kg of silver and 7 kg of copper is heated until completely melted and then cooled slowly to 900°C. Find:

(a) The mass of liquid present at 900°C.

(b) The mass of eutectic $\beta$ (silver-rich) after cooling to 700°C.

(c) The total mass of $\beta$ formed from the noneutectic $\alpha$ after it is cooled rapidly to 400°C and held until phase equilibrium again occurs.

**8-4.** Into a crucible are placed 20 kg of tin and 60 kg of lead (see Fig. 8-11). This mixture is heated until completely melted, stirred well, and cooled very slowly to 100°C. Find the following, for 100°C:

(a) The total mass of tin in the eutectic present in the crucible.

(b) The composition of noneutectic $\alpha$.

(c) The mass of noneutectic $\alpha$.

**8-5.** In the lead–tin diagram (Fig. 8-11) what is the percentage of $\beta$ in the eutectic?

**8-6.** (a) If 40 kg of tin and 60 kg of lead (see Fig. 8-11) are placed in a crucible, heated to 350°C, and slowly cooled to 180°C, what are the amounts of the microstructure(s) present?

(b) What is the composition of each phase present?

(c) If the mixture is then cooled to 50°C, what are the amounts of each phase present?

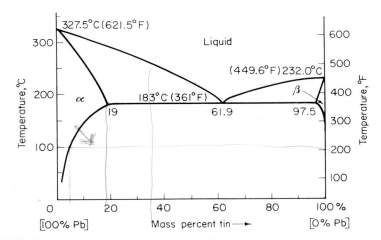

**Figure 8-11**    Lead–tin equilibrium phase diagram. [From "Phase Diagrams," *Metals Handbook: Metallography, Structures and Phase Diagrams,* 8th ed., vol. 8, T. Lyman, ed. (Metals Park, Ohio: American Society for Metals, 1973), p. 330.]

**8-7.** Compute the amount of each microstructure and the fraction of each phase present in a 0.3% carbon steel slowly cooled from the melt to each of the following temperatures:
   **(a)** 930°C.
   **(b)** 760°C.
   **(c)** 700°C.

**8-8.** A 1.00% C steel that has been cooled slowly from 1100°C and then held at 740°C for 0.5 h is air-cooled to room temperature.
   **(a)** Name the phases present.
   **(b)** Describe the microstructure.
   **(c)** Calculate the fraction of noneutectoid carbide present.

**8-9.** A mixture of 20 kg of element A and 20 kg of element B (see Fig. 8-10) is melted in a crucible and cooled slowly to 0°C.
   **(a)** Calculate the change in mass of eutectic $\alpha$ between 199°C and 0°C. Is this an increase or a decrease?
   **(b)** Describe the physical process by which such a change in mass can occur.

**8-10.** A mixture of 6.49 kg of silver and 11.63 kg of copper is cooled slowly from 1100°C to 400°C. Calculate the mass of noneutectic $\alpha$ present at 400°C.

**8-11.** A mixture of 2.45 kg of lead and 1.43 kg of tin is heated to 350°C and slowly cooled to 100°C.
   **(a)** Calculate the total mass of $\alpha$ at 100°C.
   **(b)** Calculate the mass of eutectic $\beta$ at 100°C.
   **(c)** Calculate the composition of noneutectic $\beta$ at 100°C.

**8-12.** A small, rectangular bar of 0.47% C steel has the dimensions 14.2 mm × 7.48 mm × 57.3 mm. This bar is heated to 850°C for 2 ks and then oil-quenched to 20°C. Calculate the total mass of eutectoid iron carbide in this bar at 20°C.

# 9

# Multiple Phases

The effectiveness of second phases in interfering with dislocation motion and in forcing the generation of new dislocations has a marked influence on the strength of a solid. The size, shape, and distribution of a second phase, as well as the form in which the first phase occurs, affect the ease of dislocation motion. This chapter examines the behavior of several metal alloys and provides examples of the control of strength through the control of second phases.

## PHASE BOUNDARIES

In Chap. 8 it was shown how two or more phases can be formed and why they coexist in equilibrium at moderate temperatures. A *microstructure* is a phase or combination of phases of specified geometry unique enough to warrant its being assigned a special name. Any single-phase material may thus be called a microstructure, in which case the microstructure is also a phase. Mixtures of two or more phases that are, say, layered, or with one phase spheroidal, or with one phase disk-shaped or rod-shaped, may be properly designated microstructures. If two phases are intimately mixed on a very fine scale, for example, the eutectic microstructure of Ag–Cu in Fig. 8-3(a), then the *phase-boundary area* within the *microconstituent* (i.e., the eutectic mixture) will be extensive. In contrast, a microstructure consisting of two phases, each of normal grain size, will have much less phase-boundary area.

Phase boundaries are very effective barriers to dislocation motion for at least two reasons: (1) The atoms are displaced from their equilibrium positions by the forces acting on them by unlike atoms in a different crystal at some angle to the first, in much the same way as was described earlier for grain boundaries; and (2) if a dislocation does pass through the phase boundary, it is forced to move in a crystal of different orientation and Burgers vector, and possibly with a different crystal structure. Because only the orientation difference occurs at a grain boundary in a single-phase solid, with no change in magnitude of Burgers vector or crystal structure, grain boundaries are in general less effective barriers to dislocation motion than are phase boundaries. In addition to this argument, the second phases that form in many alloy systems are chemical compounds with structures in which dislocation motion is much more difficult than in either pure constituent element. This further enhances the effectiveness of many phase boundaries as barriers to dislocation motion. This chapter discusses the marked effect of phase boundaries on strength.

## MEAN FREE PATH

Chapter 5 described the effect of dislocation glide distance on strength. As the mean glide distance increases, the number of moving dislocations required for a given strain decreases. Thus, the last dislocations move at a lower stress than would be required for the same strain but with a shorter glide distance because more dislocations would then be needed to produce the same strain. This increase in glide distance is generally associated with a reduction in strength. The glide distance for dislocation motion in the softer phase of a multiple-phase alloy varies considerably, depending on the location and orientation of each dislocation in the microstructure. When a series of random measurements of glide distance between harder phases is taken, the mean value of these measurements is termed the *mean free path.* Figure 9-1 shows the dependence of the yield strength of pearlite on the mean free path in the ferrite layers. When manganese is introduced into an iron–carbon alloy, it acts as if it is in some way equivalent to carbon, and so the fraction of carbon in the eutectoid composition is then lower than 0.77%. The data for Fig. 9-1 are for pearlite from 0.77%C down to 0.56%C, and for manganese contents up to 3.5%C (which is much higher than is normally present in steels). The range of yield strengths shown by the shaded area in Fig. 9-1 suggests that factors in addition to mean free path are influencing strength. Other strengthening mechanisms must be operating in addition to phase-boundary effects. This conclusion is generally valid for the more complex materials because most of the strengthening mechanisms discussed in previous chapters can act simultaneously with phase boundaries to increase strength. In order to determine the strengthening effect of any one variable, the other variables should be closely controlled. Because this frequently is not possible, the individual contribution of each strengthening mechanism cannot be established in many materials.

The data in Fig. 9-1 are not in convenient form for estimating the strength of pearlite that will result from a given heat treatment. A rule-of-thumb procedure, based on measurements, can be used for plain-carbon steels in the form of bars or plates on the order of 10 mm in diameter or thickness, as follows:

1. If furnace-cooled from austenite *(annealed),* the pearlite has relatively wide spacing and is termed *coarse pearlite* ($P_c$) with a Brinell hardness number (HB) of about 240. It has been observed that for ductile steels,

$$S_u \, [\text{MPa}] \approx 3.45 \, \text{HB} \qquad (9\text{-}1)$$

Thus, $S_u$ for coarse pearlite is about 830 MPa. Furnace-cooled eutectoid steel often exhibits some *spheroidization,* the formation of spheroidal carbide particles, as can

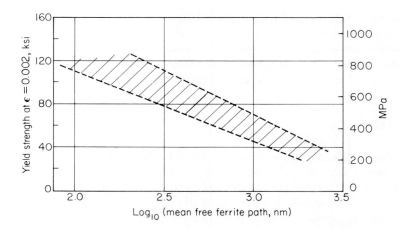

**Figure 9-1**   Variation of yield strength with mean free ferrite path in pearlite for a range of carbon and manganese contents. [From R. W. Guard, "Mechanisms of Fine-Particle Strengthening," *Strengthening Mechanisms in Solids* (Metals Park, Ohio: American Society for Metals, 1962), p. 261.]

be seen in Figs. 8-7 and 8-8. Therefore, in such cases its hardness represents not the hardness of coarse pearlite but that of a mixture of coarse pearlite and spheroidite. Spheroidite has a lower hardness (about 150 HB) than coarse pearlite for reasons given in the next section.

2. If air-cooled from austenite *(normalized),* the pearlite has a narrower, intermediate spacing and is termed *medium pearlite* ($P_m$), with a hardness of about 280 HB and $S_u \approx 970$ MPa. If a specimen of air-cooled pearlite is reheated to just below the eutectoid temperature, say 700°C, and held for a long time, say 50 ks, the layers of iron carbide will fully agglomerate into individual spheroid-shaped particles called *spheroidite.* This microstructure is similar to that just discussed under furnace-cooling conditions, except that in the present case no layered pearlite remains. Spheroidized pearlite ($P_s$) has a hardness of about 150 HB and $S_u \approx 520$ MPa.

3. If oil-quenched from austenite, the pearlite has very narrow spacing and is termed *fine pearlite* ($P_f$), with a hardness of about 380 HB and $S_u \approx 1.3$ GPa.

4. For contrast, the hardness of free ferrite alone is about 80 HB.

These approximate values of HB are listed in Table 9-1.

For faster cooling rates, and in fact even for oil quenching of some high-carbon plain-carbon steels, the formation of martensite (see Chap. 10) instead of pearlite markedly alters the final hardness. The calculations of (1) through (4) are only approximate and should be used with caution, not for design purposes, because variations in austenite grain size, cooling rate, specimen size, and composition can alter these results. For a design application, the specific properties of the steel to be used must be known.

**TABLE 9-1**   HARDNESS OF FERRITE, SPHEROIDITE, AND PEARLITE[a]

| Constituent | Abbreviation | HB, 3000 kg$_f$ | Rockwell Equivalent |
| --- | --- | --- | --- |
| Ferrite | $\alpha$ | 80 (500 kg$_f$) | 47 HRB |
| Spheroidite | P$_s$ | 150 | 80 HRB |
| Coarse pearlite | P$_c$ | 240 | 22 HRC, 100 HRB |
| Medium pearlite | P$_m$ | 280 | 30 HRC |
| Fine pearlite | P$_f$ | 380 | 41 HRC |

[a]Approximate hardness of constituents of plain-carbon steel, to be used only for rough estimate, not for design.

The data in Fig. 9-1 can be roughly approximated by the Petch equation [Eq. (6-4)], to give the yield strength of pearlite as a function of mean free ferrite path $\bar{p}$ ($\mu$m):

$$S_y \approx 200 \ \text{MPa} \ + \ (240 \ \text{MPa})\left[\bar{p} \ (\mu\text{m})\right]^{-1/2} \tag{9-2}$$

The following example suggests that the effect of the mean free ferrite path on yield strength can be approximated by Eq. (9-2).

**Example 9-1**

If the yield strength of coarse pearlite is approximately 70% of its ultimate strength, estimate the spacing of coarse pearlite.

**Solution**   From Eq. (9-1), $S_u \approx 3.45 \times 240 = 830$ MPa and $S_y = 581$ MPa, and from Eq. (9-2), $\bar{p} = 1/[(581 - 200)/240]^2 = 0.397$ $\mu$m. To check this, measure the pearlite spacing of the electron micrograph shown in Fig. 8-8, which is approximately coarse pearlite.

The inverse relationship between strength and dislocation glide distance holds for many other materials. For example, Fig. 9-2 shows the variation in strength with the reciprocal square root of the spacing of hard particles in aluminum powder that has been sintered (pressed together at high temperature) to make a single piece of metal. Mechanisms for strengthening by fine particles are discussed later in this chapter.

**Figure 9-2**   Tensile strength vs. (particle size)$^{-1/2}$ for a sintered aluminum powder. [From R. W. Guard, "Mechanisms of Fine-Particle Strengthening," *Strengthening Mechanisms in Solids* (Metals Park, Ohio: American Society for Metals, 1962), p. ?.

**Example 9-2**

After hot rolling at 900°C, plain-carbon structural-steel plates are stacked and allowed to cool. Describe two changes in this procedure that would increase their room-temperature strength.

**Solution**   Alterations in grain size or spacing of pearlite can change the strength. The grain size can be reduced by lowering the hot-rolling temperature. Hot rolling can be regarded in a simplified way as cold rolling and instantaneous recrystallization (although it actually is not), so lowering the temperature at which recrystallization occurs reduces the diffusion rate, decreases the grain size, and increases strength.

Strength increases caused by finer grain size are small compared with strength increases that result from changes such as in the pearlite spacing. Thus, the cooling rate can be increased to produce finer pearlite through cooling in still air without stacking or cooling with a cold-air blast.

## SHAPE OF HARDER PHASE

For a given amount—by mass or volume fraction—of second phase, the shape of the second phase can exert a substantial influence on the strength of the alloy. The two diagrams in Fig. 9-3 show why this can be true. In Fig. 9-3(a) shading indicates those regions where dislocation motion in a particular direction will be impeded by second-phase particles. Figure 9-3(b) shows how flattening these same particles increases the interference with dislocation motion. In this example, it is true that dislocation motion in the vertical direction would be made easier. However, in every other direction the particles interfere more; thus, the mean free path would be smaller in Fig. 9-3(b) than in Fig. 9-3(a), and hence the strength would be higher.

The limiting case of the change shown in Fig. 9-3 occurs when each particle is flattened to a layer extending fully across the grain. Pearlite is such a material, although it is not made by flattening spheroidal particles. If a layered structure is changed to a spheroidal structure, the strength should decrease. This can be accomplished by heating pearlite to produce spheroidite, as discussed earlier. This microstructure of spheroidal carbide in ferrite is also called *spheroidized cementite,* as discussed earlier, and it has a hardness of about 150 HB compared with 240 HB for coarse pearlite, for example. The microstructure of spheroidite, shown in Fig. 9-4, should be compared with the microstructure of pearlite, shown in Fig. 8-8.

Control of the shape of the second phase holds substantial promise for higher strengths in some common alloys. By cooling a eutectic alloy so as to force solidification

(a)                                                           (b)

**Figure 9-3**   Schematic diagrams showing the effect of shape of second phase: (a) spheroidal, and (b) flattened.

**Figure 9-4**    Electron micrograph of replica of polished and etched surface of spheroidite.

to progress along a nearly plane front, the characteristic shape of the eutectic microstructure can be changed. Figure 8-3(b) shows, for example, the fine lamellar shape of directionally solidified eutectic silver–copper. Microstructures such as these provide the possibility of achieving higher strengths without the difficulties of artificially dispersing long, high-strength whiskers or hard, fine particles in a metal to increase its strength (see the discussion on insoluble phases later in this chapter). The directional eutectic has the further advantage that good surface bonding exists between the two different phases. Figure 9-5 shows a directionally solidified eutectic used to exhibit flow patterns in machining.

## AMOUNT OF HARDER PHASE

If more of a harder phase is present, whether as layers, spheroids, or in some other shape, the mean free path in the softer phase will decrease and the strength will increase. Figure 9-6 shows the increase in yield strength with increasing volume fraction of fine iron particles of a range of sizes in a model experiment using mercury at a very low temperature. Such an increase is to be expected, as is the appearance of an upper limit to the strength. The sudden change in slope of the graph in Fig. 9-6 probably occurs because the stress has become high enough for the dislocations to move through the particles by shearing them or to move around the particles by other mechanisms. (Some of the possible mechanisms that achieve this are described in later sections of this chapter.)

(a)

(b)

**Figure 9-5**   (a) Microstructure of undeformed, directionally solidified silver–copper eutectic alloy. (b) Microstructure of cut chip from machining test showing discontinuous shearing. The "built-in grid" of the eutectic is deformed into the chevron pattern associated with a double shearing process. [From W. B. H. Cooke and W. B. Rice, ASME, *J. Engr. Industry*, 95 (1973), p. 844. Photographs courtesy W. B. Rice.]

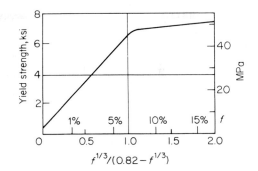

**Figure 9-6** Yield strength of frozen mercury vs. volume fraction *f* of iron particles. [From W. H. Meiklejohn and R. E. Skoda, "Dispersion Hardening," *Acta Met.* (New York: Pergamon Press, 1960), 8, p. 773. Reprinted with permission of the publisher.]

The strength of some two-phase alloys is found to be approximately proportional to the fraction of the second (harder) phase or microstructure present. Pearlite, the eutectoid mixture of alternate layers of ferrite and iron carbide in steel, behaves as if it were a separate, harder constituent when it is mixed with free ferrite. Experiments demonstrate that a reasonable estimate of hardness and strength of a *hypoeutectoid* (less than eutectoid carbon) pearlitic steel can be made by adding together the products of the mass fraction and hardness of each constituent. (The word *constituent* here refers to a component of the microstructure that is treated as a separate entity. Pearlite is such a component, but because it contains two phases, it cannot be called a phase. Sometimes the term "microconstituent" is used synonymously with constituent.) The general form of the method for calculating hardness is

$$HB \approx \text{(mass fraction pearlite)}(HB_P) + \text{(mass fraction ferrite)}(80) \qquad (9\text{-}3)$$

From Table 9-1, $HB_P$ is 380 for fine pearlite, 280 for medium pearlite, and 240 for coarse pearlite. If spheroidite of eutectoid composition is present instead of pearlite, its HB is about 150.

**Example 9-3**

Calculate the hardness and strength of a furnace-cooled 0.20% C steel.

**Solution**    The mass fraction of austenite at 728°C is approximately 0.20/0.80 = 0.25, which is then the mass fraction of pearlite at 20°C. The fraction of ferrite is 1 − 0.25 = 0.75. Furnace cooling produces coarse pearlite (HB = 240), with no effect on ferrite properties (HB = 80). Thus, the hardness of this steel is HB = (0.25)(240) + (0.75)(80) = 120 HB, and $S_u$ = 3.45 (HB) = 414 MPa. These values are very close to those measured for this steel.

## FINE PARTICLES

For a given fraction of a harder second phase, the greatest strengthening effect might be presumed to occur when the particles are as small as possible and thus have the greatest possible phase-boundary surface area (see Fig. 9-3). The limiting case for the finest possible particles is then isolated atoms. But this would be solid-solution strengthening, which was shown in Chap. 7 to have only a moderate effect on strength. Observation and experiment have demonstrated instead that the greatest strengthening effect occurs for very small second-phase particles of maximum dimension two or three orders of magnitude larger than atomic dimensions. When a dislocation intersects an array of such fine particles, it does not behave in the same way as a dislocation that encounters the large surface of a phase boundary, where the dislocation is blocked along a large part of its length. Because *fine-particle strengthening* is one of the most important and effective strengthening mechanisms, an introduction to the theory, with examples of commercial alloys, is given in the paragraphs that follow.

**Example 9-4**

Which is the most effective shape for fine particles, each having the same mass, for interference with dislocation motion: rods, thin plates, or spheres?

**Solution**   The answer depends on which dimensions are restricted. Rods are more effective if their diameter is the same as the thickness of the plates with which they are compared because they are much longer and thus have the higher surface area per unit volume. If the length of the rods is held the same as the diameter of the (circular) plates, then the diameter of the cylindrical rods will be relatively large and the plates will be more effective. In either case, for a given particular volume, spheres provide the minimum surface area, so they are least effective in interfering with dislocation motion. But because spheroidal particles can exist in stable form down to very small sizes, smaller than rods or spheres, fine spheroidal particles turn out to be among the most effective strengthening mechanisms.

The portion of a dislocation that approaches an array of fine particles is held up by the presence of these second-phase imperfections in the crystal structure in the same manner as described earlier in this chapter for phase boundaries in general. Portions of the dislocation between the fine particles are at first not held up, so the dislocation starts to bend around each particle. This situation is shown in Fig. 9-7(a), where the shaded circles represent fine particles and the shear stress $\tau$ is such as to cause the dislocation to move to the right in the figure. If further strain is to occur, the stress must be raised. (Recall the argument in Chap. 5 that the number of dislocations is very large, so the behavior described here occurs simultaneously in a small fraction, which is still a fairly large number, of this very large number of dislocations.) As the dislocation approaches this array of particles, one or more of the following can occur: (1) The dislocation forms loops around the particles and continues; (2) the dislocation moves over or under the particles by cross slip and continues; (3) the particles are sheared by the dislocation; or (4) the dislocation is blocked by the particles, while other dislocations elsewhere in the crystal move.

The first possibility was described by Orowan, in the sequence shown in Fig. 9-7. The shear stress $\tau$ required to move a dislocation through the space between particles of average spacing $f$ is estimated from dislocation elasticity calculations to be

$$\tau \approx \frac{Gb}{f} \qquad\qquad (9\text{-}4)$$

where $G$ is the shear modulus of elasticity and $b$ is the Burgers vector. As the stress increases, the dislocation can bow out between the particles [Fig. 9-7(b)]. Figure 9-8 is a thin-foil transmission electron micrograph showing bowed-out dislocations pinned by fine

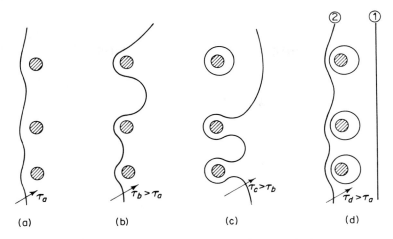

**Figure 9-7**   Orowan's model for fine-particle strengthening.

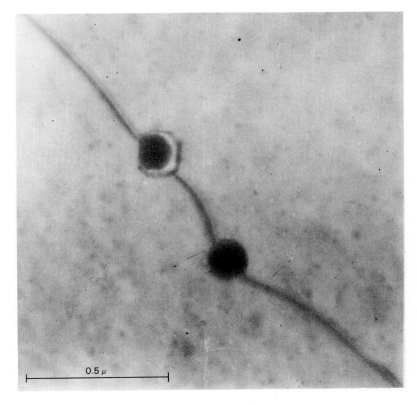

**Figure 9-8**   Dislocations in MgO bowed out after being pinned by fine particles (probably of magnesium nitride). (Photo courtesy of Gareth Thomas.)

particles in a magnesium oxide matrix. When the shear stress is sufficient to produce a radius of curvature small enough to allow the dislocation segment to pass between two particles [Eq. (9-4)], the dislocation wraps completely around the particle until it touches itself. At this instant the dislocation line snaps out into a more nearly straight line and leaves a dislocation loop around each particle. This process is shown in Fig. 9-7(c) and (d). The loops that form around the particles are called *Orowan loops*. With further increase in stress, the dislocation can either form secondary loops around each of the remaining particles or a single loop encircling two or more particles; the first case is shown here. [By a similar mechanism of bowing of a dislocation that is pinned at two points but otherwise lies out of the plane, a *Frank–Read source* can generate a large number of dislocations when the stress reaches the value of Eq. (9-4). The discussion of cross slip that follows, and Fig. 9-10, describe the essential geometry of the Frank–Read source.]

Figure 9-7(d) shows the original dislocation 1 moving off to the right; the stress required to move it, once past the row of particles shown, is lower than the stress when particles are present, but in the real case any dislocation would continuously encounter fine particles at many points along its length and would not experience a drop in stress. This model thus shows how the presence of fine particles leads to an increase in stress for dislocation motion.

**Example 9-5**

Estimate the particle spacing in an aluminum alloy having a yield shear strength of 400 MPa.

**Solution**    From Eq. (9-4), because aluminum is the matrix material, $G = 24$ GPa and $b = 286$ pm, and so $f = Gb/\tau$ = (24 GPa)(286 pm)/400 MPa = 17 nm. This agrees quite well with the particle spacing observed in aluminum alloys.

The model in Fig. 9-7 also exhibits another property observed in real alloys having fine particles: rapid strain strengthening. The Orowan loops that are left around the particles repel subsequent dislocations of like sign, and the effect is an apparent enlargement of each particle with a consequent reduction in the effective spacing between particles. The geometry faced by the new approaching dislocation, shown as dislocation 2 in Fig. 9-7(d), thus requires a higher stress for dislocation motion around the particles already circled by loops. The rate of increase in stress with strain should therefore be greater than in a similar single-phase solid. Figure 9-9 shows this effect, where the plastic stress–strain slope of the alloy with even relatively large particles is much greater than for the supersaturated single-phase alloy. (The initially higher strength of the single-phase alloy, represented by curve (a), is due to the effect of strain fields produced by supersaturation that impede dislocation motion; a single-phase alloy of Al–2 atomic percent Cu will always be supersaturated at or below room temperature. The alloy that contains particles, represented by curve (b), is not supersaturated.)

The second alternative mechanism listed earlier in this section occurs when the dislocation moves on a different slip plane and thereby bypasses the fine particles by going over or under them. This possibility is limited to a dislocation line that is of pure screw form because at moderate temperatures an edge dislocation cannot move except on its slip plane. Motion of a screw dislocation onto a different plane, called *cross slip* (see Fig. 5-

**Figure 9-9** Strain hardening of single crystals of Al–2 atomic percent Cu at −196°C: (a) supersaturated solid solution produced by quenching; (b) with fine particles of CuAl$_2$ about 2.5 $\mu$m apart. [From G. Greetham and R. W. K. Honeycombe, "The Deformation of Single Crystals of Aluminum–4.5 Per Cent Copper Alloy," *J. Inst. Metals,* 89 (1960–1961), p. 13. Reprinted by permission of the Metals Society.]

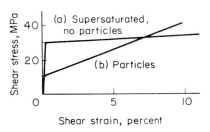

43), requires an increase in stress because the dislocation must be extended and hence has greater strain energy. Thus, cross slip can also contribute to fine-particle strengthening and, to a certain extent, to an increased strain-hardening rate because the longer dislocations will be more effective barriers to other dislocations.

**Example 9-6**

Why cannot a small portion of a screw dislocation cross slip over a particle and then continue on?

**Solution**   When only a small portion of a screw dislocation cross slips, the connecting dislocation lines are at least part edge (see Fig. 9-10). Dislocation climb (see Chap. 6) is then required for motion of these edge components in the direction of continued slip. Climb is so slow as to be virtually impossible at room temperature for most metals. Therefore, further slip can result only from the bowed-out portion in Fig. 9-10(c), and the original dislocation is stopped. This bowed-out segment can form a Frank–Read source, as discussed earlier, but unless the stress is high enough to force the dislocation to bow out to a half-circle, the requirement of Eq. (9-4), the source will not operate.

The third possibility, shearing of the fine particles, apparently occurs if the stress becomes great enough. Shearing of the particles also results in an increase in phase-boundary surface area, a kind of strain hardening should occur when this mechanism is in operation. In Fig. 9-6 the curve becomes nearly flat when the volume fraction of iron particles is large enough to raise the stress to a level where some other mechanism, such as this, apparently comes into play. If this mechanism is the shearing of the particles, it explains the sudden change in slope and the very gradual increase in yield stress with further increase in the volume fraction of iron.

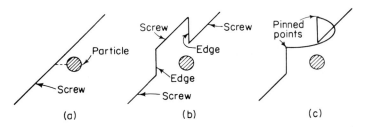

**Figure 9-10** Solution to Example 9-6, showing restricted motion of the cross-slipped portion of a dislocation.

The last possibility, that other dislocations will begin to move, is trivial because the other dislocations will sooner or later be blocked in the same way as the dislocation in Fig. 9-7(a). At that time, a further increase in stress will be required to move the dislocations by (1) looping around the particles, (2) cross-slipping over the particles, or (3) shearing the particles.

Frequently new dislocations are observed to form at fine particles. These dislocations cause much cross slip and entanglement of other dislocations in the vicinity of the particles. This produces substantial strengthening of the same magnitude as for the processes already described but in a more complex manner. Although dislocation configurations in real materials are almost always more complicated than in the simple models described here, the reasons for strengthening remain the same.

## ALUMINUM–COPPER ALLOYS

The classic case of fine-particle strengthening occurs at the aluminum-rich end of the aluminum–copper system. Discovery of this alloy, generally accepted as occurring in 1909, only a few decades after discovery of the Hall process for extracting aluminum from bauxite, drove the growth of the evolution of aircraft only a few years after the 1903 powered flight of the Wright brothers. In fact, it has recently been discovered that the crankcase of the engine of the *Wright Flyer* (the first successful powered aircraft) was made of a precipitation-hardened aluminum–copper alloy similar to that described here; this engine was thus made several years prior to the generally accepted time of discovery of this alloy.

Figure 9-11 shows about half of the aluminum–copper equilibrium diagram. (Note the similarity in general appearance to the full silver–copper system discussed in Chap. 8.

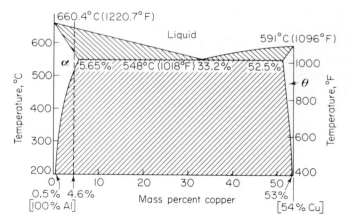

**Figure 9-11**   Aluminum-rich half of the aluminum–copper equilibrium phase diagram. [From Lowell A. Willey, "Phase Diagrams for Binary Alloy Systems," *Metals Handbook: Metallography, Structures and Phase Diagrams,* 8th ed., vol. 8, T. Lyman, ed. (Metals Park, Ohio: American Society for Metals, 1973), p. 259.]

This form is quite common, and in the Al–Cu case an intermetallic compound divides off a portion of the equilibrium diagram as if it were a separate diagram.) Consider a mixture that is 4.6% Cu by mass (shown as a dashed line), cooled slowly from the melt. At about 570°C the microstructure becomes 100% $\alpha$, which is an aluminum-rich substitutional solid solution. With further cooling, at about 510°C some $\theta$ phase begins to form at the grain boundaries. This is an *intermetallic compound* (a chemical combination of two metals) of the approximate atomic ratio of $CuAl_2$. During slow cooling through the two-phase region, $\theta$ continues to form at the grain boundaries. The final product after slow cooling is relatively weak and only moderately ductile because the brittle $\theta$ exists at the grain boundaries. Cracking can occur in the grain boundary $\theta$ after plastic deformation and will thus reduce the ductility of the matrix.

If the single-phase $\alpha$, for example at 540°C, is cooled very rapidly by water quenching to room temperature (the room-temperature equilibrium conditions are similar to the conditions at 200°C in Fig. 9-11), the $\theta$ phase will not have time to form. It has already been shown that formation of a new phase requires time and elevated temperature for diffusion of atoms. In this case, diffusion of Cu atoms into a region and Al atoms out of that region is required until there is about one Cu atom for every two Al atoms in the region where a $CuAl_2$ particle is to form. If time and elevated temperature are not provided, as is the case for water quenching, the atoms of Cu and Al must remain about where they were at 540°C. Thus, immediately after quenching, this alloy will still be a solid solution even though the phase diagram indicates that two phases will be present under equilibrium conditions. The stress–strain curve for this as-quenched alloy, aided by supersaturation effects, was shown in Fig. 9-9 and is repeated in Fig. 9-12 as curve (a). (Note that 2 atomic percent Cu in Al is 4.6 mass percent Cu because the atomic mass of Cu is 2.36 times that of Al.)

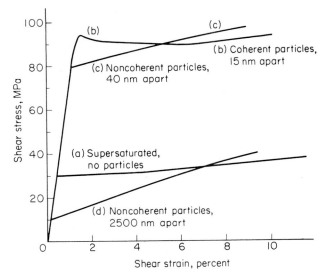

**Figure 9-12** Shear stress–strain behavior of single crystals of Al–2 atomic percent Cu (4.6% by mass) at −196°C: (a) supersaturated solid solution produced by quenching; (b) coherent platelets about 15 nm apart, by aging 170 ks at 130°C; (c) noncoherent particles about 40 nm apart, by aging 99 ks at 190°C; (d) noncoherent particles about 2500 nm (2.5 μm) apart, by aging 170 ks at 250°C. [From G. Greetham and R. W. K. Honeycombe, "The Deformation of Single Crystals of Aluminum–4.5 Per Cent Copper Alloy," *J. Inst. Metals,* 89 (1960–1961), p. 13. Reprinted with permission of the Metals Society.]

If the supersaturated solid solution is held at room temperature or above for a period of time (a process called *aging*), the system will tend to change toward formation of the equilibrium phases. Several distinct stages have been identified, from which three are selected for discussion here. Curve (b) in Fig. 9-12 exhibits the very high strength that results from formation of extremely small particles, or platelets, of pure copper one atom thick, spaced about 15 nm apart. These particles are examples of *Guinier–Preston zones* (GP zones), which form according to the previously described mechanisms. Figure 9-13 shows these platelets (the smaller ones) as detected in an electron micrograph of a thin foil of alloy of similar composition and treatment. Note how very small the spacing between particles is.

One of the reasons for the large strengthening effect of these particles is that they are *coherent* with the lattice, that is, the atoms of each platelet match up on a one-to-one basis with the surrounding atoms of $\alpha$. This is shown schematically in Fig. 9-14(a). Because the lattice constant for pure copper (361 pm) is slightly smaller than for aluminum (404 pm) and the surrounding matrix is nearly pure aluminum, the matrix is elastically strained by

0.5 $\mu$

**Figure 9-13**   Thin-foil electron micrograph of Al–4% Cu solution heat-treated and aged 36 ks at 160°C. The very fine particles in the lower part of the photograph are GP zones, while the larger particles in the upper portion are coherent particles of CuAl$_2$, an intermediate state between the GP zone and the noncoherent CuAl$_2$ particle. This remarkable photograph was taken by *dark field* electron microscopy, in which only the diffracted portion of the electron beam is imaged. [From G. Thomas, "Kikuchi Electron-Diffraction and Dark-Field Techniques in Electron-Microscopy Studies of Phase Transformations," *Trans. AIME,* 233 (1965), p. 1608. ©1965 by AIME and reprinted with its permission. Photo courtesy of Gareth Thomas.]

the particle. This strain field impedes dislocation motion, so the effective size of the particles is increased; that is, a dislocation need not intersect the particle in order to have its motion restricted. The rate of strain hardening of curve (b) in Fig. 9-12 is very low, probably because the very high initial yield stress is high enough to shear the platelets (which are only one atom thick) in a plane not parallel to the platelets.

Aging at higher temperature produces the larger particles that give curve (c) in Fig. 9-12. These particles are *noncoherent,* of approximate composition $CuAl_2$, and spaced closely enough together to retain yield strength nearly as great as for coherent platelets, with a much greater rate of strain hardening. Figure 9-14(b) shows how the strain in the

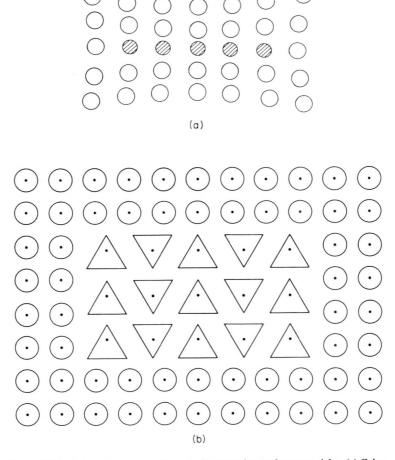

(a)

(b)

**Figure 9-14**   Schematic representation of coherent and noncoherent particles. (a) Coherent atoms of slightly smaller atomic size than the matrix. (b) Noncoherent particle of an intermetallic compound (triangles) of much larger molecular spacing than the spacing of atoms in the matrix.

surrounding matrix can be relatively small in a noncoherent particle because the mismatch between the particle molecules and the matrix atoms is so great. Although the effective size of these particles is not as much greater than their actual size as in the case of coherent particles, they are apparently stronger in shear than coherent platelets, probably because noncoherent particles are more than one atom thick.

**Example 9-7**

For what reason are larger particles less likely to be coherent?

**Solution**   Larger particle size causes the atomic mismatch at the phase boundary to be so large that the forces on the atoms at the surface of the particle become higher than the atomic bonding forces between the particle and matrix atoms. Thus, the atoms at the phase boundary move nearer to their respective equilibrium positions, with the resulting gross mismatch of unlike atoms across the phase boundary shown in Fig. 9-14(b). In small particles the cumulative effect of the size mismatch is not so great [see Fig. 9-14(a)], so coherency is possible.

Curve (d) in Fig. 9-12 shows the effect of aging at a much higher temperature, which causes the fine particles to agglomerate to form fewer and larger noncoherent particles of consequently wider spacing. The marked reduction in yield strength and substantial strain hardening follow consistently the arguments already presented. This reduction in strength with excessive aging is called *overaging*. In commercial practice, overaging can be deliberately induced in order to stabilize an alloy to resist strength reduction at elevated temperatures. The designer must then accommodate to a lower strength at the outset but obtains in exchange a somewhat higher allowable operating temperature.

Another structure involving fine particles that is of great economic importance is tempered martensite. Because of its special nature, this structure will be discussed separately in Chap. 10.

## INSOLUBLE PHASES

The chief disadvantages of the microstructures described so far are that (1) they are limited to phases that precipitate from a supersaturated solid solution, and (2) the fine particles and thin layers agglomerate at intermediate temperatures and are redissolved at higher temperatures. A much wider choice of properties ought to result from intimate mixtures of insoluble and dissimilar solids, called *composites,* to produce a solid of desirable overall properties. Nature provides many examples of structural composites, such as wood and bone, which combine rigidity and lightness of weight. Composites are covered thoroughly in Chap. 17, so the discussion here serves only as an introduction. Two of the many developments of composites are mentioned here for illustration; both depend on phase-boundary strengthening and the very small size of their constituents for their unique properties.

Very fine (less than 1-$\mu$m) thoria powder has been successfully dispersed throughout pure nickel, apparently by mixing the two in powder form and then sintering. The strength of this composite is not particularly high at intermediate temperatures, but because thoria retains its hardness up to temperatures above the melting point of nickel (1453°C), the composite retains substantial strength up to above 90% of the absolute melting temperature of nickel. This is called *dispersion strengthening* or *dispersion hardening,* terms reserved for the mechanical mixing of fine particles in a matrix in contrast to natural processes. Figure 9-15 shows the strength of this composite at high temperatures in contrast to that of two other corrosion-resistant metal alloys. Thoria–nickel has not come into general use, chiefly because it is too brittle for normal engineering applications.

Continuous graphite filaments can be manufactured with a diameter of 7 $\mu$m and a fracture strength on the order of 2 GPa. With a density of about 1800 kg/m$^3$, these filaments have a specific strength (strength/specific gravity) of about 1100 MPa compared with that for high-strength low-alloy steel (250 MPa) and for wrought aluminum alloys (about 200 MPa). However, fine filaments alone are of little use for engineering purposes unless they can be fabricated into a solid of useful shape. Thus, manufactured filamentary composites consist of high-strength filaments surrounded by some castable matrix material. One useful matrix material for graphite is high-strength epoxy (a complex polymer—see Chaps. 12 and 13). Thus, a part of useful shape, such as a thin sheet, can be made with aligned and closely spaced graphite filaments surrounded by epoxy. Figure 9-16 shows the fracture surface of such a composite. The tensile strength of this

**Figure 9-15** High-temperature strength of thoria-nickel (TD nickel) compared with two other high-temperature materials.

material parallel to the direction of the aligned filaments is about 1 GPa, less than that of the filament alone, but its overall density is less, about 1500 kg/m$^3$, so the specific strength is about 650 MPa.

Graphite–epoxy composites owe their high strength in part to the small distance across the diameter of the filament and through the epoxy between adjacent filaments. The magnitude of these dimensions is about 100 times the spacing of fine-particle-strengthened aluminum alloys, but on the other hand, the restriction of slip processes to the dimensions of each filament is a powerful deterrent to slip. We will see in Chap. 14 why the limitation of maximum possible crack size to no larger than the filament diameter is so powerful a strengthening mechanism.

High cost and low toughness will continue to restrict the use of high-performance composites. The most likely uses will probably continue to be for those special applications requiring high specific strength where cost is not a deterrent to use, where different mechanical properties are required in different directions in a part, or where unusual combinations of physical properties (such as low thermal conductivity) and strength are required. It may be expected that application of the fundamental rules of strengthening mechanisms will continue to produce better, more practical composites. (Chapter 17 covers composite materials in detail.)

**Figure 9-16**   Fracture surface of a graphite–epoxy composite, as seen in the scanning electron microscope. In this specimen the plane of fracture varies from filament to filament. Some of the filaments have pulled out of the surrounding epoxy matrix in the process of fracture, and cracking along the graphite–epoxy interface can also be seen. Both processes contribute to increased fracture *toughness* (see Chap. 14). (Photograph courtesy of W. H. Durrant.)

## REFERENCES

1. American Society for Metals, *Metals Handbook,* 9th ed., vol. 8, *Metallography, Structures, and Phase Diagrams.* Metals Park, Ohio: American Society for Metals, 1973.

2. American Society for Metals, *Strengthening Mechanisms in Solids.* Metals Park, Ohio: American Society for Metals, 1962. Chapter 9, "Mechanisms of Fine Particle Strengthening," by R. W. Guard, evaluates existing data on the subject for several materials.

3. Caddell, R. M., *Deformation and Fracture of Solids.* Englewood Cliffs, N.J.: Prentice Hall, 1980. Chapter 7 provides a thorough basic description of dislocation interaction with fine particles.

4. LeMay, Iain, *Principles of Mechanical Metallurgy.* New York: Elsevier-North Holland, 1981. Chapter 6 discusses strengthening mechanisms by the action of multiple phases.

5. McLean, D., *Mechanical Properties of Metals.* New York: John Wiley, 1962. Chapter 6 includes a thorough development of the theory of fine-particle strengthening.

6. Smith, Charles O., *The Science of Engineering Materials.* Englewood Cliffs, N.J.: Prentice Hall, 1977. Chapter 9 discusses solid-state reactions that lead to strengthening, and Chap. 10 discusses the formation and spacing of pearlite in steel.

## PROBLEMS

**9-1.** Which strengthening mechanisms can occur in
   **(a)** Pure elements?
   **(b)** Single-phase solids?
   **(c)** Multiple-phase solids?

**9-2.** One method of increasing the strength of copper–aluminum alloys is by precipitation hardening. Certain alloy compositions of copper and zinc (from 0 to 37% zinc) are called *alpha brasses.* Are these alloys susceptible to precipitation hardening? Explain your answer.

**9-3.** The *Bauschinger effect* consists of a reduction in yield strength in compression following initial yielding in tension (or in the reverse order—tension after compression). Give two explanations for the Bauschinger effect, one based on simple dislocation motion and the other based on the Orowan mechanism for fine-particle strengthening.

**9-4.** Assume that we can produce mixtures of a pure fcc metal (element Z) and a hard, brittle intermetallic compound ($\theta$) in a variety of shapes and sizes. In Fig. 9-17 are five plausible true-stress–true-strain curves (A through E) and five microstructures (1 through 5, all at 500×). Match each microstructure with the most likely stress–strain curve without duplication.

**9-5.** A metal alloy is in the form of a bar 5 mm in diameter. When furnace-cooled from the melt, it looks like microstructure (1) in Fig. 9-18 at room temperature. (All microstructures are shown at 400×; the dark phase is the harder phase.) The bar is then moderately cold-worked.
   **(a)** The metal is heated to the lowest temperature at which it is a solid single phase and then water-quenched. At the instant immediately following the quench, which microstructure in Fig. 9-18 most closely represents its probable appearance?
   **(b)** If the specimen is moderately cold-worked and heated again to the highest temperature at which it is still a solid single phase and then water-quenched, which microstructure most closely represents its probable appearance?
   **(c)** Which will have the greater yield strength, the specimen after treatment (a) or after treatment (b)?

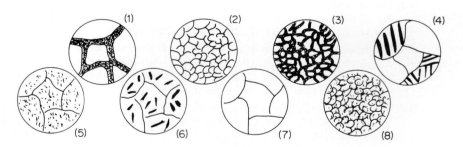

**Figure 9-17**

**Figure 9-18**

(d) Following treatment (a), the specimen is held at room temperature for 1 month. Assuming that its microstructure has changed during this period, state which microstructure in Fig. 9-18 best represents it.

(e) What is this process called?

(f) If process (d) is carried out at a moderately elevated temperature for the same time, say 80°C for 1 month, describe any differences in the appearance of the microstructure that will result.

(g) Which specimen will have the highest strength: the one shown as (1), after treatment (a), after treatment (b), after treatment (d), or after treatment (f)?

(**h**) Describe one dislocation mechanism that can explain why the strongest microstructure in part (g) is stronger than the next-strongest microstructure.

**9-6.** Calculate the room-temperature Brinell hardness of austenitized and furnace-cooled (i.e., fully annealed) plain-carbon steel of

(**a**) 0.20% C.

(**b**) 0.40% C.

(**c**) 0.70% C.

**9-7.** Calculate the Rockwell B or C hardness of plain-carbon steel of 0.30% C that is austenitized and then given each of the following treatments, each one taken separately:

(**a**) Air-cooled to 20°C.

(**b**) Furnace-cooled to 20°C.

(**c**) Oil-quenched to 20°C.

(**d**) Oil-quenched, heated to 700°C for 10 h, and air-cooled to 20°C. (Use the hardness conversions given in App. 4.)

**9-8.** You have available the following steels as bars 10 mm in diameter: 0.20, 0.40, and 0.77% C, all plain-carbon steels. These bars can be heated to any desired temperature and furnace-cooled, air-cooled, or oil-quenched. Considering each of the following requirements as a separate problem, select a steel and describe a heat treatment (within the above limitations), including temperatures, that will produce (within ± 10%)

(**a**) Tensile strength = 655 MPa.

(**b**) HB = 100.

(**c**) HB = 160.

**9-9.** A piece of 0.54% C steel 10 mm in diameter is held at 885°C for 3.6 ks and then furnace-cooled to 20°C. Its final hardness is 92 HRB. A piece of 0.30% C steel 10 mm in diameter is held at 885°C for 3.6 ks and then oil-quenched to 20°C. Its final hardness is 92 HRB. Explain, from a fundamental point of view, why these two tests produce the same hardness for different steels and cooling rates.

**9-10.** A bar of 0.72% C steel is held at 885°C for 3.6 ks and then air-cooled to 20°C, at which temperature its hardness is found to be 27 HRC. It is then reheated to 700°C for 150 ks and then air-cooled to 20°C, at which temperature its hardness is found to be 81 HRB. Explain, from a fundamental point of view, why its hardness changes.

**9-11.** A mixture of 0.26% carbon, balance iron, of total mass 947 g is furnace-cooled from the melt. For each of the temperatures 1149°C, 780°C, and 726°C,

(**a**) Calculate the mass of each phase present.

(**b**) If the phases are mixed in the form of a characteristic microstructure, name the microstructure and calculate its total mass.

**9-12.** Calculate the percentage of pearlite present in a 1.10% C steel after it is slowly cooled from 900°C to 100°C.

**9-13.** Which has the higher hardness, 0.40% C steel that has been air-cooled from 885°C or 0.40% C steel that has been oil-quenched from 885°C? Briefly explain your answer from a dislocation theory point of view.

**9-14.** Calculate the Brinell hardness number of a 0.50% C steel that has been oil-quenched from 900°C.

**9-15.** A set of schematic microstructures is given in Fig. 9-19. For each material and process listed here, write the letter of the microstructure from Fig. 9-19 that best describes the alloy after the process described. (An answer may be used more than once.)

(**a**) 2024 aluminum (UNS A92024), 510°C for 17 ks, water-quenched to 20°C, time = 0.

**(b)** 1100 aluminum (UNS A91100), 510°C for 12 ks, furnace-cooled to 20°C, time = 40 Ms.

**(c)** 0.95% C steel (UNS G10950), air-cooled from 900°C to 20°C.

**(d)** 2024 aluminum (UNS A92024), 510°C for 12 ks, water-quenched to 20°C, heated to 100°C for 3 ks, water-quenched to 20°C.

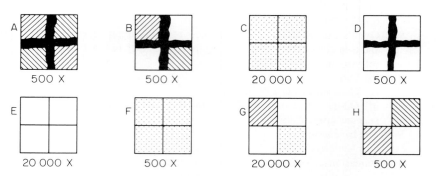

**Figure 9-19**

**9-16.** An aluminum alloy bar 6.49 mm in diameter is heat-treated so that the average spacing of $CuAl_2$ particles is 37.3 nm. How many 200-kg filing cabinets can it support simultaneously in tension before it breaks?

**9-17.** For each of the following two processes, taken separately, after the time(s) specified and at the final temperature stated:

**(a)** Name the phases present; if supersaturated, say so.

**(b)** Determine the composition of each phase.

**(c)** Describe the shape (spheroids, fine particles, flakes, etc.) or microstructure (eutectic, eutectoid, grain boundary, or free) of all phases present except the most prevalent phase. Both specimens are bars 12 mm in diameter:

**1.** Aluminum–4.8% copper, furnace-cooled from 680°C to 300°C.

**2.** Aluminum–5.5% copper, furnace-cooled from 680°C to 540°C, water quenched to 20°C, heated to 250°C for 600 ks, water-quenched to 20°C.

**9-18.** Draw schematic microstructures and label phases for a specimen of 2011 aluminum (UNS A92011) 5 mm in diameter after each of the following treatments and times taken sequentially:

**(a)** 500°C for 3 h, furnace-cooled to 20°C.

**(b)** 500°C for 3 h, water-quenched to 20°C, time = 0.

**(c)** 200°C for 1 ks.

**(d)** 200°C for an additional 7 ks.

**9-19.** In an aluminum–copper alloy of shear strength 275 MPa, how far apart are the fine particles?

**9-20.** Two tensile specimens of 2017 aluminum (UNS A92017: composition 4% Cu, 0.5% Mg, 0.5% Mn, balance Al) give the following data:

| Heat Treatment | Initial Diameter, mm | Max. Load, kN | Fracture Diameter, mm |
|---|---|---|---|
| 550°C, furnace-cooled | 12.8 | 31.1 | 9.93 |
| 550°C, water-quenched, 20°C for 100 Ms | 12.8 | 56.2 | 10.2 |

**(a)** Calculate the strength and true strain at fracture for both specimens.

**(b)** Describe precisely but briefly how the different heat treatments led to different strengths.

**9-21.** What limits the maximum temperature at which precipitation-hardened aluminum alloys can be used? Describe the mechanism involved.

**9-22.** A bar of 2024 aluminum (UNS A92024) is soaked at 510°C for 10.8 ks and then

    **(a)** water-quenched in "zero time" to 20°C and then

    **(b)** held at 371°C for 0.3 ks and water-quenched and then

    **(c)** held at 371°C for 3 ks more and water-quenched.

Draw a schematic microstructure at 20,000× magnification of the microstructure immediately after each of the three processes and label each phase with its name (not a Greek letter) or chemical symbol.

**9-23.** Compute the amount of each microstructure and the fraction of each phase present in a 0.30% C steel slowly cooled from the melt to each of the following temperatures:

    **(a)** 925°C.

    **(b)** 760°C.

    **(c)** 700°C.

**9-24.** A 1.00% C steel that has been cooled slowly from 1100°C and held at 730°C for 0.5 h is air-cooled to room temperature.

    **(a)** Name the phases present.

    **(b)** Describe the microstructure.

    **(c)** Calculate the fraction of noneutectoid carbide present.

**9-25.** A plain-carbon steel of 0.60% C is heated to 885°C for 0.5 h and then air-cooled to 20°C, yielding a hardness of 220 HB. The same specimen is reheated to 885°C for 0.5 h and then oil-quenched to 20°C; it has a hardness of 260 HB. Explain, from a fundamental point of view, why the hardness changes.

**9-26.** Describe in detail a heat treatment (without cold work) that will give a hardness of approximately 21 HRC for a bar of 1040 steel 10 mm in diameter.

**9-27.** Estimate the Brinell hardness of a steel with 0.47% carbon after it is air-cooled from 810°C to 20°C.

**9-28.** Estimate the hardness you would expect to find for a 7-mm-thick plate of 0.27% carbon steel that has been air-cooled from 875°C to 20°C.

**9-29.** A 10-kg bar of 0.33% C plain-carbon steel is air-cooled from 850°C to 20°C.

    **(a)** Calculate the mass of eutectoid iron carbide present at 20°C.

    **(b)** Calculate the mass of noneutectoid iron carbide present at 20°C.

    **(c)** Estimate the Brinell hardness of the bar at 20°C after the air cooling.

# 10

# *Martensites and Tempered Martensites*

The principal strengthening mechanisms were described in earlier chapters, but *martensite,* which in the Fe–C system is a microstructure of high hardness, is considered separately because it is both unique and important. The microstructures that can be produced from martensite are likewise important, for although they usually are not as hard as martensite itself, so-called *tempered martensite* structures can be produced with a wide range of strength and ductility that makes them very useful engineering materials. Furthermore, martensite and its products are often involved in the multiple-strengthening effects used in producing some ultrahigh-strength steels. The essential aspects of martensite strengthening are therefore presented in this chapter, accompanied whenever possible by a description of the basic strengthening mechanisms that contribute to the observed behavior. In many instances, the relative magnitudes of the effects of several mechanisms acting together are not known because each mechanism cannot be tested separately.

Martensitic microstructures can also be produced in the iron–nickel system and in nonferrous alloy systems such as copper–manganese, copper–zinc, nickel–titanium, and indium–thallium. These other martensites are of growing interest, not because of any high hardness properties but because of their unusual thermoelastic mechanical behavior: Their elastic stiffness can change markedly over a relatively small temperature range, and this behavior, together with an associated phenomenon called the *shape-memory effect* (discussed later in this chapter), can be exploited in the design of thermostatic control devices.

## MARTENSITIC TRANSFORMATION IN Fe–C

If the rate at which a small specimen of plain-carbon eutectoid steel (0.77% C) is cooled from just above the eutectoid transformation temperature (727°C) is successively increased, the resulting room-temperature microstructures will be different. In Chap. 9 pearlite spacing was shown to vary from coarse for furnace cooling to fine for oil quenching. If a still faster cooling process, such as water quenching, is used, the temperature of the specimen will drop so fast that when it reaches about 270°C, pearlite will not yet have begun to form. At about this temperature, a small portion of the austenite transforms in a very short time, on the order of 100 ns, to a new phase that is body-centered tetragonal, called *martensite*. As the specimen continues to cool, martensite continues to form from the remaining austenite in very rapid local bursts until at room temperature the structure is mostly martensite with a small amount (about 8% for a 0.77% C steel) of *retained austenite* still present. All the retained austenite at room temperature is in a metastable state, unable to transform to martensite because the temperature is too high and unable to transform to the equilibrium products, ferrite and iron carbide, because the temperature is too low for the necessary atomic diffusion. If the specimen is then cooled to the temperature of liquid nitrogen (−196°C), more retained austenite will transform to martensite until only about 3% of the structure remains as retained austenite. Thus, upon return to room temperature the 3% retained austenite will remain and will not change even if cooled again to −196°C.

The formation of martensite occurs as a *diffusionless* transformation and is therefore unique among the transformations considered in this book. It should be clear that virtually no diffusion can occur in iron in 100 ns at 270°C, so the atoms in martensite must be in almost the same positions as the atoms of the austenite from which the martensite was formed. Figure 10-1 shows that any fcc structure, in this case two fcc unit cells of austenite, can also be considered body-centered tetragonal (bct). Only the atoms of one bct unit

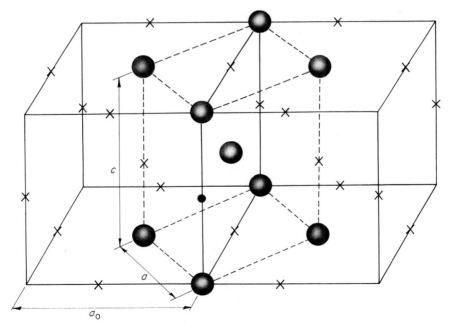

**Figure 10-1**    Two face-centered cubic unit cells of austenitic iron, shown as solid lines, with lattice constant $a_0$. Dashed lines show how atoms from two adjacent fcc cells can be viewed as body-centered tetragonal, with lattice constants $c$ and $a$. Possible interstitial lattice sites are indicated by $x$, with a carbon atom (small, black circle) shown at one of these sites.

cell are drawn in this diagram; the atoms at the remaining corners and faces of the fcc cells have been omitted. The fcc structure is ordinarily used to describe the structure of austenite because it is geometrically simpler than the bct structure, but here we wish to show how bct martensite can result from relatively small displacements of atoms, so the bct description of the crystalline structure of austenite is used.

**Example 10-1**

Find the ratio $c/a$ of the bct cell in Fig. 10-1.

**Solution**   $c = a_0$ and $a = a_0/\sqrt{2}$, so $c/a = \sqrt{2}$. This is a tetragonal structure because all angles are 90°, two of the sides of the unit cell are equal in length, and the third ($c$) is different.

If no carbon is present in austenite, it will transform to a perfect bcc structure (ferrite) at temperatures below 912°C irrespective of the cooling rate. This transformation may be considered as a collapsing of the bct unit cell in Fig. 10-1 until $c = a = 286$ pm. It is accompanied by a slight change in the effective atomic radius of iron from the fcc structure to the bcc structure.

When carbon is present in the austenite, it is found at the middle of a few of the edges of the fcc unit cells and at the centers of the unit cells (see Fig. 10-1). These two positions are crystallographically equivalent because any iron atom can be considered to be either at a corner or on a face of a unit cell. Carbon atoms cannot occupy a very large proportion of the sites in equilibrium austenite; if they did, the resulting structure would be severely strained and thus unstable. Figure 10-1 shows the possible sites for interstitial carbon in face-centered cubic iron (austenite), with a carbon atom situated at one of these sites.

When austenite containing some carbon attempts to transform (at high cooling rates) to the bcc structure of pure ferritic iron, the carbon acts to prevent full contraction of the $c$-direction. Figure 10-2 shows a unit cell of martensite in which $c > a$ because a carbon atom

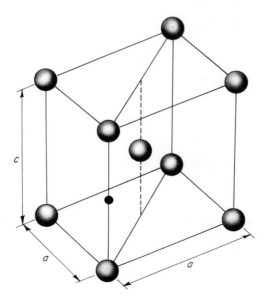

**Figure 10-2**   Martensite unit cell, where $c > a$. Spheres are iron atoms; the small, black circle is an interstitial carbon atom.

is situated along one of the edges parallel to the *c*-direction. (If there is no carbon present, then the austenite will transform to ferrite; ferrite might thus be thought of as martensite of zero carbon content.) The carbon atom is in the same position in the lattice that it occupied in the austenite lattice in Fig. 10-1; this emphasizes the diffusionless character of the martensite transformation. The presence of interstitial carbon along *c*-axis edges causes the *c*-direction to be slightly larger than the *a*-direction. The effect of the amount of carbon on the lattice parameters is shown in Fig. 10-3. As would be expected, an increase in carbon content increases *c/a*. All unit cells of a given martensite crystal have parallel *c*-axes. Once the *c*-direction is established, those carbon atoms situated in sites other than along the *c*-direction shift very rapidly into the nearest *c*-direction sites, jumping a distance that is less than *c*.

**Example 10-2**

Calculate the number of empty sites that occur for every full site in 0.77% C martensite.

**Solution**   Let *N* be the number of atoms of an element and *W* the weight of each element present. Then,

$$\frac{N_{Fe}\left(\text{atomic wt.}\right)_{Fe}}{N_C\left(\text{atomic wt.}\right)_C} = \frac{W_{Fe}}{W_C}$$

$$\frac{N_{Fe}}{N_C} = \frac{W_{Fe}\left(\text{atomic wt.}\right)_C}{W_C\left(\text{atomic wt.}\right)_{Fe}}$$

$$= \frac{99.23\% \times 12.0}{0.77\% \times 55.85} = 27.7$$

In each unit cell of martensite (see Fig. 10-2) there are two atoms of iron and $(2 \times {}^1\!/_2) + (4 \times {}^1\!/_4) = 2$ possible whole interstitial sites for carbon, considering only the potential sites that are parallel to the *c*-direction. (Note that all the sites are shared with one or three other unit cells.) Then for 0.77% C in Fe, $N_{Fe}$ = number of interstitial sites, and

$$\frac{\text{No. interstitial sites}}{\text{No. atoms of carbon}} = 27.7$$

$$\frac{\text{No. empty sites}}{\text{No. atoms of carbon}} = 27.7 - 1 = 26.7$$

It is not obvious from the example how such a small fraction of interstitial sites actually containing carbon atoms can cause all unit cells of iron to be tetragonal. Most unit cells contain no carbon at all. Yet X-ray diffraction studies of martensite crystals have

**Figure 10-3**   Variation in lattice parameters with carbon content of martensite. [From C. S. Roberts, "Effect of Carbon on the Volume Fractions and Lattice Parameters of Retained Austenite and Martensite," *Trans. AIME*, 197 (1953), p. 203. © 1953 by AIME and reprinted with its permission.]

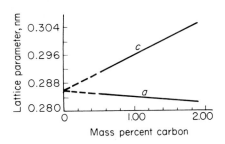

given the data represented in Fig. 10-3. Thus, we may conclude that the local straining by one carbon atom is so severe that its influence extends to all unit cells in the vicinity. The presence of one carbon atom along an edge parallel to the *c*-direction not only elongates the unit cell but also probably causes some local warping of the unit cell out of a perfect tetragonal shape. The presence of such large, highly localized strains in martensite must contribute substantially to its high hardness.

The manner in which martensite plates form from austenite is a complex process of local shearing, compression, and extension. Martensite in ferrous alloys normally occurs as very small, nonparallel, lens-shaped crystals. Figure 10-4(a) is a photograph of the characteristic appearance of martensite (in this case Fe–Ni martensite) with some austenite that has not transformed. In the course of formation of a martensite crystal, the matrix surrounding it is inevitably strained because iron–carbon martensite forms with an increase of specific volume. If the surrounding matrix is already martensite, it will probably be subjected to an increase in local stresses, owing to the nonparallel occurrence of the martensite plates in the matrix. If the surrounding matrix is austenite, it will probably deform plastically to accommodate the formation of the martensite plate. Thus, the retained austenite may be locally strain-hardened as a result of the martensite transformation.

(a)

(b)

**Figure 10-4**    (a) Optical micrograph of martensite and retained austenite (white background) in Fe–32% Ni, cooled to –195°C. [From V. F. Zackay, ed., *High-Strength Materials* (New York: John Wiley & Sons, 1965), p. 404. Reprinted with permission of the publisher. Photograph by D. W. Stevens, courtesy of W. C. Leslie.] (b) Martensite plates in 60 Mn–40 Cu alloy heat-treated at 400°C for 8 h and water-quenched. In contrast to Fe–Ni martensite (where the needles form in a random fashion), here the plates are roughly parallel over large distances. This permits easy, reversible movement along the length of the needles in shape-memory alloys. Related nonferrous alloys display very high acoustic damping for similar reasons. Curly dislocation lines may be seen in the background of the picture. (Courtesy J. A. Hedley.)

In those nonferrous alloy systems in which martensitic microstructures can be produced, the crystallographic changes are not necessarily the same as those occurring in the Fe–C system (i.e., not always fcc $\Rightarrow$ bct). In the case of Cu–Mn and In–Th the changes are fcc $\Rightarrow$ fct; in Ni–Ti the change is bcc $\Rightarrow$ orthorhombic. The shear deformations associated with these nonferrous martensitic transformations are much smaller than in steels, and the cell dimensions of the martensite phases are not greatly different from those of the parent matrix phases. Consequently, the differences in size between transformed and untransformed regions can be accommodated elastically within the body, so such martensitic structures can respond reversibly to changes in applied stress and/or temperature [Fig. 10-4(b)]. This allows these alloys to sometimes display the *shape-memory* (or *marmem*) effect whereby, for example, a bent strip of the alloy in its martensitic state curls and uncurls as the temperature is altered.

**Example 10-3**

A trick metal spoon curls up when used to stir hot tea. Give a possible explanation for this observation.

**Solution**   It is likely that the spoon is made from a shape-memory metal such as one of the nonferrous martensitic memory alloys. The behavior described can be caused to occur in the following way: The spoon, in a curled-up shape, is heat-treated in some appropriate manner in order to get it into the state where it has a latent memory. Such heat treatment might be a rapid quench to room temperature of the sort used in Fe–C martensites, but some nonferrous martensites can be produced in other ways; details do not matter here. The quenched structure will consist of the two phases of some transformed martensite and some untransformed matrix (analogous to retained austenite in the Fe–C system).

If the curled-up spoon is then flattened out at room temperature, it stays flat. At first sight, it might be thought that the spoon remains flat because the unbending strains have exceeded the yield strain. However, this is not the case because in the martensitic state these alloys have relatively large elastic limits ($e \approx 0.02$ and greater). But if the elastic limit has not been exceeded, why does the spoon not resist permanent flattening and spring back? The reason is bound up in the peculiar behavior of the martensitic structures.

In the act of flattening, the acicular phase boundaries between the martensite induced by heat treatment and the untransformed matrix move so as to consume matrix material and convert it to more martensite. Such boundaries can move easily because of the comparatively small differences in the size of crystal cells in the transformed and untransformed phases in these particular memory alloys. In the flattened condition, there is thus a greater volume of martensite than in the curled state. Internal elastic stresses (akin to residual stresses) are induced when the martensite boundaries are caused to move by some external loading. Such internal stresses attempt to restore the original shape of the body on unloading, when the martensite interfaces move back and the stress-induced martensite reverts to the matrix phase. Under a certain range of temperature (exploited in this spoon example), the elastic restoring forces are insufficient on their own to reverse the deformation, so that the spoon remains flat under the action of the internal stresses. However, an appropriate increase in temperature will cause the spoon to "remember" its original shape and so curl back up as a result of the changes in resistance to movement with temperature to the martensite–matrix phase boundaries. (Such resistance to movement is analogous to Peierls–Nabarro forces in dislocation mechanics.) The martensite is thereby reversibly restored, and hence the spoon bends up when used to stir hot tea.

Because the resistance to movement of the phase boundaries decreases with increasing temperature, attempts to flatten the spoon in a somewhat higher range of temperatures will not succeed. However, the martensitic material reversibly deforms over quite large strains, as mentioned earlier, and this type of response in these solids is termed *superelastic* behavior.

Polymer spoons can also display similar large recoveries over certain ranges of temperature as a result of relief of residual stresses produced in forming processes. Vacuum-formed objects made from inappropriate sheet material have been known to flatten out if exposed to sunlight over long periods. Heat-shrinkable plastics, which are irradiated and quenched polymers, work for reasons similar to the martensitic memory metals: They are deformed (expanded) when hot and then quenched while still in the expanded state. On reheating, they remember their original, smaller shape and thus shrink tightly around whatever is being covered.

The amount of carbon present in Fe–C austenite influences the temperatures for the *start* of the martensitic transformation, called $M_s$ and for the *finish* of the transformation, called $M_f$. These are the temperatures for 1% and 99% martensite, respectively. Figure 10-5 shows the drop in these temperatures with increasing carbon content. This drop might be expected from energy considerations: With higher carbon content there is more strain energy in the martensite that is to be formed, so to provide the necessary driving force to form martensite the transformation must take place at a lower temperature where the free energy difference between fcc and bct is greater. Thus, instead of transformation to the desired low-energy bcc ferrite, the absence of diffusion causes the fcc austenite to transform to a lower-energy but still highly strained martensite. Note that for carbon con-

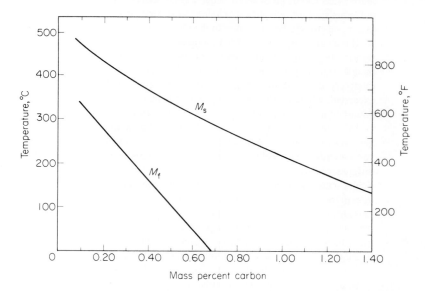

**Figure 10-5**   Effect of carbon content on temperatures for start ($M_s$) and finish ($M_f$) of martensite transformation. [From A. R. Troiano and A. B. Greninger, "The Martensite Transformation," *Metals Handbook*, T. Lyman, ed. (Metals Park, Ohio: American Society for Metals, 1948), p. 265.]

tents above about 0.65%, some retained austenite is present following quenching to room temperature.

**Example 10-4**

One step in the heat treatment of some high-strength steels is to cool them to a very low temperature well below room temperature. Why is this done?

**Solution**   This step usually follows a quenching process, which is intended to produce as much martensite as possible. Higher-carbon steels have substantial amounts of retained austenite at room temperature (Fig. 10-5), so further cooling is necessary to transform this austenite to martensite. If the part is made to close tolerances and may subsequently be exposed to temperatures much lower than the original quench temperature, the increase in specific volume resulting from the additional martensite formation during subcooling could lead to dimensional misfit and failure. Because maintenance of very low temperatures entails an additional processing cost, such treatment is restricted to steels having very high hardness or strength and requiring close tolerances, where the extra expense is justified.

## MECHANICAL PROPERTIES OF MARTENSITE

The hardness of iron–carbon martensite depends on the amount of carbon present. Figure 10-6(a) gives the Rockwell C hardness of thin wafers (about 2 mm thick) of plain-carbon steel that have been heated into the austenite range and then water-quenched to room temperature. If the quenching is fast enough to form all martensite, the hardness can be reliably predicted from the carbon content of the steel. With ordinary quenching techniques, it is not possible to obtain martensite in plain-carbon steels of less than about 0.20% C. (Reasons for this limitation are discussed later in this chapter.) With the carbon content above about 0.80%, a saturation in the HRC hardness occurs. That this does not give a true picture of the resistance to penetration can be deduced from the observation that the Vickers pyramid hardness (obtained with a diamond-shaped indenter, with hardness related inversely to the area of indentation), shown in Fig. 10-6(b), continues to rise with increasing carbon. The HRC measurement of depth of indentation is apparently not sensitive to changes in the martensite in this range. The continuing increase in hardness shown in Fig. 10-6(b) is consistent with the argument that more carbon strains the lattice more and thus inhibits dislocation motion more effectively.

Quenching steel more than 2 mm thick usually results in less than 99% martensite (for reasons that are discussed later in this chapter), with consequent lower hardness. For example, Fig. 10-6(c) shows the range of hardness with carbon content when 80% martensite is produced. This hardness range can be considered typical of usual quenching procedures.

The strength of martensite is often much lower than would be predicted from the equation $S_u$ [MPa] $= 3.45 \times$ HB. For example, Fig. 2-25 shows two instances where martensite produced by water quenching, with about 600 HB, has a strength of only about 700 MPa instead of the predicted 2 GPa. On the same graph is another water-quenched martensite specimen, lower in carbon, with about 310 HB and 830 MPa, which is much closer to the predicted behavior. This fickle behavior is related to the kind of fracture that occurs in martensite. Whereas pearlitic steel usually deforms plastically and then fractures by ductile

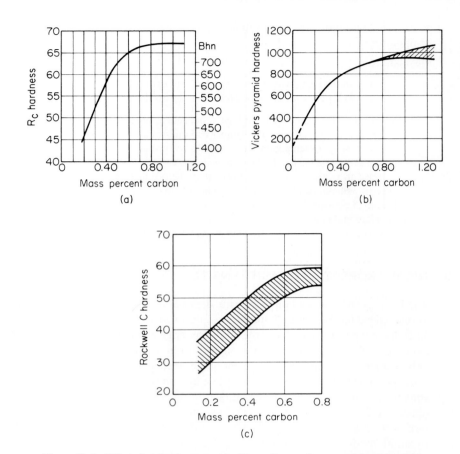

**Figure 10-6**   Effect of carbon content on hardness of iron-carbon martensite. (a) and (b) show the hardness for 99.9% martensite. [From Edgar C. Bain, *Functions of the Alloying Elements in Steel* (Metals Park, Ohio: American Society for Metals, 1939), p. 36.] Graph (c) shows the range of hardness obtained for 80% martensite. [From ASM Committee on Hardenability, "The Selection of Steel for Hardenability," *Metals Handbook: Properties and Selection of Metals,* 8th ed., vol. 1, T. Lyman, ed. (Metals Park, Ohio: American Society for Metals, 1961), p. 189.]

tearing, martensite fractures before plastic flow sets in, while the load is still increasing along a nearly elastic curve. Such *brittle* behavior can result in wide variations in strength.

Figure 10-7 contrasts true stress–strain curves for a 0.40% C[*] low-alloy steel after each of several different heat treatments. Curve (a) has the substantial deformation that is

---

[*]The American Iron and Steel Institute (AISI) and the Society of Automotive Engineers (SAE) designate plain and low-alloy steels by use of a simple code in which the first two digits indicate the extent of alloy content (where 10 refers to plain-carbon steels) and the remaining digits indicate the amount of carbon present, in hundredths of 1%. Hence, 1040 is a plain-carbon steel of about 0.40% C, and 4140 is a low-alloy steel of 0.40%C (AISI-SAE tables give the alloy composition range). The Unified Numbering System (UNS) for metals and alloys has been adopted by the American Society for Testing and Materials (ASTM), SAE, AISI, and others. Since UNS designations are coming into common use, both the AISI-SAE and UNS designations are frequently listed together in this book.

characteristic of a specimen composed of ferrite and pearlite. Curve (b), for which only the fracture point is shown, exhibits the usually small deformation that precedes fracture in as-quenched martensite. The energy required to fracture each specimen is related to the area under each curve in Fig. 10-7. Because the area under curve (b) is very small, as-quenched martensite is considered brittle. The area under curve (a) is large, indicating that a microstructure of ferrite and pearlite is usually ductile at room temperature.

**Figure 10-7**  True-stress–true-strain curves at room temperature for tensile specimens (9.07 mm initial diameter) of a low-alloy steel of 0.40% C (AISI 4140, UNS G41400), initially heated into the austenite range: (a) furnace-cooled (ferrite and coarse pearlite); (b) oil-quenched (martensite); (c) oil-quenched and then heated to 230°C for 1.8 ks (stress-relieved martensite); (d) oil-quenched, 430°C for 1.8 ks (tempered martensite); (e) oil-quenched, 590°C for 1.8 ks (tempered martensite); (f) oil-quenched, 700°C for 36 ks (mostly spheroidite). (Data by William G. Ovens.)

The low tensile strength of as-quenched martensite is partly due to residual stresses that occur during quenching. These large residual stresses result from the nonuniform local strains during the transformation from austenite to martensite. When a specimen of martensite containing residual stress is loaded in tension, the residual tensile stresses can add to the applied tensile stresses to cause fracture at a lower applied tensile stress than would be required to cause fracture in the absence of residual stresses. If much plastic deformation occurred in the martensite before fracture, the residual stresses would be greatly reduced and the load for fracture would be increased. Thus, it is the brittle nature of the fracture of martensite that is indirectly responsible for the frequent lowering of its strength. Oil quenching, with the lower residual stresses that result from a slower cooling rate, is therefore preferable to water quenching if the steel is such that martensite can be produced at the slower cooling rate.

The hardness test thus provides a truer measure than the tension test of the ability of martensite to resist slip. A conical or spherical hardness indenter causes compressive stresses in all three directions (triaxial compression) in the specimen. Slip occurs because the compressive stresses are not of the same magnitude in all three directions. Tensile fracture cannot occur without tensile stresses, so in materials without flaws or cracks there is no immediate possibility that fracture can interfere with plastic deformation in the hardness test. Furthermore, the hardness test is not substantially influenced by residual stresses because the large plastic deformation that occurs in the presence of the triaxial compressive stresses wipes out most of the residual stresses.

**Example 10-5**

Construct Mohr's circle for stress for (a) a uniaxial tensile test, and (b) a hardness test where $\sigma_1 > \sigma_2 > \sigma_3$ and all are compressive. Select magnitudes of shear stress for slip ($\tau_{slip}$) and tensile stress for fracture ($\sigma_f$) so that fracture occurs before slip in (a) and slip occurs before fracture in (b).

**Solution**   The construction is shown in Fig. 10-8. (See Chap. 2 for the Mohr's circle construction if review is required.) The radius of the largest circle for each case gives the maximum shear stress. Thus, for the tensile test, if $\sigma_f < 2\tau_{slip}$, tensile fracture will occur. In the hardness test, slip will always occur first. It should therefore be clear from this construction how plastic deformation can occur in the hardness test of very brittle metals.

Heating martensite to the range 150–260°C for 2 ks apparently allows the residual stresses from quenching to be relieved by local slip. Some changes on a very fine scale also occur in the microstructure of martensite, but these do not substantially reduce the hardness and may aid in the relief of residual stresses. Curve (c) in Fig. 10-7 shows the marked increase in strength that occurs in oil-quenched AISI-SAE 4140 steel after *stress relieving* at 230°C. In this temperature range the residual stresses resulting from rapid nonequilibrium cooling and the austenite-to-martensite transformation are substantially reduced with negligible change in the microstructure. This results in an increase in strength over that of as-quenched martensite, which fails prematurely with virtually no plastic deformation. In order to discriminate here between these two similar-appearing microstructures, the stress-relieved microstructure is called *stress-relieved martensite.* The strength of stress-relieved martensite is much closer than that of as-quenched martensite to the value predicted from the hardness test, although the fracture strain remains small. In

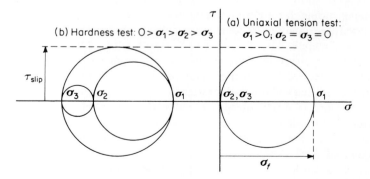

**Figure 10-8**   Solution to Example 10-5 showing in a simple way why a brittle material does not fracture under moderate triaxial compression.

Fig. 2-25 four of the five highest strengths are for steels that have been quenched and then tempered in the range 200–260°C.

## TEMPERED MARTENSITE

When quenched specimens of steel are heated to temperatures above about 260°C, the excess carbon in the martensite precipitates out in a series of steps that terminate with formation of very fine particles of $Fe_3C$ in a matrix of bcc ferrite. The ferrite can be thought of as martensite with almost all of the carbon removed, as was pointed out earlier in this chapter. Because this process takes place at relatively low temperatures, diffusion is quite limited. Carbon can thus diffuse only very short distances, so the $Fe_3C$ forms as very fine particles uniformly distributed throughout the ferrite. The resulting microstructure gives an ideal combination of high strength and ductility, which is here called *tempered martensite* (although no martensite is present when the precipitation of $Fe_3C$ is complete). Note that "to temper" means "to moderate"; in this case the terminology refers to the reduction of brittleness.

Curves (d) and (e) in Fig. 10-7 show the range of mechanical behavior that can be expected of tempered martensite of UNS G41400. The hardness of tempered martensite is more closely related to strength than was observed for as-quenched martensite. Most of the data in Fig. 2-25 in the 900–1700 MPa strength range are for tempered martensite. Proper selection of tempering temperature can provide (within limits) a wide range in strength, hardness, and ductility because the higher tempering temperature allows a higher rate of diffusion with a consequent larger size and wider spacing of the $Fe_3C$ particles. Consequently, the strength and hardness of tempered martensite depend on both the percentage of carbon and the tempering temperature.

Figure 10-9 shows the variation in hardness with tempering temperature for steels with a range of carbon content. If lengthy tempering occurs at just below the eutectoid transformation temperature, as for curve (f) in Fig. 10-7, the $Fe_3C$ forms as very large spheroidal particles of extremely wide spacing. The mixture of ferrite and this form of $Fe_3C$ is virtually the same as spheroidite (see Chap. 9), so this microstructure can be considered a mixture of free ferrite ($\alpha$) and spheroidite ($P_s$ of 0.77% C). This structure has lower strength and higher ductility than any tempered martensite or pearlitic microstructure. Curve (f) in Fig. 10-7 exhibits this structure's characteristic properties. Microstructures of ferrite + spheroidite can also be produced by first forming ferrite + pearlite and allowing the $Fe_3C$ to agglomerate into large spheroids at about 700°C, as described in Chap. 9. The test results shown in Fig. 10-9 for tempering at 700°C do not represent the hardness of spheroidite (e.g., for 0.74% C steel) because the samples were tempered for only 1.8 ks. Complete spheroidization, forming larger $Fe_3C$ particles, requires conditions on the order of 10 h (36 ks) at 700°C.

The ductility of tempered martensite in low-alloy steels increases and the strength decreases with increasing tempering temperature, except in the range of about 260–400°C. Tempering in this range decreases notch ductility (as measured in a special notched-impact test; see Chaps. 2 and 14) to values lower than are achieved through tempering at either

**Figure 10-9** Hardness at 20°C as a function of tempering temperature (all tempered for 1.8 ks) for steels of a range of carbon and alloy content. Water quenching (plain-carbon steels) was in ice water; oil quenching (low-alloy steels) was at 20°C. [Data on plain-carbon steels by Roger A. Heimbuch. Data on low-alloy steels from American Society for Metals, *Source Book on Industrial Alloy and Engineering Data* (Metals Park, Ohio: American Society for Metals, 1978).]

260 or 400°C. This loss of ductility, also known as *500°F embrittlement* or *tempered martensite embrittlement*, is accompanied by a reduction in strength below that from tempering at 500°F (260°C), so there are no compensating gains in mechanical properties to warrant tempering in the range of about 260–400°C. (For consistency with the use of degrees Celsius, 500°F embrittlement may also be termed *250°C embrittlement*, because 250°C is a rough approximation of 500°F.) The cause of this embrittlement is associated with the diffusion of iron carbide and possibly ferrite platelets to prior austenite grain boundaries, with consequent grain-boundary fracture.

In some low-alloy steels that have been quenched to produce martensite, tempering in a temperature range above 550°C promotes precipitation of very fine particles, which are usually carbides formed with some of the alloying elements. Important carbide-forming elements are titanium, vanadium, molybdenum, and tungsten. These fine particles act to inhibit dislocation motion and therefore can reverse the trend toward lower strength with higher tempering temperature. This is called *secondary hardening*. This behavior can be seen in a mild way in Fig. 10-9, where the hardness of 8660 steel (UNS G86600) at the higher tempering temperatures, resulting from carbide precipitation, not only exceeds the hardness of the 0.74% C steel but also exceeds the hardness of the 1.03% C steel. Secondary hardening is much more marked than this in certain other low-alloy steels.

**Example 10-6**

How well do the tempered martensites hold their strength at elevated temperatures?

**Solution**   Not very well, in general, because the fine particles of $Fe_3C$ in ferrite agglomerate in a manner analogous to $CuAl_2$ in aluminum (see Chap. 9), and the strength drops. Secondary hardening provides an alternative mechanism for maintaining strength at higher temperatures, and high-temperature materials make extensive use of stable fine particles.

## HARDENABILITY

Tempered martensite possesses a good balance between high-strength and moderate ductility, and it is used where the requirement for high strength justifies the greater processing cost than for pearlitic steels. Tempered martensite can be made only from martensite. Likewise, stress-relieved martensite of high carbon content is the only practical structure having hardness in the 55–65 HRC range. The conditions that promote the formation of martensite are therefore of great engineering interest, and a substantial amount of knowledge on the subject is available in the form of articles, books, and handbooks. *Hardenability* is a measure of the ease with which martensite can be formed in a given steel; this property can be measured quantitatively in several satisfactory ways.

The time available for transformation of austenite to pearlite becomes shorter when transformation occurs at temperatures progressively further below the eutectoid temperature; the time reaches a minimum at about 550°C, and at lower temperatures diffusion is so slow that pearlite does not form at all. If a specimen of eutectoid steel is cooled from austenite so fast that when it reaches 550°C no pearlite has formed, the microstructure will remain austenite during further rapid cooling until it reaches the $M_s$ temperature, which is about 270°C for plain-carbon steels with 0.77% C (see Fig. 10-5). Martensite then begins to form from the austenite, and at room temperature this steel will be mostly martensite with a small fraction of retained austenite. The ease of forming martensite is thus seen to be related solely to the ease of avoiding the formation of pearlite, and thus it bears no relation to the speed of the austenite-to-martensite transformation which is always very rapid.

The effects of cooling rate and temperature on microstructure are shown in Fig. 10-10 for small pieces of eutectoid plain-carbon steel cooled as described in the preceding paragraph. When time is plotted on a logarithmic scale, continuous cooling curves are nearly straight lines. The dashed line labeled "Cooling rate A" represents the fastest cooling rate that still results in all pearlite at room temperature. Specimens cooled at "Cooling rate B" and at faster rates result in all martensite (plus retained austenite) at room temperature, in the manner described in the preceding paragraph. Cooling rates between curves *A* and *B* produce a mixture of microstructures because the transformation of austenite to pearlite does not have enough time to reach completion at high temperature. The lower edge of the shaded ($\gamma + P$) region, labeled $P_{finish}$ in Fig. 10-10, represents the finish of pearlite formation. Nothing further happens for cooling rates between curves *A* and *B* until the temperature reaches the $M_s$ line and martensite starts to form from the remaining untransformed austenite. At room temperature, after cooling at a rate between *A* and *B,* four phases are present: austenite, ferrite and carbide (together as pearlite), and martensite.

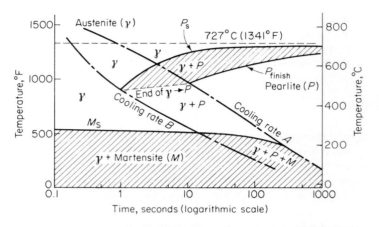

**Figure 10-10** Continuous cooling transformations of a eutectoid plain-carbon steel.

### Example 10-7

How can four phases be present in Fig. 10-10 when the equilibrium-phase diagram gives only two, ferrite and carbide?

**Solution** The equilibrium diagram applies only to equilibrium and near-equilibrium cooling, certainly not to fast quenching where diffusion is limited. Thus, the phases present after fast cooling are neither limited to nor described by the equilibrium diagram. In this case, a diagram such as Fig. 10-10 tells what phases are actually present.

The final microstructure after quenching therefore depends on two factors: (1) the rate at which austenite transforms to pearlite, a factor that is a property of the steel and can be specified by the limiting cooling rates *A* and *B* that span the cooling-rate range from all pearlite to no pearlite, respectively; and (2) the actual rate at which a given grain of steel is cooled. The first of these factors is just another way of describing hardenability, for if the transformation of austenite to pearlite can be suppressed (i.e., if the whole shaded region ($\gamma + P$) is moved to the right, so cooling rate *B* is also moved to the right in Fig. 10-10), the limiting cooling rate that can just produce martensite will be slower. A slower cooling rate is easier to achieve, so such a steel has a higher hardenability. Parts that are thick, say 200 mm, where the cooling rate in the middle of the cross section is slower than at the outside, can be transformed to martensite only if a high-hardenability steel is used. Figures like Fig.

10-10 are variously called time–temperature–transformation (*TTT*) curves, *S* curves, or Bain curves.

Many alloying elements have been found that increase hardenability; carbon (higher carbon measurably increases hardenability in plain-carbon steels), manganese, molybdenum, chromium, silicon, and nickel are common examples. Alloying elements other than carbon have negligible influence on the mechanical properties of martensite. The primary purpose of these elements is to slow the formation of pearlite by interfering with diffusion and thus to enable martensite to be produced at slower cooling rates, with the nose of the diagram being shifted to the right. Steels possessing this feature have higher hardenability. Another purpose of alloy additions, particularly of strong carbide-forming elements such as titanium, vanadium, and molybdenum, is to promote secondary hardening during tempering, as described in the preceding section. When the alloying elements that are added total no more than 5–10% of the mass of iron present, the steel is called a *low-alloy steel.* Such a steel should not be confused with an *alloy steel,* which usually has over 10% alloying elements and often has a markedly different microstructure—such as all stable austenite at room temperature—as well as unique properties such as high corrosion resistance. Some of these high-alloy steels are discussed in Chap. 11. The line of demarcation between plain-carbon and low-alloy steels remains fuzzy because most commercial plain-carbon steels have a greater mass of alloying elements other than carbon present. For example, the standard composition of a nominally 0.40% plain-carbon steel has 0.60–0.90% Mn, an element that is very effective in increasing hardenability but has little effect in these amounts. (The small amount of manganese in plain-carbon steels acts preferentially to combine with sulfur as MnS and thus prevents the formation of brittle FeS.)

The continuous cooling–transformation diagrams of two common low-alloy steels are shown in Figs. 10-11 and 10-12 for a wide range of cooling rates. (The cooling rates are also given as distances from the quenched end of an end-quenched bar, a procedure that is described in the following paragraphs.) Figures 10-11 and 10-12 contain two more structures than the plain-carbon eutectoid steel in Fig. 10-10: ferrite and bainite. Ferrite is absent from Fig. 10-10 because no free ferrite is produced during quasi-equilibrium cooling of eutectoid steel. And bainite is a microstructure of very fine particles of carbide in a matrix of ferrite which forms directly from austenite. Bainite and tempered martensite are thus of similar microstructure, but tempered martensite is formed only from the tempering of martensite and bainite is formed directly from the cooling of austenite.

Figure 10-12 shows how bainite can be formed in a low-alloy steel by continuous cooling. It is possible to produce bainite in plain-carbon steel by *isothermal transformation,* which simply means transformation at constant temperature. (Contrast this with the continuous cooling transformation in Figs. 10-10–10-12.) For example, if a plain-carbon eutectoid steel is quenched very rapidly from 800°C to 400°C, it will remain austenite. (We can use Fig. 10-10 to establish this, as the quenching process up to this point is continuous cooling; however, from this point on we cannot use Fig. 10-10 because we will no longer be cooling.) If the specimen is held at 400°C for, say, 1000 s, the austenite will transform to ferrite and carbide in the form of bainite. This occurs because (1) the temper-

**Figure 10-11**    Continuous cooling transformations for AISI 4140 steel (UNS G41400).
Cooling rates are at 700°C. A, austenite; F, ferrite; P, pearlite; B, bainite; M, martensite.
[From Homer Research Laboratories, Bethlehem Steel Co., reprinted in *Source Book on
Industrial Alloy and Engineering Data* (Metals Park, Ohio: American Society for Metals,
1978), pp. 125, 129.]

ature is too low to allow sufficient diffusion to produce pearlite, and (2) the temperature is
above the $M_s$ and so the austenite cannot transform to martensite.

**Example 10-8**

How would you determine the time required to produce all bainite at 400°C?

**Solution**    A series of tests on very small specimens should be run: (1) Quench from 800°C to
400°C; (2) hold at 400°C for time *t;* (3) water-quench to 20°C; (4) measure hardness. If any
austenite remains at time *t,* it will be transformed to martensite when water-quenched (except
for a fraction of retained austenite), giving higher hardness than for all bainite. For times
greater than the time *t* required just to produce all bainite, no martensite is produced and the
hardness should remain substantially constant. The bainite can be established by metallographic examination. This procedure is used to establish *isothermal transformation diagrams*
for steels.

**Figure 10-12** Continuous cooling transformation for AISI 4340 steel (UNS G43400). Cooling rates are at 700°C. A, austenite; F, ferrite; B, bainite; M, martensite. [From Homer Research Laboratories, Bethlehem Steel Co., reprinted in *Source Book on Industrial Alloy and Engineering Data* (Metals Park, Ohio: American Society for Metals, 1978), pp. 125, 129.]

The amount of nonmartensite transformation products is indicated in Fig. 10-11 for a range of cooling rates for AISI 4140 steel at several levels from 1% to 90%; the balance, which is austenite, will transform to martensite on further cooling to 20°C. Thus, a cooling rate of 108°C/s at 700°C leads to about 99% martensite, a cooling rate of 18.3°C/s leads to slightly more than 50% martensite, and a cooling rate of 2.8°C/s yields less than 10% martensite. By contrast, in Fig. 10-12 for AISI 4340 steel, a cooling rate of 2.8°C/s at 700°C produces more than 95% martensite. A cooling rate that produces more than 95% martensite in 4340 steel produces less than 10% martensite in 4140 steel. Thus, AISI 4340 has a much higher *hardenability* than AISI 4140 (a result of the presence of nickel in 4340 in addition to the chromium and molybdenum of AISI 4140).

The grain size of the austenite also influences hardenability. If the grain size is small, the large surface area of grain boundaries will mean large surface energy, and there will also be more grain-boundary sites for the nucleation of pearlite. For both reasons, fine-grained austenite forms pearlite faster than coarse-grained austenite and hence has a lower hardenability.

The second factor cited previously that strongly influences the final microstructure is the actual rate of cooling of a given grain of steel. The specimens in Fig. 10-10 were very small, so the cooling rate was nearly the same throughout. As the thickness increases, so does the difference in cooling rate between the center of a piece of steel and the surface. It is therefore possible to have a quenched piece of steel where the cooling rate at the surface is faster than rate *B* in Fig. 10-10, producing martensite (and retained austenite), while the cooling rate at the center is slower than rate *A*, producing all pearlite (and ferrite, if the steel is below 0.77% C). Thus, the dimensions of the piece and the quenching medium (water, oil, or air) control the cooling rate at a given point in the steel.

The *Jominy* end-quench hardenability test was developed to measure with a single test the hardness, and hence the microstructure, that results from a wide range of cooling rates. This test consists of water quenching one end of a specimen [1 in. (25.4 mm) in diameter and 4 in. (102 mm) long] that has been heated to about 40°C above the lowest temperature at which the steel is all austenite. The physical setup for this test is shown in Fig. 10-13. It can be seen that the cooling rate at the quenched end is very rapid and progressively decreases with distance from the quenched end. (Details can be obtained from Ref. 1.) Two parallel

**Figure 10-13**   Jominy end-quench hardenability test.

flats, 0.015 in. (380 $\mu$m) deep, are ground on the cylindrical surface for hardness measurements. The Rockwell C hardness along the flats is measured and plotted vs. distance at $1/16$-in. (1.6-mm) intervals from the quenched end of the bar. Such *Jominy hardenability curves* are shown in Fig. 10-14 for a 0.40% C plain-carbon steel (1040) and the 0.40% C low-alloy steel (4140) used for the tensile results given in Fig. 10-7 and included in the hardness results in Fig. 10-9. The higher-hardenability 4340 steel is also included in Fig. 10-14. The shaded areas represent the usual ranges of hardness that result from the allowable variations in carbon content and alloy content for these steels. Note that the three steels have about the same hardness at the quenched end because the hardness of the 100% martensite is controlled only by the 0.40% carbon content. The cooling rate at the quenched end is fast enough to form martensite in most plain-carbon and all low-alloy steels.

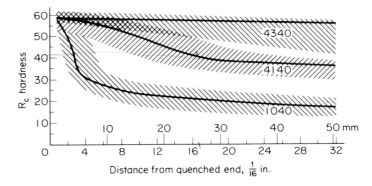

**Figure 10-14**   Jominy hardenability curves for three 0.40% C steels: AISI 1040 (UNS G10400), AISI 4140 (UNS G41400), and AISI 4340 (UNS G43400). Shaded regions show ranges; solid curves show typical values.

A comparison of Jominy curves for different steels is based on little or no variation in thermal properties such as thermal conductivity and specific heat. For plain-carbon and low-alloy steels there appears to be no substantial change in these properties from those of iron. Thus, at any particular distance from the quenched end, the cooling rate in the Jominy test is the same for all these steels. Each point therefore experiences a unique cooling rate which can be plotted as the dashed lines in Figs. 10-11 and 10-12. The drop in hardness for greater distances from the quenched end corresponds to the decreasing amount of martensite that results as the cooling rate is decreased from curve *B* to curve *A* and slower in Fig. 10-10, and from *Jominy distances* ranging from $1/8$ in. to 3 in. in Figs. 10-11 and 10-12.

**Example 10-9**

Why would a designer specify a particular low-alloy steel, for example, AISI 4140, in place of a plain-carbon steel such as AISI 1040 when both can be quenched to about 58 HRC (see Fig. 10-14)?

**Solution**   If a part is small enough to be quenched in water to the hardness desired, then plain-carbon steel may be adequate. However, in practice even in this case an oil-quenched low-alloy steel is usually specified because high residual stresses and the consequent possibility of cracking can result from the high cooling rate from water-quenching. For larger parts, it is impossible to achieve very much martensite in regions away from the surface of AISI 1040 steel, so to obtain mostly martensite a higher-hardenability steel must be used. Compare the typical hardness values for the three steels in Fig. 10-14 at a Jominy distance of 16 ($^{16}/_{16} = 1$ in., a cooling rate of 5.6°C/s at 700°C): 21 HRC for 1040, 42 HRC for 4140, and 57 HRC for 4340. Note that high-hardenability steel has one important disadvantage besides higher cost: Welding requires special precautions to avoid cracking because so much martensite forms during cooling. Thus, the designer should not specify a steel of higher hardenability than is needed.

## MECHANICAL PROPERTIES OF TEMPERED AISI 4140

Most mechanical properties are changed by tempering, as discussed previously and shown in Figs. 10-7 and 10-9. Figure 10-15 shows several property changes from tempering AISI-SAE 4140 steel over a range of temperatures following oil quenching from 840°C. The trade-offs from tempering can be seen in Fig. 10-15, where one gains ductility in exchange for reductions in strength and hardness. The dip in Charpy impact energy in the temperature range 200–400°C, already described as 250°C embrittlement, usually leads to a choice of tempering either above or below this temperature range. Note that the data in Fig. 10-15 are from room-temperature tests; the abscissa represents the *tempering* temperature, not the testing temperature.

## HIGH-STRENGTH LOW-ALLOY AND DUAL-PHASE STEELS

High-strength low-alloy (HSLA) steels have been developed in recent years to provide steels having strengths substantially greater than those of plain-carbon steels at a cost substantially less than for low-alloy steels such as were discussed earlier.

In the planning of the 800-mile-long Trans-Alaska oil pipeline in the late 1960s, there was a requirement for bigger-diameter pipelines having better mechanical properties than could be provided by the then standard linepipe steels (0.22-0.25% C plain-carbon steels with 1% Mn or more, having strengths up to 420 MPa). These steels are difficult to weld in thicknesses greater than 10 mm under normal conditions, let alone in the severe climate of Alaska, and an early specification asked for 16-mm-wall-thickness pipe. Fully killed normalized (air-cooled) steels could partly satisfy the requirements for oil and gas transmission lines. These steels contain small additions of aluminum, which combine with nitrogen to form aluminum nitride particles during the solidification of the steel or during the subsequent normalizing treatment. The particles prevent austenite grain growth during the normalizing treatment by pinning the austenite grain boundaries. The fine-grained austenite transforms to a fine-grained ferrite and hence produces a steel of enhanced yield strength and resistance to brittle fracture. Unfortunately, the fully killed normalized steels have relatively high carbon-equivalent values, so problems of weld cracking remained.

**Figure 10-15**  Mechanical properties as a function of tempering temperatures for oil-quenched AISI 4140 (UNS G41400) bars of 12.7 mm diameter. "Area (%)" is cold work at fracture, $r_f$. "Elongation (%)" is nominal strain to fracture $e_f$. [From Republic Steel Corp., "Alloy Steels," *Source Book on Industrial Alloy and Engineering Data* (Metals Park, Ohio: American Society for Metals, 1978), p. 36.]

A most important development came with the introduction of "controlled rolled" steels, in the manufacture of which attention is paid to accurate control of rolling temperatures, with accelerated cooling of heavy plates by air or water. This technology is particularly successful with *microalloying additions* in the steel, such as Nb, V, and Ti. The roles of microalloying elements are relatively complex, but essentially the niobium, which starts to form precipitates of Nb(CN) at temperatures below 1200°C, is added to control recrystallization during rolling. Vanadium remains in solution above about 900°C but then precipitates out in the form of fine V(CN) particles during the austenite–ferrite transformation. The major effect of niobium is therefore to promote a fine grain size, while vanadium gives a precipitation-strengthening reaction.

The majority of HSLA steels (now used for offshore applications, bridges, and general purposes, as well as pipelines) have ferrite–pearlite structures. Owing to their poor hardenability, quenching and tempering of HSLA steels produces various nonmartensitic structures (predominately bainite with varying proportions of ferrite depending upon the steel composition and cooling rates). The helpful role played by the microalloying elements is to keep the austenite grain size small, giving improved mechanical properties. Bainitic HSLA steels are available with proof strengths up to even 900 MPa.

In order to permit easy formability, the carbon content of HSLA steels has been reduced considerably, and some "acicular ferrite" or bainitic steels have only 0.03% C. Even so, yet improved press-formability is required for automobile components, and this has led to the development of *dual-phase steels*. These are produced by annealing fairly low-carbon steels in the two-phase ($\alpha + \gamma$) region and then quenching to convert the higher-carbon $\gamma$ to martensite. The microstructure of dual-phase steels is almost the same as that of a particulate metal matrix composite (see Chap. 17) having about 20% of very hard, relatively coarse martensitic particles distributed in a soft ferrite matrix. Dual-phase steels are strong, tough, and readily formable, and find extensive application in the automobile industry. The usual discontinuous yielding of mild steels is replaced by continuous yielding at low offset yield stresses followed by a high work-hardening rate. As a consequence good stretch formability may be achieved. The dual-phase structure can also be produced by controlled cooling directly from hot rolling, but this route is more difficult to control in the steelworks.

## ULTRAHIGH-STRENGTH STEELS

Many of the strengthening mechanisms that have been described can be combined to produce metals of unusually high strength. Two examples are the nickel–martensite maraging steels and the ausforming of certain high-hardenability steels.

Martensites of iron and nickel can be produced in a manner similar to martensites of iron and carbon, although the cooling rates required are much slower and the final structure is less hard. Figure 10-4(a) shows the appearance of iron–nickel martensite. A high-nickel steel that has been vacuum-melted (for control of impurities) and that will produce iron–nickel martensite upon air cooling may have, for example, the following composition: 17.5% Ni, 12.5% Co, 3.75% Mo, 1.8% Ti, 0.15% Al, balance Fe, plus very small amounts of C, Si, and Mn. After air cooling to form martensite, the ultimate tensile strength is approximately 1 GPa. When heated to 480°C for 11 ks, a very fine precipitate forms from the alloying elements, increasing the room-temperature yield strength to 2.4

GPa while retaining the original martensite. This is called *maraging*. Note the similarity of this process to secondary hardening, where fine particles are formed on heating after quenching. The chief difference between maraging and secondary hardening is that in maraging the martensite is stable and does not transform to tempered martensite. The practical advantages of a maraging steel are that the martensite can be formed with slow cooling, and this relatively soft structure can then be cold-worked or machined. The subsequent aging process takes place with virtually no dimensional changes and at a temperature that is so low that the quality of the finished surface is not destroyed. The final structure represents a combination of retained austenite, iron–nickel martensite, and finely dispersed alloy particles. This steel has excellent fracture toughness.

*Ausforming* combines cold work and tempered martensite strengthening in a unique way. It is possible to deform some high-hardenability steels while they are still austenite in a temperature range below the eutectoid temperature, so recrystallization does not occur and neither pearlite nor bainite is formed. After being cooled to form martensite and then tempered, these steels can have extremely high strengths. For example, a 5% Cr steel, Vascojet 1000, air-cooled from 1000°C to 20°C, deformed 90% at 590°C, and tempered at 565°C, has the following room-temperature properties: $S_u$ = 2.6 GPa, $S_y$ = 2.3 GPa, $r_f$ = 40%, $\varepsilon_f$ = 8%, and 63 HRC. This ausforming steel retains acceptable properties to 550°C, where, for example, its strength is 1.8 GPa.

Although the reasons for this exceptional behavior are not fully understood, the "cold-worked" tempered martensite microstructure may have the fine particles of carbide preferentially located near the dislocations formed during ausforming so as to be unusually effective in preventing dislocation motion. It is also possible that aging effects are important here because fine alloy–carbide particles have been found to influence the hardness in some instances for ordinary martensites.

# REFERENCES

1. American Society for Metals, *Metals Handbook,* 9th ed., vol. 1, *Properties and Selection: Irons and Steels.* Metals Park, Ohio: American Society for Metals, 1978.

2. American Society for Metals, *Metals Handbook,* 9th ed., vol. 2, *Properties and Selection: Nonferrous Alloys and Pure Metals.* Metals Park, Ohio: American Society for Metals, 1979.

3. American Society for Metals, *Source Book on Industrial Alloy and Engineering Data.* Metals Park, Ohio: American Society for Metals, 1978. A book of useful engineering data, mostly as tables and graphs, of metals and alloys of industrial importance such as iron and steel, stainless, refractory metals, aluminum, copper, magnesium, and titanium.

4. Birchenall, C. Ernest, *Physical Metallurgy.* New York: McGraw-Hill, 1959. Pages 290–309 give details on the formation of martensite, tempered martensite, and bainite.

5. Reed-Hill, Robert E., *Physical Metallurgy Principles.* Princeton, N.J.: Van Nostrand, 1964. Chapter 15 is particularly valuable for its detailed description of the martensite transformation. See also Chap. 16, "The Iron–Carbon System," and Chap. 17, "The Hardening of Steel," which provide much more detail than is possible here.

## PROBLEMS

**10-1.** What microstructures result from each of the following treatments, taken independently, of a previously annealed 0.77% C steel?

(a) Heat to 740°C, water-quench.

(b) Heat to 820°C, air-cool.

(c) Heat to 150°C, water-quench.

(d) Heat to 760°C, furnace-cool.

(e) Heat to 800°C, water-quench; reheat to 480°C for 2 ks, air-cool.

**10-2.** For each of the following treatments, taken consecutively, calculate (or estimate, if necessary) the Brinell hardness number. Start with a bar of 0.60% C steel 10 mm in diameter of unknown history:

(a) Heat to 850°C for 2 ks, air-cool to 20°C.

(b) Heat to 815°C for 2 ks, water-quench to 20°C.

(c) Heat to 120°C for 4 ks, oil-quench to 20°C.

(d) Heat to 480°C for 2 ks, air-cool to 20°C.

(e) Heat to 700°C for 40 ks, water-quench to 20°C.

(f) Heat to 815°C for 2 ks, furnace-cool to 20°C.

**10-3.** Find the tensile strength of 0.40% C steel after each of the following heat treatments, taken independently:

(a) Heat to 850°C, water-quench to 20°C, and stress-relieve (no residual stresses present).

(b) Heat to 730°C, air-cool to 20°C.

(c) Heat to 730°C, water-quench to 20°C, and stress-relieve (no residual stresses present).

(d) Heat to 870°C, quench in molten salt to 740°C, oil-quench to 20°C.

**10-4.** Estimate the maximum tensile load that initially annealed bars of 0.30% C steel 6 mm in diameter can support after the following heat treatments, each taken separately:

(a) Furnace-cool from 850°C.

(b) Oil-quench from 850°C.

(c) Water-quench from 850°C and stress-relieve.

**10-5.** When you heat a piece of cold-worked 0.40% C steel to 850°C and water-quench it, it becomes harder. When you heat a piece of cold-worked alpha brass to 500°C and water-quench it, it becomes softer. Explain why similar heat treatments produce opposite effects in the two alloys.

**10-6.** Small pieces of plain-carbon steels are given the treatments listed below. What are the resulting microstructures and hardnesses?

| No. | Percentage of Carbon | Austenitizing Temp., °C | Quenching Medium |
|-----|----------------------|-------------------------|------------------|
| 1   | 0.77                 | 760                     | Oil              |
| 2   | 0.77                 | 760                     | Water            |
| 3   | 0.20                 | 760                     | Water            |
| 4   | 0.40                 | 840                     | Water            |

**10-7.** The surface microstructure of a bar of 0.40% C steel 6 mm in diameter, water-quenched from 730°C, contains one-half ferrite and one-half martensite. (This is an example of a *duplexed* or *dual-phase* microstructure.)

(a) What is the carbon content of the martensite immediately after quenching?

(b) What is the carbon content of the ferrite immediately after quenching?

(c) Estimate the surface hardness of this bar. Show how you obtained your answer.

**10-8.** A ball-bearing company makes high-precision bearing races (the running surfaces for ball bearings) of AISI-SAE 52100 steel (UNS G52986), a low-alloy steel of 1.00% C. These are first rough-machined, then austenitized and oil-quenched, and then cooled to −75°C and warmed to room temperature before the final precision-grinding operation. What might occur if the −75°C cooling is omitted? Explain why.

**10-9.** Tempering of martensite at 430°C produces a precipitate, and the final hardness is lower than the hardness before tempering. In contrast, precipitation in 5% Cu–Al alloys at 150°C (age hardening) *increases* the hardness. Explain this difference in the behavior of the two alloys.

**10-10.** A production part 75 mm in diameter made of AISI-SAE 1040 steel (UNS G10400), water-quenched from 850°C, is hard enough at the outside surface but too soft at the center.

(a) What changes would you recommend?

(b) In practice, parts of this size are virtually never water-quenched from austenite. Why not?

**10-11.** The surface of a bar of AISI-SAE 1040 steel (UNS G10400) 50 mm in diameter, water-quenched from 870°C, is near 43 HRC.

(a) If this bar were 100% martensite, what would be its hardness?

(b) Why is this bar not 100% martensite?

**10-12.** A part made of AISI 4140 steel (UNS G41400) fractured in service under impact loading. Several other identical structures are to be constructed, and the owner wants to avoid future fractures. These parts had been oil-quenched from 840°C, tempered at 480°C, and air-cooled, resulting in 1.1 GPa strength. The designer had specified this treatment so these parts would be extra strong—to be on the "safe" side although the design stress, including all safety factors, was only 0.96 GPa. What would you recommend that the owner do to reduce the chances for failure of the structures that are still to be built?

**10-13.** Is it possible for two steels to have the same hardability but different hardnesses? Explain very briefly.

**10-14.** Two complete Jominy curves, for steels *A* and *B,* are given in Fig. 10-16.

(a) Which steel has the higher hardenability, *A* or *B?*

(b) Which would probably be better for use in knife blades?

(c) Which will give more uniform strength after quenching?

(d) Which has the higher percentage of alloying elements other than carbon?

(e) Which will have more pearlite present at the center of a bar 40 mm in diameter that has been water-quenched from 900°C?

(f) Which contains harder martensite?

**Figure 10-16**

**10-15.** Name or describe each of the phases present and the microstructure in which these phases are found after the treatment given:

(a) 0.20% C–99.8% Fe, slowly cooled from the melt.

(b) 1.20% C–98.8% Fe, slowly cooled from the melt to 750°C and then water-quenched.

(c) Alloy (b), given the same treatment and then reheated to 430°C for 2 ks.

(d) AISI 1040 steel (UNS G10400), quenched from 840°C to 480°C, held for 600 s, and then oil-quenched.

(e) AISI 1095 steel (UNS G10950), quenched very rapidly from 870°C to 150°C.

(f) Alloy (e), given the same treatment and then reheated to 315°C for 2 ks.

**10-16.** Isothermal (constant-temperature) transformation data for very small pieces of 0.77% C plain-carbon steel initially at 800°C are given below for two bath temperatures, each followed by water quenching to 20°C.

| Bath Temp. 675°C | | Bath Temp. 370°C | |
|---|---|---|---|
| Time, s | HRC | Time, s | HRC |
| 5 | 63 | 5 | 63 |
| 10 | 63 | 10 | 61 |
| 15 | 53 | 15 | 52 |
| 20 | 43 | 60 | 50 |
| 30 | 27 | 100 | 45 |
| 60 | 24 | 150 | 43 |

Name the microstructure(s) and phase(s) present at the end of

(a) 5 s at 675°C.

(b) 5 s at 370°C.

(c) 15 s at 675°C.

(d) 15 s at 370°C.

(e) Explain, using the fundamental ideas of strengthening mechanisms, why the hardness after 60 s at 675°C and water quenching is lower than the hardness after 150 s at 370°C and water quenching. Start by describing the final microstructures of these two specimens.

**10-17.** For each treatment listed (assuming bars 6 mm in diameter), write the number of the schematic microstructure in Fig. 10-17 that most accurately depicts the final microstructure. A microstructure may be used in more than one answer. In all cases $\alpha$ and $\beta$ are used to indicate any possible solid solutions, and $\theta$ is used to indicate an intermetallic compound or harder phase of some other type. Note magnifications.

(a) 70% Cu–30% Zn, cooled slowly from the melt, heated to 315°C for 2 ks, oil-quenched.

(b) 5% Cu–95% Al, water-quenched from 510°C, heated to 260°C for 4 ks, oil-quenched.

(c) 5% Cu–95% Al, cooled slowly from the melt to 20°C.

(d) 99.6% Fe–0.4% C, cooled slowly from the melt to 20°C.

(e) 99.6% Fe–0.4% C, water-quenched from 840°C.

(f) 99.6% Fe–0.4% C, oil-quenched from 840°C and then heated to 700°C for 20 h.

(g) 99.6% Fe–0.4% C, water-quenched from 840°C and then heated to 425°C for 2 ks.

(h) AISI 4140 steel (UNS G41400), oil-quenched from 840°C.

(i) 99% Fe–1 % C, cooled slowly from 870°C to 740°C and then water-quenched.

(j) AISI 1040 steel (UNS G10400) from the outer surface of an end-quenched Jominy bar at position $J = 24$.

**10-18.** Taking into account the functional requirements and general material characteristics and properties, match each of the following materials with the most appropriate example of use.

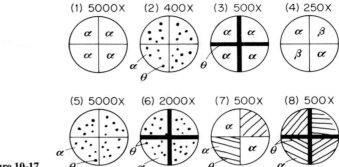

**Figure 10-17**

| Material | Examples of Use |
|----------|-----------------|
| (1) Aluminum alloy | (a) Die casting, cooking utensils, hydraulic tubing |
| (2) Brass | (b) Common nails |
| (3) 1010 steel | (c) Roller bearings |
| (4) 1040 steel | (d) Gears, crankshafts, low-quality hand tools |
| (5) 1080 steel | (e) Radiator cores, springs, tubes |
| (6) 10120 steel | (f) High-quality wrenches, hammers, pliers, axes |
| (7) 4140 steel | (g) Moderate, uniform-strength shafts, high-quality bolts |
| (8) 52100 steel | (h) Shear blades, knife edges, files |

**10-19.** An AISI 1040 steel (UNS G10400), oil-quenched from 850°C, has a hardness of 28 HRC; the same heat treatment for an AISI 4140 steel (UNS G41400) gives a hardness of 56 HRC.
  **(a)** Describe the differences in microstructure.
  **(b)** Show how these differences lead to the difference in hardness.
  **(c)** Explain why the 4140 steel gave a different structure after the same heat treatment.

**10-20.** An AISI 4140 steel (UNS G41400) bar 12.8 mm in diameter is oil-quenched from 870°C, reheated to 540°C for 2 ks, and then air-cooled to 20°C. Calculate the maximum load that this bar can support in tension.

**10-21.** You are given a piece of AISI 1050 steel (UNS 10500) in the shape of a short bar 8 mm in diameter.
  **(a)** Describe a heat treatment, giving exact temperatures, times, and cooling rates (e.g., quenching in oil or water, air cooling, furnace cooling), for achieving the maximum possible hardness in this steel without changing its chemical composition. What value of hardness can you achieve?
  **(b)** Describe a heat treatment in the same way as in part (a) for achieving the *lowest* possible hardness for this steel. What hardness would you expect?

**10-22.** Tensile tests on three specimens of AISI 4140 steel (UNS 41400) gave the following data (initial diameter of each is 9.07 mm):

| No. | Heat Treatment | HRC | Max. Load, kN | Fracture Diameter, mm |
|-----|----------------|-----|---------------|------------------------|
| 1 | Oil-quench (OQ) | 53 | 73.4 | 9.04 |
| 2 | OQ, 233°C for 2 ks | 48 | 113 | 6.50 |
| 3 | OQ, 566°C for 2 ks | 27 | 66.7 | 5.87 |

(a) Describe the final microstructure of all three specimens and give the fractions of each phase present.

(b) Explain why the maximum load for specimen 2 is greater than for specimen 1. Be specific as to cause.

(c) Explain from a fundamental point of view why the true strain at fracture of specimen 3 is greater than that for specimen 2.

**10-23.** Draw the approximate curve of Charpy V-notch energy measured at 20°C vs. tempering temperature for AISI 4140 steel (UNS G41400) that has been first oil-quenched from 850°C. Explain all parts of this curve.

**10-24.** An AISI 4140 steel (UNS 41400) is continuously cooled from 820°C to 20°C at a rate so as to give a hardness of 41 HRC. Name all phases present at 20°C.

**10-25.** A plain-carbon steel of 0.38% C is heated to 820°C for 2 ks, oil-quenched to 20°C, heated to 517°C for 2.5 ks, water-quenched to 20°C, subcooled to −60°C for 10 ks, and air-heated to 20°C. Estimate its final HB.

**10-26.** A plain-carbon steel of 0.77% C is heated to 760°C for 2 ks, cooled to 350°C in 1 s, held at 350°C for 1 ks, and oil-quenched to 20°C. Name the microstructure(s), sketch the microstructure(s), and label the phase(s) that result.

**10-27.** An unknown steel of unknown heat treatment gives the following hardness data: as received, 27 HRC; heated to 885°C and water-quenched, 56 HRC; heated to 885°C and oil-quenched, 49 HRC.

(a) What is the approximate carbon content?
(i) 0.20% C.
(ii) 0.40% C.
(iii) 0.80% C.

(b) What is the alloy content?
(i) Plain-carbon steel.
(ii) Low-alloy steel.

(c) What was the as-received condition?
(i) Annealed.
(ii) Normalized.
(iii) Quenched to martensite.

**10-28.** You are provided with three specimens of steel of identical size: 10 mm × 20 mm × 4 mm thick. Each is initially tested for hardness, then heated to 840°C for 0.5 h, and then water-quenched to 20°C. The final hardness is then measured. Results are as follows:

| Specimen | Initial Hardness, HRC | Final Hardness, HRC |
|----------|-----------------------|---------------------|
| A        | 20                    | 62                  |
| B        | 20                    | 46                  |
| C        | 20                    | 66                  |

(a) Describe the prior heat treatment of each specimen before you measured the initial hardness.

(b) Describe the microstructure of each specimen and name all phases present in each specimen at the time you measured the initial hardness.

(c) Give the chemical composition (as much about it as you know) of each specimen.

**10-29.** A bar of AISI-SAE 4140 steel (UNS G 41400) of dimensions 11.3 mm × 17.6 mm × 37.4 mm is heated to 810°C for 2 ks and then oil-quenched to 20°C.

(a) Name the microstructure(s) present.

(b) Estimate the Rockwell C hardness of this bar following the above heat treatment and state the source of your information.

This bar is then heated to 400°C for 2 ks and air-cooled to 20°C.

(c) Name the microstructure(s) present.

(d) Estimate the Rockwell C hardness of this bar following the second heat treatment and state the source of your information.

# 11

# *Stainless and Heat-Resistant Steels*

Stainless steels were developed chiefly in the twentieth century to provide a low-cost alloy of iron that is resistant to many types of corrosion. In addition, some alloys of this class of steels maintain their strength and corrosion resistance at elevated temperatures. This chapter covers the general classes of stainless and heat-resistant steels; it describes their strengthening mechanisms and examines the most common causes of their embrittlement or loss of corrosion resistance. Because it is impossible to cover the large number of different alloys thoroughly in a single chapter, only a few of the more common stainless steels are described as examples.

For reasons described later in this chapter, stainless steels are essentially iron–chromium alloys that may or may not contain other elements such as carbon, nickel, and molybdenum. These elements can cause the stable crystal structure of the iron to be either fcc or bcc, and so we use the terms "austenite" and "ferrite," respectively; however, the austenite and ferrite do not have the same composition as in Fe–C. Here we are not dealing with a simple Fe–C alloy as in previous chapters, so we should not be surprised that the behavior of Fe–Cr alloys is markedly different.

## STAINLESSNESS

P. Monnartz and W. Borchers first discovered in 1911 that the corrosion rate of iron–chromium steels drops to virtually zero when the chromium content reaches 12%. This phenomenon can be seen in Fig. 11-1, taken from later work by R. Franks. In a 10-year (315-Ms) outdoor test, mild steel corroded away at the rate of about 680 fm/s, Fe–4% Cr at about 140 fm/s, and Fe–12% Cr at virtually a zero rate. Above 12% chromium, *passivity* occurs. This is the formation of a surface condition that inhibits corrosion completely. For this reason, virtually all stainless steels have a chromium content of at least 12%.

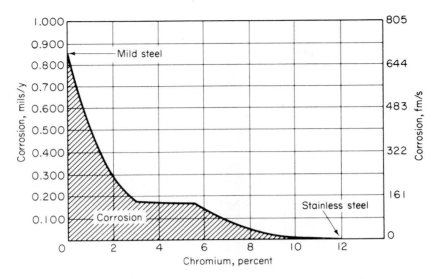

**Figure 11-1** Corrosion rate for Fe–Cr alloys in an outdoor industrial atmosphere, measured for 10 years (315 Ms). [From R. Franks, "Chromium Steels of Low-Carbon Content," *Trans. ASM,* 35 (1945), p. 616, disc. p. 636. Copyright American Society for Metals. Reprinted with permission.]

## IRON–CHROMIUM–CARBON EQUILIBRIUM

Figure 11-2 is a pseudobinary equilibrium diagram of iron and chromium developed by Zapffe (Ref. 7). It is "pseudo" because the carbon content varies to match the different alloys in the figure. Nevertheless, Fig. 11-2 (combined later with Fig. 11-3) shows clearly why the different classes of stainless steels behave in different ways.

The small region in the upper left of Fig. 11-2, which represents austenite (Fe–Cr), shows the limits of austenite in a pure Fe–Cr alloy of composition up to 12.7% Cr. However, with small amounts of carbon present, this *austenite loop* expands to about 18% Cr (for about 1% C content) to form the larger loop labeled "austenite (Fe–Cr–C)." Alloys of this composition range (12–18% Cr) can usually be air-cooled to form martensite and are thus the *martensitic stainless steels*. In general, the alloys of this class having higher Cr also have higher C because increased Cr alone leads to a reduction in the range of the austenite loop, which must be compensated for by increased C, which increases the size of the austenite loop. Thus, a balance is required between chromium and carbon contents.

Fe–Cr steels above 12.7% Cr and with zero carbon are ferritic at all temperatures, as are Fe–Cr–C steels ranging from 12.7 to 18% Cr if the carbon content remains low enough. Above 18% Cr, the alloys of higher carbon are usually ferritic at 20°C. Thus, low-carbon alloys ranging from 12 to 30% Cr are the *ferritic stainless steels*. Ferritic stainless steels always have lower carbon contents than martensitic stainless steels of the same chromium content.

**Example 11-1**

On what basis would a designer decide between martensitic and ferritic stainless steel?

**Figure 11-2**   Pseudobinary equilibrium diagram for Fe–Cr, with carbon content varying to balance chromium. [From Carl A. Zapffe, "Constitution of Stainless Steels," *Stainless Steels* (Metals Park, Ohio: American Society for Metals, 1949), p. 114.]

**Solution**   Ferritic stainless steel can be cold-worked at 20°C, so it is appropriate for applications requiring cold forming and moderate stress levels. Martensitic stainless steel cannot be cold-worked, but it is much stronger and harder and thus useful for structural applications.

## IRON–CHROMIUM–NICKEL EQUILIBRIUM

Addition of more than about 6% nickel to an Fe–18% Cr alloy leads to a stable austenite region at progressively lower temperatures with increasing nickel, as shown in Fig. 11-3. In fact, the transformation from austenite to ferrite on cooling is so sluggish that it normally does not occur for nickel contents above about 8% in the Fe–18% Cr alloy, for example. Thus, we have another method, the addition of nickel, for extending the austenite range of Fe–Cr alloys, but in this case the austenite is stable at room temperature. These alloys are the *austenitic stainless steels*. Because of the stability of the austenite, they cannot normally be heat-treated to increase strength.

The only other austenite-promoting element of consequence is manganese, which can be used in place of some of the nickel to produce a stable austenitic alloy. We will consider these alloys as part of the austenitic class.

## MARTENSITIC STAINLESS STEELS (Fe–Cr–C)

This class has a chromium content from about 12 to 18% and a carbon content from about 0.15 to 1.20%, depending upon the chromium content. In general the carbon content must increase with the chromium content in order to permit expansion of the austenite loop.

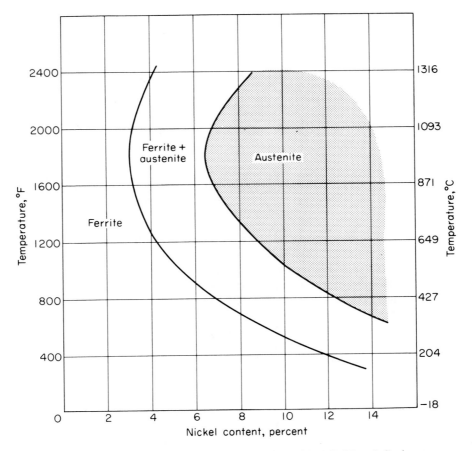

**Figure 11-3**  The effect of nickel content on an alloy of Fe–18% Cr. [From J. Gordon Parr and Albert Hanson, *An Introduction to Stainless Steel* (Metals Park, Ohio: American Society for Metals, 1965), p. 27.]

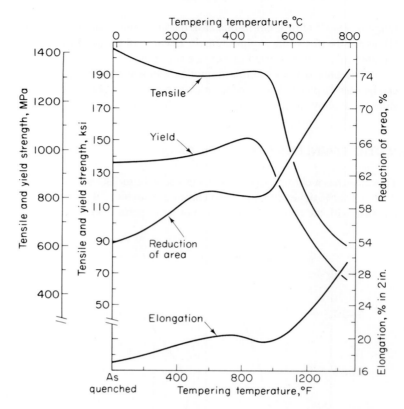

**Figure 11-4**    Mechanical properties of Type 410 after austenitizing, quenching, and tempering. [From *The Making, Shaping and Treating of Steel,* 8th ed., H. E. McGannon, ed. (Pittsburgh, Pa.: U.S. Steel Corp., 1964). Copyright 1964 by United States Steel Corporation.]

From this point onward, there will be extensive discussion of the effects of chemical composition on behavior. For simplicity we shall often omit the % sign when giving compositions; it should always be clear from the context.

One of the most used martensitic stainless steels is alloy Type 410 (UNS S41000). (Throughout this chapter we will designate this alloy by its traditional number, 410, but the new UNS designation is provided for ready reference. Compositions of the common stainless steels are given in App. 7.) The composition of Type 410 is 11.5–13.5 Cr and 0.15 max C. (Most of the common stainless steels have composition limits ranging as follows: 1–2 max Mn, 0.5–1.5 max Si, 0.040–0.060 max P, and 0.030 max S. These limits are omitted in this discussion; see App. 7 for exact limits.) Type 410 is used as a general-purpose, heat-treatable stainless steel for such applications as machine parts, pump shafts, cutlery, jet engine parts, screws, and valves. Most martensitic stainless steels can be air-cooled from austenite to produce martensite, but the austenitizing temperature for the 12–14% Cr grades must be closely controlled to avoid the solid–liquid two-phase region above the austenite loop (see Fig. 11-2). The recommended austenitizing temperature range for Type 410 is 930–1010°C, followed by air- or oil-quenching. The part can then be either stress-relieved at 200–370°C or tempered.

Typical properties of Type 410 following quenching and stress relieving are 1.3 GPa strength, 1.0 GPa yield strength, 375 HB, 15% elongation at fracture, and 54 J Izod impact energy.

The mechanical properties of alloy Type 410 after tempering over a range of temperatures are given in Figs. 11-4 and 11-5. Note the severe temper embrittlement that occurs in the 370–600°C range for this alloy. As this temperature range also leads to decreased corrosion resistance, recommended tempering temperatures are in either the 200–370°C range or the 560–610°C range. As is usual for martensitic steels (see Chap. 10), tempering in the lower range gives higher strength, hardness, and lower toughness, while tempering in the upper range gives the opposite.

Although retained austenite is not a serious problem in this alloy because of its relatively low carbon content, martensitic stainless steels with higher chromium and carbon must be subcooled immediately after quenching. Cooling to about −75°C forces the retained austenite to transform to martensite and avoids later service embrittlement or dimensional changes that might occur if much retained austenite can transform.

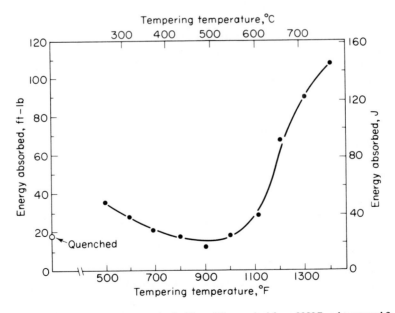

**Figure 11-5**   Izod impact test results for Type 410 quenched from 980°C and tempered 3 h at indicated temperature. [From W. C. Rion, Jr., *Stainless Steel Information Manual for the Savannah River Plant,* vol. I, Properties, USAEC Report DP-860 (July 1964), NASA N64-33060. Reprinted with permission.]

The alloys of this class with higher chromium and carbon are particularly susceptible to *hydrogen embrittlement* (see Chap. 18). Hydrogen can enter the steel during melting or heat treating or during such processes as electroplating. It can cause premature
cracking and unexpected failure of a part. Hydrogen-generating processes should be
avoided or subsequent heating in the range 200–370°C used to drive the hydrogen from
the metal.

Martensitic stainless steels of higher chromium (approaching 18% Cr) and carbon
contents to 1.20% have tensile strengths up to 1.9 GPa, yield strengths to 1.8 GPa, and
hardness to 600 HB. They are also more corrosion-resistant and more expensive than 12%
Cr martensitic steels.

## FERRITIC STAINLESS STEELS (Fe–Cr)

When the carbon content is reduced for any given chromium content, the austenite loop
contracts and the alloy becomes partially or completely ferritic when in the 1000°C range.
Steels with carbon content lower than that of martensitic stainless steels, but with
chromium content in the range 12–30%, are usually ferritic.

The most commonly used ferritic stainless steel is alloy Type 430 (UNS S43000). Its
typical composition is 16–18 Cr and 0.12 max C. (Note that it has both higher Cr and lower
C than the martensitic Type 410 discussed in the previous section.) This alloy is very commonly used in corrosion-resistant nonstructural applications requiring cold forming, such
as kitchen sinks, decorative trim, annealing baskets, and nitric acid tanks. While more
expensive (because of higher Cr content) than the martensitic stainless steels, the ferritic
stainless steels are more corrosion-resistant. Heat treatment consists of annealing at
700–790°C followed by air cooling or water quenching. This lower annealing temperature
avoids the formation of austenite. (See Fig. 11-2.)

Rapid cooling after annealing is essential to avoid what was traditionally called
*885°F embrittlement,* which we shall call *475°C embrittlement* following the suggestion of
Parr and Hanson (Ref. 5). When held in the temperature range of 450–550°C, ferritic stainless steels experience a loss in toughness and an increase in hardness. The loss in toughness is so severe that the steels are useless unless the embrittlement is removed by holding
at 600°C and quenching. The mechanism is not well understood, being attributed either to
formation of a coherent precipitate or to atomic-level ordering, but the effect is more
marked with the higher-chromium grades. Slow cooling through the 450–550°C range is
thus to be avoided. For this reason, the ferritic stainless steels are not generally welded,
unless it is possible to remove the effects of 475°C embrittlement by subsequent heating
and quenching. As other satisfactory weldable grades of stainless steel are available, the
ferritic grades are usually used only for unwelded parts.

Another problem that precludes welding of the ferritic grades is grain growth. As
temperature increases, so does grain size. In the martensitic grades, reheating leads to a
phase change to austenite, with consequent fine grain size. But because the ferritic grades
cannot be heated to force a phase transformation, they can achieve grain-size reduction
only by cold work and recrystallization. This is usually impractical, so for this reason as

well, the ferritic grades are not usually welded. Grain growth can be controlled somewhat by the addition of nitrogen.

Typical mechanical properties of Type 430 are tensile strength 480 MPa, yield strength 275 MPa, and 155 HB in the annealed condition, and tensile strength 620–900 MPa and yield strength 450–850 MPa in the cold-worked condition. Nominal strain to fracture of 20% for 51-mm gage length in the annealed condition drops to 16% in the cold-worked condition. It is clear that the ferritic stainless steels are not usually the choice for structural applications because of their low strength. The higher-Cr members of this class have cold-worked strengths lower than those of Type 430, but they are more corrosion-resistant.

Two other new problems occur in the ferritic grades: *sigma-phase embrittlement* and *sensitization*. In Fig. 11-2, at 46% Cr and temperatures below 821°C, the *sigma phase* occurs. This phase is an intermetallic Fe–Cr compound that is hard and brittle, and extends to include Cr contents in the 20–30% range. Fortunately, sigma forms slowly, and so it is not generally present unless the alloy is held between 500 and 800°C for many hours. Sigma forms first at grain boundaries and leads to an embrittled part. It can be dissolved by holding at 900°C and quenching.

*Sensitization,* while more significant in the austenitic stainless steels, does occur in the ferritic grades. In alloys with chromium contents above 16%, rapid cooling after holding at temperatures above 925°C leads to accelerated grain-boundary attack when in a corrosive medium and eventually to failure. As sensitization is more of a problem with austenitic stainless steels, it is discussed further in the next section. Sensitization in ferritic stainless steels can be avoided by not heating to 925°C (note the low annealing temperature range of 700–790°C), which means that welding is undesirable. We now have three reasons for not welding the ferritic grades: 475°C embrittlement, grain growth, and sensitization.

## AUSTENITIC STAINLESS STEELS (Fe–Cr–Ni)

Figure 11-3 suggests that a very high nickel content is required for austenite to be stable at 20°C, but this is misleading. The austenite-to-ferrite transformation in Fe–Cr–Ni alloys is very slow, and in fact an 18% Cr alloy with 8% Ni remains stable as austenite at 20°C. Most of the austenitic grades are based on variations of this basic "18-8" austenitic stainless steel. It cannot be hardened by heat treatment, but some alloys show remarkable strengthening by cold work, having strain-hardening exponents $n$ (in $\sigma = K\varepsilon^n$) of 0.4–0.5. This leads to difficulty in mechanical working, which is due to the rapid increase in flow stress for small deformations (see Chap. 2). The austenitic grades of 16–30% Cr and 6–20% Ni are the most corrosion-resistant of all the stainless steels and do not suffer embrittlement at very low temperatures. They are generally nonmagnetic and for this reason are used in some electrical applications.

The basic 18-8 alloy is either Type 301 (UNS S30100) or Type 302 (UNS S30200). Type 301 has typically 16–18 Cr, 6–8 Ni, and 0.15 max C. Type 302 is the same except that the Cr limit is 1% higher and the Ni limit is 2% higher. Here we consider these a single alloy type. This stainless steel is annealed in the range 1000–1120°C and then air-cooled or water-quenched. In this condition Type 301 has tensile strength 760 MPa, yield strength 275 MPa, and 60% elongation at fracture over a 51-mm gage length. The influence of cold work on Type 301 (and some other stainless steels) is shown in Fig. 11-6. This remarkable increase in strength with cold work is largely attributable to the transformation of unstable austenite to martensite as a result of the cold work. The balance of the strength increase results directly from plastic deformation. (Note that stainless steel introduces no new strengthening mechanisms but combines them in new ways.) Types 301 and 302 are widely used where parts are cold-formed and require higher corrosion resistance, strength, and ductility than provided by ferritic Type 430.

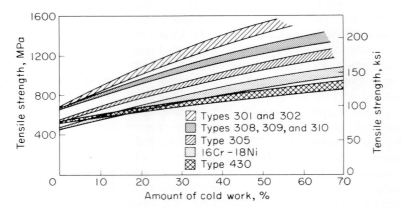

**Figure 11-6** Effect of cold rolling on tensile strength of selected stainless steels. [From ASM Committee on Wrought Stainless Steels, "Wrought Stainless Steels," *Metals Handbook*, vol. 3, 9th ed., David Benjamin, ed. (Metals Park, Ohio: American Society for Metals, 1980), p. 33.]

**Example 11-2**

Compare the costs of the common grades of martensitic, austenitic, and ferritic stainless steels.

**Solution** Remember that iron is by far the cheapest metal in any steel. Chromium and nickel are much more expensive. Thus, 18 Cr–8 Ni austenitic stainless steel (Types 301 and 302) is the most expensive, ferritic grades such as the 17% Cr Type 430 are less expensive because of the absence of nickel, and martensitic 12.5% Cr Type 410 is the cheapest.

The chief problem with the austenitic grades is sensitization, already mentioned as a minor problem with the ferritic grades. When heated in the range 425–800°C, chromium carbide precipitates at the grain boundaries. This is called sensitization because it leads to local depletion of the chromium content to below 12% Cr in the vicinity of the grain boundaries, with a resulting loss of passivity and thus sensitivity to a corrosive environment. The alloy then experiences rapid corrosion at the grain boundaries and can literally fall apart. Sensitization is reversible if, prior to service, the part is heated to about 1100°C, soaked long enough to redissolve the chromium carbides, and then water-quenched. If possible, the temperature range of 425–800°C should be avoided altogether, but this is not possible during processes such as welding, so special measures must be taken.

Many of the austenitic alloys were developed in order to control the problem of sensitization. The two most effective means for controlling sensitization are (1) keeping the carbon content very low so that little chromium is lost to chromium carbide, and (2) introducing other elements that can preferentially and harmlessly combine with the carbon before the chromium does, thus effectively tying up the carbon.

Alloys Type 304 (UNS S30400) and Type 304L (UNS S30403) reduce but do not eliminate the sensitization problem through limiting carbon to low levels. The composition of Type 304 is 18–20 Cr, 8–10.5 Ni, 0.08 max C. Type 304L differs significantly only in having an extra-low carbon content, a maximum of 0.03%. These grades are used where welding precludes the use of Types 301 and 302, as in welded cryogenic (very-low-temperature) vessels. Comparison of the compositions of Types 302 and 304 shows that the only differences are that Type 304 has 1% more Cr and half the C. The establishment of two separate alloys having progressively lower carbon content demonstrates the importance of the sensitization problem. The mechanical properties of Types 304 and 304L are close to each other and have only slightly lower values than those of Type 302; for Type 304, tensile strength is 580 MPa, yield strength is 290 MPa, and elongation at fracture is 55%. The substantial differences in strain hardening between Types 301 and 304 can be seen in Fig. 11-7. Clearly, the higher Cr and Ni and lower C contents of Type 304 permit less strain-induced transformation of austenite to martensite, with consequently less strain hardening.

Type 347 (UNS S34700) controls sensitization by small additions of niobium. Its composition is 17–19 Cr, 9–13 Ni, 0.08 max C, and Nb = 10 × C min. The intent of the

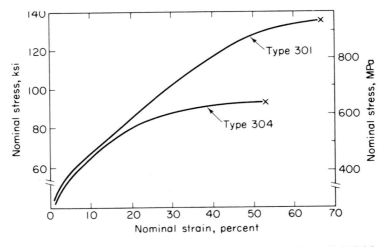

**Figure 11-7**    Stress–strain curves for austenitic Types 301 and 304. [From K. G. Brickner, "Stainless Steels for Room and Cryogenic Temperatures," *Met. Eng. Quart.* (May 1968). Copyright American Society for Metals. Reprinted with permission.]

composition is to be reasonably sure that enough niobium (which is a strong carbide former) is present to tie up the carbon before it can combine with any chromium. Notice that, except for a small decrease in chromium and an increase in nickel, the Nb requirement is the only change from the composition of Type 304. Type 321 (UNS S32100) is virtually identical to Type 347 except that titanium is used as a stabilizer instead of niobium. Annealed mechanical properties of Types 347 and 321 are very similar to those of Type 304. Since titanium reacts rapidly with oxygen at high temperature, Type 321 cannot be used for electric arc welding. Types 347 and 321 are used for high-temperature welded parts such as aircraft exhaust manifolds, boiler shells, pressure vessels, and welded tank cars for chemicals (347).

Type 316 (UNS S31600) has molybdenum added to expand the passivity range, which thus reduces pitting and increases resistance to corrosion from neutral chloride solutions (such as sea water). Thus, Type 316 is one of the more popular of the stabilized austenitic stainless steels. Type 316L (UNS S31603) has been used extensively for prosthetic implants in the human body. The composition of Type 316 is 16–18 Cr, 10–14 Ni, 0.08 max C, and 2–3 Mo; Type 316L has 0.03 max C. The room-temperature mechanical properties of Type 316 are virtually identical to those of Type 304.

The low-temperature mechanical properties of the austenitic grades are unique among the stainless steels. For example, Fig. 11-8 shows the influence of temperatures down to 20 K on some properties of Types 304 and 304L.

For some of the austenitic alloys that contain some ferrite when welded, sigma formation can occur through extended heating in the range 500–800°C. The sigma can be redissolved in the same way as for the ferritic stainless steels, by soaking at 900°C and quenching.

**Example 11-3**

Why are the austenitic stainless steels widely used?

**Solution**   Although they are the most expensive of the stainless steels, the austenitic grades can be (moderately) cold-worked, many can be welded, they have excellent corrosion resistance, and their strength, ductility, and elevated temperature properties are good.

## PRECIPITATION-HARDENED STAINLESS STEELS

Developed during the 1940s, the precipitation-hardened stainless steels make possible precision fabrication of a corrosion-resistant part of high strength and hardness. None of the other grades allows this: The martensitic stainless steels suffer dimensional changes when undergoing the martensite transformation; the ferritic grades are never very strong; and the austenitic grades achieve their highest strength as a result of cold forming.

The principles of precipitation hardening, as already described in Chap. 9 for the Al–Cu system, apply equally well to Ni–Al and Ni–Ti: The nickel-rich ends of both of these systems exhibit decreasing solubility with decreasing temperature in the same fashion as the aluminum-rich end of the Al–Cu system (Fig. 9-11). Thus, if either Al or Ti is present in a martensitic stainless steel that contains Ni, fine particles of intermetallic com-

**Figure 11-8**   Change in mechanical properties with decreasing temperatures for cold-worked Types 304 and 304L. [From *Chromium–Nickel Stainless Steel Data,* Sect. I: Wrought Stainless Steels, Bulletin C. "Mechanical and Physical Properties of Austenitic Chromium-Nickel Stainless Steels at Subzero Temperatures," The International Nickel Co., Inc. (1963). Reprinted courtesy of The International Nickel Company, Inc.]

pounds of Ni–Al or Ni–Ti will be formed upon aging at intermediate temperatures. This aging takes place at temperatures well below the austenitizing temperatures required to produce martensite, so the hardening and strengthening that result from the formation of fine particles during aging can take place after fabrication, forming, and machining.

A common precipitation-hardened stainless steel is 17-7 PH (UNS S17700) of the following composition: 16–18 Cr, 6.5–7.75 Ni, 0.09 max C, and 0.75–1.50 Al. This alloy is austenitized at 954 ± 14°C and air-cooled to 20°C. In order to force the transformation of as much austenite to martensite as possible, it is then cooled to below –68°C for 29 ks. The aging (hardening) temperature required for fine-particle precipitation of Ni–Al depends on the strength level desired. Figure 11-9 shows the variation with age-hardening temperature of strength and elongation for aging 17-7 PH for 3.6 ks. For example, actual recommended treatments are as follows: For a minimum tensile strength of 1.45 GPa, age at $510 \pm 6°C$ for 3.6 ks; for strength of 1.03 GPa, $593 \pm 6°C$. Thus, 17-7 PH, for 510°C aging, can have a yield strength of 1.5 GPa and a hardness of 47 HRC. Since it can be cold-drawn prior to this final heat treatment, it is useful for such applications as springs, knives, and pressure vessels.

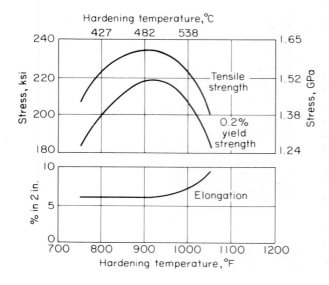

**Figure 11-9**   Tensile and yield strengths, and nominal strain at fracture as a function of aging temperature for 17-7 PH (UNS S17700). The specimens were first heated at 954°C for 600 s, air-cooled, liquid-cooled to −73°C for 29 ks; age-hardened for 3.6 ks at indicated temperatures. [From ASM Committee on Heat Treating of Stainless Steel, "Heat Treating of Stainless Steel," *Metals Handbook,* 8th ed., vol. 2, T. Lyman, ed. (Metals Park, Ohio: American Society for Metals, 1964), p. 251.]

The precipitation-hardening stainless steels have slightly better corrosion resistance than the martensitic stainless steels because of generally higher chromium content. At high strength levels, the precipitation-hardening steels are susceptible to stress-corrosion cracking in some environments (see Chap. 18), and they are notch sensitive. Thus, in practice these steels are often deliberately overaged to lower strengths in order to avoid stress-corrosion problems and to increase their notch toughness.

## ELEVATED TEMPERATURE PROPERTIES OF STAINLESS STEELS

Many of the stainless steels of the four classes described in the foregoing sections retain their corrosion-resistant properties to elevated temperatures. Some of them also retain excellent strength at elevated temperatures. For these reasons, stainless steels are frequently used for service above room temperature where plain-carbon steels are inadequate. Figure 11-10 shows the short-term strength of stainless steels, taken from an ordinary tension test, over a range of temperatures from 300° to 800°C. The precipitation-hardening grades are clear choices at temperatures up to about 500°C, above which the austenitic grades are superior.

One measure of the high-temperature performance of materials is their resistance to *creep*. (This is discussed in Chaps. 12 and 13.) Creep is the slow deformation of a material under stress at elevated temperature that occurs with virtually no strain hardening. The

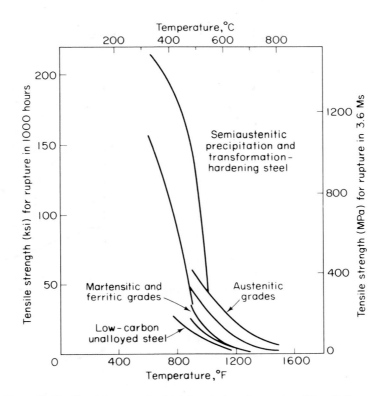

**Figure 11-10** Short-term strength of stainless steels vs. temperature. (From T. D. Parker, "Strength of Stainless Steels at Elevated Temperature," *Met. Eng. Quart.,* May 1968. Copyright American Society for Metals. Reprinted with permission.)

*creep rate* in this case is defined as the total nominal strain divided by the time a specimen is exposed to a fixed tensile stress and temperature. (In general, creep rate is the instantaneous slope of the strain vs. time curve; see Chap. 12.) Figure 11-11 shows the stress that will produce a fixed creep rate in a number of different stainless steels. Compositions and other properties of the numbered alloys can be found in App. 7 and in Refs. 4–6. Type 316 has become one of the most commonly used stainless steels for elevated-temperature service, at least in part because of the high resistance to chloride environments resulting from the 2–3% molybdenum in its composition.

## HEAT-RESISTANT STEELS

In addition to conventional grades of stainless steels for use at elevated temperatures, there is a special group of oxidation- and corrosion-resistant steels called *heat-resistant steels.* This means that these steels are capable of resisting operating temperatures above 650°C. In general, they are an extension of the ordinary grades of stainless steels into higher alloy

**Figure 11-11**     Stress required to produce a creep rate of 0.1%/1000 h ($2.78 \times 10^{-10}$/s) vs. testing temperature. [From J. Gordon Parr and Albert Hanson, *An Introduction to Stainless Steel* (Metals Park, Ohio: American Society for Metals, 1965), p. 38.]

contents and less iron. Thus, there are alloys of Fe–Cr (10–30% Cr), Fe–Cr–Ni (>18% Cr and >7% Ni), and Fe–Ni–Cr (>25% Ni and >10% Cr), as well as nickel-base and cobalt-base alloys. Carbon content is usually higher than in the ordinary stainless steels because of the need for some strength at high temperatures, but heat-resistant steels are not selected for high strength at high temperatures. (Chapter 16 discusses such materials.)

The most widely used heat-resistant stainless steel is an austenitic grade designated Type HH when cast and Type 309 when wrought (UNS S30900). Its composition is 22–24 Cr, 12–15 Ni, and 0.20 max C. It is used for many high-temperature applications, particularly because of its excellent oxidation resistance to 1100°C. Its strength at elevated temperatures is exceeded by that of several other stainless steels (see Fig. 11-11), so its chief applications are in low-stress parts: annealing trays, exhaust manifolds, and radiant tubes.

## S U M M A R Y

Iron becomes corrosion-resistant when it contains more than 12% Cr. The wide range of stainless-steel alloys represents a solution to the problems of embrittlement, sensitization, high-temperature failure, low strength, low formability, and high cost. No single alloy provides the best solution to all the problems.

## R E F E R E N C E S

1. American Society for Metals, *Metals Handbook,* 8th ed., vol. 1, *Properties and Selection of Metals*. Metals Park, Ohio: American Society for Metals, 1961, pp. 437–636.

2. American Society for Metals, *Metals Handbook,* 8th ed., vol. 2, *Heat Treating, Cleaning and Finishing.* Metals Park, Ohio: American Society for Metals, 1964, pp. 243–270.

3. American Society for Metals, *Metals Handbook,* 9th ed., vol. 3, *Properties and Selection: Stainless Steels, Tool Materials and Special-Purposes Metals.* Metals Park, Ohio: American Society for Metals, 1980.

4. American Society for Metals, *Source Book on Stainless Steels.* Metals Park, Ohio: American Society for Metals, 1976.

5. Parr, J. Gordon, and Albert Hanson, *Introduction to Stainless Steel.* Metals Park, Ohio: American Society for Metals, 1965.

6. Peckner, D., and I. M. Bernstein, *Handbook of Stainless Steels.* New York: McGraw-Hill, 1977.

7. Zapffe, Carl A., *Stainless Steels.* Metals Park, Ohio: American Society for Metals, 1949.

## PROBLEMS

**11-1.** Explain why it is often desirable for stainless steels with a higher chromium content to have a higher carbon content.

**11-2.** What is the *essential* requirement for "stainlessness" in an iron-base alloy?

**11-3.** Describe the chief advantages of martensitic stainless steel over austenitic stainless steel.

**11-4.** How would you specify change(s) in the composition of a ferritic stainless steel so that it becomes a martensitic stainless steel?

**11-5.** Figure 11-3 shows that an iron alloy of, for example, 18% Cr and 8% Ni is ferritic at temperatures below about 300°C. Why is an 18-8 steel called an austenitic steel?

**11-6.** Which stainless steel has the best corrosion resistance? Why is it not used for all applications?

**11-7.** Which stainless steel has the highest hardness? What are its concomitant disadvantages?

**11-8.** Why are some martensitic stainless steels subcooled below room temperature after heat treatment? Which martensitic stainless steels are so treated?

**11-9.** What limitations does 475°C embrittlement in ferritic stainless steels impose on its use?

**11-10.** What is the most important advantage of precipitation-hardened stainless steel over martensitic stainless steel?

**11-11.** Explain the reason(s) for the expansion of the basic 18 Cr-8 Ni austenitic stainless (e.g., Types 301 and 302) composition into 304, 304L, 316, 321, and 347 (which are the most common of this set).

**11-12.** For what reason(s) do austenitic stainless steels have large values of $n$ in $\sigma = K\varepsilon^n$?

**11-13.** What is sigma-phase embrittlement? In which stainless steel(s) can it occur?

**11-14.** Describe two advantages of ferritic over austenitic stainless steels.

**11-15.** Speculate as to why an austenitic stainless steel, Type 309, is probably a reasonable choice for heat-resistant applications. Describe the disadvantages of **(a)** ferritic, **(b)** martensitic, **(c)** precipitation-hardening stainless steels for heat-resistant applications.

**11-16.** Describe the most important problems and disadvantages associated with the use of stainless steel for automobile fenders and include in your discussion each of the reasonable possible choices for this application.

**11-17.** Steels of Types 410 and 440C are martensitic, whereas Type 430 is ferritic.

     **(a)** Compare the compositions of these three steels and describe why each has the microstructure it has.

     **(b)** Which has the highest hardness? Why?

**11-18.** Type 440A stainless steel has 0.60–0.75% carbon and 16–18% chromium, among other elements. In contrast, Type 442 stainless steel has 0.20 max. carbon and 18–23% chromium. Describe the reason(s) for this marked difference in required carbon content and the differences in final microstructure that result.

**11-19.** Two stainless steels have the following compositions:

|   | C | Mn | Cr | Ni | Other | Balance |
|---|---|-----|------|------|--------|---------|
| A | 0.02 | 1.50 | 18.5 | 10.7 | — | Fe |
| B | 0.06 | 1.50 | 18.5 | 10.7 | 0.42 Ti | Fe |

The only differences in composition are that steel A has 0.02% C and no Ti, and steel B has 0.06% C and 0.42% Ti. Explain precisely the purposes of these differences in composition.

# 12

## Rate- and Temperature-Dependent Mechanical Properties: Creep and Viscoelasticity

No mention was made in Chap. 2 of the possibility of basic strength properties being rate-dependent. There is no reference to time in $\sigma = K\varepsilon^n$, and by implication, sudden application of a stress that is then kept constant produces the corresponding strain "in no time." On testing machines, we do not expect a different answer for extension under a given load if we leave the load on and go back the next day. For most of the materials so far considered, such a description of mechanical behavior is broadly correct, and room-temperature strengths and hardnesses are "fixed." There are exceptions, such as in steel prestressing wires used under high stresses for the reinforcement of concrete; we also saw in Chap. 9 that the strength of precipitation-hardened alloys during aging varies with time. Even then, those alloys chosen for use at room temperature do not appreciably change their strengths under ambient conditions. Reasons for such deliberate choices are obvious: It would be foolish in an engineering design to allow the strength of parts to change as time goes by.

Nevertheless, it was implied in Chap. 4, when discussing the structure of solids, that in the cooling down from liquid to crystals or liquid through molasses to glassy solids, the mechanical properties vary with temperature and rate of deformation—particularly at elevated temperatures approaching the melting/freezing range of crystals or at temperatures above the glass-transition temperature for amorphous solids. In broadest terms, materials are stiffer and stronger at lower than at higher temperatures. If experiments, using conventional testing machines, demonstrate that the common engineering solids are comparatively unaffected by changes in temperature and by changes in speed of deformation under ambient conditions, room temperature must be far removed from their melting points or glass-transition temperatures. Thus, steels melt at about 1450°C, which is much greater than room temperature (about 23°C); brasses melt at about 900°C; many aluminum alloys melt at about 600°C; glass fuses at about 1100°C; and many ceramics and minerals melt at over 2000°C. On the other hand, the melting points of other metals and alloys are much closer to room temperature; for example, lead melts at 327°C, tin at 232°C, general-purpose solder (50% Pb/50% Sn) at 216°C, and Wood's metal (50% Bi/25% Pb/12.5%

Sn/12.5% Cd) at 66°C, and, of course, mercury (which is a metal) has a melting point of −39°C, so it is liquid at room temperature. Most common polymers have glass-transition temperatures $T_g$ comparatively close to room temperature (e.g., for nylon, 60°C; for polystyrene, 100°C); indeed, for some, room temperature is above $T_g$ (e.g., for low-density polyethylene, −20°C; for polypropylene, 0°C); elastomers by definition have $T_g$ below room temperature, that of polyisobutylene (butyl rubber) being −70°C. Butter melts at 31°C.

We should not be surprised if the mechanical properties of all these solids are markedly affected by temperature and speed of deformation. Some toffee behaves in a brittle fashion when hit with a hammer but is pliable in the hands when deformed slowly. Silly putty (silicone) responds quite differently when deformed quickly rather than slowly, as does chewing gum. Even some of the solids that are far removed from their melting points at room temperature can have different strengths under impact or dynamic conditions (i.e., under very high rates of strain). At higher temperatures, the strengths of all solids become more and more rate dependent.

An appreciation of this behavior is necessary either because we are looking for solids with which to make items that have to operate at high temperatures (e.g., gas turbines) or because we want to deform things more easily (as in the "hot working" of metals). The traditional room-temperature description of solids, as given in Chap. 2, is inadequate for such purposes, and we must broaden our coverage to include stress–strain curves over wide spectra of temperature and rates of deformation. Furthermore, we may have to introduce new methods of testing and new types of descriptions of mechanical behavior. In this way, so-called *creep testing* and the phenomenon of *relaxation* play their part.

It is important to realize from the outset that "high temperature" in this context is defined relative to the melting temperature or glass-transition temperature of the solid concerned, so room temperature may be "high" for some solids. It is useful for this field of study to introduce the concept of *homologous temperature,* defined as the ratio of the temperature of interest to the melting temperature, where the temperatures are in degrees Kelvin (K). Thus, zinc at room temperature is at (23 + 273)/(419 + 273) = 0.43 of the melting point; this may be written 0.43 $T_m$ and called "0.43 homologous." Low-carbon steel is at (23 + 273)/(1450 + 273) = 0.17 $T_m$. Because homologous temperatures serve merely to bracket temperature ranges, the choice of liquidus or solidus for alloys is not critical.

## STRAIN RATE

Whenever strengths depend on time, the concept of strain rate is a convenient way of bringing time into the description of mechanical behavior. The average strain rate in a tensile test is the strain divided by the time taken to reach that strain level. This time depends on the velocity of the testing machine crosshead or on how rapidly the load is applied by the testing machine, or in an engineering design by how quickly a bridge, for example, becomes strained by a train running over it. It is just as appropriate to talk about load rates or stress rates, but it is common practice to use strain rates, and of course it is possible to convert from one to another.

Both engineering (nominal) strains and true (logarithmic) strains were employed in Chap. 2, so we have

$$\text{Nominal strain rate} = \frac{e}{t}$$

$$\text{True strain rate} = \frac{\varepsilon}{t}$$

The units of both are "per second," or $s^{-1}$, as strain is dimensionless. In the elastic range, both formulas give the same value for strain rate because in that range of deformation $\varepsilon = \log(1 + e) \approx e$. In the plastic range, however, there can be wide differences between $\varepsilon$ and $e$, and because $\varepsilon$ is considered to be the more meaningful measure of strain, we limit our discussion to rates of true strain (and, if necessary, to rates of true stress).

To account for possible variations in strain rate during a test, we can rewrite the foregoing as

$$\text{Instantaneous strain rate} = \frac{d\varepsilon}{dt} = \dot{\varepsilon}$$

where the fluxional (dot) notation is used for derivatives with respect to time. Because

$$d\varepsilon = \frac{dL}{L}, \quad \dot{\varepsilon} = \left(\frac{1}{L}\right)\frac{dL}{dt} = \frac{V}{L} \tag{12-1}$$

where $V$ is the crosshead velocity of the testing machine. Furthermore, because $L = L_0 \exp \varepsilon$, where $L_0$ is the initial gage length now stretched to length $L$,

$$\dot{\varepsilon} = \frac{V}{L_0 \exp \varepsilon} \tag{12-2}$$

For small strains (including the elastic region) $\exp \varepsilon = 1$, so

$$\dot{\varepsilon} = \frac{V}{L_0} \tag{12-3}$$

Note that Eqs. (12-1) and (12-2) suggest a constant strain rate for small strains but a diminishing strain rate in tension as plastic deformation continues. In compression, increasing strain rates through the plastic range are suggested because $L$ decreases as $\varepsilon$ increases.

These expressions assume *uniform deformation*, so that they cannot be applied when barreling occurs in compression, and particularly when necking occurs in tension (where the deformation becomes concentrated in a small zone). The differences in strain rate within and outside necked regions play important roles in Lüders band formation, "polymer drawing," and superplasticity, which are discussed later.

A test piece of gage length 75 mm, strained by a crosshead moving at 200 $\mu$m/s, has $\dot{\varepsilon} = (200 \times 10^{-6})/(75 \times 10^{-3}) \approx 3 \times 10^{-3}\ s^{-1}$. Traditional laboratory testing machines (whatever their load capacity) produce strain rates no greater than about $10^{-1}\ s^{-1}$ on test pieces of traditional size. Special machines are required to produce greater strain rates.

**Example 12-1**

Determine the average elastic strain rate for a material with a 0.2% offset yield strength of 150 MPa if loaded at a constant load rate of 3 kN/s. The gage section is 11.3 mm in diameter.

**Solution**   The cross-sectional area is $\pi(11.3)^2/4 = 100$ mm$^2$. Therefore, the yield load is $(150 \times 10^6)(100 \times 10^{-6}) = 15$ kN. The time taken to reach yield load is $(15 \times 10^3)/(3 \times 10^3) = 5$ s. Thus,

$$\dot\varepsilon = \frac{0.002}{5} = 400 \times 10^{-6} \text{ s}^{-1}$$

**Example 12-2**

If the material in Example 12-1 follows $\sigma = 230 \, \varepsilon^{0.3}$ MPa, determine the strain rate in the plastic range up to ultimate.

**Solution**   The load at ultimate is

$$[230 \, (0.3)^{0.3} \times 10^6](100 \times 10^{-6})\exp(-0.3) = 11.9 \text{ kN}$$

because the area at ultimate is $A_0 \exp(-n)$. The time required to reach maximum load, at a loading rate of 3 kN/s, is thus 4 s. Therefore, neglecting the yield strain, the average strain rate between yield and ultimate is 0.3/4 = 0.075 s$^{-1}$. Notice that the strain rate has increased by two orders of magnitude.

Exercise 12-2 well illustrates how difficult it is to conduct a constant strain rate test through both the elastic and the plastic regions. During the work-hardening stage after the onset of gross yielding, the test piece is less stiff and the load varies much more slowly than in the elastic region. Much greater strain increments are obtained for the same stress increments. In order to complete tests in reasonable amounts of time, deformation rates often have to be increased after yield. One way of picking up the yield point on old-fashioned hydraulic testing machines was to note when the load needle failed to keep up with the "pacing disk" that was set for some chosen load rate and that this effect occurs because of changes in stiffness of the test piece at yield.

Despite these changes in rate of deformation during testing, it turns out that the room-temperature $\sigma$–$\varepsilon$ curves of most traditional engineering metals are *not* much affected by the value of $\dot\varepsilon$, provided that the strain rate is *less* than about $10^{-1}$ s$^{-1}$. The values of $\sigma_y$ and $\sigma_u$ do not depend very much on how fast the test is performed or on how much it is speeded up after yield. At higher rates of strain, changes in the level of the room-temperature $\sigma$–$\varepsilon$ curve can occur for some alloys. Since such higher strain rates can be produced only with special testing machines, information on strength and hardness of solids at low homologous temperatures is not always available.

## BEHAVIOR OF METALS

If $\sigma$–$\varepsilon$ tests are performed at different strain rates and temperatures, it is found that both the shape and the level of the curves are affected but in opposite senses; that is, an increase in strain rate increases stress levels and sometimes work-hardening rate, but when deformation is carried out at elevated homologous temperatures, the whole level of the $\sigma$–$\varepsilon$ curve drops.

If we can still describe the true stress–strain behavior by an equation of the form $\sigma = K\varepsilon^n$, it means that $K$ must vary with both rate and temperature. The values quoted in Chap. 2 and App. 3 merely reflect temperatures around 23°C and strain rates around $10^{-4} \sim 10^{-2}$ s$^{-1}$. Figure 12-1 shows how $K$ depends on $\dot{\varepsilon}$ at various temperatures for annealed 2024 aluminum. Clearly, temperature always seems to affect $K$, but strain rate becomes significant only at temperatures greater than 200°C (which is about $0.58T_m$); for temperatures less than that, $K$ is essentially independent of $\dot{\varepsilon}$ for this aluminum alloy, being, for example, about 330 MPa at 28°C.

The logarithmic scales in Fig. 12-1 suggest that

$$K = C\dot{\varepsilon}^m \tag{12-4}$$

where $C$ (the intercept at $\dot{\varepsilon} = 1$) decreases as the temperature $T$ increases. The slope of the plot, $m$, is called the *strain-rate sensitivity* and changes with temperature, as shown by the increasing slopes with increasing temperature.

To compare the level of stress at two strain rates, we can write

$$\frac{\sigma_2}{\sigma_1} = \left( \frac{\dot{\varepsilon}_2}{\dot{\varepsilon}_1} \right)^m$$

where the subscripts 1 and 2 refer to the different rate conditions.

In Fig. 12-1, $m = 0.005$ (a very shallow slope) at room temperature. Increasing the strain rate by a factor of $10^4$ produces an increase in flow stress of less than 5%. This is why precipitation-hardened aluminum alloys, copper alloys, and magnesium alloys are considered rate-insensitive at room temperature. Other materials, particularly annealed bcc

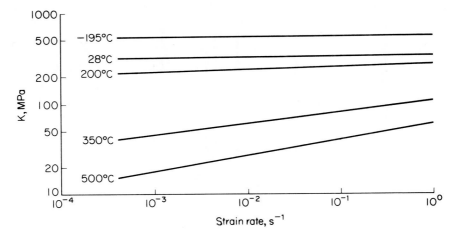

**Figure 12-1**   Variation of $K$ in $\sigma = K\varepsilon^n$ with strain rate and temperature for 2024 aluminum. [After D. S. Fields and W. A. Backofen, *Trans. ASM*, 55 (1959), p. 950.]

metals, aluminum, and titanium, show greater rate sensitivity. For example, annealed 1010 steel having a yield strength of 210 MPa at $\dot{\varepsilon} = 10^{-3}$ s$^{-1}$ has a flow stress of 450 MPa at $\dot{\varepsilon}$ = $10^2$ s$^{-1}$. This corresponds to $m = 0.06$, from $450/210 = (10^2/10^{-3})^m$. Such greater strain-rate sensitivity disappears the more work-hardened the material. When cold-rolled to a strain of $\varepsilon_{cw} = 1$, 1010 steel changes its yield strength only from 630 MPa at $10^{-3}$ s$^{-1}$ to 735 MPa at $10^2$ s$^{-1}$; again, high-purity aluminum cold-rolled to a 50% reduction in area has yield strengths of 98 MPa at $\dot{\varepsilon} = 10^{-3}$ s$^{-1}$ and 133 MPa at $10^3$ s$^{-1}$. For both, $m$ is then roughly 0.01 to 0.02.

At higher homologous temperatures, $m$ increases for all metals, some results being shown in Fig. 12-2. At 600°C copper may have an $m$ of 0.1, and the flow stress is increased by 10% when the strain rate is merely doubled.

**Example 12-3**

The range of crosshead speeds on some testing machines can be altered by three orders of magnitude. What is the minimum temperature at which pure aluminum test pieces will show yield strengths differing by a factor of 2 between the slowest and fastest available crosshead speeds?

**Solution**   At some chosen fixed yield strain, yield strengths differing by a factor of 2 means that the $K$ values must differ by the same factor. Thus, in Eq. (12-4), if $K_2 = 2K_1$, $\log 2 = m$ $\log(\dot{\varepsilon}_2/\dot{\varepsilon}_1)$. Assuming that $\dot{\varepsilon}$ is directly proportional to crosshead velocity and using the fact

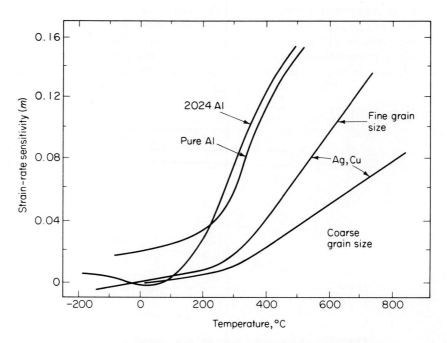

**Figure 12-2**   Effect of temperature on strain-rate sensitivity index ($m$).

that the fastest speed is 1000 times the slowest, we have $\log 2 = m \log 10^3$, and so $m = 0.1$. Inspection of Fig. 12-2 shows that such a sensitivity is displayed at slightly below 400°C.

The value of $C$ in Eq. (12-4) decreases as the temperature increases, and semilogarithmic plotting of $C$ and $T$ shows that $C$ varies according to $\exp(1/T)$. As discussed later, the form of this temperature dependence is a manifestation of *thermal activation* on an atomic level. Thus, in general,

$$\sigma = K_0 \ exp\left(\frac{1}{T}\right)\dot{\varepsilon}^m \varepsilon^n \qquad (12\text{-}5)$$

This is the full version of the simple $\sigma = K\varepsilon^n$ empirical relation; both $m$ and $n$ vary with temperature, $m$ increasing and $n$ decreasing as the temperature is raised.

In $\sigma$–$\varepsilon$ experiments at low homologous temperatures (say, less than $0.5T_m$), rapid application of fixed stresses produces rapid strains, which are achieved very quickly as shown in Fig. 12-3(a). At higher temperatures these strains do not stop suddenly but increase slowly in the time following the application of stress and go over into slow plastic flow, as in Fig. 12-3(b). This flow is called *creep*. Sometimes the creep stops, and further loads have to be added to cause further deformation. Even so, the supplementary creep components of strain can be appreciable. On other occasions, particularly at high homologous temperatures, creep strains never settle down and rapid plastic flow continues for a short or long time until the test piece breaks, *without the need for additional loads*. For example, Fig. 12-4 shows the effect of a constant force of 89 N (produced by a hanging mass of 9.08 kg) on a Pb–Sn solder wire 3.2 mm in diameter at room temperature, which is $0.65T_m$ for this alloy. Note the rapid extension in the early stages, followed by a deceleration to a nearly steady rate of creep extension, followed by a final gradual acceleration to fracture.

Under different loads the creepy Pb–Sn alloy behaves differently. For larger loads the creep rate increases, and fracture occurs in a shorter time (Fig. 12-4). At very high loads the transition between the deceleration and acceleration phases is sometimes only a point of inflection; at lower loads than shown in Fig. 12-4 the extension tends asymptotically to a fixed value, with no final fracture. In general, however, there is a more-or-less straight extended portion of the creep curve between the deceleration and

**Figure 12-3**  Strains resulting from the rapid application of a fixed stress: (a) for $T < 0.5\ T_m$ and (b) for $T > 0.5\ T_m$, showing occurrence of creep strains in the higher temperature range.

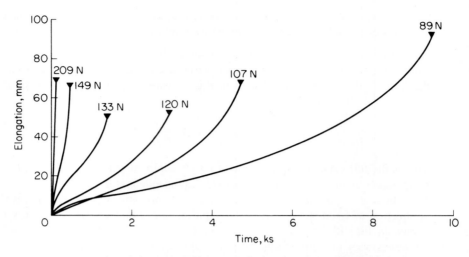

**Figure 12-4**  Creep of commercial Pb–Sn solder wire of initial diameter 3.2 mm and length approximately 90 mm, at 20°C. The axial force applied to each specimen is given. (Data courtesy of W. H. Durrant.)

acceleration parts. Accordingly, creep curves are conventionally subdivided into four parts:

1. A sudden extension upon initial application of load
2. A period of deceleration (called *transient* or *primary* creep)
3. A period of constant creep rate *(steady-state* or *secondary* creep)
4. A period of acceleration *(tertiary* creep), which ends with fracture.

In most cases tertiary creep follows necking, with its decrease in cross-sectional area, internal hole formation, and consequent increase in stress, akin to conventional $\sigma$–$\varepsilon$ curves. For reasons of practical convenience, tensile creep tests are performed with the load and not the stress being kept constant, although tertiary creep can usually be eliminated under the correct conditions.

The effect of increased temperature on creep curves is similar to that of increased load (or stress): Extensions occur more quickly at higher homologous temperatures. On the other hand, this same solder wire would have a conventional $\sigma$–$\varepsilon$ curve, which would be essentially time-independent, at sufficiently low temperatures (say, at −100°C).

Whenever the time effects of creep become appreciable, conventional stress–strain tests are not usually performed, and dead-load tests as just described are used to characterize mechanical behavior (i.e., strain as a function of time at constant stress). There is an overlap temperature range between the two regions, when a $\sigma$–$\varepsilon$ curve *can* describe the behavior, provided that adequate attention is paid to rate and temperature effects on $K$, as described earlier in this section. Then creep is beginning to be felt, but creep strains settle down after a time; this is before creep entirely dictates mechanical behavior.

Orowan has linked the $\sigma$–$\varepsilon$ curve in this region with transient creep, as shown in Fig. 12-5. He suggested in the construction of the $\sigma$–$\varepsilon$ curve that the stress is applied in small increments $\Delta\sigma$. During each corresponding time interval $\Delta t$, creep occurs as in a conventional creep test under the same current average stress level. The $\sigma$–$\varepsilon$ curve is then the plot of the accumulated sum of the stress increments vs. the accumulated sum of the creep strain increments. In Fig. 12-5, $\Delta\sigma/\Delta t$ is constant for purposes of illustration; that is, it is a constant-stress-rate test.

The decrease in creep rate in the transient part is caused by work hardening, but competition can arise from thermal softening. Steady-state creep occurs when the two effects are balanced.

Creep occurs in metals because mechanisms come into play that allow easier flow but are absent at low homologous temperatures. Dislocation movement by climb (described in Chap. 6) is one such mechanism. Dislocations piled up against a barrier on some slip plane can move on if they can somehow jump over the barrier by moving to the next slip plane above. This can come about by migration of vacancies (holes or missing atom sites) to the "bottom" of the extra half-plane of the dislocation, taking out one row of atoms and effectively moving the base of the dislocation up onto the slip plane above. There are many naturally occurring vacancies in crystals, and diffusion of vacancies is a

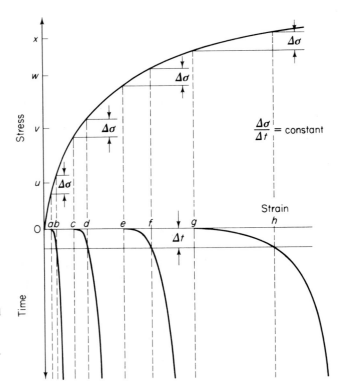

**Figure 12-5** Orowan's description of the connection between transient creep at various stresses and the monotonic stress–strain curve. At every time increment $\Delta t$, transient creep strains $ab$, $cd$, $ef$, $gh$, . . ., are produced by average stress levels $Ou$, $Ov$, $Ow$, $Ox$, . . . . The monotonic $\sigma$–$\varepsilon$ curve is the accumulated sum of all the stress increments $\Delta\sigma$ vs. the accumulated sum of all the creep-strain increments $\Delta\varepsilon$.

process helped by increasing temperature (i.e., it is thermally activated), so it contributes more to creep as the temperature rises.

Another event that is thermally activated is atomic mobility. At low homologous temperatures, atoms may be thought of as fixed in the crystal lattice, but as the temperature rises they can change places with one another. This process, called *self-diffusion,* comes into play significantly at temperatures around $0.5\ T_{\mathrm{m}}$ and higher. Atomic mobility is particularly marked in grain boundaries (where the crystal lattice positions are mismatched), more than in the grains themselves. The enhanced atomic mobility of grain boundaries at high homologous temperatures makes them behave like a viscous fluid, and the associated grain-boundary sliding contributes significantly to creep strains. The inverse exponential temperature terms in Eq. (12-5) represent the collective effects of thermal activation.

Empirical relations between creep strain and time, as may be determined from experiments in which masses are hung on wires, take on the following form (among others):

$$\dot{\varepsilon}_{\mathrm{transient}} = A_1 t^{-2/3} \qquad (12\text{-}6)$$

where $A_1$ is some constant depending on the stress; the rate decelerates as time goes by. The constant is a function of stress and temperature as first studied by E. N. da C. Andrade in 1910. It will be shown later how transient creep can be linked to viscoelasticity and how steady-state creep can be linked to non-Newtonian flow.

**Example 12-4**

Describe the influence of grain size on creep resistance.

**Solution**   Clearly, if grain-boundary sliding is a significant contributor to creep strains, the latter are reduced if the regions in which sliding can take place are kept as small as possible, that is, if the grain boundary area is small. Thus, other things being equal, larger grain sizes have smaller creep rates.

## BEHAVIOR OF AMORPHOUS SOLIDS

Amorphous materials stiffen and harden progressively as the temperature drops, unlike crystalline materials, whose mechanical properties change markedly at the melting/freezing point. For noncrystalline solids, as the temperature drops there is a gradual transition from liquid behavior through molasses-type behavior to that of elastic solids. Liquids often display simple Newtonian viscosity in which (for the purposes of this book) shear stress $\tau$ is proportional to the *rate* of shear strain $d\gamma/dt$ or $\dot{\gamma}$, that is,

$$\tau = \eta\dot{\gamma} \qquad (12\text{-}7)$$

The constant of proportionality $\eta$ is called the *coefficient of viscosity* or the *viscosity* of the liquid. Note that here the response is related to the *rate* of displacement, whereas for materials displaying simple elastic or plastic behavior the response is related to the displacement (strain) itself.

As the temperature drops and the liquid thickens toward an amorphous solid, the shear stress ceases to be directly proportional to the strain rate, and instead we have non-proportional (non-Newtonian) behavior given by

$$\tau = \eta(\dot\gamma)^q \tag{12-8}$$

This expression, like the linear Newtonian relationship, may be integrated to give the magnitude of the strain at any time:

$$d\gamma = \left(\frac{\tau}{\eta}\right)^{1/q} dt$$

$$\gamma = \int \left(\frac{\tau}{\eta}\right)^{1/q} dt \tag{12-9}$$

The applied $\tau$ may vary with time (as in a simple tension test where the load, and hence stress, is increased as the test proceeds). Integration of Eq. (12-9) would require an expression for $\tau$ as a function of $t$ in order to evaluate the change in strain over some period of time. Note, however, that even with $t$ constant, $\gamma$ *increases with time* according to

$$\gamma = \left(\frac{\tau}{\eta}\right)^{1/q} t \tag{12-10}$$

(We assume that $\gamma = 0$ at the reference observation starting time when $t = 0$.)

This is similar to the creep behavior of metals, such as given by the simple Andrade relation for transient $\dot\varepsilon$ [Eq. (12-6)] when integrated. The magnitude of induced strain depends on $\tau$, of course (greater stresses producing greater strains), but it particularly depends on $1/q$. When this index is very small, $\gamma$ is small even if $\tau$ is large; that is, the amorphous material is not very rate dependent. On the other hand, when $1/q$ is large, great strains are produced in short times even by very small stresses.

### Example 12-5

An amorphous solid follows $\tau = 800\ \dot\gamma^2$ Pa at some temperature above $T_g$. What strain is produced 20 s after a uniform shear stress of 80 Pa is applied to a block of the solid?

**Solution**  From Eqs. (12-8) and (12-10),

$$\gamma = \left(\frac{80}{800}\right)^{1/2}(20) = 6.32$$

Note that if the index were 0.2 instead of 2, $\gamma$ would be $2 \times 10^{-4}$, which is considerably smaller.

In all these expressions, the strains are irreversible upon removal of the stress, and even after some period following the removal of $\tau$, the amorphous body retains its deformed shape. However, at lower temperatures (particularly below $T_g$ for polymers) elasticity comes into play. Then the deformation at any instant has two components: an irreversible flow component (as already discussed) and a reversible elastic component. The relative magnitude of the component deformations or strains depends on the character of the solid and the temperature. At lower and lower temperatures, the flow component diminishes and the amorphous material behaves more and more like a time-independent elastic body with all deformation reversible. The range of mechanical properties shown by different polymers at room temperature—some, such as polymethylmethacrylate (PMMA), being virtually elastic in response and others, such as low-density polyethylene, being very creepy—reflects the relative importance of the flow and elastic components of strain. We have

$$d\gamma_{\text{total}} = d\gamma_{\text{flow}} + d\gamma_{\text{elastic}} \qquad (12\text{-}11)$$

Here $d\gamma_{\text{flow}}$ comes from $\tau = \eta(\dot{\gamma})^q$ and $d\gamma_{\text{elastic}}$ comes from $\tau = G\gamma$, where $G$ is the shear modulus for the amorphous solid. Methods of measuring these mechanical properties are discussed later.

In many cases of interest, the magnitude of the recoverable elastic strains is small in comparison with that of the irreversible strains. This applies to metal deformation problems in which the plastic strains swamp the elastic strains, and it applies to the behavior of amorphous solids above $T_g$. Sometimes, however, $d\gamma_{\text{elastic}}$ and $d\gamma_{\text{flow}}$ are comparable in magnitude. This is particularly true for polymers below $T_g$ because, as discussed in Chap. 13, their elastic moduli ($G$ in $\tau = G\gamma$ or $E$ in $\sigma = E\varepsilon$) are much smaller than those of metals, and the associated elastic strains at a given stress level are greater, becoming comparable with the flow components of strains. Thus, around $T_g$ the subject of *viscoelasticity* (i.e., mechanical behavior with combined flow and elastic components) is important.

In the transient creep of metals, where the total strains are small, reversible and irreversible components are comparable in magnitude, so the mechanics of primary creep and viscoelasticity are similar. In steady-state creep, however, the elastic component of total strain can usually be ignored, so secondary creep relations may be thought of as rather complicated non-Newtonian flow.

Dividing Eq. (12-11) by time, we obtain

$$\dot{\gamma}_{\text{total}} = \dot{\gamma}_{\text{flow}} + \dot{\gamma}_{\text{elastic}}$$
$$= \left(\frac{\tau}{\eta}\right)^{1/q} + \left(\frac{1}{G}\right)\dot{\tau} \qquad (12\text{-}12)$$

If at $t = 0$ a stress $\tau^*$ is applied to a body following Eq. (12-12) and then kept constant for some time $t^*$, there will be an instantaneous strain $\tau^*/G$ which will remain unchanged throughout $t^*$ (since elastic deformations occur without reference to time), followed by a steadily increasing flow component given by $\gamma_{\text{flow}} = (\tau^*/\eta)^{1/q}t$ during the period $0 < t < t^*$. This is shown in Fig. 12-6 for $q = 1$. The rapidity with which $\gamma$ increases with $t$ depends on $q$: For $q < 1$, $1/q$ is large and the straight line $AB$ is steep; the opposite is true for $q > 1$ and $1/q$ is small. The straight line $AB$ for Newtonian viscosity with $q = 1$ lies in between. If the stress $\tau^*$ is removed at time $t^*$, $\gamma_{\text{elastic}} = \tau^*/G$ is recovered instantaneously. However, the accumulated flow component remains. Clerk Maxwell first discussed this type of behavior

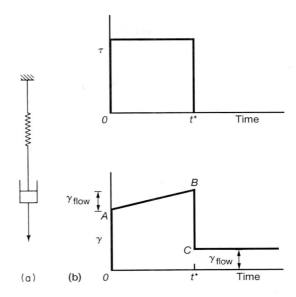

**Figure 12-6** (a) Mechanical analog, consisting of a spring and dashpot in series, which describes a Maxwell solid for $q = 1$. (b) Strain (or displacement) vs. time response to a step application and removal of stress (or load).

with combined elastic and Newtonian components of deformations, and so a material following Eq. (12-12) is called a *Maxwell solid.*

There is a mechanical analog for a Maxwell solid which consists of an elastic spring in series with a dashpot (i.e., a viscous damper or type of shock absorber); see Fig. 12-6(a). The same load (stress) is experienced by both the spring and the dashpot, and the total deflection (strain) is the sum of the two components. A dashpot can be made from a piston containing holes, sliding in a bath of oil; it is viscous and "takes time to get going," so in Fig. 12-6(b) *OA* reflects the zero-time spring elongation under stress, and the (straight) path *AB,* the subsequent extension of the dashpot piston rod. The spring returns on removal of the load, moving the dashpot bodily with it (*BC*); however, the dashpot piston rod end remains in its extended position and can return to the starting position only if a load is applied in the opposite direction from before.

Some polymers do not behave as if they are Maxwell solids. Rather, they exhibit *delayed elasticity*; that is, instead of attaining an elastic strain instantaneously, there is some sort of retardation which delays the process. Long polymer chains, tangled up like spaghetti, must uncurl as they are deformed, and this can produce delayed elastic effects.

Another mechanical analog, first described by Voigt, that displays such an effect is shown in Fig. 12-7(a). This time the spring and dashpot are in parallel rather than in series. Thus, they share a common displacement (strain) rather than a common load or stress. We have, at some strain $\gamma$,

$$\gamma_{total} = \gamma_{elastic} + \gamma_{flow}$$
$$= G\gamma + \eta(\dot{\gamma})^q \tag{12-13}$$

For a constant stress $\tau^*$ applied for a time $t^*$, Eq. (12-13) becomes a linear differential equation for the simplified case of $q = 1$ (i.e., for a Newtonian dashpot). The solution is

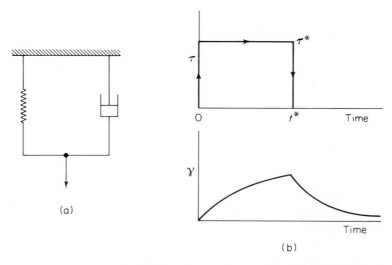

**Figure 12-7**   (a) Mechanical analog, consisting of a spring and dashpot in parallel, which describes a Voigt solid. (b) Strain (or displacement) vs. time response to a step application and removal of stress (or load).

$$\gamma = \left(\frac{\tau^*}{G}\right)\left[1-\exp\left(\frac{-Gt}{\eta}\right)\right] \tag{12-14}$$

which describes how the strain $\gamma$ grows with time $t$ for the duration of $\tau^*$. Figure 12-7(b) displays the behavior, which is an exponential increase toward the elastic strain ($\tau^*/G$) that would have been attained instantaneously in the absence of the retarding dashpot. Removal of $\tau^*$ produces an exponential return to zero strain.

A combination of the Voigt and Maxwell solid behaviors into a four-element model [Fig. 12-8(a)] embodies many of the essential characteristics of the mechanical response of polymers: instantaneous deformation, deformation decaying to a constant rate, followed by some permanent deformation after stress removal. The total strain in the combination during the application of a constant $\tau^*$ is

$$\gamma_{\text{total}} = \frac{\tau^*}{G_1}+\frac{\tau^* t}{\eta_1} + \frac{\tau^*}{G_2}\left[1-\exp\left(\frac{-G_2 t}{\eta_2}\right)\right] \tag{12-15}$$

where subscripts 1 and 2 are used to differentiate between the two spring and dashpot systems. Figure 12-8(b) shows the complete stressed-unstressed strain-time history for the four-element model. Precise simulation of real polymers usually requires much more complex models, but all the essential features are present in this *Voigt-Maxwell model*.

It should be emphasized that the springs and dashpots have no physical meaning[†]; rather, microstructural characteristics of polymers lead to the same sort of response as

---

[†]Fracture in polymers is not, as a student once asserted, when the piston falls out of the dashpot.

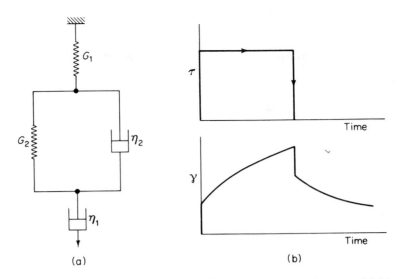

**Figure 12-8**   (a) Maxwell and Voigt analogies combined into a four-element model. (b) Strain or displacement vs. time response to a step application and removal of stress (or load).

these conceptual analog models. Values for the $G$'s and $\eta$'s of the analog models would be obtained such as by curve-fitting experimental $\gamma$ vs. $\tau$ data to Eq. (12-15). The microstructural behavior of creepy metals leads to a similar state of affairs, except that in this case the elastic and recoverable components of strain are usually negligible because of the relatively large elastic moduli of metals, even at high homologous temperatures.

### Example 12-6

A tensile force of 500 N was applied to a polymer rod, 5 mm × 5 mm × 500 mm long. The rod stretched to 508 mm very quickly and then continued to lengthen so that by the same time on the next day it was 550 mm long. Another day later, it was 570 mm long. Determine the parameters $E_1$, $E_2$, and $\eta_2$ of a three-element tensile analog that describes this behavior. [The three-element model consists of a spring and a dashpot in parallel coupled to another spring in series, i.e., the arrangement in Fig. 12-8(a) but with $\eta_1$ omitted.]

**Solution**   In terms of tensile stresses and strains, from Eq. (12-15), rewritten in tensile nomenclature, for the case of no $\eta_1$ element,

$$\varepsilon = \frac{\sigma}{E_1} + \frac{\sigma}{E_2}\left[1 - \exp\!\left(\frac{-E_2 t}{\eta_2}\right)\right]$$

At time zero,

$$\varepsilon = \frac{508 \text{ mm} - 500 \text{ mm}}{500 \text{ mm}} = 0.016 = \frac{\sigma}{E_1} = \frac{500\,\text{N} / (25 \times 10^{-6})\,\text{m}^2}{E_1}$$

$$E_1 = 1.25 \text{ GPa}$$

Twenty-four hours later,

$$\varepsilon = \frac{550 \text{ mm} - 500 \text{ mm}}{500 \text{ mm}} = 0.1$$

$$= 0.016 + \left[\frac{500 \text{ N}/(25 \times 10^{-6})\text{m}^2}{E_2}\right]\left[1 - \exp\left(\frac{86.4 \text{ ks} \times E_2}{\eta_2}\right)\right]$$

Forty-eight hours later,

$$\varepsilon = \frac{570 \text{ mm} - 500 \text{ mm}}{500 \text{ mm}} = 0.14$$

$$= 0.016 + \left[\frac{500 \text{ N}/(25 \times 10^{-6}) \text{ m}^2}{E_2}\right]\left[1 - \exp\left(\frac{173 \text{ ks} \times E_2}{\eta_2}\right)\right]$$

For convenience, set

$$\exp\left(\frac{86.4 \text{ ks} \times E_2}{\eta_2}\right) = Z$$

Then dividing the relations for 48 h and 24 h gives

$$\frac{(6.2 \times 10^{-9})E_2 / \text{MPa}}{(4.2 \times 10^{-9})E_2 / \text{MPa}} = \frac{1 - Z^2}{1 - Z} = 1 + Z$$

$$1.476 = 1 + Z$$

$$0.476 = Z$$

$$0.476 = \exp\left(\frac{-E_2 \times 86.4 \text{ ks}}{\eta_2}\right)$$

$$E_2 / \eta_2 = 8.59 \times 10^{-6} / \text{s}$$

Substituting back into the equations gives $E_2 = 125$ MPa and $\eta_2 = 14.5$ TPa-s.


## STRESS RELAXATION

Creep is the increase in $\gamma$ with $t$ at constant $\tau$, as illustrated by a mass hanging on a wire or a hot turbine blade stretching under centrifugal force. The only way in which such an increase in $\gamma$ can be eliminated (for a material in a given state) is to reduce $\tau$ as $t$ increases. In a Maxwell solid, for example, $\dot{\gamma}_{\text{total}} = 0$ if $\gamma_{\text{total}}$ remains constant, so

$$\left(\frac{\tau}{\eta}\right)^{1/q} + \left(\frac{1}{G}\right)\dot{\tau} = 0 \tag{12-16}$$

For $q = 1$ the solution of this differential equation is

$$\tau = \tau^* \exp\left(\frac{-Gt}{\eta}\right) \tag{12-17}$$

where $\tau^*$ is the initial stress that produced a strain $\gamma^*$ that is now remaining constant. Equation (12-17) gives the manner in which $\tau^*$ must be decreased to keep $\gamma^*$ fixed, as shown in Fig. 12-9. When a time-dependent body is strained and the strain then held (as by driving the crosshead of a screw-driven testing machine to some new position), the stress itself naturally decays according to some relationship like Eq. (12-17). Internal flow takes place so as to decrease the stress, the process being called *stress relaxation*. It is characteristic of all creepy materials.

Note that when $t = \eta/G$ in Eq. (12-17), $\tau = \tau^* \exp(-1) = 0.37\tau^*$; $\eta/G$ is called the *relaxation time* (i.e., the time required for the stress to relax to $1/e$ of its initial value). It is similar to time constants of alternating-current circuits or vibrating springs (hence, again, the spring-dashpot analogs of viscoelastic behavior).

### Example 12-7

Show how the value of $\eta/G$ in a Maxwell solid can predict, for a given loading duration, whether the behavior will be predominantly elastic, predominantly flow, or a mixture of both.

**Solution**   $\gamma_{\text{elastic}} = \tau/G$; $\gamma_{\text{flow}} = (\tau/\eta)t$. Thus, $\gamma_{\text{flow}} = [(G/\eta)t]\gamma_{\text{elastic}}$. When $\eta/G$ is large, $\gamma_{\text{flow}} \ll \gamma_{\text{elastic}}$; when $\eta/G$ is small, $\gamma_{\text{flow}} \gg \gamma_{\text{elastic}}$ for the same $t$. When $\eta/G$ is large, a long time elapses before much relaxation takes place, that is, the solid is hardly rate dependent at all; the opposite effect occurs for small $\eta/G$. When $\eta/G$ and $t$ are comparable, the mixed behavior of the Maxwell model comes into play.

The ratio of relaxation time to observation time ($\eta/Gt$) is sometimes called the *Deborah number*. It is zero for ideal fluids, infinite for ideal solids, and unity at $T_g$.

The presence of creep and/or relaxation in solids causes difficulties in the experimental measurement of elastic moduli. For example, if a mass is hung on a wire in order to determine a pair of $\sigma$–$\varepsilon$ values that will give $E$ by division, it is clear that in the pres-

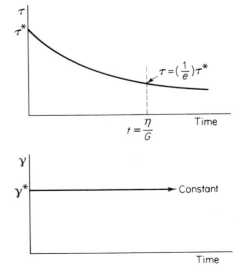

**Figure 12-9**   Stress relaxation where stress falls with time at constant strain.

ence of creep the value of $\varepsilon$ depends on how soon it is measured after application of the stress. That is, in Fig. 12-4 any value on the rising strain curve might be taken. Similar considerations apply to a soft (e.g., hydraulic) testing machine (see Chaps. 2 and 14). Or, in a hard machine where a fixed strain might be put on, the associated stress relaxes away from the value first seen after attainment of the chosen strain level, along the lines of Fig. 12-9.

It is customary, therefore, in polymers to talk of *relaxed* and *unrelaxed moduli.* Unrelaxed moduli are those obtained from instantaneous readings of stress and strain and are the same as "dynamic" or high-strain-rate moduli, such as relate to the passage of sound waves through a polymer where the molecules do not have time to creep about. The moduli of Voigt–Maxwell models and the like are strictly unrelaxed moduli. Relaxed moduli, on the other hand, come about from dividing the stress by that strain attained some considerable time after first stressing. They are thus much smaller than unrelaxed or dynamic moduli. As described in Chap. 13, the elastic properties of some polymers are given in terms of a so-called 10-s modulus, which is stress divided by the strain reached 10 s after first loading.

### Example 12-8

A bar of low-density polyethylene is suspended from two fine wires 25 mm apart and forced to vibrate by sending pulses down one of the wires. The time taken for a sound wave to travel along the bar is measured by picking up the rod vibration in the second wire. If that time is 54 $\mu$s and the density of the polymer is 920 kg/m$^3$, calculate the dynamic modulus of the polymer.

**Solution** The velocity of sound in a solid is given by $\sqrt{E/\rho}$ where $E$ is the modulus and $\rho$ is the density. Thus, if the time taken to travel 25 mm is 54 $\mu$s, the velocity must be (25 $\times$ 10$^{-3}$)/(54 $\times$ 10$^{-6}$) = 463 m/s. Therefore, 463 = $\sqrt{E/920}$, from which $E = 197$ MPa.

### Example 12-9

Contrast the relative effects of strain rate and temperature on the moduli of metals and polymers.

**Solution** At a given temperature, particularly in the range below half-homologous, the elastic moduli of most metals are not much affected by rate. In contrast, the elastic moduli of polymers can be quite rate dependent, the extremes being given by the relaxed and unrelaxed values. For example, at room temperature ($<T_g$) PMMA displays approximately 6 GPa in tests, the durations of which are microseconds, but only about 3 GPa in slower tests taking 10 s or so. Other glassy polymers (and semicrystalline polymers) behave similarly, their moduli changing by up to a factor of 2. Some rubbery polymers of exceptionally high molecular weight flow over long periods of time, which means that their moduli tend to zero. Occasionally the application of a high strain rate to a rubbery polymer can transform the material through the $T_g$ into the glassy region, in which low-rate rubbery moduli of less than 2 MPa can become high-rate glassy moduli of perhaps 2 GPa in magnitude.

## CREEP AND HARDNESS

When a hard indenter is pressed into the surface of a solid to form a plastic indentation, the mean pressure $p$ over the indentation may be written

$$p = c\sigma_y$$

where, as explained in Chaps. 2 and 19, $c$ is a proportionality factor that depends on the shape of the indenter ($2.5 < c < 3.0$ most often) and $\sigma_y$ is the uniaxial compressive yield stress as locally augmented by the indentation process itself. That is, using $\sigma = K\varepsilon^n$, we have

$$p = (cK)\,\varepsilon_{av}^n \tag{12-18}$$

where $\varepsilon_{av}$ is the average plastic strain produced in the vicinity of the indenter. For ball indenters, $\varepsilon_{av} = 0.2(d/D)$, where $d$ is the diameter of the projected impression and $D$ is the diameter of the indenter. In such cases the representative strain increases as the ball is pushed further in, which is physically sensible. With pointed indenters, however, where the indentations are geometrically similar whatever their depth, the representative strain is a fixed value; for example, a Vickers pyramid produces a strain of 0.08. This strain corresponds to an equivalent ($d/D$) for a ball indentation of 0.4, or an impression of diameter rather less than the radius of the ball; this is typical of recommended hardness testing practice.

When the yield stress is rate and temperature dependent, the apparent hardness changes, as the longer the load is left on, the bigger the size of the indentation. Using Eq. (12-5), we may rewrite Eq. (12-18) as

$$p = cK_0 \, \exp\!\left(\frac{1}{T}\right) t^{-m} \, \varepsilon_{av}^{m+n} \tag{12-19}$$

Thus, the hardness diminishes with $t$, giving linear logarithmic plots of $p$ vs. $t$, as shown in Fig. 12-10. It is remarkable in Fig. 12-10 that the creep process during indentation is not greatly dependent on indenter shape, except for the 60° cone. It turns out that, apart from the 60° cone, the deformation zones below all the other indenters resemble a radial compression (Fig. 12-11), and the reason that the behavior appears similar is that the change of hardness with time is a measure of the rate of progress of the elastic–plastic boundary "hemisphere" into the solid. Becasuse this plastic zone increases continuously as time progresses, it is not surprising that a relationship exists between transient creep and hardness via Eq. (12-5), which is itself related to Eq. (12-6) and Fig. 12-5.

The time dependence of indentation hardness at high homologous temperatures can be used as a simple means of selecting creep-resistant alloys.

In the case of solids with relatively low moduli, such as polymers, indentation pressures are lower than would be expected from either $p = c\sigma_y$ or Eq. (12-19), and elastic recovery of the indentations (principally in the depth of indentation) affects the results. This happens because the elastic strains in the region of the indentation are no longer negligible compared with the plastic strains; only when the strain imposed by the indenter is 5–100 times greater than the yield strain does the simple $p = c\sigma_y$ relation hold. Even so, as shown by K. L. Johnson, proper $c$ factors can be developed for smaller ratios of applied strain to yield strain.

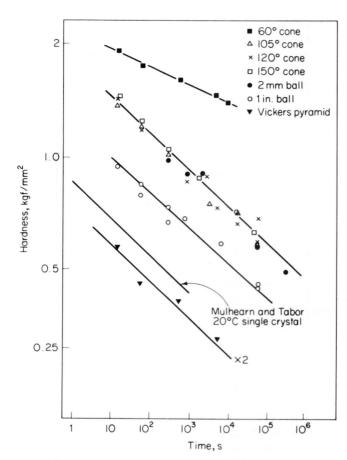

**Figure 12-10** Variation of hardness with loading time for indium polycrystalline specimens at room temperature. Except for the 60° cone, the slope is the same for conical, pyramidal, and spherical indenters. (The behavior is also similar to that observed by Mulhearn and Tabor on a single crystal using a spherical indenter.) Evidently, the creep process during indentation is not greatly dependent on indenter shape. [From A. G. Atkins, A. A. dos S. Silverio, and D. Tabor, "Indentation Hardness and the Creep of Solids," *J. Inst. Metals,* 94 (1966), p. 369. Reprinted with permission of the Institute of Materials.]

## POLYMER DRAWING, LÜDERS BANDS, AND METAL SUPERPLASTICITY

In simple tension tests of common engineering metals, elongations are limited by necking. The true strain at ultimate when necking commences is numerically equal to the work-hardening index $n$, as shown in Chap. 2. For mild steel, $n = 0.1 \sim 0.25$, so the uniform elongation at the maximum load is no more than 25%; even with brass, for which $n = 0.4 \sim 0.5$, the elongation is limited to 50%. Thereafter deformation is concentrated in the necked regions leading to fracture. There are circumstances involving some metals and some polymers under which a neck forms but does not give immediate fracture. Rather the neck travels along the test bar in a stable fashion, the thinned-down neck being "fed" material from the adjacent shoulders (Fig. 12-12). The drawn material in the neck is usually very strong and often anisotropic; it is also highly fibrous in nature. Some high-strength nylon fibers are produced in this fashion, as are glass filaments by "hot-drawing" glass

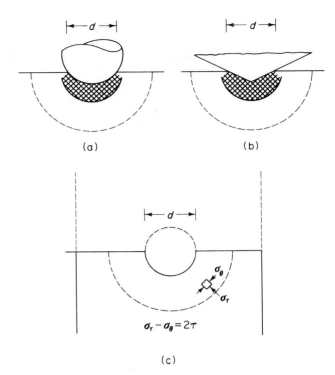

**Figure 12-11**  Plastic deformation mode: (a) for a spherical indenter; (b) for a conical indenter; (c) representing the equivalent plastic expansion of a hemispherical cavity into a semi-infinite solid. [From A. G. Atkins, A. A. dos S. Silverio, and D. Tabor, "Indentation Hardness and the Creep of Solids," *J. Inst. Metals,* 94 (1966), p. 369. Reprinted by permission of the Institute of Materials.]

rods. Chewing gum behaves similarly. In so-called *dieless drawing* of hot metals, a sharp temperature gradient is maintained at one section of the rod under tension, and significant reductions in area can be obtained in that region.

An explanation for this type of behavior is available when strain-rate sensitivity is taken into account. For material inside and outside a neck Eq. (12-4) may be rewritten, for constant $K$,

$$\frac{\dot{\varepsilon}_w}{\dot{\varepsilon}_s} = \left(\frac{\sigma_w}{\sigma_s}\right)^{1/m} = \left(\frac{\text{cross-section shoulder}}{\text{cross-section neck}}\right)^{1/m}$$

**Figure 12-12**  Polymer drawing, that is, the stable growth of a neck in high-density polyethylene.

where the subscripts w and s refer to neck and shoulder, respectively.

Suppose, for example, that at an early stage of necking the cross section in the necked region is 1% less than elsewhere. Then, $\sigma_w = (1/0.99)\sigma_s$ and $\dot{\varepsilon}_w = (1.01)^{1/m}\dot{\varepsilon}_s$.

For many common engineering materials at room temperature, $m = 0.005$ (Fig. 12-2), so

$$\dot{\varepsilon}_w = (1.01)^{200}\dot{\varepsilon}_s = 7.3\dot{\varepsilon}_s$$

This means that the necked regions stretch much more rapidly and that little deformation occurs in the shoulders; in fact, these elastically unload after ultimate in the usual type of tension test. By the time a 2% difference in cross-sectional area develops, $\dot{\varepsilon}_w = 52\,\dot{\varepsilon}_s$, so the process is self-accelerating. However, were $m$ to have a much larger value, such as 0.5, we would have for a 1% area difference,

$$\dot{\varepsilon}_w = (1.01)^2\,\dot{\varepsilon}_s = 1.02\,\dot{\varepsilon}_s$$

That is, the rate of deformation in the necked region is only 2% greater than elsewhere. This means that a great deal of deformation will continue in the shoulders despite the attempted localization of the deformation at the first place the change in areas took place. Even when the neck has developed to the extent that its cross section is down to 90% of the shoulders, its rate of deformation is only $(1/0.9)^2 = 1.23$ or 23% greater. With such slow localization of flow, a very large amount of deformation continues throughout the piece.

At neck initiation, $dP = 0$, as discussed in Chap. 2. The subsequent load drop bottoms out, however, and $dP = 0$ again when the neck stabilizes and draws out. Both instances correspond with $d\sigma/d\varepsilon = \sigma$, but the types of tangency are different (Fig. 12-13). The first has the $\sigma$–$\varepsilon$ curve "inside" the tangent line, and, as with traditional engineering metals under tension at room temperature, this corresponds with the strain $\varepsilon_A$ at neck initiation. The second has the $\sigma$–$\varepsilon$ curve "outside" the tangent line and corresponds with stable neck propagation at that fixed value of true strain $\varepsilon_B$. It should be noted that extension of the $\sigma$–$\varepsilon$ curve temporarily stops at $\varepsilon_B$ because the growth of the stable neck does *not* further strain the material in the neck. Only when the neck would have run to both gripped ends of the test piece is the rest of the curve picked up. In contrast, the curve of steady load vs. extension (on the chart of the testing machine) goes on for as long as the neck grows. Since a specimen 100 mm long has a shorter $\Delta L$ region than a specimen 1 m long, conversions of $\Delta L$ into $e$ values or into "extension ratios" ($L_f/L_0$) are quite meaningless because of the same objections that were raised in Chap. 2 to the use of percentage elongation without reference to gage length in tension tests. Even so, many polymer data are still published in terms of extension ratios or as "many hundreds of percent" extension.

There is a similarity between regions of *discontinuous yielding* in ductile metals and the growth of stable necks in polymers. In a highly polished steel sample, discontinuous yielding is seen to be accompanied by a series of striations that ripple along the gage section in the period *ab* in Fig. 2-20. Such *Lüders bands*, as they are called, may be thought of as a series of stable necks. Careful measurement shows many pairs of tangency points sat-

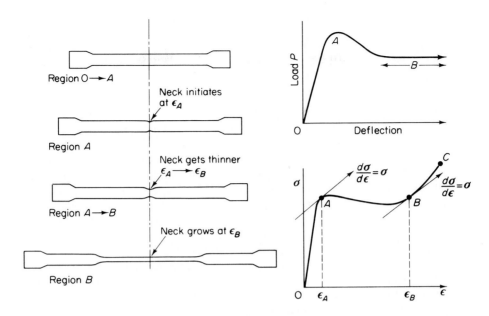

**Figure 12-13** Mechanics of (isothermal) stable and unstable necks; both occur at $d\sigma/d\varepsilon$ = $\sigma$, but the types of tangency are different. Unstable necks at $A$ have the $\sigma-\varepsilon$ curve "inside" the tangent line. Stable necks at $B$ have the curve "outside" the tangent line.

isfying $d\sigma/d\varepsilon = \sigma$ in the discontinuous yielding region at the $dP = 0$ local maxima and minima. The strain-hardening part of the $\sigma-\varepsilon$ curve beyond initial yielding up to ultimate and then to fracture follows upon exhaustion of the supply of Lüders bands. Note that, strictly speaking, the same problems as occur in polymers arise in regard to the meaning of strain in terms of $\Delta L$ during Lüders band propagation, except that whereas stable necks in polymers occur at large strains (see Fig. 13-9), discontinuous yielding in ductile metals takes place at very low strain levels, so the resulting errors in converting from true to nominal strains are small. That is, the strain range $ab$ in Fig. 2-20(c) is, say, 0.001 ~ 0.02, but the strain range $\varepsilon_A \sim \varepsilon_B$ in Fig. 12-13 may be 0.1 ~ 0.3. Note that not all polymers display stable necks and that many have $P-\Delta L$ and $\sigma-\varepsilon$ curves like those for brass in Fig. 2-2; details are given in Chap. 13.

**Example 12-10**

Lüders bands were observed in the $1 \times 10 \times 100$-mm gage length of a polished low-carbon steel test piece. The extension of the test piece was 100 $\mu$m. Determine the extent of Lüders band propagation in the gage length, given that Lüders band strain is approximately 0.02 within the band.

**Solution**   The apparent strain of the test piece is $(100 \times 10^{-6})/(100 \times 10^{-3}) = 0.001$. The local strain in a Lüders band is given as 0.02. If the parts of the gage section outside the bands have essentially zero strain, the extension of 100-$\mu$m strain is produced by some regions with zero strain and the band regions with 0.02 strain. To simplify matters, let us assume that the Lüders bands are perpendicular to the applied load (they usually are not), in which case the accumulated length of gage section taken up by the Lüders bands must be $(100 \times 10^{-6})/0.02 = 5$ mm. That is, of the 100-mm gage length, all the Lüders ripples add up to 5 mm.

In principle, the values of $\varepsilon_A$ and $\varepsilon_B$ for polymers and Lüders strains may be deduced by differentiating a relation such as $\sigma = K_0 \exp(1/T) \dot{\varepsilon}^m \varepsilon^n$, which additionally takes account of the possibility of the bar heating up from the work of deformation. They are not predicted from differentiation of $\sigma = K \varepsilon^n$ because that is a single-valued function describing the work hardening after initial yield (i.e., the region *buf* in Fig. 2-20).

Great strain-rate sensitivity is also displayed by some metals under certain conditions, where it is known as *superplasticity*. The circumstances that promote this behavior are those often associated with large rates of creep in metals, namely, high homologous temperatures, low strain rates, and a very fine grain size. Indeed, a good superplastic metal is the antithesis of a good creep-resistant alloy. It appears that quasi-viscous mechanisms such as grain-boundary sliding may contribute significantly to the strain-rate sensitivity and account for an appreciable fraction of the total deformation.

Reduction in area of over 90% (and associated tremendous elongations) have been reported for superplastic alloys in tension tests performed under the correct conditions. For example, at 300°C (approximately 0.8 $T_m$), an 80 Zn–20 Al alloy displays $m \sim 0.6$ when $10^{-3} < \dot{\varepsilon} < 10^{-1}$ s$^{-1}$. This means that formability limits of superplastic sheet metals can be increased to such an extent that it is possible to use forming techniques normally employed only with glass and plastics (Chap. 13). Figure 12-14 is a diagram (to scale) showing the tensile elongation and forming capability of the Zn–Al superplastic alloy. Care must be taken to ensure that the material remains superplastic during forming. Grain growth (which would reduce the grain-boundary area) must be avoided. Two-phase (duplex) microstruc-

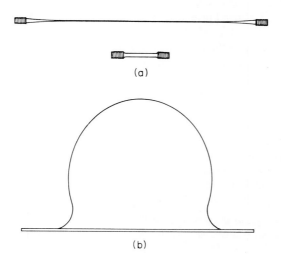

(a)

(b)

**Figure 12-14**    (a) Diagram of a tensile bar of 80 Zn–20 Al extended 10 times its original length at 350°C and 200 $\mu$m/s. (b) Bulged sheet of 80 Zn–20 Al alloy. Sheet thickness 0.75 mm; diameter of hole through which it was bulged is approximately 100 mm. [After W. A. Backofen, I. R. Turner, and D. H. Avery, *Trans. ASM* 60 (1964), p. 980.]

tures are especially useful in this regard; thus, eutectic Pb–Sn is another superplastic solid under the right conditions.

## REFERENCES

1. Alexander, J. M., and R. C. Brewer, *Manufacturing Properties of Materials.* London: Van Nostrand, 1963.

2. Cottrell, A. H., *The Mechanical Properties of Matter.* New York: John Wiley, 1964.

3. Johnson, K. L., "The Correlation of Indentation Experiments," *Journal of the Mechanics and Physics of Solids,* 18 (1970), p. 115.

4. McClintock, F. A., and A. S. Argon, *Mechanical Behavior of Materials.* Reading, Mass.: Addison-Wesley, 1966.

5. McCrum, N. G., B. E. Read, and G. Williams, *Anelastic and Dielectric Effects in Polymeric Solids.* New York: John Wiley, 1967.

## PROBLEMS

**12-1.** The data for $\sigma$ vs. $\varepsilon$ over a range of $\dot{\varepsilon}$ in Fig. 12-15 were obtained on low-carbon steel. Determine the parameters in the relation $\sigma = \text{constant} \times \varepsilon^n \dot{\varepsilon}^m$.

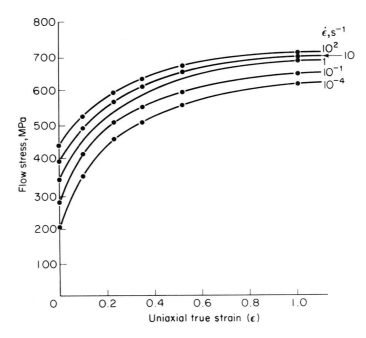

**Figure 12-15**

**12-2.** For the two-element model in Fig. 12-16 calculate the required minimum value of $\eta$ in order to limit the total strain of the system in 10 Ms to 0.003.

$E = 35.4$ GPa

$\eta$

$\sigma = 83.9$ MPa     **Figure 12-16**

**12-3.** The load displacement behavior of a certain polymer is described by a spring and dashpot in series as shown in Fig. 12-17. For the force application $P$ as shown from time $t_1$ to time $t_4$, plot the corresponding displacement $x$ within that time interval.

**Figure 12-17**

**12-4.** A steel beam subjected to bending loads in the laboratory deforms almost instantaneously when a load is applied, and the deformation remains constant with time. A lead–tin solder wire in tension deforms by a small amount when a load is applied and continues to deform with time until fracture occurs. Why does the lead–tin solder behave differently from the steel?

**12-5.** In a constant-load tensile test on 50 Pb–50 Sn solder at room temperature the specimen exhibits initial behavior of the form $\varepsilon = C[1-\exp(-Et/\eta)]$, where $\varepsilon$ is the strain, $C$ is a constant, $E$ and $\eta$ are constants, and $t$ is time. Then, for most of the test, the strain approximates the form $d\varepsilon/dt$ = constant. Explain the reasons for the *change* in behavior in terms of basic changes that occur in the structure of the metal.

**12-6.** The deflection–time curve of the lead–tin alloy in Fig. 12-4 can show a slightly upturned shape throughout the "steady-state creep" region. Why?

**12-7.** A lead–tin wire, initially 100.0 mm long and 3.18 mm in diameter, is loaded in tension with a constant force of 118 N at 20°C. After 1.74 ks its length is 119.0 mm, and 0.90 ks later its length is 129.5 mm.

    **(a)** Estimate the magnitude of the steady-state viscosity $\eta_1$ (in Pa-s) if the specimen is considered a four-element Voigt–Maxwell model.

    **(b)** Is this a liquid or a solid?

**12-8.** Results for a creep test of Pb–Sn solder are given in Fig. 12-18. Calculate $\eta_1$ and $E_1$ if we assume this specimen approximates the Voigt–Maxwell four-element model shown.

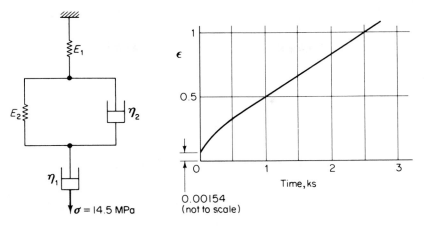

**Figure 12-18**

**12-9.** The strain–time curve for a viscoelastic polymer under a tensile stress of 100 MPa is shown in Fig. 12-19. Regarding the polymer as the Voigt–Maxwell model shown,
  (a) Calculate the value of $E_1$.
  (b) Calculate the value of $\eta_1$.

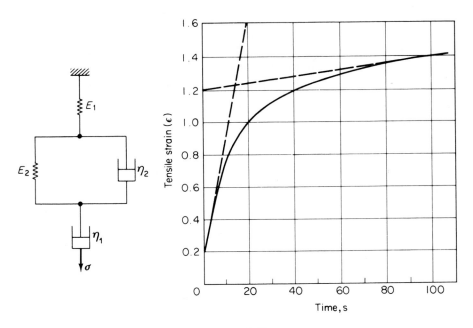

**Figure 12-19**

**12-10.** The strain–time curve for a four-element Voigt–Maxwell model is given in Fig. 12-20. Calculate the strain after 1 ks for a model identical in every respect except that $E_1$ is one-fourth the $E_1$ of the model shown.

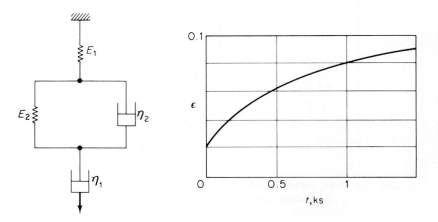

**Figure 12-20**

**12-11.** The relative ductility or brittleness of silly putty depends on the rate at which it is stressed. Which of the three models shown in Fig. 12-21 best describes the behavior of silly putty? Explain why you reject the other models.

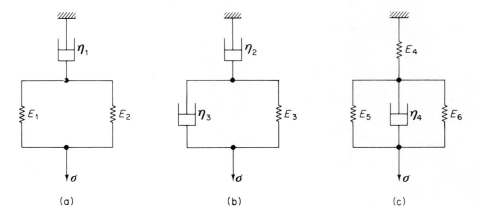

(a)          (b)          (c)

**Figure 12-21**

**12-12.** A four-element viscoelastic model produces the strain–time curve in Fig. 12-22. Draw to *reasonably accurate scale* the strain–time curve, on the same coordinates, for tripling $E_1$, halving $\eta_1$, and lowering $\sigma_A$ by 20%.

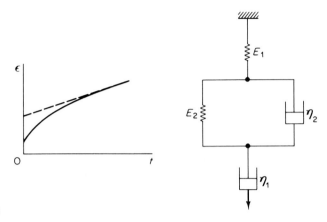

**Figure 12-22**

**12-13.** A round, tensile test piece of some polymer necked (as a metal often necks in tension), but instead of necking on down and fracturing, the polymer neck stabilized and "drew out" at constant diameter. Explain why this happens with reference to the load extension or stress–strain curve for the polymer. The following loads and diameters applied to such a stable neck test:

|                     | Diameter, mm | Load, N |
| ------------------- | ------------ | ------- |
| Initial condition   | 10           | 0       |
| Neck initiates      | 8            | 1000    |
| Neck grows steadily | 6            | 560     |

Sketch this behavior on a true-stress–true-strain ($\sigma$–$\varepsilon$) plot, marking off important values of $\sigma$ and $\varepsilon$ and indicating any significant aspects of the diagram.

**12-14.** The relaxation time of polyisobutylene is about 200 s at 50°C. Determine the relative magnitudes of the elastic and flow components of strain under loads of duration 1 s, 10 s, 100 s, and 1000 s. At 50°C the same polymer is held at a constant strain for 10 ks. Sketch the percentage fall in the applied stress from its initial value over the 10-ks time interval.

**12-15.** A series of ball indentation experiments under a load of 5 kg$_f$ were carried out at room temperature on a block of tin. The load was left on for various lengths of time, and the following diameters of the hardness impressions were produced: 1.13 mm after 10 s; 1.26 mm after 100 s; 1.39 mm after 1000 s; and 1.51 mm after 3200 s. Obtain from these data an estimate for the strain-rate sensitivity of tin at room temperature.

**12-16.** A strip is reduced in thickness, with no sideways spread, from $t_{in}$ to $t_{out}$ by pulling through tapered dies (sheet drawing). If the speed of the strip exiting the die is $V_{out}$, show that the mean strain rate for the process is $2V_{out}(1-r)\tan\alpha/t_{out}$, where $\alpha$ is the semi-angle of the die and $r$ is the percentage reduction in area for the drawing process.

Determine the mean strain rate when $V = 1$ m/s, $r = 20\%$, $\alpha = 8°$, and $t_{out} = 2$ mm. Comment on the magnitude of this strain rate as compared with quasi-static rates of testing, and using the data in Fig. 12-15, determine by how much the working loads in the process are increased when rate effects are taken into account.

**12-17.** A bar of polyethylene 10 mm in diameter reduces uniformly to 9.5 mm diameter and then forms a stable neck 8.2 mm in diameter that propagates down the bar. Show that percentage elongation is an unsuitable parameter for characterizing the neck formation and provide a sensible parameter.

**12-18.** A four-element Voigt–Maxwell model for deformation of a polymer has elements $E_1$ and $\eta_1$ in series with $E_2$ and $\eta_2$ in parallel.

    **(a)** Draw the strain–time curve for this system with the correct shape but without numbers; label it "curve (a)."

    **(b)** Increase $\eta_2$; on the same coordinates as curve (a) draw the new curve with a dotted line; label it "curve (b)."

    **(c)** Decrease $E_1$, using original values of $E_2$, $\eta_2$, and $\eta_1$, and draw a new curve with long dashes on the same coordinates; label it "curve (c)."

**12-19.** The model of polymer behavior shown in Fig. 12-23 applies to polymer X. If we wish to utilize a 127-mm-long part made of polymer X that is to be loaded to an axial tensile stress of 67.3 MPa, calculate the required magnitude of $\eta_2$ that will just allow a total deflection of 1 mm in 1 year. $E = 16.2$ GPa and $\eta_1 = 17.2$ PPa-s.

**Figure 12-23**

**12-20.** The elements of a Voigt solid have the following properties: $\eta_A = 43.6$ GPa-s and $G_B = 833$ MPa. A second solid has the same Voigt elements as the first plus a viscous element $\eta_C = 67.2$ GPa-s in series with the Voigt elements. If each solid is subjected separately to a shear stress of 97.6 MPa, calculate the difference in the total shear strain $\gamma$ for the two solids after 372 s in each case.

**12-21.** A four-element model of a polymer Y has the following tensile constants:

Voigt elements: 18.7 GPa and 24.6 TPa-s

Maxwell elements: 12.4 GPa and 672 GPa-s

If a specimen of polymer Y that is 173 mm long is subjected to a tensile stress of 142 MPa, calculate the *change* (increase or decrease?) in total elongation (mm) after 1.00 ks that will result if the value for the viscous Maxwell element is doubled.

**12-22.** Calculate the strain in the viscoelastic model in Fig. 12-24 at the end of 15 s if $\eta_1 = 45.3$ GPa-s, $\eta_2 = 27.9$ GPa-s, $E = 694$ MPa, and $\sigma = 72.1$ MPa.

**12-23.** When it is difficult to test prototype structures or vehicles for structural integrity or "crashworthiness," models are tested instead. Show that in impact tests at the same impact velocity, the strain rate is greater in the model structure than in the prototype.

**Figure 12-24**

**12-24.** The yield stress of low-carbon steel is rate dependent. What effect does this have on the energy absorbed by a prototype structure predicted from the energy measured in an experiment on a small-scale model of the structure?

# 13

# Polymers and Design with Time-Dependent Solids

The mechanical properties of polymeric solids (commonly referred to as plastics) have been touched on in earlier chapters. It was shown that a whole range of stress–strain–time responses is possible, depending on conditions, from very creepy behavior to stiff elastic behavior, with the possibility of a rubbery range between. The structure of the polymer and its makeup, together with temperature and rate of deformation, determine to a great extent the mechanical behavior.

Many polymers are now accepted as regular engineering materials and given the loose description *engineering plastics*. The term is clearly artificial because applicability depends on circumstances, but it probably originated to distinguish those polymers that can be substituted satisfactorily for metals in certain designs from those with inadequate properties. When polymers are used in design, they fulfill a need that either cannot be met with traditional materials or cannot be met as cheaply otherwise. The mass production of intricate, accurate parts by injection molding is one feature of thermoplastics of which advantage is often taken.

Examples of plastics used successfully in engineering contexts are polyvinyl chloride (PVC) in large-diameter pipes, polyolefins in small-diameter pressure pipes, nylons and acetals in small gear wheels, polypropylene (PP) in hinges, and high-density polyethylene (HDPE) in artificial hip joints. In most of these applications the choice of material is determined by some particular virtue sufficiently valuable to offset the relatively low moduli and strengths. Other load-bearing products designed in polymers include bottle crates, vegetable trays, and other stacking containers injection-molded from polypropylene or high-density polyethylene; pallets thermoformed from acrylonitrile butadiene styrene (ABS) sheet or molded from HDPE or PP; and chairs made by a variety of techniques from materials that include PP, PVC, ABS, polystyrene, acrylic, and HDPE.

The limitations imposed by creep on the wide application of polymers led to intensive research programs in the 1950s and 1960s aimed at increasing the temperatures at which polymers can be used without sacrifice of mechanical properties. Research was spurred by aerospace and defense needs. Even so, few of the newly developed materials have really reached significant production levels (in tonnage) in comparison with polyethylenes, polyvinyl chlorides, polypropylenes, and ABS. Of those that have, nylons and polyacetals account for a significant part of the consumption of "new" engineering plastics, followed by polycarbonate. Of the rest of the newer materials, only modified polyphenylene oxide (PPO) (e.g., GE Noryl) and thermoplastic polyesters [polybutylene terephthalate (PBT) and polyethylene terephthalate (PETP)] are used in reasonable amounts; others, such as fluorocarbons, polysylphones, and polyamides, are expensive and thus find only specialized applications.

In engineering design with polymers, it would be ideal if the region where the behavior is primarily elastic, with small rate effects, could always be used. Conventional strength-of-materials design calculations could then be employed to a reasonable approximation. But other factors (corrosion resistance, low mass, cost, etc.) may suggest the use of polymers whose behavior would be time dependent in use. Then, such polymer parts must be stressed merely to levels that produce only tolerable creep. Examples of such calculations (which also apply, of course, to creepy metals) are given later.

Although polymer parts are often cast (molded) directly to shape, some polymers may be subjected to the same sort of mechanical working processes (e.g., rolling, drawing, and extrusion) as metals, resulting in interesting changes in mechanical properties. Before considering mechanical behavior and design with time-dependent solids, however, some consideration must be given to the molecular structure of typical polymers and how structure influences mechanical properties.

## POLYMER STRUCTURE

Polymerization produces giant long molecules, often joining together thousands of *monomer units,* like a length of beads on a string. The molecular forces that hold individual links together in the chain are primary valence bonds (Chap. 4), and it is secondary bonding that attracts separate polymer chains to one another. Although these secondary cohesive bonds are relatively weak, there are so many of them between the many backbone strands that a tightly cohering mass results. Differences in the strength of secondary bonding produce different types of mechanical behavior in polymers. For example, weak secondary bonding gives a rubbery and stretchy polymer; stronger interchain cohesive forces give stiffer and stronger plastics. The substitution of different groups in the polymer chain produces stronger secondary bonding and thus polymers with altered mechanical properties. For example, acrylonitrile is essentially polyethylene [$(CH_2)_n$], but with alternate H—C—H units replaced by H—C—N units.

In addition to the foregoing simple *linear polymers,* it is possible for chains to grow out sideways from the main-chain backbone as branches do on a tree. Furthermore, the branches may link across to other main chains, giving *ladder* or *network* polymers (Fig. 4-3). Under the circumstances, main chains are held together not only by secondary valence forces but also by the stronger primary forces along the "rungs" of the (three-dimensional) ladder. This, in effect, produces a single, strong, giant molecule.

As explained in Chap. 4, the more molecules interfere with one another in the liquid state, the more viscous the liquid and the less the chance that a crystalline solid with a clear-cut freezing/melting point will be formed on cooling. Thus, branched and network polymers are more "amorphous" than simple linear polymers. Polymers of the class called *thermosetting plastics* start out in the liquid state but solidify under the influence of heat and/or a catalyst hardener by forming cross links between linear main chains, which reduce the mobility of these strands. One example of a thermoset is epoxy glue. Once formed, thermosets tend to char and degrade on heating. In contrast, polymers such as simple linear polymers and glasses soften to a toffee consistency on heating and return to their earlier state on cooling. They are called *thermoplastics*.

The properties of polymers depend strongly on their detailed structural shape, where the chemistry of the monomer and repeat patterns are paramount. Particularly important is chain geometry. Chains of mers (a *mer* is one basic molecular unit) are usually represented in textbooks as being planar, but they are actually three-dimensional. The bonding directions of a carbon atom may be thought of as acting through the corners of a tetrahedron, so even a molecularly simple chain zigzags in space. For example, the chemical structure of polyethylene $(CH_2)_n$, is usually drawn as in Fig. 13-1(a), whereas it really looks more like Fig. 13-1(b). Polymethylmethacrylate is similar to polyethylene, but instead of being simple hydrogen atoms, the side groups are more bulky methyl $(CH_3)$ and $(COOCH_3)$ groups (Fig. 13-2). In truth, PMMA is not very complicated compared with many other polymers.

Chains in general wander in many directions and become raveled and matted, and the usual analogies for polymers are tangled string or masses of spaghetti. It follows that the mechanical properties of polymeric materials are markedly dependent on the flexibility of the joints in these chains and the ease with which the chains can slide over one another

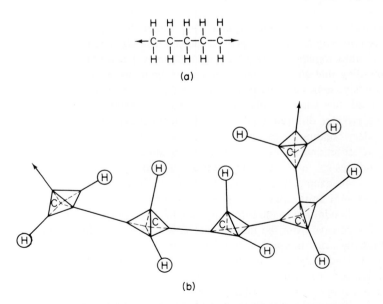

(a)

(b)

**Figure 13-1** (a) Planar and (b) three-dimensional representations of polyethylene; bonding directions of carbon atoms act as if through the corners of a tetrahedron.

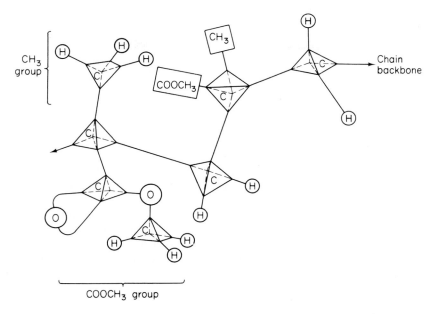

**Figure 13-2**  Polymethylmethacrylate molecule, consisting of methyl (CH₃) and (COOCH₃) side groups on the carbon backbone chain.

during deformation; time-dependent behavior results, in part, from the uncurling of long, tangled chains, which gives delayed elastic effects. The behavior is further modified when the chains are cross-linked to one another as in network polymers.

Again, the behavior is modified if there are crystalline regions in the polymers. Although most bulk polymers are amorphous, it is possible for regular arrays of touching parallel chains to order themselves and form a lattice in certain regions of a polymer, especially when the polymer consists of short, simple, linear chains or has symmetrical mers with no lumpy side groups and no cross linking to other chains.

**Example 13-1**

The volume vs. temperature plot for a certain polymer cooled at two widely different rates is shown in Fig. 13-3. At room temperature, which condition, A or B, is likely to be more brittle?

**Solution**  Condition B is amorphous, and condition A is crystalline, because of the distinct change in volume at a single (melting) temperature. Condition A is less likely to show viscoelastic behavior and thus is likely to be less ductile, hence more brittle.

Of the common polymers, polyethylene, with its simple structure of chains less than about 100 mers long, has a strong tendency to crystallize, achieving perhaps 80% overall crystallinity. Others, such as nylon, propylene (of the so-called isotactic type), and PVC, are partially crystalline (say 50–70%), whereas polymethylmethacrylate, with lumpy side groups, is virtually impossible to crystallize. Note that the more complex the molecular chain system, the less chance there is that crystalline regions (having a regular and

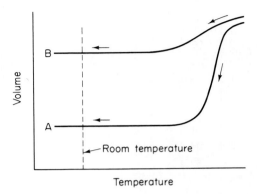

**Figure 13-3** Volume vs. temperature plots of two polymers (Example 13-1).

repeated three-dimensional pattern) will easily be found. For similar reasons, branched and network polymers do not readily crystallize. As a rule, partially crystalline polymers are opaque or at most translucent. Amorphous polymers, on the other hand, are usually transparent. This provides a rough but convenient means of identification: Remember that window glass is transparent and glasses are amorphous. In a bulk polymer containing microcrystalline regions, a single molecular chain may extend through both the crystalline and noncrystalline zones.

The arrayed regions are called *crystallites,* and it has been found that the parallel chains are not simply linear but fold back and forth in pleats (called a Chinese cracker structure by analogy with the firework), as shown in Fig. 13-4. Here again, the simple planar drawing is in reality a closely packed, three-dimensional figure. The platelike folded areas are called lamellae and extend individually for about 10 nm. Structured aggregates of crystallites occur on a macroscopic scale, and lamellae appear in plan view to flow on edge radially outward from small regions like spokes in a wheel. The resulting rosette structure is called a *spherulite* (Fig. 13-5).

**Figure 13-4** "Chinese cracker" arrangement of crystallites in crystalline regions of polyethylene.

**Figure 13-5**   Spherulite in polyethylene, which is a structured aggregate of crystallites on a macroscopic scale. (Photograph courtesy of G. S. Yeh.)

Polymers that have some crystallinity *whiten* on straining; that is, the highly stressed areas change color. The drawn necks in high-density PE are very white; if a hardness ball is pressed into a lump of PVC, the distribution of nonuniform strains is seen. The probable cause of whitening is the mechanical breakup of spherulites leading to light scattering. When a whitened region is warmed, the color disappears and the process is reversed.

## BEHAVIOR RELATIVE TO GLASS-TRANSITION TEMPERATURE

As described in Chap. 4, the mechanical properties of amorphous polymers may be conveniently split into two principal regimes: above $T_g$ and below $T_g$. Above $T_g$ time-dependent creepy behavior is evident, whereas as the temperature drops below $T_g$, time and rate effects gradually become less marked and creep is low (Fig. 4-4 and Chap. 12). Near $T_g$ the behavior is often called *leathery.* The glass transition is caused by changes in the molecular mobility of the main polymer backbone chains; diminished creep below $T_g$ comes about from a stiffening of these main chains in comparison with their flexibility above $T_g$. Any small creep and rate effects that are present below $T_g$ arise from molecular mobility of polymer branch side chains. At low enough temperatures, a "secondary transition" may occur, after which the mobility of even these side chains is restricted. At temperatures below the secondary transition, the polymer is then essentially elastic and brittle, like some time-independent metals and ceramics.

Figure 13-6(a) shows the changes in modulus that occur with changes in molecular mobility at the secondary and primary transition points as the temperature varies. The modulus falls by a factor of 3–5 as side-chain mobility comes into play, and by a further factor of 10–500 as main-chain mobility increases [the ordinate in Fig. 13-6(a) is a logarithmic scale]. The existence and position of the transitions and their magnitudes depend on the detailed molecular structure, on the conditions of measurement, and indeed on the history of the sample. In a way, this statement is no different from the observation that recrystallization temperatures in metals depend on the amount of prior cold work. In the case of partially crystalline polymers, it is only the amorphous part that may show such transitions. Consequently, the decreases in modulus are less marked for them, and modu-

**Figure 13-6**   Schematic changes with temperature of (a) modulus and (b) damping capacity for polymers. Temperatures at which the damping peaks occur usually coincide with the inflection in the modulus–temperature curve; the glass-transition temperature $T_g$ is usually defined as the intersection of tangents, so the primary and secondary transition temperatures differ somewhat from the temperatures of maximum damping.

lus–temperature plots for crystalline polymers often show only minor drops before melting. Example 12-9 concerns the relative effects of strain rate and temperature on the moduli of metals and polymers.

Other properties, such as toughness and damping capacity, also change at transition temperatures. It is a feature of these materials that their ability to damp out vibrations increases markedly near the transition temperatures when compared with their damping capacity at temperatures much below and above. For example, the oscillations of a torsional pendulum, the suspension of which is made from a polymer displaying transitions, die away much more quickly at these transition temperatures, and plots of damping capability have associated peaks, as shown schematically in Fig. 13-6(b). The bounce of rubber balls, measured at different temperatures, is the least at temperatures near their primary and secondary transition temperatures because of the high damping and low restitution. The temperatures at which the damping peaks occur usually coincide with the points of inflection of the modulus–temperature curve; as $T_g$ is conventionally defined in terms of the intersection of the tangents at "bends" in the modulus–temperature curve, $T_g$ and the temperature at peak damping are not the same. For many polymers the difference is on the order of 25°C.

Phenomenologically, the mechanical behavior of a creeping polymer above the secondary transition, and certainly above the primary glass transition, is similar to that of a creeping metal. However, because the elastic moduli of polymers are in general much smaller than those of metals, the recoverable elastic strains of polymers may be comparable in magnitude with the irreversible creep strains, which is not the case for most creeping metals of engineering importance. The combination of viscous sliding and normal elastic behavior leads to *viscoelasticity:* Large amounts of deformation can occur for quite low stresses, as described in Chap. 12.

Generally speaking, the more complicated the molecule and the more cross linking present, the higher the glass-transition temperature. The following are the approximate $T_g$ values for some common polymers: polyisobutylene (butyl rubber), −70°C; polyethylene, −20°C; polypropylene, 0°C; nylon, 60°C; PVC, 80°C; polymethylmethacrylate, 100°C; polystyrene, 100°C; PTFE, 125°C; polycarbonate, 150°C. Thus, at room temperature, nylon, PVC, PMMA (Plexiglas), polystyrene, PTFE, and polycarbonate display quasi-elastic behavior, with comparatively small time effects, whereas all the rest display marked viscoelasticity. Obviously, if nylon were warmed above its respective $T_g$, it would become creepy. Conversely, the behavior of polypropylene much below 0°C is time independent, and so on. The glass-transition temperature of polymers may be reduced by adding a plasticizer, which is a liquid in which the polymer is soluble but that does not combine with the polymer at the molecular level. For example, the glassy polymer cellulose nitrate is softened ($T_g$ is lowered) by the addition of camphor, to give celluloid; toughness is produced at the expense of rigidity. Plastic raincoats are made from plasticized PVC. Another technique is to mix two polymers together to obtain a *copolymer* whose $T_g$ lies between those of its components. The corresponding mechanical behavior is affected by changes in $T_g$, so polymers can be tailored to suit particular applications.

**Example 13-2**

Select $T_g$ relative to 23°C for the following applications: (a) a child's rubber ball for maximum bounce, (b) a polymer for electrical insulation in the Arctic, (c) rubber fenders for a boat dock in Miami, (d) a polymer with the highest possible room-temperature strength, (e) rubber for automobile tires.

**Solution**   (a) << 23°C (b) << 23°C (c) ~23°C (d) >> 23°C (e) << 23°C

## ELASTICITY

Because along most polymer chains the backbone is formed of carbon bonds, one might think that polymers should have the same stiffness as other substances with carbon–carbon bonds such as diamond. The Young's modulus of diamond is about 760 GPa, but diamond has a specific gravity of about 3.5, whereas that of many polymers is about unity. So, making allowances for the differences in density (and hence fewer bonds per unit volume), we might say that polymers should have moduli of perhaps (1/3.5)(760) = 207 GPa.

In practice, the moduli of typical polymers are about 700 MPa. Only with certain natural polymers in fiber form, such as the cellulose molecule in wood, is the modulus at all close to the expected value. Why?

The reasons for the discrepancies should be clear from the clues just given:

1. Diamond has an ordered crystal structure.
2. Fibers, although amorphous, are usually oriented with the chain backbones along the fibers.
3. Regular, amorphous tangles of string or spaghetti have no orientation whatever.

Polymer chains zigzag in space and, like coil springs, must be more flexible than the same length of chain bonds stretched straight. The exceptional deformability of the rubber molecule is due to the helical structure of the isoprene carbon chain. In gutta-percha, another form of the rubber polyisoprene, the carbon chain is zigzag rather than helical, and the substance is much less deformable. A coil spring made from metal wire is much more compliant than the same length of straight wire. Even in a highly crystalline polymer, where a fairer comparison could be made between theory and practice, the Chinese cracker structure of lamellae can never be as stiff as a straight chain.

Wide-angle X-ray examinations show that the rolling process orients molecular chains nearly parallel to the rolling direction, and improvements in polymer stiffness along the rolling direction can be marked. For example, the Young's modulus of undeformed polypropylene is perhaps 710 MPa; a rolling reduction in area of 87% at about 120°C can increase the modulus more than sixfold, precise values depending on strain rate and temperature (cold work also being possible). The Dupont Company sells material of this type for strapping. Other polymers, such as polyoxymethylene and nylon-66, also improve in modulus under similar processing conditions. In the case of polyethylene, cold rolling up to a 25% reduction in area produces a decrease in modulus at first, followed by a subsequent increase (Fig. 13-7). (Note also the effect of the speed of the testing crosshead on the results.) The reduction is probably associated with a breakup of the crystalline spherulite structures without observable chain orientation, the latter coming into play only at higher amounts of cold work.

**Figure 13-7** The effect of cold work on the elastic modulus of high-density polyethylene at two crosshead speeds. [From R. M. Caddell, T. Bates, and G. S. Yeh, *Mat. Sci. Engr.,* 9 (1972), p. 233. Reprinted with permission.]

Other mechanical properties (particularly strength) are affected over these different ranges of prestrain, as discussed later (Fig. 13-13), and other polymers may be affected differently.

**Example 13-3**

The following values of elastic modulus $E$ were obtained by measuring the deflection of a polyethylene cantilever beam ($\delta = PL^3/3EI$) with a constant load on its end after the times indicated.

| Time, s | 30 | 60 | 90 | 120 |
|---------|----|----|----|-----|
| $E$, MPa | 24 | 22 | 20 | 19 |

Explain why the calculated values of $E$ change with time.

**Solution** As time passes, the viscoelastic deformation increases (Chap. 12). Thus, the tip deflection of the cantilever is greater for the same load, so in $\delta = PL^3/3EI$ the apparent $E$ decreases with time.

The example indicates a need to standardize rates of loading, time under load, and so on; hence, it is customary with polymers to quote a *10-second* modulus, which relates deflection and load 10 s after reasonably rapid application of the load. The data in Fig. 13-7 are typical 10-s moduli.

## BRITTLE BEHAVIOR AND CRAZING

Polymers much below their $T_g$ behave in a brittle fashion, with virtually linear stress–strain curves up to fracture. Rate effects are not very marked. At room temperature polystyrene and PMMA are typical examples. Both these solids display a characteristic called *crazing* (Fig. 13-8).

**Figure 13-8**   Crazes in tensile test piece of polystyrene. [From R. P. Kambour, *Macromol. Rev. I,* 1 (1973). Reprinted with permission. Photograph courtesy R. P. Kambour.]

Crazes, which were first observed in 1949 by J. Sauer and coworkers in PS, appear to be tiny, widely distributed, cracklike defects which produce bright reflections. They are *not* true cracks, however, but rather regions of highly plastically deformed material interspersed with voids. The microstructure of a craze resembles that of a sponge and when stressed appears as a set of oriented microfibrils of diameter 20 ~ 40 nm elongated in the direction of maximum tensile stress. The formation of crazes in glassy polymers is the localized plastic response to stress concentrations at stresses lower than the overall yield stress (see next section); unless suppressed by compressive hydrostatic stress fields, crazes often form at a tensile stress level of about $\sigma_y/2$. The formation of crazes is linked to the tensile hydrostatic component of stress and cavitation processes.

That crazes are not simple cracks is confirmed by the fact that test pieces cut from highly crazed material show respectable strengths and do not, as might be expected, have low strength. In real cracks in glassy polymers, crazed structures are often found in the highly deformed region around the crack tip (see Chap. 14).

## YIELD AND FLOW STRESS

There is some confusion in terminology associated with the mechanical behavior of polymers, a result of the lower moduli of polymers compared with that of, say, metals and of the phenomenon of the drawing of stable necks that sometimes occurs for polymers (Chap. 12). It is necessary to spell out these differences fully to avoid misunderstanding the polymer strength data given in handbooks and scientific literature.

Engineers have traditionally differentiated between yield stress and ultimate tensile strength in metals, but in polymers these terms are sometimes confused.

The common definitions of yield strength $S_y$ and tensile strength $S_u$ of ductile metals are illustrated in Fig. 13-9 with reference to a load-extension (or nominal-stress–nominal-strain) curve for an annealed low-carbon steel. For all practical purposes, yield strength, elastic limit, and limit of proportionality are given by the same stress $S_y$, where the associated strain is most often less than about 0.2%. There is essentially no change in cross-sectional area between $O$ and $Y$, so nominal stress and true stress are the same at yield.

The maximum load point $U$, at which an unstable neck initiates, gives the ultimate tensile strength (or tensile strength) $S_u$. The associated strain is numerically equal to the work-hardening index $n$ (Chap. 2) and might be quite considerable, say, $n = 0.2$ or about a 20% reduction in area. There is a marked but uniform reduction in cross-sectional area between $Y$ and $U$; the true stress at ultimate $\sigma_u$ is thus greater than $S_u$. The unstable neck at $U$ *always* leads to fracture in the region where it first develops.

The flat portion of the curve at $Y$ is really a series of ripples in the load trace, with the associated propagation of Lüders bands (see Chap. 12). With certain materials and testing arrangements, a stress spike can sometimes occur at $Y$, giving an upper- and a lower-yield point. For design purposes, $S_y$ as shown in Fig. 13-9 is usually the important strength parameter. As discussed in Chap. 2, $S_y$ may be increased by cold working; when the amount of prestrain exceeds $n$, the yield and maximum load points coincide, with immediate necking.

Most other metals display nominal stress–strain curves like the one shown in Fig. 13-10 where there is a gradual transition from elastic to plastic behavior. The flat region at $Y$ in Fig. 13-9 is not in evidence, and the yield point is defined by an offset (proof) method. The associated yield strain is still on the order of 0.2%. The reduction in area to ultimate can be as high as 40% for these metals, such as some brasses and stainless steels, which have $0.35 < n < 0.5$. Furthermore, $S_y$ can again be increased by cold working.

A typical load extension (or nominal stress–strain) plot for polymers such as polyethylene, nylon, and polyvinyl chloride, which "draw" at room temperature, is shown in

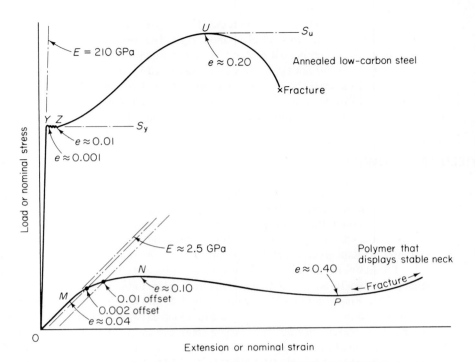

**Figure 13-9** Nominal stress–strain curves for a polymer that displays stable neck propagation ("cold drawing") as compared with an annealed plain low-carbon steel that exhibits a pronounced yield point. [From C. S. Lee, R. M. Caddell, and A. G. Atkins, *Mat. Sci. Engr.*, 18 (1975), p. 213.]

the lower half of Fig. 13-9; this behavior was discussed in Chap. 12 where the reasons for the phenomenon in terms of strain-rate sensitivity were investigated. There is a departure from linearity at *M* (where the strain may be about 1%), and the load curve rises to a local maximum at *N* (where the strain may be 10%), at which point the stable neck initiates. The load then falls as the neck is reduced in cross-sectional area, until stability is reached and the neck propagates along the test piece at the essentially constant load *P*. Subsequently, after the neck has propagated along the length of the test bar, the load increases again and fracture eventually ensues under a rising load, a situation akin to failure in brittle materials. The particular polymer fracture load may be less than or greater than the load at *N*, depending on the interaction of the geometry of the propagated neck and the original shoulders of the test piece. When the drawn section is cut out and a new test piece made from it, the strength of this highly oriented material becomes an extrapolation of the *true-stress–true-strain* curve beyond *P*.

    After the local load maximum at *N*, measurements and definitions of strain in terms of *length*, such as the nominal strain given by $(L - L_0)/L_0$ or the "draw ratio" $L/L_0$, are quite meaningless because the reduction is nonuniform. Clearly, *any* value is possible depending on the reference gage length (the percentage elongation in tensile tests of metals is likewise

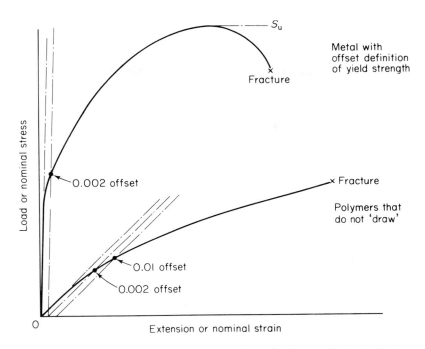

**Figure 13-10** Nominal stress–strain curve for a polymer that does not display localized necking (i.e., does not "cold-draw") as compared with a ductile metal that does not exhibit a pronounced yield point. [From C. S. Lee, R. M. Caddell, and A. G. Atkins, *Mat. Sci. Engr.*, 18 (1975), p. 213.]

meaningless without reference to the starting gage length). It is most sensible beyond $N$ in Fig. 13-9 to measure the strain in terms of percentage reduction in area $(A_0 - A)/A_0$ or true strain $\varepsilon = \ln(A_0/A) = 2 \ln(d_0/d)$. Then during most of the stable neck propagation, changes in true stress and true strain are minor until the neck has propagated to the shoulders. It is worth noting that if the initial neck formation is very sudden, it may be very difficult experimentally to obtain points between $N$ and $P$. Thus, in Fig. 13-14 (later), it is impossible with micrometer calipers to pick up the neck-formation strains over the range 0.05–0.50 for the as-received polycarbonate. The region $NP$ in Fig. 13-9 and its growth of *one* stable neck corresponds to the region $YZ$ on the low-carbon steel curve where there are many local maxima and minima in the load, corresponding to the propagation of many Lüders bands.

The deformation and strains at $N$ and $P$ (Fig. 13-9) in ductile polymers that draw are much greater than those at $Y$ and $Z$ in low-carbon steel. Moreover, the range of strain between $M$ and $N$ is extremely large compared with the corresponding regions for ductile metals that occur near $Y$. The point at which permanent deformation sets in is somewhere between $M$ and $N$. The curvature over a large range of strains between $M$ and $N$ and the local maximum in load at $N$ have led some workers to call $N$ the ultimate tensile strength for polymers. Reference to Fig. 13-9 shows that this is erroneous if we mean that ultimate tensile strength is followed by an unstable neck and fracture as is the case with ductile metals. However, a point such as $N$ may indicate the maximum *load,* and in this context the use of the term "tensile strength" is consistent with a long-standing definition. However, it is clear that $N$ and $P$ do not represent upper- and lower-yield points in the traditional sense.

Others identify $N$ as the local yield point, and the similarity of events between $N$ and $P$ and those between $Y$ and $Z$ makes such an approach justified. However, the strains at $N$ are much larger than at $Y$, and from a design point of view $M$ may be more meaningful as a limiting strain; even then, deformations are greater than traditional metal yield strains. However, the principal objection to using $N$ as a yield value relates to those polymer load-extension curves, such as in Fig. 13-10, that show no zero-slope load maximum. The off-set definition of yield allows a consistent method of determination whatever the shape of the load-extension diagram.

Polymers that do not draw (such as polymers containing cold work) have continuously rising load–extension (nominal stress–strain) curves, similar to the one in Fig. 13-10 for metals, but the strains are larger and the moduli smaller.

**Example 13-4**

Tangled string and spaghetti are analogies for long-chain polymers. Explain the mechanical behavior for which each model is appropriate.

**Solution**   The tangled or knotted string analogy describes rubbery behavior, where the material may be stretched appreciably but then becomes stiff and cannot be moved further without breaking some of the string (compare to a rubber band). The spaghetti analogy describes the situation where continuous appreciable sliding is possible among the chains, and so viscous flow comes into play as well as elasticity, which it eventually swamps at higher and higher temperatures.

Viscous sliding is more inhibited the longer the chain length of simple linear polymers, and it is particularly restricted when there is cross linking between the chains. Thus, the spaghetti model reverts to tangled string. The vulcanization of rubber, where sulfur is added to produce cross linking, allows a range of properties to be produced ranging from soft, compliant natural rubber to harder, stiffer tire rubbers.

Appendix 8 lists some representative mechanical properties for various polymers. The two groupings in the table relate to thermoplastics and thermosets. These limited data must be used with caution and are merely indicators of mechanical properties; most thermoset data happen to relate to reinforced (high-impact) grades. Processing conditions, strain rates, temperature, number and type of side branches, and so on, affect the results. The glass-transition temperatures are only guides; in the case of partially crystalline polymers, $T_g$ relates to the amorphous phase. Unlike the case for metals, it is not really possible to compare, except in broadest terms, a given polymer (of which there may be numerous grades) with another polymer based on these few data. The trade names given are in no way exhaustive, different suppliers having other names. Furthermore, manufacturing processes have changed over the years so that, for example, commercial PMMA is slightly different from what it formerly was. Similarly, the designations of high- and low-density versions of PE have lost much of their meaning.

Elastomers are not covered in App. 8. Their properties vary widely, with elastic moduli between 1 and 4 MPa and tensile strengths between about 5 and 30 MPa. The various names and types of synthetic rubber include SBR (styrene butadiene copolymer), butyl rubber (isobutene and isoprene), neoprene (chloropene), silicone rubbers (substituted silicones), urethane rubbers (polyester urethanes), and thiokol (polysulfides).

Both polymeric and nonpolymeric additives, such as asbestos, cellulose, fabric, flock, glass, graphite, and wood flour, are used in commercial polymers in amounts up to 50%. These are over and above the small amounts of additives employed to protect against environmental attack or as pigments and dyes. Polymeric additives are often used in the form of fine dispersions of rubbery particles in a matrix of stiffer polymer; this composite structure promotes resistance to crack propagation and can give high-impact strength, ABS being the classic example. In a different manner, up to 15% of PTFE can be added to nylon to improve wear resistance. Nonpolymeric additives increase the moduli of polymers but can sometimes reduce the impact strength. The coefficient of thermal expansion is often reduced, and this is of benefit not only in engineering design but also in reduced shrinkage during the solidification and cooling stages in molding. The most successful type of additive for engineering plastics is a nonpolymeric fiber (carbon, glass, etc.). Chapter 17 concentrates on such filament-reinforced composites. Figure 13-23 (later) shows how the incorporation of glass fibers affects creep data for polymers (in this case polypropylene).

## EFFECT OF MECHANICAL PROCESSING ON PROPERTIES

If polymers are rolled, drawn, or extruded, their properties can be changed. This behavior is similar to that of ductile metals, except that the trends are not always the same. The response of high-density polyethylene sheet subjected to rolling at room temperature may be used as an illustration. "Rolling" is used here in the same sense as rolling of metals and should not be confused with the calendaring process used to produce the sheet from the raw polymer in the first place. Polyethylene is a material that in the as-received condition displays stable neck growth with stress-whitening in the neck. The tensile strength at neck initiation is about 21 MPa, when the associated true strain is about 0.1. The first difference between the rolling of a ductile metal and of this polymer is the amount of elastic recovery on exit from the mill; the thickness of the rolled strip is greater than the roll gap setting. The recovery takes place over a period of time but eventually stabilizes out. Figure 13-11(a) shows the thickness recovery after 1 week for a PE sheet of starting thickness 3.3 mm rolled with rolls 75 mm in diameter to different nominal percentage reductions in area (as defined by the roll gap setting). As the absolute recovery does not increase beyond about 25% cold work, this means that the actual cold work is directly correlated with the apparent cold work in the manner of Fig. 13-11(b). With other polymers the recovery shows a general correlation with $T_g$, being smaller the higher the glass-transition temperature. This is sensible in terms of recovery and relaxation behavior.

Tensile tests on PE specimens containing up to 50% cold work show that the onset of necking and stress-whitening is retarded until larger strains (Fig. 13-12) for greater cold work; note that higher strain rates reduce these strains at ultimate load. However, the ten-

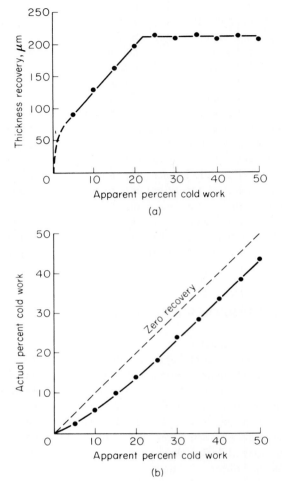

**Figure 13-11** (a) Thickness recovery after 1 week as a function of apparent cold work. (b) Comparison of actual vs. apparent percent cold work of high-density polyethylene. [From R. M. Caddell, T. Bates, and G. S. Yeh, *Mat. Sci. Engr.*, 9 (1972), p. 223. Reprinted with permission.]

sile strengths (maximum load divided by starting gage cross-sectional area) in such tests on PE stay comparatively constant for a given testing strain rate; the true stress at ultimate is, of course, much greater for larger cold work because the uniform area reduction to ultimate is greater, as shown in Fig. 13-13. Data for two crosshead speeds varying by two orders of magnitude are shown, out to ultimate load only in this figure; that is, the end points correspond with $d\sigma/d\varepsilon = \sigma$. This behavior should be contrasted with the behavior of ductile metals (Chap. 2), for which prior cold work increases the tensile strength and reduces the true strain at ultimate. The effect of cold work on the elastic modulus of rolled PE has been mentioned earlier, and the change in behavior at 25% cold work seems to relate to the breakup of crystallites at heavy reductions in area.

What does happen for increasing prestrains is that the load drop (Fig. 13-9) associated with neck formation, which bottoms out when the neck stabilizes and begins to draw,

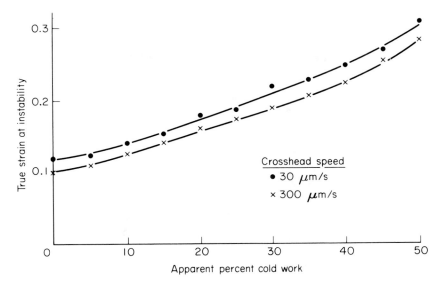

**Figure 13-12**    Effect of cold work on the true strain at tensile instability for two strain rates. [From R. M. Caddell, T. Bates, and G. S. Yeh, *Mat. Sci. Engr.*, 9 (1972), p. 223. Reprinted with permission.]

gets progressively smaller. The drawing load in as-received PE might be about 0.6 of the maximum load, whereas at 50% cold work it might be 0.8. With some other polymers, necking and load drop are eliminated at smaller amounts of prestrain, for example, at about 30% cold work for polycarbonate and polysulfone and about 40% cold work for ABS and PPO when unidirectionally rolled. Delay or elimination of necking is extremely useful in deep-drawing operations in sheet, which may thereby be taken to much larger strains.

Extrusion and drawing have similar effects. In the case of polycarbonate, cold working by extrusion eliminates necking, and the tensile strength thereafter increases from 66 MPa in the as-received condition through 81 MPa at 18% reduction to about 120 MPa at heavy reductions. The fracture strains continuously decrease with increasing cold work, but the true stress at fracture increases at first and subsequently drops. In Fig. 13-14, points satisfying the necking criterion $d\sigma/d\varepsilon = \sigma$ may be located only on the as-received $\sigma$–$\varepsilon$ curve, which is therefore the only condition that displays stable neck drawing. Note, as mentioned earlier, that no data were recorded over the strain range 0.05 ~ 0.50 in this case because of the rapidity with which the specimen necked.

After extrusion, polycarbonate, nylon 6-6, and PE are all weaker in compression than before, and pronounced directional effects are introduced by the forced flow through the throat of the die, with axial material being weaker than material transverse to the die. This may be associated with buckling of oriented chains.

Extrusion increases the elastic modulus from 2.4 GPa for as-received rod through 2.7 GPa for an 18% reduction in area up to about 3.6 GPa for a 64% reduction in area. This is of interest to designers, but, unfortunately, small amounts of cold working (18%) pro-

**Figure 13-13**  Typical true-stress–true-strain behavior of polyethylene cold-rolled to different amounts and tested at two strain rates. [From R. M. Caddell, T. Bates, and G. S. Yeh, *Mat. Sci. Engr.,* 9 (1972), p. 223. Reprinted with permission.]

duce substantial decreases in 1% offset yield stress, from 63 MPa to 46 MPa. Although additional cold work increases the yield stress above this value, it never goes above the original strength; for example, at 64% reduction it is about 62 MPa. However, if the cold-worked material is heat-treated at 100°C for a few hours, the yield strength may be increased to about 76 MPa with only a slight reduction in modulus. Heat-treating of polymers is a relatively unexplored area, but it may provide a means of obtaining specially desired combinations of mechanical properties, namely, in the foregoing case, higher strengths combined with higher stiffnesses.

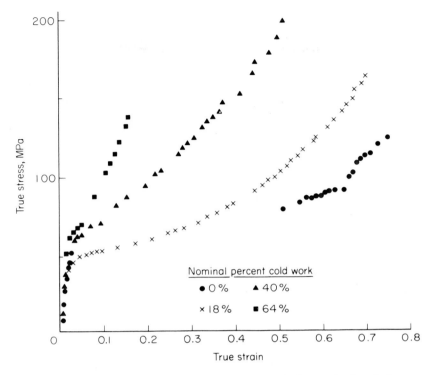

**Figure 13-14**   Tensile true-stress–true-strain curves of PC for four levels of nominal percent cold work by extrusion. [From C. S. Lee, R. M. Caddell, and A. G. Atkins, *Mat. Sci. Engr.*, 18 (1975), p. 213.]

## FLOW UNDER COMBINED STRESSES

As described in Chap. 2, many stress situations in practice do not conform to the simple tests performed in the laboratory. Hence, there is a need to study the behavior of materials when stressed in two or more directions and to relate the results to the simple measurements for application to other cases. Sheets of material may be subjected to biaxial tension by pressurizing one side with a fluid, similar to blowing up a balloon. In the laboratory, hydraulic machines can bulge disks into dome shapes, producing the results for various polymers shown in Fig. 13-15. The four sectioned PVC specimens at the top illustrate the sequence of doming that accompanies increasing pressure. Note the peculiar "Asian hat" shape of the PE specimen at the right; this is typical of the latter stages of bulging of this polymer after a stable neck forms.

At the pole of the dome there is balanced tension, where the stress, as in surface tension bubbles, is given by

$$\sigma = \frac{pR}{2t}$$

**Figure 13-15**    Bulged specimens of ABS/PVC alloy, PVC, and PE (left to right). Sectional specimens at center top show bulging sequence under increasing pressure. (Photograph courtesy of W. H. Durrant.)

where $p$ is the pressure, $R$ is the radius of the dome at the pole, and $t$ is the thickness of the sheet [Fig. 13-16(a)]. Using ideas of superposition, we may view the balanced tensile stress as a hydrostatic tension coupled with simple compression of magnitude $pR/2t$ through the sheet thickness [Fig. 13-16(b)]; the algebraic sum of the stresses in each direction gives the original equal biaxial tension. Thus, a bulge test gives the same results as a *simple compression test,* providing that hydrostatic stresses do not affect flow. Devices exist for measuring the strain through the thickness, so equivalent compressive stress–strain curves may be generated from bulging experiments, as shown for various polymers in Fig. 13-17. Superimposed on the figure are simple uniaxial tensile stress–strain data (see Fig. 13-14 for other PC data). The agreement is very good, suggesting that for these polymers uniaxial tensile and compressive behavior are similar. Note that bulging experiments produce equivalent compressive stresses unaffected by friction; in this they differ from conventional compression tests, which have platen friction.

The rise and fall in the curves display the characteristics discussed earlier, particularly the growth of stable necks which develop into thin disks at the pole (Fig. 13-15).

Figure 13-18 shows the true stress–strain behavior of high-density PE subjected to uniaxial and balanced biaxial tension (equivalent to uniaxial compression). Here there is *not* the same agreement between bulge data and tension data. (Note that the strains in this figure extend to much larger values than in Fig. 13-13, which went only as far as neck initiation.) The reason concerns the fact that flow in many polymers is really *pressure dependent*; that is, compaction makes it more difficult for them to flow in tension or hydrostatic tension makes it more difficult for them to flow in compression. The assumption about equivalent stress states is not valid in such cases; hence the discrepancy in Fig. 13-18 for PE, in which the effect is more marked than for the polymers in Fig. 13-17.

Experiments investigating the pressure dependency of flow in polymers may be performed under combined stress states by means of thin-walled tubes subjected to combinations of axial loading, internal or external pressure, and torsion. *Yield criteria* (i.e., which

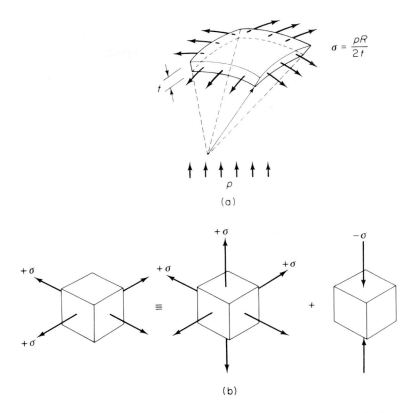

**Figure 13-16** (a) Balanced biaxial tension produced at pole of dome in sheet bulged over circular hole. (b) Equivalent stress state showing that bulging is equivalent to frictionless simple compression.

describe those combinations of $\sigma$ and $\varepsilon$ on various planes and in various directions that will cause yielding) may be established from the results of such experiments.

## DESIGN WITH TIME-DEPENDENT SOLIDS (POLYMERS AND METALS)

When deformations due to creep are only minor and when a part has to operate under load for, say, less than 1 ks (about $^1/_4$ h), the modulus measured in a tensile test at constant strain rate may be used in standard strength-of-materials formulas in order to arrive at suitable proportions. Many room-temperature plastics design calculations can be performed in this way, and the usefulness of the data may be modestly extended in scope to cover a range of test temperatures but always when the time scale is short term. For longer durations of loading, where time-dependent effects are more marked, consideration must be given to

**Figure 13-17**   True-stress–true-strain behavior of PC, PVC, and an ABS/PVC laminate subjected to uniaxial and balanced biaxial tension. [From R. Raghava, R. M. Caddell, L. Buege, and A. G. Atkins, *J. Macromol. Sci. Phys.*, B6, 4 (1972), p. 655. Reprinted with permission of Marcel Dekker.]

the mechanical behavior of metals and polymers expressed in terms of *creep, relaxation,* and *stress rupture.* As explained in Chap. 12, creep relates to the increase in strain with increasing time at constant stress; relaxation relates to the decrease in stress at constant strain with increasing time; and stress rupture data give the times for which a body can sustain a given stress before breaking. These three types of information are appropriate to different design criteria as follows:

1. Creep strain data are appropriate for situations where the maintenance of accurate size is important (e.g., turbine blades, casings, and fittings).
2. Stress relaxation data are appropriate for situations where stress levels must be maintained (e.g., bolts in tension, washers, and flanges in compression).
3. Stress-rupture data are appropriate for situations where breakage must be avoided but where shape stability is not vital (e.g., bottles, pipes, pressure vessels, and tubes).

Figure 13-19(a) shows room-temperature tensile creep curves for a PVC compound, and Fig. 13-19(b) gives the stress-relaxation cross plot derived from the creep data. It is also possible to perform the other cross plot of stress vs. strain at constant times, but as it is not so useful in design, it is not shown here. Similar sets of curves apply at other tem-

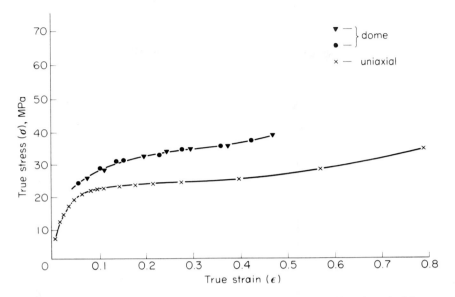

**Figure 13-18**   True-stress–true-strain behavior of high-density PE subjected to uniaxial and balanced biaxial tension. [From R. Raghava, R. M. Caddell, L. Buege, and A. G. Atkins, *J. Macromol. Sci. Phys.*, B6, 4 (1972), p. 655. Reprinted with permission of Marcel Dekker.]

peratures. The fact that the creep strain–time curves at low stress levels are nearly horizontal for short loading times in Fig. 13-19(a) is the justification for using constant rate moduli in strength-of-materials design stress calculations when the time scale is short.

From creep curves such as in Fig. 13-19(a) may be derived *creep modulus* data. The creep modulus is defined as stress/creep strain at some chosen strain level and may be plotted against time in order to display changes in stiffness. Figure 13-20 shows creep moduli for a different plastic (an acetal copolymer) at two temperatures. The 10-s modulus mentioned earlier is but one point on such curves.

When the right-hand end failure points are taken from creep curves such as those in Fig. 13-19(a) and plotted as breaking stress vs. time to failure at that stress, stress-rupture curves are produced. Figure 13-21 gives such curves for a high-density polyethylene compound. When using such data for polymers, care must be taken to note that precise values depend on processing conditions and other factors, so no single failure curve is characteristic of a particular material. A variation on the stress-rupture approach is to postulate different "failure" criteria (such as some maximum allowable strain) connected with the evidence of optical failure by crazing, or avoidance of excessive deflections, and so on.

Similar types of creep, relaxation, and stress-rupture curves are available for creepy metals, as shown, for example, in Fig. 13-22, where the data are for a chrome-molybdenum steel at 600°C.

**Figure 13-19**  Tensile creep at 20°C of a PVC compound. [From S. Turner, *Plast. Rub. Mat. Appl.*, 2, 3 (1977), p. 90. Reprinted with permission.]

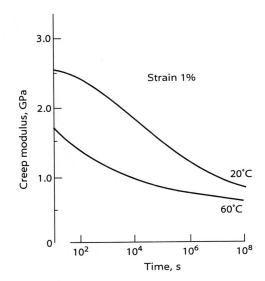

**Figure 13-20** Creep modulus vs. logarithmic time for an acetal copolymer. [From P. C. Powell, *Engineers Digest,* 38, 6 (1977), p. 13. Reprinted with permission.]

As mentioned earlier in this chapter, the incorporation of fibrous additives affects the creep behavior of polymers. Figure 13-23 gives data for a plain and a fiber-reinforced polypropylene. It is seen that the presence of fibers changes the magnitude of the strain response to a stress but does *not* change the basic time dependence to any significant degree. In contrast, fibers reduce the creep sensitivity to temperature quite effectively,

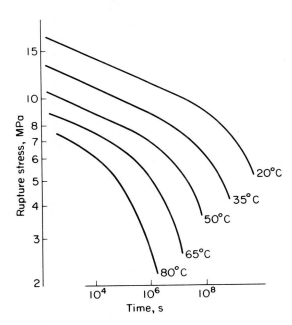

**Figure 13-21** Creep rupture data for an HDPE pipe compound. [From P. C. Powell, *Engineers Digest,* 38, 6 (1977), p. 13. Reprinted with permission.]

**Figure 13-22** Creep-rupture data for 2.25% Cr–Mo steel at 600°C. Test results from five countries embracing a total of 26 casts. Scatter of data falls within 20% of mean line. [From M. P. E. Desvaux, *Engr. Mat. Design* (1976), p. 38. Reprinted with permission.]

mainly because the modulus of the fibers remains constant while the modulus of the matrix falls, slowly in some regions but by perhaps two orders of magnitude, as the temperature rises through the glass–rubber transition region (the region associated with main-chain mobility). Thus, the effective reinforcing efficiency of the fibers increases as the temperature rises, though this is true only where the coupling between fibers and matrix is good; where the coupling is poor, the mechanical properties of a fiber-reinforced grade are worse than those of an unreinforced grade at high temperatures because the matrix expands away from the fibers and the overall performance becomes that of a matrix with holes.

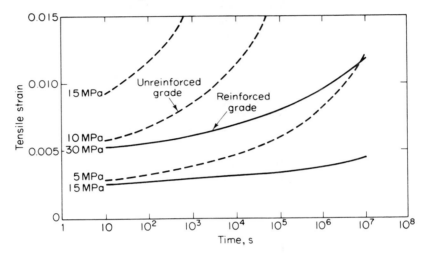

**Figure 13-23**   The effect of reinforcing fibers on the tensile creep of polypropylene at 23°C. [From P. L. Clegg and S. Turner, *Metal. Mat. Technol.*, 7, 12 (1975), p. 618. Courtesy of *The Metallurgist and Materials Technologist.*]

The time dependence of mechanical properties introduces several complications into the design process for thermoplastics, such as calculation difficulties when the applied forces fluctuate. Even in simple cases, the fact that elastic moduli and yield stresses have been replaced by viscoelastic functions means that the designer must choose the appropriate datum from an array of values rather than use a simple established constant. An associated difficulty is comparing the cost-effectiveness of different materials. For instance, the relative cost per unit stiffness of two materials with rather different viscoelastic behavior depends on whether service stresses act for short or long periods and on the service temperature. The choice of polypropylene or high-density polyethylene turns on their similar yet distinctly different viscoelastic characteristics. The short-term modulus values of the two classes of material are similar at room temperature (the actual magnitudes depending on the make and grade), but the modulus ratio changes and favors polypropylene for stresses of long duration and for elevated temperatures. The superior load-bearing capability of polypropylenes under stresses of long duration is part of the reason why beer crates and soft-drink crates are molded predominantly out of polypropylene in Great Britain and predominantly out of high-density polyethylene in most of the rest of Europe; in British

breweries, crates are stacked very high to meet seasonal peak demand, whereas in conti-
nental breweries the stacks are generally lower because the demand is steady.

Moment-by-moment design calculations for creepy solids under varying stresses are
really very complicated, as the compliances vary with time and stress, so the strain
response depends on the history of loading. For related reasons (which are beyond the
scope of this book), proper stress-relaxation data are *not* strictly the same as cross-plotted
creep data. For most practical purposes, however, the differences are not marked unless a
polymer is being used near its transition points (e.g., the glass transition), where moduli
and Poisson's ratios may change by an order of magnitude within a short period of time.

In order to simplify safe design procedures, a "worst-case" principle is often
employed, which amounts to using the least stiffness shown by the body in strength-of-
materials calculations. The steps in the procedure are

1. Identify the maximum service temperature and the maximum (continuous) duration
   for which the maximum load is applied.
2. Calculate the maximum stress(es) in the proposed design.
3. Read, interpolating as necessary, the strain from the appropriate creep curve relevant
   to the worst case.
4. Calculate the creep modulus.
5. Use this creep modulus in the formulas for elasticity to predict deflections, deformation, stability, and load relaxation.

For example, a polyacetal cantilever is 100 mm long and 6 mm wide and carries a
constant force of 10 N at the free end. What beam depth for a rectangular cross section is
required so that the tip deflection does not exceed 2 mm over 2 years at 20°C? We require a
modulus $E$, which depends on strain, time, and temperature, to use in the strength-of-materials expression for cantilever deflection, namely, $\delta = PL^3/3EI$, where $\delta$ is the deflection, $P$
is the load, $L$ is the length of the beam, and $I = th^3/12$, the second moment of area (moment
of inertia) of a section of width $t$ and depth (height) $h$. Assuming that strains in bending are
small, we can use a 1% strain creep modulus taken from Fig. 13-20. After 2 years (63 Ms)
at 20°C the modulus is about 850 MPa. Thus, $2 \times 10^{-3} = 10 \times (100 \times 10^{-3})^3/3 (850 \times 10^6)I$,
from which $h = \sqrt[3]{12I/t} = 15.8$mm. This gives a beam $L/h$ ratio of 6, which is arguably on
the borderline of strength-of-materials applicability. Nevertheless, it is clear that the beam is
overstiff in the short term in order to be adequately stiff after long periods.

**Example 13-5**

A polyacetal beam spring of rectangular cross section consists of a quarter-circle of radius 30
mm, width 6 mm, and depth 3 mm. It is deflected by a fixed amount, 3 mm. In what manner
does the thrust exerted by the deformed spring vary with time and temperature? (The tip thrust
of a spring with such a geometry is given by $P = 4\delta EI/\pi R^3$, where R is the radius of the arc.)

**Solution**   From Fig. 13-20 the 1% creep moduli after 100 s, 100 ks, and 100 Ms are 2.4
GPa, 1.5 GPa, and 830 MPa, respectively, at 20°C, and 1.3 GPa, 830 MPa, and 630 MPa,
respectively, at 60°C. Substitution of the fixed deflection and geometry of the beam in the
expression for thrust gives $P = 1.91 \times 10^{-9}E$. Thus, at 20°C, the respective tip forces are 4.6
N, 2.86 N, and 1.59 N; and at 60°C, they are 2.5 N, 1.59 N, and 1.2 N. Clearly, if a thrust
such as 4.5 N is required over long periods, design based on short-term data is inadequate.

**Figure 13-24**   Ashby fracture mechanism map for 304 stainless steel showing fields of dominant micromechanisms of fracture at different temperatures and stresses. (M. F. Ashby and coworkers. Reprinted with permission.)

Creep rupture design calculations proceed along the following lines. Suppose that an extruded high-density polyethylene pipe of mean diameter 100 mm is required to withstand water at 65°C for 500 h at a maximum pressure of 600 kPa. What is the minimum thickness of the pipe wall? Using Fig. 13-21 as an indicator of behavior, we find that the expected failure stress after 500 h (1.8 Ms) at 65°C is about 4.5 MPa. If the design factor of safety is 1.3 (typical of HDPE water pipe applications), the design working stress is 4.5/1.3 = 3.46 MPa. Using the "thin-walled cylinder" hoop stress formula for the pipe [i.e., $\sigma = pD/(2t)$, where $\sigma$ is the hoop stress, $p$ is the fluid pressure, $D$ is the pipe diameter, and $t$ is the wall thickness], we have $3.45 \times 10^6 = (600 \times 10^3)(100 \times 10^{-3})/2t$, from which $t = 8.67$ mm.

A design life of 50 years is required for underground gas piping made from polymers. Stationary gas turbines with metal components used under steady conditions (for pumping oil and gas, for example) work for at least 20,000 h before major overhauls; and civil aero-engines, with regular thermal cycling, go for about 4000 h between repairs. Very often the rea-

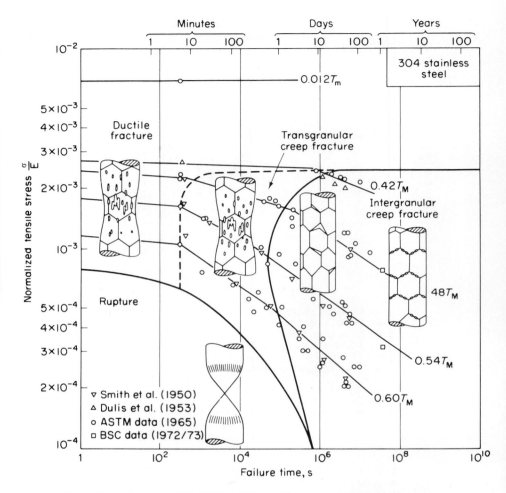

**Figure 13-25**    Cross plot of Fig. 13-24 as tensile stress vs. time to failure with lines of constant homologous temperature superimposed. (M. F. Ashby and coworkers. Reprinted with permission.)

son for taking out of service is not creep but other effects (such as erosion of the blades), so many components are reused and therefore are designed for a cumulatively longer creep life.

Because of the length of time required and costs involved in acquiring creep data, there is a temptation to extrapolate test results taken over a few hundred or a thousand hours out to $10^5$ (which is about 11.4 years). Extrapolation can be a dangerous procedure, creep rupture data, for example, perhaps not falling on a curve extrapolated from short times, thereby giving a false indication of component life. The consequence of poor design of boiler or turbine components based on unproved data can be expensive failures. For power-plant steel, "safe" extrapolation never uses more than a factor of 3 between the duration of actual data and the required design stress.

M. F. Ashby and coworkers have constructed *fracture-mechanism maps* which are relevant to this matter. In their usual presentation the maps are diagrams with tensile stress as one axis and temperature as the other, showing the fields of dominance of a given micromechanism of fracture. Figure 13-24 is a typical diagram, in this case for 304 stainless steel; further details on the mechanisms of fracture will be given in Chap. 14. Superimposed on the fields are contours of constant time to fracture. The maps may be cross-plotted as tensile stress vs. failure time with lines of constant temperature superimposed (Fig. 13-25). The validity of extrapolation of short-term results at constant temperature to long times depends on whether the lines of constant temperature in figures like Fig. 13-25 have "unknown" downward kinks at long times. Such dangerous occurrences have been observed in some materials such as Type 316 stainless steel and some superalloys.

## REFERENCES

1. Alfrey, Turner, and E. F. Gurnee, *Organic Polymers.* Englewood Cliffs, N.J.: Prentice Hall, 1967.

2. Harris, B., and A. R. Bunsell, *Structure and Properties of Engineering Materials.* London: Longmans, 1977.

3. Kambour, R. P., "A Review of Crazing and Fracture in Thermoplastics," *Journal of Polymer Science: Macromolecular Reviews,* 7 (1973), p. 1.

4. Lever, A. E., and J. Rhys, *The Properties and Testing of Plastics Materials.* London: Temple Press, 1962.

5. Powell, P. C., "Designing Load-Bearing Plastics Products," *Engineers' Digest,* 38, 6 (1977), p. 13.

6. Tebbatt, T. E., "Plastics in Engineering," *Engineering Materials and Design,* 9 (1976), p. 36.

7. Treloar, L. R. G., *The Physics of Rubber Elasticity,* 3rd ed. Oxford: Clarendon Press, 1975.

8. Ward, J. M., *Mechanical Properties of Solid Polymers.* New York: Wiley-Interscience, 1971.

9. Williams, J. G., *Stress Analysis of Polymers.* New York: John Wiley/Halstead, 1973.

## PROBLEMS

**13-1.** Poly($\alpha$-methylstyrene) exhibits a higher glass-transition temperature than does polystyrene. Explain. (The structures of these two polymers are shown in Fig. 13-26.)

**Figure 13-26**    (a) Polystyrene mer    (b) Poly($\alpha$-methylstyrene) mer

**13-2.** Describe briefly how increasing each of the following can change the strength of a polymer:
   **(a)** Crystallinity.
   **(b)** Cross linking.
   **(c)** Branching.

**13-3.** Thin samples of translucent penton (chlorinated polyether) on copper strips are heated on a hot plate and then cooled either by water quenching or by air cooling. Qualitative indications of the ductility and hardness of the samples cooled in different ways may be obtained by bending the supporting copper strips. Describe any likely differences in mechanical properties and explain their origins.

**13-4.** Using the four-element model of a polymer, describe the effect of
   **(a)** Increasing strain rate on measured elastic modulus.
   **(b)** Increasing temperature on measured elastic modulus.
   **(c)** Long-term constant stress on deformation.

**13-5.** On the coordinates in Fig. 13-27 is drawn the curve for the four-element model of a polymer. Draw on these same coordinates the curve for the same polymer at a slightly lower temperature (say, 20°C lower) and at the same applied stress.

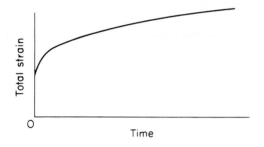

**Figure 13-27**

**13-6.** A cantilevered polyethylene beam exhibited an elastic modulus under a 200-g load of 29.6 MPa after 20 s, whereas under a 500-g load for the same time the elastic modulus was 27.7 MPa. Why are the moduli different?

**13-7.** The rebound height of bounced rubber balls reached a minimum when tested over a wide temperature range. With what property, independently obtained, is this minimum best correlated?

**13-8.** What is the approximate relationship between the temperature (call it $T_a$) of maximum energy absorption in a polymer and its glass-transition temperature $T_g$?

**13-9.** Describe briefly two important changes in the mechanical properties of polymers that occur
   **(a)** when the temperature is changed from 20°C below the glass-transition temperature $T_g$ to 20°C above $T_g$, and
   **(b)** when the temperature is changed from 150°C above the glass-transition temperature to 10°C below $T_g$.

**13-10.** The minimum rebound for a ball of butyl polymer occurred at about −18°C. The minimum rebound for a natural rubber ball is at about −40°C.
   **(a)** Which do you think is the better bouncing ball at 21°C? Why?
   **(b)** Which ball has the higher glass-transition temperature?

**13-11.** The rebound height (in mm) of rubber balls dropped from 500 mm is measured as follows:

| | Material | $T_g$, °C | –196°C | 0°C | 21°C | 100°C |
|---|---|---|---|---|---|---|
| A | SBR | –63 | 420 | 120 | 250 | 350 |
| B | Polysar butyl | –40 | 350 | 5 | 10 | 220 |

(a)  Which rubber will have the higher rebound at 50°C, A or B?

(b)  Which rubber will have the higher rebound at –40°C, A or B?

(c)  Which rubber would you expect to have the higher strength at –50°C? Why?

**13-12.** Two bending tests were made 300 s apart on a polyethylene cantilevered beam. The results are plotted in Fig. 13-28; the same load (250 g) was used each time. Explain:

(a)  the difference in deflection, and

(b)  the difference in the shape of the curve. Consider fundamental mechanisms.

**Figure 13-28**

**13-13.** Cis-polybutadiene is used for superballs, styrene butadiene is used for regular car tires, natural rubber is used for truck and plane tires, and butyl rubber is used for inner tubes. Explain why these particular rubbers are chosen for these jobs.

**13-14.** Why do polypropylene trash can lids crack in winter in certain parts of the United States?

**13-15.** Explain why thermoplastics (particularly those reinforced with fibers or particles) can be used satisfactorily in the severe environment of engine compartments of vehicles even though the conditions are beyond their traditional performance ranges.

**13-16.** A PVC component is bolted onto a metal frame at 20°C and thereby squashed to a 2% compressive strain. If the torque applied to secure the part were $T$, what torque would be required just to begin to compress the PVC component

(a)  1 day later?

(b)  1 year later? (Use the data in Fig. 13-l9.)

**13-17.** A pipe made of polystyrene has outer diameter 75 mm and inner diameter 70 mm. It is used to transport a fluid under different service loading conditions. When used in a flat, horizontal position, the service life of the pipe is very adequate, yet in another application requiring a small bend, an identical pipe fails in too short a time. Explain the most likely reason for this difference, including the probable mechanisms that lead to failure. (Note Prob. 16-13 on the use of Weibull statistics for the creep rupture lifetime of polyethylene pipe.)

# 14

# *Fracture*

We have all broken things. Sometimes we want fracture to occur, as when we split wood, drill holes, plow fields, or tear paper. Often we do not want fracture to occur, as when stones hit car windshields, china vases fall to the floor, bolts snap off upon tightening, or ships break in two.

In this chapter, we explore different types of fracture and the factors affecting the manner in which things break. Explanations are given for the observed discrepancies between ideal fracture strengths and those measured in practice. The concept of fracture toughness is introduced, along with the salient features of the mechanics of cracking. It is pointed out that fracture behavior can depend markedly on the method of testing or in-service conditions, so simple classifications of "brittle" and "ductile" must be used with caution. Brittle fracture in large, fabricated structures made of materials that behave in a ductile fashion in the laboratory is discussed, and it is shown that stresses for cracking in such large structures can sometimes be much smaller than design stresses based on yielding.

## FRACTURE CLASSIFICATIONS

Broadly speaking, fractures may be classified according to the amount of energy required: *brittle fracture* requires little energy, and *ductile fracture* requires much energy. In brittle fracture, the broken parts usually may be refitted together to the original dimensions, with essentially no permanent deformation, the behavior being elastic. In ductile fracture, before rupture there is considerable plastic deformation that is not recovered, and often it is impossible to refit the severed parts.

Consider the simple tensile fractures shown in Fig. 14-1, which can be produced in the laboratory. Figure 14-1(a) shows the response of a rod of glass, cast iron, or blackboard chalk at room temperature. The test piece breaks with great suddenness, and the fractured surface is *normal* to the applied *tensile* stress. On the stress–strain diagram, *OA* is essentially straight; there may be a small curvature in the region of *A*, but permanent deforma-

308

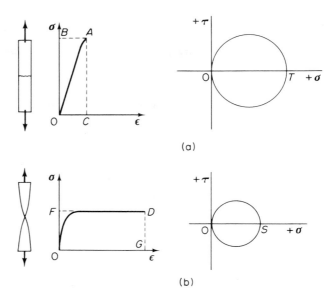

**Figure 14-1** Simple tensile fractures and associated Mohr's circles: (a) perfectly brittle, produced by normal stresses; (b) perfectly ductile, produced by shear stresses.

tion is extremely small. We can define $OB$ as the brittle fracture stress $\sigma_f$ (represented by $OT$ on the Mohr's circle diagram) and $OC$ as the fracture strain $\varepsilon_f$ (which is mostly elastic). Steels at low temperatures fracture in a brittle manner, individual crystals failing by *cleavage*, that is, splitting of atom planes. In contrast, Fig. 14-1(b) shows what can happen if rods of a soft metal such as lead or indium are pulled at room temperature. Intense *shear* slides off planes of atoms (in the classic concepts of movement of dislocations) until only a vanishing cross section is left. There is permanent deformation preceding rupture throughout the test piece, as can be seen by the plastic deformation in path $OD$, and it is impossible to remake the original rod. The tensile stress at fracture $\sigma_f$ is $OF$, and the fracture strain $\varepsilon_f$ is $OG$ and is chiefly plastic. The driving stress in this instance is the shear fracture stress, given by the radius of the Mohr's circle, by $OS/2$.

In compression, similar effects can be seen with concrete and rocks on the one hand and ductile metals on the other, except that the overall behavior is affected by friction at the loading platens of the testing machine.

Most commonly experienced fractures involve two- or three-dimensional stress states. Then, ideally brittle fractures *seem* to take place along the trajectories of maximum principal tensile stress. For example, if a hard ball is pressed into a block of glass, fracture occurs along a cone, the semiangle of which is related to the position of a peak in tensile stress [Fig. 14-2(a)]. Damage caused by projectiles hitting glass windows or protective shields is similar. Ductile fracture under complex stresses can often be difficult to interpret. The commonest example known to students is the "cup-and-cone" fracture of a typical tensile test on a ductile metal [Fig. 14-2(b)]. Once the neck forms, the stress system ceases to be uniaxial tension and changes to triaxial tension. The material above and below the neck is under lower stress than the neck itself and acts to prevent radial contraction of the metal in the neck. Under these conditions, three-dimensional tensile stresses *inside* the neck even-

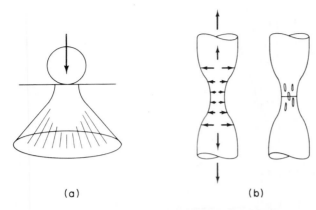

(a)                                                                    (b)

**Figure 14-2** Fractures produced under three-dimensional stress states: (a) brittle cone cracks in glass under a ball indenter; (b) cup-and-cone fracture in the tensile test of a moderately ductile metal.

tually open up internal holes around inclusions or hard second-phase particles, and when the inside area is reduced by the joining of these holes, the rest slides off around the edges. The cup is very rough from the internal necks and holes, and the cone is smooth from sliding off.

**Example 14-1**

Construct the Mohr's circle for torsion testing and demonstrate why materials that are brittle in tension sometimes yield rather than crack in torsion.

**Solution**    The Mohr's circle for torsion, which is pure shear, is a circle centered on the origin of stress. The Mohr's circle for simple tension is shown in Fig. 14-3, with maximum tensile stress given by $OA$. Brittle fracture is related to the maximum tensile stress in a system. For the same maximum tensile stress, the size of the circle for torsion is twice as big as the simple tension circle (when the radius $OB$ of one is numerically the same as the diameter $OA$ of the other). Then, however, the shear stresses in torsion are much greater than in simple tension ($OB = 2LM$). Consequently, in torsion the material shear yield stress may be reached before its tensile fracture stress. Furthermore, the mean hydrostatic stress in torsion is "less tensile" than the mean stress in tension [in torsion the hydrostatic stress is $OA + (-OC) + 0 = 0$; in tension it is $(OA + 0 + 0)/3 = OA/3$].

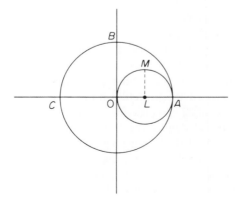

**Figure 14-3** Solution for Example 14-1, showing Mohr's circles for tension and torsion.

## FACTORS AFFECTING THE TYPE OF FRACTURE

The sort of fracture produced in a solid depends very much on conditions. The more significant controlling factors are as follows:

A decrease in *temperature* below half the absolute melting temperature tends to promote brittleness in most metals and ceramics, as does an increase in *strain rate*; polymers do not always follow this trend (see Chap. 13). The state of stress is important, and although in-service situations may be complex, the mean hydrostatic stress component [i.e., $\sigma_m = (\sigma_1 + \sigma_2 + \sigma_3)/3$], sometimes called the *triaxiality,* is a good indicator to the type of fracture obtained. *Tensile* values for $\sigma_m$ encourage brittle fracture. In contrast, very brittle substances such as stone can sometimes be made to flow plastically under large, *compressive* $\sigma_m$. The effects of residual stresses (e.g., resulting from manufacturing processes or nonuniform heating and cooling) and prestrain on fracture can also be thought of in terms of their influence on promoting triaxiality.

Microstructure is an important factor in fracture. Broadly speaking, a coarse grain size encourages brittleness at temperatures below the recrystallization range. Likewise, aging and irradiation cause brittleness. However, some alloying elements tend to promote ductile behavior, in part because of their influence on the conditions for starting very small cracks. For those microstructural variables that are well understood, the one common factor causing brittleness appears to be the encouragement of void formation and growth of microcracks. Hostile environments produce fractures that would not take place under clean conditions. Aspects of this problem, such as stress-corrosion cracking and hydrogen embrittlement, are discussed in Chaps. 18 and 19.

The interplay of strain rate and temperature influences fracture processes; that is, high temperatures and low strain rates generally promote ductile behavior in metals, but low temperatures and high strain rates generally encourage brittleness. The brittle stress–strain curves in Fig. 14-l(a) gradually change to the ductile stress–strain curves in Fig. 14-l(b) as the temperature is increased and/or the strain rate is decreased. If a third axis of increasing temperature and decreasing strain rate is added to the normal $\sigma$–$\varepsilon$ plot, we obtain the set of curves shown schematically in Fig. 14-4. Ashby fracture mechanism maps were mentioned in Chap. 13 (Figs. 13-24 and 13-25). The fracture stress can be projected into the stress–temperature or stress–strain-rate plane, which shows the changes in $\sigma_f$. Similarly, the fracture strain can be projected into the strain–temperature or strain–strain-rate plane. A similarly shaped curve (with values increasing as the temperature is raised or the strain rate reduced) is obtained if the areas under the separate $\sigma$–$\varepsilon$ curves are plotted against temperature–strain rate; as shown in Chap. 2, such areas are a measure of the energy per unit volume involved in flow and fracture. The change from brittle to ductile fracture in given types of test pieces, that is, from low $\varepsilon_f$ and relatively low fracture energies to high values, takes place over a fairly narrow span of temperature called the *transition temperature range* (Fig. 14-5).

Notch-impact tests, rather than tensile $\sigma$–$\varepsilon$ curves, are conventionally used to measure the transition temperature range. Notches produce large tensile $\sigma_m$ and hence encourage brittle behavior and reduce the plastic deformation preceding fracture. The energy of a heavy pendulum, swinging under standardized conditions, that is absorbed in fracturing

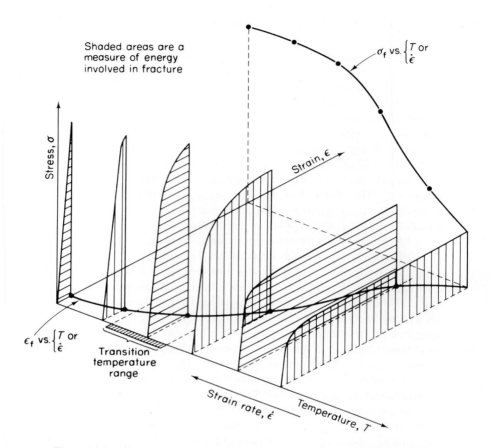

**Figure 14-4**    Relations between stress, strain, strain rate, and temperature in fracture. Brittle stress–strain curves transform to ductile curves at increasing temperature or decreasing strain rate (and vice versa) for most materials. Shaded areas under $\sigma$–$\varepsilon$ curves are some measure of the energy involved in fracture. Change from brittle to ductile behavior takes place over a reasonably narrow span of temperature, called the transition temperature range; the position of this temperature range can vary with type of test piece and conditions of measurement.

the notched specimen is measured. When the specimen is broken in three-point bending, the test is called a Charpy measurement; if the specimen is mounted as a notched cantilever, it is referred to as an Izod test (see Chap. 2). The energies thus determined are a rough measure of the area under a tensile $\sigma$–$\varepsilon$ curve as distorted by the complex nature of the deformation at the notch.

      The simple interpretation of the impact data is that even under severe triaxial conditions ductile behavior occurs at temperatures greater than the transition temperature range. In general terms, therefore, the safe operating temperatures for structures must be those that are higher than the transition temperature. It must be emphasized, however, that notch-

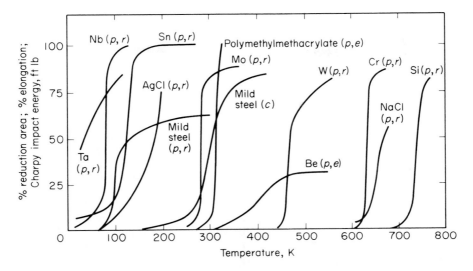

**Figure 14-5**  Transition temperature ranges for different materials (*p*, plain tension; *r*, reduction in area; *e,* elongation; *c*, Charpy V-notch impact energy). [From A. H. Cottrell, *Mechanical Properties of Matter* (New York: John Wiley & Sons, 1964), p. 358. Reprinted with permission of the publisher.]

impact tests, such as the Charpy and Izod tests, are qualitative and produce no design values analogous to yield stress, for example. These tests are useful in providing a guide to the relative ranking of various materials, chiefly steels, in their resistance to failure at discontinuities and changes in section, but design against cracking really ought to be related to the material property called *fracture toughness*, dealt with in the following sections, which can give working design stresses in the presence of cracks.

First, however, let us compare fracture strengths measured in practice with those predicted by theory.

## IDEAL FRACTURE STRENGTH
## AND MEASURED VALUES

Can we make theoretical predictions of fracture strength? How do the experimental values for $\sigma_f$ and $\varepsilon_f$ live up to our expectations? If the actual behavior differs, why? In an unstressed situation the repulsive and attractive forces between atoms in a crystal balance out to zero at a particular equilibrium spacing of the atoms. If a crystal is strained, however, the net force resisting the external pull increases from zero as the atoms are moved away from their equilibrium locations. Eventually a maximum force is reached, and then the force between atoms falls off at greater spacing, as shown in Fig. 14-6. The ideal strength of the material in terms of cohesion is related to that maximum. Typically, the curve for applied force, or stress, versus distance looks as shown in the figure. Although $\sigma$

**Figure 14-6** Forces or stresses generated in a crystal when the atoms are moved away from their equilibrium positions. $\sigma$ is not periodic, but the region of interest can be approximated by $\sigma = \sigma_{max} \sin(2\pi x/\lambda)$.

is *not* periodic, we can approximate the part of the curve of interest to us as $\sigma = \sigma_{max} \sin(2\pi x/\lambda)$, where $\lambda$ is an artificial wavelength and $x$ is the distance the atoms move from their equilibrium interatomic spacing $a_0$ during a tensile pull. For small $x$, the crystal strain increments are $dx/a_0$, and in the elastic region around the origin, $d\sigma/(dx/a_0) = E$, so $d\sigma/dx = E/a_0 = (2\pi/\lambda)\sigma_{max} \cos(2\pi x/\lambda)$. Because $\cos(2\pi x/\lambda) \approx 1$ for small $x/\lambda$, $\sigma_{max} = \lambda E/2\pi a_0$. We do not really know what $\lambda$ is, but a fair guess is that $\lambda \approx a_0$. Hence,

$$\sigma_{max} = \frac{E}{2\pi} \qquad (14\text{-}1)$$

say, $E/10$. Note the similarity between this calculation for pulling atoms apart and the calculation in Chap. 5 for sliding atoms over one another; in that case, $\tau = \tau_{max} \sin(2\pi x/b)$ really is periodic.

Measured values of $\sigma_f$ are usually about $E/10^4$ for most materials unless they are in fiber or "whisker" form, in which case the measured values have been known to approach $E/15$ (Chaps. 16 and 17). In the same way that crystal dislocations were introduced to explain the discrepancy between theoretical and actual shear strengths, the presence of *microcracks* or *flaws* is postulated to explain the discrepancies between theoretical and actual brittle strengths. Statistically, a flaw of given size is less likely to occur in a small lump of material, which explains the *size effect* according to which thin fibers are stronger than thick rods. Flaws are not figments of the imagination and have been observed in, for example, glass and low-carbon steel.

Flaws act as stress concentrators, locally raising stresses to the theoretical level, even though away from the flaw stresses may be low. The bonds adjacent to the crack tip can then be broken, and the crack opens. This is shown schematically in Fig. 14-7. All missing bonds go to the two atoms labeled $B$ and $C$, giving rise to a *stress concentration*. The crack eventually spreads, and the process repeats itself. Obviously, cracking is easiest when the applied stress is perpendicular to the crack, again leading to the concept of maximum tensile stresses for brittle fracture.

### Example 14-2

When an elastic material is strained, strain energy is stored in the stretched chemical bonds; on fracture this energy is released. Chemical energy is no different from this "explosive energy" during fracture. Why are tensile tests therefore not explosive?

**Figure 14-7**   Stress concentration produced by flaw; all missing bonds are imposed on the atoms located around $B$ and $C$.

**Solution**   Real materials achieve only a fraction of their theoretical strength because of microflaws. The amount of energy released on fracture is thus only a fraction of the energy that would be produced by an equivalent mass of explosive. In the case of fibers and whiskers, which do achieve a high fraction of their theoretical strength, there can be an explosion on testing and the fibers vanish into dust, not merely breaking into a few pieces. Only the fact that their volume is small prevents their testing from being dangerous.

## MECHANICS OF CRACKING

What happens when we load a piece of material containing a flaw? Can we predict when fracture will take place? Why do some materials yield before cracking and others crack without yielding? Can we prevent cracks from initiating and propagating?

Although many objects break with a bang and fly apart on fracture, cracking is not always necessarily catastrophic. Sometimes when an ice-covered pond is walked on, cracks propagate slowly; the growth of fatigue cracks is not fast; and the log of one ship records that members of the crew put a chalk mark each day at the tip of a growing crack in the deck. The precise response depends on the size of the piece of material, the method of loading, temperature, strain rate, environment, and so on.

Consider the problem of crack initiation and propagation in a system where the cracking loads produce general stresses in the system that are nowhere greater than the yield stress. That is, the system behaves essentially as an elastic, reversible body: There is no gross distortion, and the broken bits may be refitted back together. Brittle materials behave in this fashion, although as we shall see, materials commonly thought of as ductile *can* follow this type of behavior under certain circumstances.

During loading, before the crack runs, the load deflects the piece of material containing the crack and puts *elastic strain energy* into the material—the same as energy in a spring. The magnitude of the strain energy is given by the area under the load–extension line. When a critical state is reached, the crack begins to extend; the process of cracking may be violently fast or slow. During the propagation of the crack, the strain energy of the system changes, and kinetic energy can be generated if the crack runs very quickly. If all the interchanges of energy are balanced out, an amount will be found lost after fracture. In other words, *work must be done to open up the crack.* The specific work of cracking, the work done in developing the crack per unit area, is called *the fracture toughness R* of the solid. It is a material property, like $E$ or yield strength, and likewise depends on rate, temperature, and environment.

### Example 14-3

Calculate the fracture toughness $R$ for the case where 40 mJ of energy is required to create a crack 14 mm long in a plate 4 mm thick.

**Solution**

$$R = \frac{40 \text{ mJ}}{(14 \text{ mm})(4 \text{ mm})} = 714 \text{ J/m}^2$$

Note that the conventional interpretation of "crack area" used in the calculation is the thickness times the growth in crack length. The actual total new crack area, on both halves of the cracked part, is twice this value.

Consider quasi-static cracking, cracking that is so slow that kinetic energy is negligible. Think of a split beam of material put into a testing machine, as shown in Fig. 14-8, with a crack (say, a fine saw cut) of initial length $L_1$ and area $A_1$. A typical plot of load $P$ versus crosshead displacement $\delta$ is $Oa$ prior to any propagation of the crack. The slope of $Oa$ gives the stiffness of the cracked test piece (the reciprocal of the stiffness slope is called the compliance). Longer starter cracks make the body more compliant; that is, the slopes of lines such as $Oa$ become smaller as the crack length or area becomes bigger. (Conversion of $P$ into stress and $\delta$ into strain would *not* give Young's modulus for any of the lines. The loading and unloading behavior is nevertheless elastically reversible up and down lines such as $Oa$.) Imagine that the load increases to point $b$ in this way [Fig. 14-8(b)], and then cracking starts.

Perhaps the crack propagates in such a way that $\delta$ remains constant at $\delta_b$, that is, the load drops down to $c$ when the crack increases its area from $A_1$ to $A_2$ (which, since $A = Lt$, where $t$ is the thickness of the sheet, is the same as the crack lengthening from its starting length $L_1$ to $L_2$). This situation is shown in Fig. 14-8(c). The strain energy before the crack ran was the area $Obe$; after the crack has lengthened to $L_2$, it is area $Oce$ because the elastic stiffness line with longer crack $L_2$ is given by $Oc$. No extra work has been fed to the system because the load has not moved its point of application, that is, $\delta_b$ is fixed. If kinetic energy is negligible, the area $Obc$ has fed the crack with the work necessary to open it up. This toughness work is equal to $R(A_2 - A_1) = Rt(L_2 - L_1)$ if $R$ is the fracture toughness. Thus, $R = (\text{area } Obc)/(A_2 - A_1)$.

Suppose, on the other hand, the crack propagates at constant load $P_b$, the displacement approaches $\delta_d$ as the crack lengthens from $L_1$ to $L_2$ [Fig. 14-8(d)]. Again, the energy at the start is area $Obe$; at the finish it is $Odf$ because $Ocd$ is still the elastic stiffness line when crack length is $L_2$. This time, however, because $P_b$ has moved through a distance $(\delta_d - \delta_b)$, external work has been done, represented by area *bdfe,* and energy has been fed into the system (i.e., to the crack *and* to the test piece).

**Figure 14-8**   (a) Elastic load–displacement $P$–$\delta$ plot and energy interchanges during crack propagation. *Ob* is the stiffness of the body containing a crack of area $A_1$ (or length $L_1$ in a plate of uniform thickness); *Od* is the stiffness of the body containing a crack of area $A_2$ (or $L_2$), where $A_2 > A_1$. (b) Shape at the start of crack growth when $L = L_1$. (c) Shape at the end of crack growth (when $L = L_2$) if propagation occurs at constant deflection $\delta_b$. (d) Shape at the end of crack growth (when $L = L_2$) if propagation occurs at constant load $P_b = P_d$. [The cracked body shown is called a *double cantilever beam* test piece in fracture testing; see Fig. 14-11.]

Then, cracking work = areas *(Obe + bdfe)* – area *Odf* = area *Obd*. And $R(A_2 - A_1)$ = area *Obd*, so $R = $ (area *Obd*)/$(A_2 - A_1)$.

Notice that the constant-load case and the constant-deflection case approach the same value of work done to open the crack, as the increments of crack opening or deflection are made smaller. The work *Obd* for the constant-load case differs from the work *Obc* only by the amount *bdc*, which disappears as *bd* and *bc* approach zero.

In fact, propagation in the split beam [so-called double cantilever beam (DCB)] test piece occurs under decreasing load with increasing displacement. That is, the load path cuts *ocd* somewhere between *c* and *d*. For many geometries of cracked bodies, the actual situation usually lies between the two extremes shown in Fig. 14-8 and follows a curve such as *bh* in Fig. 14-9. Then, as shown by C. Gurney, the work done in opening up the

**Figure 14-9**  $P-\delta$ relations during cracking for many cracked bodies. *bjkh* ... is the toughness locus, and the work area *Ojk* is the work done in propagating the crack from an area corresponding to stiffness *Oj* to an area giving stiffness *Ok*. The shape of *bjkh* ... varies with cracked-body geometry and loading conditions (see Fig. 14-14).

crack from the crack area corresponding to point *j* to the crack area corresponding to *k* is the sector *Ojk*.

Sometimes (but not often) crack propagation takes place under *increasing* load, in which case the likes of *bjkh* point upward; the conical fracture ball indentation example shown in Fig. 14-2(a) happens to be one of these cases. All the foregoing arguments about segmental work areas apply as before.

Thus, if we have load-extension crack-length data we can find the toughness. Conversely, if we know something about a material's fracture toughness, it is possible to predict the likelihood of cracking in a given situation or, in other words, to predict the *strength of cracked parts*.

**Example 14-4**

A cracked plate of a certain material was stressed until it began to crack at a constant load of 500 N. Just before the crack started to run, the deflection of the plate was 2.5 mm; when the crack stopped, it was 3 mm.
(a) Calculate the work done in propagating the crack.
(b) Calculate the fracture toughness of the material if the crack area increased by 1200 mm² during propagation.

**Solution**    The toughness–work area during propagation is a triangle, similar to *Obd* in Fig. 14-8, with base $3.0 - 2.5 = 0.5$ mm and height 500 N. (a) The work is thus $500(0.5 \times 10^{-3})/2 = 125$ mJ. (b) The fracture toughness is thus $(125 \times 10^{-3})/(1200 \times 10^{-6}) = 104$ J/m².

The energy balance, which has just been described graphically, can be developed mathematically as follows.

Consider events during an increment of cracking where the external displacement of the cracked structure is $\Delta\delta$ and the associated increment of the crack area is $\Delta A$ (Fig. 14-10). The external load does work $P\,\Delta\delta$, and the crack absorbs energy $R\,\Delta A$. The strain energy $\Lambda$ of the structure changes by $\Delta\Lambda$.

The strain energy is given by a triangular area in Fig. 14-9 such as $Ojn = 1/2(P\delta)$ at any instant, so $\Delta\Lambda = \Delta 1/2(P\delta)$.

Thus, equating the rate of doing external work to that of doing internal work for quasi-static cracking and neglecting kinetic effects, we have

$$P\,\Delta\delta = \Delta\Lambda + R\,\Delta A = \Delta\left(\frac{P\delta}{2}\right) + R\,\Delta A \qquad (14\text{-}2)$$

**Figure 14-10** Events during an increment of cracking in which the displacement changes by $\Delta\delta$ and the crack area increases by $\Delta A$.

Hence,

$$P\,\Delta\delta = \tfrac{1}{2}P\,\Delta\delta + \tfrac{1}{2}\delta\,\Delta P + R\,\Delta A$$

so

$$P\,\Delta\delta - \delta\,\Delta P = 2R\,\Delta A$$

Dividing both sides by $P^2$, we obtain

$$\frac{d}{dA}\left(\frac{\delta}{P}\right) = \frac{2R}{P^2}$$

that is,

$$P^2_{\text{crack}} = \frac{2R}{d/dA\,(\delta/P)} \tag{14-3}$$

Equation (14-3) describes the relationship among $P$, $\delta$, and $A$ during cracking; that is, it is the algebraic expression for a line such as *bjkh* in Fig. 14-8, which is the locus of all $(P, \delta)$ necessary just to keep cracking going. The quantity $d/dA(\delta/P)$ is the rate of change in compliance of the cracked structure with respect to increase in crack area and is given from the slope of an experimentally determined compliance vs. crack area curve; it generally changes with $A$ for most cracked bodies.

Equation (14-3) is often written $P^2_{\text{crack}} + 2G_c/[d/dA(\delta/P)]$, where $G_c$ is called the *critical strain energy release rate*. $G_c$ is a measure of the work available from the body to feed the crack were the crack to propagate. In stable cracking, $G_c = R$ at initiation, that is, for the first incremental sector area at *b* in Fig. 14-9. The idea of strain energy being released is readily understood from the constant-displacement case in Fig. 14-8, as *Oce* after propagation is less than *Obe* before propagation. However, in the constant-load case shown in the same figure, the strain energy after propagation is *greater* than that before *(Obe)*, so the use of the term "release rate" is misleading: when $\Lambda$ increases at the same time that energy has been fed into the crack, it occurs as a result of the contribution of the external work term. If $\Lambda$ were constant along a path such as *bjkh* in Fig. 14-9, the path would be hyperbolic ($P\delta$ = constant). Therefore, strictly speaking, strain energy is released during cracking only when *bjkh* falls at a faster rate than an hyperbola.

The compliance of a structure can be measured for different starter crack lengths and $d/dA(\delta/P)$ developed graphically, or sometimes analytical expressions can be used. For example, consider the splitting in the middle of a thin, slender beam of thickness $t$ and depth $2h$, as shown in Fig. 14-11(a). Let us assume that the load–deflection response can be represented by strength-of-materials beam theory. Then, for a cantilever beam,

$$\text{Deflection} = \frac{PL^3}{3EI}$$

where $L$ is the length, $E$ is Young's modulus, and the second moment of area (moment of inertia) is $I = th^3/12$. This is the deflection of one arm of the test piece, so the observed deflection in splitting the beam $\delta$ is twice this value. Thus,

$$\delta = \frac{2PL^3}{3EI} \quad \text{or} \quad \frac{\delta}{P} = \frac{2L^3}{3EI} \tag{14-4}$$

The compliance is thus proportional to $L^3$, and a plot of experimental compliances vs. crack area looks something like Fig. 14-11(b). Because $dA = tdL$, $d/dA(\delta/P) = (1/t)d/dL(2L^3/3EI) = 2L^2/EIt$, which is an expression for the slopes in Fig. 14-11(b) at various $A$ or $L$.

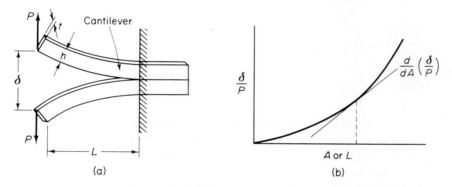

**Figure 14-11**    (a) Splitting a slender beam by opposing forces; (b) the associated compliances ($\delta/P$) for various crack areas (lengths).

Therefore, in Eq. (14-3), for this situation,

$$P_{\text{crack}}^2 = \frac{2R}{d/dA(\delta/P)} = \frac{REIt}{L^2} \tag{14-5}$$

If the material cracks at constant $R$, $P$ will decrease as the crack becomes longer and falls along a constant $R$ locus such as $bjkh$ in Fig. 14-9.

### Example 14-5

Two strips of metal are glued together as shown in Fig. 14-12. A patch of glue was omitted over the central section. The joint is pulled apart by equal and opposite central forces as shown. At what load will the glued joint start to fail? You may assume beam theory for the metal strips, which have $E = 210$ GPa. The fracture toughness of the glue is 300 J/m$^2$.

**Solution**   Both strips above and below the unglued section will deform as *encastré* beams, the deflection of each being $PL^3/192EI$. Consequently, the compliance of the glued two-beam structure is given by $\delta/P = L^3/96EI$. Using $dA = tdL$, we have $d/dA(\delta/P) = L^2/32EIt$, and in Eq. (14-3), $P_{crack}^2 = 64REIt/L^2$. Thus,

$$P_{crack}^2 = \frac{(64\times300)(210\times10^9)(10\times10^{-3})^2(1\times10^{-3})^3}{12(60\times10^{-3})^2}$$

so
$$P_{crack} = 97\ N$$

**Figure 14-12**   The glued joint discussed in Example 14-5.

Another case, of great classic interest because it was the first fracture mechanics problem to be solved, is the so-called Griffith cracking problem, which involves a small, elliptical slit of length $2L$ in a very large plate loaded by a uniform tensile stress $\sigma$ on a pair of opposite boundaries. Figure 14-13 shows the geometry and symbols used in the following analysis.

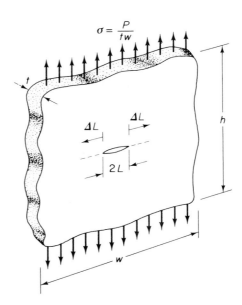

**Figure 14-13**   The Griffith cracking problem of a small elliptical slit in a very large plate loaded at its remote boundaries.

It may be shown that the stiffness $P/\delta$ of such a cracked plate (in terms of the equivalent remote boundary load $P = \sigma wt$ and the remote boundary displacement $\delta$) is given by

$$\frac{P}{\delta} = \frac{Etw}{h(1 + 2\pi L^2/hw)} \tag{14-6}$$

where $E$ is the Young's modulus of flawless material. Then, because the crack area in this case is given by $2Lt$, $dA = 2tdL$ and

$$\frac{d}{dA}\left(\frac{\delta}{P}\right) = \frac{4\pi hL}{2t^2 w^2 hE} = \frac{2\pi L}{t^2 w^2 E}$$

Hence, in Eq. (14-3), for the load to initiate cracking,

$$P^2_{\text{crack}} = \frac{2Rt^2 w^2 E}{2\pi L}$$

or

$$\sigma^2_{\text{crack}} = \frac{ER}{\pi L} \quad \text{or} \quad \sigma_{\text{crack}} = \sqrt{\frac{ER}{\pi L}} \tag{14-7}$$

Equation (14-7) is the Griffith equation for brittle fracture, written in terms of the fracture toughness $R$. In Griffith's experiments at the end of World War I, the measured strengths of glass rods were correlated with surface flaws of different sizes (i.e., different $L$). For small $L$, $\sigma$ was very high; and in almost perfect fine filaments of glass with very small $L$, $\sigma$ approached the theoretical value for the strength of glass (approximately $E/10$).

Note the general form of the Griffith equation: Fracture depends on applied stress *and* crack length, not merely stress alone as yielding does. The expression suggests that a critical stress or load must be attained before cracking is initiated. The critical value squared varies directly with the modulus and toughness and inversely with the crack length. The latter suggests a *critical crack length* below which the crack will not run for a given applied load or stress. Equivalently, the stress that causes a known crack to propagate in an object is hence a measure of the strength of that cracked object.

### Example 14-6

The strength of beer bottle glass is about 170 MPa. What is the Griffith flaw size associated with this strength? For glass, use $E = 70$ GPa and $R \approx 20$ J/m$^2$.

**Solution**    Using $\sigma^2_{\text{crack}} = ER/\pi L$ and noting that $E \approx 70$ GPa with $R \approx 20$ J/m$^2$ for glass, we have $170^2 \times 10^{12} = (70 \times 10^9)(20/\pi L)$. Hence, $L \approx 15 \times 10^{-6}$, or the Griffith flaw size is about 30 $\mu$m.

The use of the expression for a small crack in a large plate for beer bottle geometry is justified in this case by the smallness of $L$.

## CRACK TIP PLASTICITY

No molecular or atomistic interpretation has been developed thus far for $R$. What is happening at the crack tip? Where is the work of fracture going? Why do we have to supply energy at all? Conversely, why does a cracked part have any strength at all?

Clearly both brittle and ductile fracture involve separation of surfaces and *creation of new surfaces* (i.e., the atoms labeled $B$ and $C$ in Fig. 14-7 become exposed as the crack develops).

In 1920, A. A. Griffith first recognized that the exposure of interior atoms when new surfaces are created by the act of crack spreading requires work to be done to give these new surfaces their *surface free energy*. Surface free energy, analogous to surface tension in a liquid, is the work done in breaking the bonds between a unit area of atoms, thereby giving two mirror surfaces. Equivalently, we can say that atoms in the surface have "unbalanced" bonds which require more energy than interior atoms. (See the discussion in Chap. 5.)

The work must have something to do with the area under the force or stress versus atom displacement curve given in Fig. 14-6. The work to cause fracture is

$$\int_0^{\lambda/2} \sigma_{\max} \sin\left(\frac{2\pi x}{\lambda}\right) dx = \frac{\lambda \sigma_{\max}}{\pi} \tag{14-8}$$

This work is shared between the two surfaces, the surface energy of *each* of which has the symbol $\gamma_s$. (Note that in this case this is total work referenced to *both* surfaces; $R$ is defined with reference to *one* side of a double surface, i.e., a crack. See Exercise 14-3.) Therefore,

$$2\gamma_s = \frac{\lambda \sigma_{\max}}{\pi} \tag{14-9}$$

that is,

$$\sigma_{\max} = \frac{2\pi\gamma_s}{\lambda}$$

or, from the development preceding the approximation of Eq. (14-1),

$$\sigma_{\max} = \frac{\lambda E}{2\pi a_0}$$

Multiplying the two preceding equations, we get

$$\sigma_{\max} = \sqrt{\frac{E\gamma_s}{a_0}} \tag{14-10}$$

where $a_0$ is the equilibrium spacing of atoms. Note that Eq. (14-10) can be derived from Eq. (14-7) if $R = 2\gamma_s$ and $L = 2a_0/\pi$. Thus, Eq. (14-10) gives the theoretical maximum stress required to *initiate* a crack when no crack exists other than the equilibrium spacing of atoms. All these expressions for $\sigma_{\max}$ amount to the same thing in the end and suggest that $\sigma_{\max}$ ought to be approximately $E/10$.

For example, for steel at very low temperatures, the following values probably apply: $a_0 = 0.2$ nm, $\gamma_s = 1$ J/m$^2$, $E = 210$ GPa. So,

$$\sigma_{max} = \sqrt{\frac{(210 \times 10^9)1}{0.2 \times 10^{-9}}} \approx 32 \text{ GPa} \approx \frac{E}{7}$$

At this juncture, having been introduced to $\gamma_s$, the reader may be wondering what the difference is between $\gamma_s$ and $R$ (apart from the "definition" of dividing by twice the crack area or by only the area). If, when a crack opens, we create new surfaces to each of which $\gamma_s$ must be furnished, isn't that the lost energy of fracture?

To investigate this issue, think of fracture as a process of breaking a series of miniature tensile specimens ahead of the crack (Fig. 14-14). Each specimen must be taken to its fracture strain before the crack can move on. Obviously, materials differ in this imitation tensile behavior. Some involve large plastic strains with correspondingly large amounts of irreversible work before they break; others merely snap elastically with little strain prior to fracture [Fig. 14-l(a)].

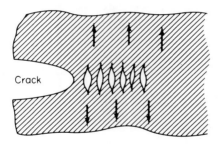

Crack

**Figure 14-14**   Oversimplified idea of fracture as a process of breaking a series of miniature tensile test pieces in the path of the crack.

With the latter brittle materials such as glass (which Griffith used for his experiments), it is found that $R$ and $\gamma_s$ have similar orders of magnitude. Hence the Griffith equation (14-7) is more commonly written with $2\gamma_s$ replacing $R$:

$$\sigma_{crack} = \sqrt{\frac{ER}{\pi L}} = \sqrt{\frac{2E\gamma_s}{\pi L}} \approx \sqrt{\frac{E\gamma_s}{L}}$$

With typical ductile or quasi-brittle materials, measured values for $R$ are very much larger than the surface free energy $\gamma_s$; in fact, $R = 1000\gamma_s$ or greater. This means that work much beyond that necessary for merely the creation of new surfaces is required for crack propagation in these materials. Where does this work go? From the model just suggested it was postulated and then demonstrated by Orowan and Felbeck that there exists a *plastic zone* around most crack tips in which the strains are locally brought up to $\varepsilon_f$ or thereabouts as the crack propagates (remember that stress and strain are concentrated at crack tips; Fig. 14-15). The plastic zone is part and parcel of the cracking process, and cracking in these materials is impossible without it. The work

**Figure 14-15** Orowan plastic zones at tips of cracks. In this illustration the zones of irreversible deformation, which form a "boundary layer" contiguous with the propagating crack, are shown for a crack in polycarbonate, but they were first explored in low-carbon steels. The vertical lines were drawn on the (photograph to indicate the progress of the crack.)

involved in plastically deforming this volume of material is the cause of $R$ being much greater than $\gamma_s$ .*

The size of the crack tip plastic zone is quite small, and it must be emphasized that it is only a *local* region of plasticity; during cracking the rest of the material is often far below yielding. This is very important because parts are often designed on the basis of the average stress at regions remote from the crack (the presence of which may not be known).

---

*The symbol $\gamma$ (without a subscript) is sometimes used to denote the *work of fracture,* which is obtained by taking the *total* area under a load–deflection curve during cracking (such as that in Fig. 14-9) and dividing by twice the crack growth area; that is, the load–deflection diagram is considered one big segmental work area. In these circumstances, $R = 2\gamma$, but unlike the Gurney method, this procedure is not capable of picking up possible variations in $R$ along the path of cracking and/or variations in $R$ with velocity, or whether some of the work has been dissipated in plastic flow remote from the crack tip.

We can think of the plastic zone as disappearing to zero in the most brittle materials and of there being a range of zone sizes up to very large in the most ductile materials.

Take care to note, however, that the idea of "little tensile test pieces" is oversimplified; real cracking is more complex, but the model serves to illustrate some salient points.

**Example 14-7**

Estimate the size of the plastic zone at the tip of a crack in low-carbon steel for which $R = 100$ kJ/m$^2$.

**Solution**    The plastic work done per unit volume up to some strain level is the area under the $\sigma$–$\varepsilon$ curve up to that strain level (Chap. 2), which, if $\sigma = K\varepsilon^n$, becomes $K\varepsilon^{n+1}/(n + 1)$. For a typical low-carbon steel, $K \approx 1$ GPa, $n = 0.15$, and the strain to fracture in a tensile test is approximately 0.6. Hence the work per unit volume to fracture is $10^9(0.6)^{1.15}/1.15 \approx 500$ MJ/m$^3$.

If we assume that the shape taken by the plastic zone at the tip of a crack is something like a cylinder of radius $r$, the volume of material newly plastically deformed when the crack moves by unit distance will be given by a cylinder with a crescent-shaped cross section. When the unit crack advance is less than $r$, the area of the crescent-shaped cross section is small compared with $\pi r^2$. The crescent area is thus approximately $(1/r)(\pi r^2) = \pi r$ when $1 < 2r$.

Consequently, for a unit depth of cylinder perpendicular to the plane of a plate (as in Fig. 14-15), the plastic work done if the newly deformed zone is strained to $\varepsilon_f$ is $500\,\pi r$ MJ. But for unit crack advance across a crack front of unit depth, the work required is numerically equal to $R$, or 100 kJ/m$^2$ in this case. If it is justifiable to equate the two works, we have $r = (100 \times 10^3)/(500 \times 10^6)\pi \approx 65$ $\mu$m. This dimension is, in fact, reasonably typical of the thickness of strained boundary layers found contiguous with cracks in some steels.

## BRITTLE CRACKING IN LARGE STRUCTURES

At the time Griffith did his work, it was not recognized that brittle fracture is possible in materials such as plain-carbon steel that are normally considered ductile. Of course, it was known that at low temperatures (below the transition range measured with Charpy specimens) steel does become brittle, but during World War II welded ships started to break by cracking even though their strength-of-materials design stresses were considerably below yield and even though metal taken afterward from near the crack had ductile $\sigma$–$\varepsilon$ behavior with respectable values for $\varepsilon_f$. What was going on? Steel is not supposed to be brittle like glass or cast iron, the temperatures were not all that cold, and evidently most of the ships were really not overstressed. The failed ships had neither yielded all over nor necked before cracking! (Some recent evidence indicates that even the *Titanic* might not have sunk following its collision with an iceberg in 1912 if its steel had not been susceptible to brittle fracture.)

It is now realized that fabricated items (ships, bridges, oil storage tanks, pipelines, airplanes, etc.) are never perfect. Somewhere a weld might not be as good as it ought to be; perhaps a metallurgical error has allowed a fissure to be rolled over in the hot mill; glued joints may not be up to specification. Consequently, we now must identify any macroflaws, or even really large cracks, in large fabricated structures (e.g., hatches in ships' decks have even been shown to be bad stress concentrators on occasion). Hence, if the conditions are favorable, cracking can initiate at these flaws and break the structure, even though stresses away from the crack may be small and *below the yield value.*

When are the conditions "favorable"? In traditional design, the stresses to which a structure is subjected are calculated from the working loads on the assumption that the structure is "perfect," with no flaws, no preexisting cracks, no locally brittle areas caused by wrong heat treating, and so on. These design stresses are usually arranged to be well below the yield stress. However, in the presence of a flaw, the applied stress necessary to open up the flaw and propagate it could be less than the design stress. The stress $\sigma$ that must be applied to the structure to initiate cracking can be obtained from an appropriate formula such as Eq. (14-5) or (14-7). Most often it takes the general form $\sigma^2 \propto ER/L$, where $L$ is the crack length, the constant of proportionality depending on specimen shape and crack orientation.

Consider two versions of the structure, the large prototype itself and a model that might be tested in the laboratory. For simplicity, consider a geometrically similar model where all dimensions (height, width, thickness, crack length) are $1/\lambda$ of those of the prototype; $\lambda$ is called the *scaling factor*. Using subscripts m and p, for model and prototype, we find that the stresses needed to cause cracking in the two sizes of structure are

$$\sigma_m^2 \propto \frac{E_m R_m}{L_m} \quad \text{and} \quad \sigma_p^2 \propto \frac{E_p R_p}{L_p}$$

If the fracture toughness and Young's modulus of the material are assumed to be the same in the laboratory and in the fabricated structure, then

$$\sigma_p^2 = \left(\frac{L_m}{L_p}\right)\sigma_m^2 \quad \text{or} \quad \sigma_p = \left(\frac{1}{\lambda^{1/2}}\right)\sigma_m \qquad (14\text{-}11)$$

Because $\lambda = L_p/L_m$, we see that the stress necessary to cause cracking in the prototype is less than the stress required to do the same thing in the model. For example, if $\lambda = 25$, the prototype cracking stress is one-fifth of that required in the model. No such scaling effect appears for yielding, flow stresses being independent of the size of the body. Thus, cracking may occur before yielding in a large structure, whereas in the laboratory a model of the same structure may yield before cracking (Fig. 14-16). Consequently, what might appear to be a safe, ductile material in the laboratory may perform quite differently and

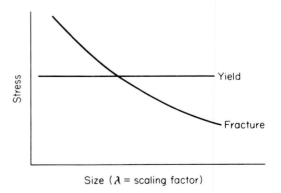

**Figure 14-16** Scale effects in brittle fracture. The larger the structure, the smaller the stress required to cause cracking, but no size effect enters calculations for yielding. Thus, a material that yields in the laboratory (small $\lambda$) may crack when fabricated into a large structure (large $\lambda$).

unsafely in a large fabricated structure. Hence, the small tensile bars taken from locations adjacent to the cracks in welded ships showed ductile behavior. This is the origin of brittle fracture in large structures. In a way, it is a logical extension of old ideas on stress concentration in the design of machine parts; it was known that even in "perfect" steel, sharp corners in small pieces led to failure; now we look at sharp corners on a gross scale.

In the simple derivation of Eq. (14-11), it was assumed that the scaling factor $\lambda$ was the same for length, height, width, and crack length. Clearly, this is not always so: Ships' steel can be tested in the laboratory in its full thickness, but the areal extent of hull or decking cannot be duplicated; again, annular cross sections of pipelines can be tested in the laboratory, but practical lengths of about a kilometer cannot, and so on. Even so, it is possible to perform the more complicated calculations for scaling laws with four different $\lambda$'s, and it is usually found that $\sigma_p < \sigma_m$.

The possibility of brittle fracture in normally ductile materials is confusing if it is assumed that fracture in ductile materials must involve gross yielding (as in a tensile test of a ductile solid). Such ductile fractures can occur, of course, as in metal forming when we try to push the process to its limit (e.g., in rolling or deep-drawing car bodies, as discussed earlier) and in opening cans of beer, sardines, and so on, but these fractures are expected. It is the fact that we are *not* expecting fracture when the design stress is below yield in a ductile material that makes it dangerous.

**Example 14-8**

A cylindrical pressure vessel 6 m long, 3 m in diameter, and with a 20-mm-thick wall exploded at 5 MPa. The steel from which it was made had the following properties: $E = 210$ GPa, $\sigma_y = 1.4$ GPa, $R = 130$ kJ/m$^2$. Show why failure was not expected on the basis of strength-of-materials design calculations but that failure was compatible with the existence of a longitudinal welding crack about 120 mm long.

**Solution**    The hoop stress in the cylinder is given by

$$\frac{pD}{2t} = \frac{(5\times10^6)3}{2(20\times10^{-3})} = 375 \text{ MPa}$$

This is between one-third and one-quarter of the yield stress, so failure should not be expected.

Using the small-crack-in-a-large-sheet expression for cracking stress, we have

$$\sigma_{\text{crack}} = \sqrt{\frac{ER}{\pi L}} = \sqrt{\frac{(210\times10^9)(130\times10^3)}{\pi\times60\times10^{-3}}} = 380 \text{ MPa}$$

As this is comparable to the design hoop stress, the failure of the pressure vessel is not unexpected.

The foregoing exercise indicates the importance of knowing whether flaws are present in a built-up structure and whether they are likely to be dangerous or not. It is customary nowadays to use stringent crack-inspection procedures (such as X-ray, ultrasonic, or other nondestructive methods) in an attempt to identify flaws before a critical part is used. If we want to use materials at a high strength level, yet clearly want cracking stresses to be greater than the design stress, we must be able to locate very small cracks because the crack initiation stress is inversely proportional to flaw size. Conversely, if the inspection equipment is reliable only for a certain lower $L$, this sets an upper value on the safe working stress in a component of given geometry.

Sometimes the cracking stresses associated with flaws are found to be very high, in which case it is occasionally permissible to leave the flaws; at other times the cracking stresses are unacceptably low and unsafe, so the component or structure must be repaired and the flaws eliminated or reduced in size. Bad welds, for example, may be ground away and the cavity refilled with fresh metal.

## FRACTURE MECHANICS

The loads or stresses that cause a crack to propagate have so far been determined from the energy and compliance method [Eq. (14-3)]. Alternatively (and nowadays much more commonly), they may be arrived at from considerations of the detailed stress fields around the tip of a flaw and in the cracked body generally. This second line of attack is the discipline called *fracture mechanics*. G. R. Irwin realized that relations connecting the applied stress and the size and orientation of a flaw can be written in terms of the applied *stress-intensity factor K,* which is an alternative parameter to $G_c$ or $R$ and is related to them, as will be explained presently. For example, a fracture mechanics expression for the stress-intensity factor at the tips of the crack in a center-cracked panel of finite width (Fig. 14-17) is

$$K = \sigma\sqrt{\pi L}\sqrt{\frac{2b\ \tan^{-1}(\pi L/2b)}{\pi L}} \qquad (14\text{-}12)$$

**Figure 14-17**  A center-cracked panel of finite width $2b$ with crack length $2L$ loaded remotely in the axial direction perpendicular to the crack.

It will be seen from the dimensions of this expression that the stress-intensity factor $K$ has the peculiar units of N/m$^{3/2}$, which also may be written as Pa$\sqrt{m}$. The *critical stress-intensity factor* $K_c$ with the same odd units is a combination of the two more easily understood properties $E$ and $R$, namely, $K_c^2 = ER$. $K$ has the form it has from the mathematics of the nonuniform stress fields around the crack tip; it must not be confused with the stress-*concentration* factor, which relates to a multiplying factor (dimensionless) by which the local average elastic stress at a change in cross section is increased above the average stress.

Experiments show that cracking occurs when $K$ [determined from formulas such as Eq. (14-12) for the test piece being employed] attains a critical value, called $K_c$, which is characteristic of the material. In fact, it may be shown that $K_c^2 = ER = EG_c$, which is the product of two separate physical properties $E$ and $R$ of a material (hence the peculiar units for $K$), so the attainment of a critical $K_c$ for propagation is the same as satisfying an energy-derived equation such as Eq. (14-5) or (14-7) at the material's $R$ level. The link between the two methods is that the distribution of stresses and strains gives an integrated strain energy for the cracked body, which, in the same manner as described for $G_c$, feeds the crack at the correct rate at first propagation. The $d/dA(\delta/P)$ term in Eq. (14-3) has been already worked out and is implicit in the fracture-mechanics expression for a particular geometry. Thus, the fracture-mechanics formula for the DCB specimen is written as $P = (K/L)(It)^{1/2}$; compare Eq. (14-5).

Notice that when Eq. (14-12) is rewritten for small $L/b$ [when $\tan(\pi L/2b) \approx \pi L/2b$], with $K_c = \sqrt{ER}$ also, we have

$$\sqrt{ER} = \sigma\sqrt{\pi L} \quad \text{or} \quad \sigma = \sqrt{\frac{ER}{\pi L}}$$

Not unexpectedly, this is the same as the Griffith expression (14-7) for a small center crack in a large plate. In general,

$$\sqrt{ER} = K = \sigma\sqrt{\pi L}(Y) \tag{14-13}$$

where $Y$ depends on cracked-body geometry and type of loading; values for $Y$ for numerous bodies of different shapes, crack orientations, and types of loading may be found in handbooks. Equation (14-13) is the same as Eq (14-3): The $d/dA(\delta/P)$ is incorporated in the $Y$ factor.

Fracture toughness *(R or $G_c$)* and critical stress-intensity factor $K_c$ are now routinely measured for material specifications and quoted in codes of practice for design for use with fracture-mechanics formulas. The designer utilizes fracture-toughness concepts by first calculating the value of the applied $K$ as a function of stress, crack length, and geometry, using solutions already worked out and available in handbooks. The value of $K_c$ for the material being used in the design is then found from published data (or measured), and the calculated $K$ is equated to $K_c$. This provides the safe value of stress for a specified value of toughness.

## PLANE STRAIN $K_{Ic}$ TESTING

Fracture mode can be markedly influenced by the state of stress in the region of fracture. An element of material subjected to the stress system shown in Fig. 14-18(a) contains stresses that lie only in the *x-y*-plane and is thus subjected to a condition of *plane stress*.

Refer to Fig. 2-11 for examples of plane stress in thin sheets. Assuming that $\sigma_x$ and $\sigma_y$ are not equal, the Mohr's circle for plane stress has the form of Fig. 14-18(b). In this case,

$$2\tau_{max} = \sigma_1 = \sigma_{max}$$

The maximum shear stress is thus relatively large, and general plastic deformation is quite likely.

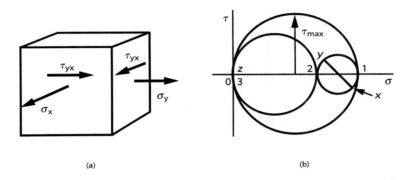

(a)  (b)

**Figure 14-18** (a) Plane stress; (b) Mohr's circle analysis for case (a), where $\sigma_x > \sigma_y$.

*Plane strain* occurs when the stress system leads to zero elastic strain in one direction, for example the $z$-direction. If $v$ is Poisson's ratio, plane strain requires that

$$\varepsilon_z = \frac{1}{E}\left[\sigma_z - v(\sigma_x + \sigma_y)\right] = 0$$

$$\sigma_z = v(\sigma_x + \sigma_y) = \frac{2v(\sigma_x + \sigma_y)}{2} = 2v\ (\text{average of }\sigma_x \text{ and } \sigma_y)$$

For $v \sim 0.3$,
$$\sigma_z = 0.6\left(\frac{\sigma_x + \sigma_y}{2}\right)$$

One way to achieve this condition is for all three stresses to be tensile, as in Fig. 14-19(a), a condition called *triaxiality* of stress. The Mohr's circle for this condition is shown in Fig. 14-19(b), and it is clear that the maximum shear stress is much less than half the maximum normal stress, as was the case for the condition of plane stress discussed earlier. Thus, a plane strain condition inhibits plastic deformation.

Whether the local stress system is plane stress or plane strain depends on

- its thickness
- its yield strength $S_y$
- the elastic stress intensity factor $K$.

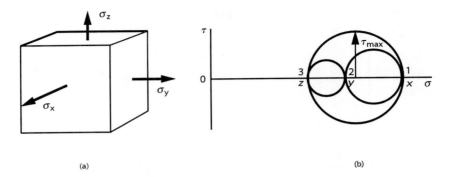

**Figure 14-19**  (a) Plane strain; (b) Mohr's circle analysis for case (a) where $\sigma_x > \sigma_y$.

For plane stress, the radius $r_p$ of the plastically deformed zone that penetrates in from each outer surface of a plate in the vicinity of the tip of a crack is

$$r_p = \frac{1}{2\pi}\left(\frac{K}{S_y}\right)^2$$

Thus, for a plate of thickness $B$, if

$$B = 2r_p = \frac{1}{\pi}\left(\frac{K}{S_y}\right)^2$$

then plane stress exists throughout virtually the full thickness of the plate.

Alternatively, if the plate is substantially thicker than $2r_p$, then mostly plane strain conditions hold, which restrict the radius of the *plane strain plastic zone* $r_p$ to a smaller size,

$$r_p = \frac{1}{6\pi}\left(\frac{K}{S_y}\right)^2$$

Thus, the radius of the plastic zone in the central region of a plate substantially thicker than twice the plane stress $r_p$ is one-third the radius of the plastic zone at both surfaces of the plate. This demonstrates that the thickness of the plastically deformed layer on the fracture surface is always greater in the vicinity of the outer surface of a thick plate than in the vicinity of its midthickness.

The material property $K_{Ic}$ characterizes the resistance of a material to fracture in a neutral environment in the presence of a sharp crack under severe tensile constraint, which means plane strain conditions. (The I in the subscript refers to mode I fracture, which occurs under ordinary tension in the presence of a partial transverse crack.) From consideration of the influence of plane strain discussed earlier, increasing thickness of a part must lead to a decrease in $K_{Ic}$ because a progressively greater portion of the fracture occurs under plane strain conditions as plate thickness increases. The value of $K_{Ic}$ becomes asymptotic to a minimum value with increasing thickness; this minimum value is called the *critical plane strain stress-intensity factor* or *critical plane strain fracture toughness* because deformation in the thickness direction at the tip of a crack is constrained by the surrounding elastic material so most of the strain occurs in the two directions that lie in the plane of the plate. Because use of the lowest (plane strain) value of toughness leads to a conservative design for thinner parts, the plane strain $K_{Ic}$ is usually used in design, and national standards have been developed for measuring $K_{Ic}$. However, if a design involves a thinner part and a reliable value of $K_{Ic}$ is known for this thickness, the designer can use this higher value for a more efficient design.

The differences in behavior are shown in the types of fracture surfaces produced. Plane-strain fractures tend to be flat but are accompanied by *shear lips* (compare tensile cup-and-cone fractures), as in Fig. 14-20, when plane-stress edge effects come into play (note the sideways contraction of the specimen on the right-hand side of the figure). In practice, most fractures are "mixed mode" in character. These effects have long been known in the fracture surfaces of Charpy specimens.

**Figure 14-20** Fracture surfaces of Charpy V-notch test specimens of a very brittle steel (left) and a moderately ductile steel (right). The initial machined notch lies along the bottom edge of each specimen in the photograph. The restriction of plastic flow prior to fracture led to plane-strain conditions in the brittle steel, whereas the ductile specimen of the same dimensions allowed sufficient plastic flow to permit conditions of plane stress near the free edges of the cross section. The 45° shear fracture that results, called a *shear lip,* and the associated plastic contraction of the specimen, are evident along the three sides of the fracture. (Photograph courtesy of W. H. Durrant.)

The fractography of brittle and ductile fractures in general is discussed in Chap. 19.

To ensure that the bulk of the thickness of a plate is subjected to plane-strain conditions, the American Society for Testing and Materials (ASTM) standards specify a minimum thickness $B$ of a fracture toughness test piece given by

$$B \geq 2.5 \left( \frac{K_{Ic}}{S_y} \right)^2 \tag{14-14}$$

where $S_y$ is the 0.2% offset yield strength. This thickness is thus about 16 times the plane-stress plastic zone radius and about 47 times the plane-strain plastic zone radius. This is an empirical rule for metals, experiments showing that the fracture toughness attains its smallest value at this thickness and greater. For polymers, it is found that $B \geq 5(K_{Ic}/S_y)^2$ for plane strain to apply.

Precracked three-point bend test pieces that conform to the size limitations shown in Fig. 14-21 have been found experimentally to produce consistent measurements of $K_{Ic}$ (Ref. 1). Because sharp rather than blunt cracks require smaller loads to cause fracture, and

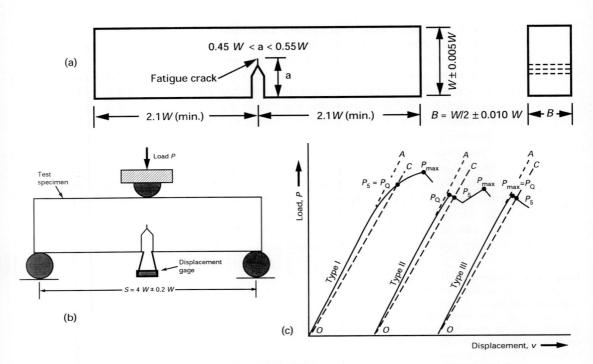

**Figure 14-21**    (a) ASTM standard fracture toughness bend specimen; (b) standard bend fixture; (c) types of load–displacement diagrams. [American Society for Testing and Materials, "Standard Test Method for Plane-Strain Fracture Toughness of Metallic Materials." Designation: E399-90. (ASTM, 1990).] Copyright ASTM. Reprinted with permission.

because we are looking for the lowest value of toughness, the tip of the crack of nominal length $a$ in Fig. 14-21(a) is created by fatigue loading the specimen after the sharp notch is machined. Because precise control of the crack length is impossible, the total crack length $a$ is allowed to range from 0.45 to 0.55 times the specimen width $W$. The test is conducted in a three-point bending fixture as shown in Fig. 14-21(b). As the load $P$ increases, the displacement gage measures the increase in the opening of the notch at the bottom (tensile) edge of the specimen. Obviously, because $K_{Ic}$ is not known until after the test is conducted, yet $B$ must be selected before the test is begun, the test must involve a preliminary selection of $B$. After the test, a tentative value (called $K_Q$) that has been measured can be used in Eq. (14-14) to establish whether $B$ was sufficiently large for the test to be valid. If not, a new thicker specimen must be made and a new test conducted. When a sufficiently large $B$ has been reached and other criteria described below have also been met, then we have obtained an accurate value for $K_{Ic}$.

The load vs. displacement record should have a shape similar to one of the three curves shown in Fig. 14-21(c). To determine the value of $P_Q$ for any of the three types of curves, construct a line with a slope 0.95 times the slope of the elastic portion of the curve. Thus, line $OC$ has a slope 0.95 times the slope of line $OA$. [In Fig. 14-21(c), the slope of

*OC* is somewhat less than this for clarity.] The intersection of *OC* with the *P–δ* curve at load $P_5$ establishes the value of $P_Q$ for type I; for types II and III, where *P* reaches a value greater than $P_5$ before point $P_5$, the earlier $P_{max}$ is used for $P_Q$.

After it is established that all the dimensions of the test specimen fall within the tolerances of Fig. 14-21(a) and (b), the next check to establish whether the test is valid is as follows.

1. Calculate $P_{max}/P_Q$; if $P_{max}/P_Q > 1.10$, the test is not valid. The tentative value $K_Q$ is determined as

$$K_Q = \left( \frac{P_Q S}{BW^{3/2}} \right) Y\left( \frac{a}{W} \right)$$

All these values have already been defined and measured except $Y(a/W)$, which is derived from the fracture-mechanics analysis for this geometry:

$$Y\left(\frac{a}{W}\right) = \frac{3(a/W)^{1/2}[1.99 - (a/W)(1-a/W)(2.15 - 3.93 a/W + 2.7a^2/W^2)]}{2(1+2a/W)(1-a/W)^{3/2}}$$

In order to simplify calculation of $K_Q$, $Y(a/W)$ is tabulated as follows:

| a/W | Y(a/W) | a/W | Y(a/W) |
|---|---|---|---|
| 0.450 | 2.29 | 0.500 | 2.66 |
| 0.455 | 2.32 | 0.505 | 2.70 |
| 0.460 | 2.35 | 0.510 | 2.75 |
| 0.465 | 2.39 | 0.515 | 2.79 |
| 0.470 | 2.43 | 0.520 | 2.84 |
| 0.475 | 2.46 | 0.525 | 2.89 |
| 0.480 | 2.50 | 0.530 | 2.94 |
| 0.485 | 2.54 | 0.535 | 2.99 |
| 0.490 | 2.58 | 0.540 | 3.04 |
| 0.495 | 2.62 | 0.545 | 3.09 |
| — | — | 0.550 | 3.14 |

Calculate the value of $K_Q$ from the previous equation. Then apply the final check to establish whether the test is valid.

2. Calculate $2.5(K_Q/S_y)^2$. If this is less than the value of *B*, then $K_Q = K_{Ic}$. Otherwise the test is not valid.

Other types of specimens may be used for $K_{Ic}$ testing, such as a "compact tension" test piece, which is a sort of "stubby" DCB specimen. Further details may be found in national standards.

**Example 14-9**

An ASTM E399 fracture toughness test of a metal X is conducted with a specimen of the following dimensions: thickness = 3.15 mm; length between supports = 26.0 mm; height = 6.30 mm; and crack length = 3.37 mm. Metal X has an elastic modulus of 169 GPa and a yield strength of 127 MPa. The load–displacement results are shown in Fig. 14-22.

**Figure 14-22**  Example 14-9.

(a) Calculate from these data the critical mode I plane-strain stress-intensity factor.

(b) Is this result valid?

(c) Show why or why not.

**Solution**   The trial value $K_Q$ is calculated from

$$K_Q = \left( \frac{P_Q S}{BW^{3/2}} \right) Y \left( \frac{a}{W} \right)$$

$P_Q$ is obtained from the load–displacement curve in Fig. 14-22 using the criteria defined in Fig. 14-21(c). A line drawn with a slope 0.95 times the slope of the straight-line portion $OP$ intersects the curve to the right of the first maximum load at 150 N; therefore, this is a curve of type II, and $P_Q = 150$ N. A check against $P_{max} \leq 1.10$ $P_Q$ gives a maximum allowable value of 165 N; the observed $P_{max}$ is clearly less than 165 N, so $P_Q = 150$ N is valid.

The remaining values for the equation are obtained from the dimensions shown in Fig. 14-21(a) and (b): thus, $B = 3.15$ mm, $W = 6.30$ mm, $a = 3.37$ mm, $S = 26.0$ mm. A check for the tolerance for $B$ is $B = W/2 \pm 0.010$ $W = 3.15$ mm $\pm 0.063$ mm and shows that $B$ is within tolerance. A check for the tolerance for $S = 4W \pm 0.2$ $W = 25.2$ mm $\pm 1.3$ mm shows that $S$ is within tolerance.

$Y(a/W)$ is found from $a/W = 0.535$, so $Y(a/W)$ is found from the table to be 2.99. Therefore,

(a)

$$K_Q = \left[ \frac{(150 \text{ N})(26.0 \text{ mm})}{(3.15 \text{ mm})(6.30 \text{ mm})^{3/2}} \right] 2.99 = 7.40 \text{ MPa} \sqrt{\text{m}}$$

(b) and (c): In addition to the checks already noted above, from Eq. (14-14),

$$B \geq 2.5 \left( \frac{K_Q}{S_y} \right)^2 = 2.5 \left( \frac{7.40 \text{ MPa} \sqrt{\text{m}}}{127 \text{ MPa}} \right)^2 = 8.50 \text{ mm}$$

But $B = 3.15$ mm, so the thickness is too small to provide adequate plane-strain conditions. The test must be repeated with larger values of $B$ until the condition in Eq. (14-14) is met.

## MAXIMUM TENSILE-STRESS FRACTURE CRITERIA

A criterion for fracture based on the attainment of some characteristic maximum tensile stress has been used with a certain degree of success, particularly in very brittle solids. It should be clear, however, that quantitative criteria for fracture based merely on some maximum tensile stress (or combination of stresses) address only part of the necessary and sufficient conditions for fracture. No consideration is given to the energetics of crack propagation. Although maximum stress theories may be intuitively appealing and *may* work reasonably well under the circumstances, they should not be expected to apply to large structures containing critical flaws.

    Consider the tensile strength of brittle materials such as glass, ice, ceramics, rocks, and epoxies. It is given by the fracture load divided by the tensile bar cross-sectional area, or if determined by bending, from the simple beam theory outer-fiber stress (the latter bending strength is called the *modulus of rupture* in the ceramics field). The cracking load in such tests is really the solution of Eq. (14-3) or (14-13) for the particular specimen geometry and the material's natural distribution of flaws (from which fracture starts). The cracking loads are likely to be quite different in different geometries. It is not surprising, therefore, that different values for tensile strength of the same material are obtained from different tests irrespective of the scatter in data resulting from the statistical distribution of natural flaws in these solids (Chap. 16). Consequently, in maximum-stress cracking theories, what is the "real" reference tensile strength to be used? Because it is the toughness value rather than the tensile stress that is the material property, it seems more satisfactory to work in terms of the fracture resistance and Eq. (14-3) or (14-13) when all the information is available.

## GENERALIZED YIELDING, COD, AND THE *J*-INTEGRAL

In the split-beam example in Fig. 14-11, it was assumed that the stresses at regions remote from the crack tip were considerably below the yield strength of the material, so all irreversible deformation was concentrated in zones contiguous with the crack tip. Such irreversible work is, of course, the origin of the material fracture toughness. If the arms of the split beam are not sufficiently deep, however, stresses greater than the yield stress can be produced, so irreversible work (which has nothing to do with the process of cracking) must also be performed. The approximate maximum stress in the arms is given by engineers' bending theory as

$$\sigma = \frac{My}{I}$$

From Eq. (14-5), the moment required to produce cracking is

$$P^2L^2 = M^2 = REIt$$

Thus, the depth of the one arm when $\sigma = \sigma_y$ (the yield stress) during cracking is

$$h = \frac{3ER}{\sigma_y^2} = 3\left(\frac{K_c}{\sigma_y}\right)^2$$

In other words, if $h < 3ER/\sigma_y^2$, the arms will yield before the body cracks. Look at the intersection of the "yield" and "fracture" lines in Fig. 14-16, which illustrates such a "deformation transition" between elastic fracture and plastic flow. Note this criterion for DCB specimen minimum arm *height* as a multiple of $(K_c/\sigma_y)^2$ and the ASTM minimum *thickness* requirement Eq. (14-14). The proportions of the remaining ligament of length $(W - a)$ in the path of the crack in Fig. 14-21(a) are also chosen to avoid yielding up there.

Depending on the complex interaction of work hardening and the loads required to propagate the crack subsequently, a load–deflection plot under these circumstances might appear as in Fig. 14-23. Appreciable curvature exists, and the unloading line does not go back to the origin; the split beam remains yawed open, with permanent deflection $Oq$, because of the *generalized yielding*. Clearly it is difficult from the $P–\delta$ diagram to differentiate between the cracking work and the yielding work performed away from the crack tip (hence one criticism of the use of $\gamma$, "the work of fracture," from the total area under a $P–\delta$ plot).

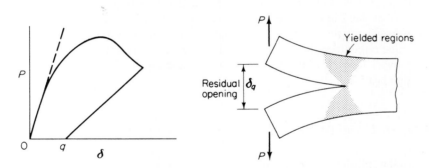

**Figure 14-23**   *P–δ* diagram for combined flow and fracture, where yielding remote from the crack faces occurs during (or at least before) cracking. Unloading lines do not return to the origin.

The mechanics of cracking preceded by and accompanied by generalized yielding is not yet properly formulated. One approach uses the Wells–Cottrell concept of *critical crack tip opening displacement* (COD or CTOD) which recognizes that cracks initiate and propagate at their own "natural sharpness." An appreciation of the idea comes from the "miniature tensile test pieces" at the tip of a crack shown in Fig. 14-14, where the COD is related to the plastic stretch needed to cause fracture. It may be shown that

$$R \approx \sigma_y \text{ (COD)}$$

so COD is yet another alternative to $R$, $G_c$, or $K_c$ as a measure of toughness. Like those parameters, COD may change with change in constraint (plane stress). The *COD design curve* relates working stresses, crack lengths, and toughness rather like Eq. (14-3) or (14-13).

Another approach to postyield fracture employs a critical value of the so-called *J*-integral $J_c$. Curvature in the $P–\delta$ plot in Fig. 14-23 is caused by irreversible plasticity, giving residual displacements on unloading. It is possible to have nonlinear $P–\delta$ plots which are, in fact, reversible and on unloading return to the origin of displacement. The curvature in such $P–\delta$ plots is caused by *geometric* nonlinearities. (It happens when domes are loaded and even in the simple DCB test piece in Fig 14-11 if fracture does not occur until large displacements are reached; the effect is produced in many specimens if they are loaded out of plane, by putting nuts and bolts through the loading pin holes and pulling on one side.) Fracture under reversible nonlinear conditions is governed by *nonlinear elastic fracture mechanics* (nlefm). In Fig. 14-9, the nonlinear stiffness lines *Ob, Oj, Ok*, and so on, would be curved, but the work done $R \, \Delta A$ in propagating the crack from the crack area corresponding to point *j* to the crack area $\Delta A$ corresponding to *k*, is still the curvilinear sector *Ojk*. There are also formulas that relate cracking load, crack length, and toughness under these conditions. It may be shown that, even though the $P–\delta$ curvature in Fig. 14-23 is caused by plasticity and not by reversible geometric nonlinearities, the same mechanics apply at least up to the first fracture. That is, nlefm may be used to predict loads and displacements at ductile fracture initiation. $J_c$ is nothing more than $R$ in nlefm, although it is often presented as an arcane contour integral which in turn is really Eq. 14-2 couched in different terms for nonlinear conditions.

## NOTCH BRITTLENESS
## AND TOUGHNESS/STRENGTH RATIOS

It is often stated that the area under a tensile curve out to fracture is a measure of fracture toughness. Although with test pieces of identical size in the laboratory a relative ranking between materials can be obtained (in the same way that impact data are used), the differences in behavior mentioned in the foregoing discussion, as well as the odd scaling effects, are not always brought out by merely looking at areas under stress–strain curves. The classification of materials as brittle or ductile on the basis of some arbitrary laboratory test may therefore be misleading. In Charpy tests, for example, marked differences in the transition temperature range for steels are produced by varying the type and depth of the notch. Many solids can be made to behave in a brittle manner in one type of test but in a ductile manner in some other test. We have seen that fracture behavior may be altered by changing specimen thickness in tests that measure $R$: Thick test pieces have predominantly small crack tip plastic zones, and the applied cracking stresses are much lower than the material yield stress, whereas thin specimens have large-necked zones and often yielded regions remote from the crack tip. Again, fracture scaling laws that favor cracking instead of yielding in large pieces complicate our simple use of the words "brittle" and "ductile."

Perhaps we should classify solids in terms of a toughness/strength ratio rather than merely toughness or strength alone. A useful concept in this connection is *notch brittleness*. Consider pulling a notched piece of a material usually considered to be ductile, such as a flat tensile bar with a saw cut (Fig. 14-24). The behavior might be brittle with easy cracking if the conditions are favorable. On the other hand, the stress on the so-called net section (i.e., the section of material left where the notch has been made and on which the greatest average stress in the system occurs) may get as high as the yield strength of the material before cracking commences. This suggests that yielding, rather than cracking, would occur at the net section in the same way that yielding took place in the arms of the split beam. However, we must remember the arguments about a triaxial state of stress in the neck of a tensile bar when the shoulder material constrains material from flowing into the neck and hence prevents elongation. Something similar happens in the notched sheet, and the result is that we have to apply *more* than the simple yield stress in order to obtain plastic flow. It can be shown that the average net stress necessary to promote plastic flow in the presence of deep notches (where the stress in the shoulders is much lower) can be as high as $3\sigma_y$; the situation is a sort of "tensile" hardness test (Chap. 2).

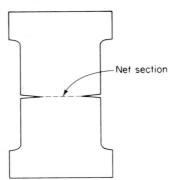

**Figure 14-24**   Tensile test of a notched specimen to assess notch brittleness.

A *notch-brittle* material is one that is ductile in an ordinary laboratory test but brittle in notched tests with large plastic constraint factors (i.e., in tests with deep notches). Hence, if $\sigma_{crack}$ is the boundary cracking stress and $\sigma_y$ the yield stress, we have (1) a *simply brittle* material if $\sigma_{crack} < \sigma_y$, (2) a *notch-brittle* material if $\sigma_y < \sigma_{crack} < 3\sigma_y$, and (3) a *simply ductile* material if $\sigma_{crack} > 3\sigma_y$. Because the cracking stress is given by some formula such as $\sigma^2_{crack} = K_c^2/\pi L Y^2$ [Eq. (14-13)], where $Y$ is a constant for the specimen geometry, the performance of a material in this test depends on the relative values of $\sigma_y$, and $3\sigma_y$. That is, a simply brittle solid has $K_c/\sigma_y < Y\sqrt{\pi L}$; a simply ductile material has $K_c/\sigma_y > 3Y\sqrt{\pi L}$; and a notch-brittle solid has $K_c/\sigma_y$ between these values. Knowing the appropriate $Y\sqrt{\pi L}$ for a chosen type of test piece permits the value of $K_c/\sigma_y$ for the solid to be bracketed, and different types of notched specimens can, in principle, refine the answer. However, it is customary nowadays to actually measure $K_c$ or $R$ independently from standard toughness tests, as described earlier in this chapter, rather than to use notch-brittleness tests to characterize the behavior.

Nevertheless, the toughness/strength ratio given by $K_c/\sigma_y$ or $ER/\sigma_y^2$ is a most useful classification for materials. It has already appeared in this chapter in connection with transitions from cracking to yielding and in the ASTM thickness requirement for plane-strain fracture. Appendix 9 gives some representative values of $K_c/\sigma_y$. We see a gradual ranking, from materials traditionally considered brittle to those commonly considered ductile. However, note that although a material possessing a high ratio is more prone to display ductile behavior before fracturing, such materials are not immune from brittle fracture in very large fabricated structures. The Liberty ships and T-2 tankers of World War II were made from steel having $K_c/\sigma_y$ probably about $0.5\sqrt{m}$ at 20°C, but on the order of $0.1\sqrt{m}$ at −10°C.

**Example 14-10**

Explain why notch-sensitivity data for low-carbon steels are often erratic.

**Solution**    One type of test piece that may be used to investigate notch brittleness is the round tensile bar containing deep peripheral V-notches. The cracking expression for such a test piece is given approximately by $\sigma_n^2 \approx 2ER/\pi D$, where $\sigma_n$ is the net section stress at cracking and $D$ is the gage section diameter. For bars 10 mm in diameter, with $E = 210$ GPa, if we take a constant $R$ at about 130 kJ/m² (i.e., $K_c = 165$ MPa$\sqrt{m}$), $\sigma_n \to 1.3$ GPa; for 20-mm bars the cracking stress is 930 MPa, and for 30-mm bars, 750 MPa. The yield strength range of common steels is about 200 ~ 500 MPa, and the constrained yield strength range, given by $3\sigma_y$, is then 600 ~ 1500 MPa. Clearly, there are steel test pieces for which $\sigma_n$ and $3\sigma_y$ will be comparable (10-mm steel bars of yield point approximately 400 MPa, 20-mm bars of approximately 300 MPa, and 30-mm bars of approximately 250 MPa). Consequently, some tensile bars may crack, while other nominally identical specimens may flow; similar behavior is found with Charpy and Izod steel specimens. Notch brittleness under these conditions is sensitive to slight variations in composition, previous heat treatment, mechanical work, and temperature. Change in specimen size evidently may move the solid from the boundary between notch sensitivity and simple ductility and may also change ranking comparisons made with other materials having the changed test piece size.

Note the "fracture scale effect" in the results; even the material of lowest yield strength will crack if the test piece is made big enough.

## FRACTURE IN VERY THIN SHEETS

If a notched sheet of material is very thin, another type of fracture can occur. As before, the stress becomes concentrated at the tip of the notch. The piece will not be ripped brittly because the toughness of thin sheets is greater than that of thick sheets, but as the load is increased, shear stresses may be produced that cause slip at 45° in the *thickness* of the sheet rather than across the width. It can be shown (Example 14-11) that the average stress required to do this in a thin sheet is less than $\sigma_y$. Then, out-of-plane tearing can easily take place and produce complete fracture (Fig. 14-25). It is a locally ductile fracture, which could not take place in a thick sheet because the area of the shear planes would be so large that the normal stress required would probably be sufficient instead to cause generalized yielding across the width.

Such fractures should be familiar to the reader when observed in aluminum foil, chocolate wrappers, and so on. They occur because the material is thin. Since aircraft skins and pipelines are also thin, this sort of fracture can be dangerous.

Although produced by slip, and thus locally ductile, out-of-plane tear fractures are comparatively low-energy fractures. Because they are not accompanied by generalized yielding throughout the sheet, such fractures should really be classified as brittle.

**Figure 14-25**   Out-of-plane tearing in thin sheets at stresses below the yield stress, caused by slip-off through the thickness of the sheet.

**Example 14-11**

Explain in terms of dislocation theory why out-of-plane tearing fractures in thin sheets are possible at stresses below the generalized yield stress. Why are such fractures not possible in thick sheets?

**Solution**   The tearing fracture in thin sheets occurs because, literally, dislocations go through the (thin) thickness of the sheet out to the other side, and slip-off results. The situation is analogous to a screw dislocation: The original set of dislocations, sufficient to produce complete slide-off at the start of the fracture, remains sufficient to produce slide-off all the way along merely by gliding forward. In thick sheets, however, incremental growth of the crack requires more edge dislocations to be injected and activated in the slip lines. The yield zones must extend over bigger and bigger proportions of the cross section for the crack to advance. This is much more difficult to achieve than slip-off. It is important to realize that tearing fractures succeed because the *plastic* displacement at the tip of the crack is accommodated by *elastic* displacements elsewhere. There is no need for plastic zones to span the entire cross section (see Fig. 14-25).

## REFERENCES

1. ASTM. *1991 Annual Book of ASTM Standards,* American Society for Testing and Materials, Philadelphia, Pa. See No. E399, "Standard Test Method for Plane-Strain Fracture Toughness of Metallic Materials."

2. Atkins, A. G., and Y.-W. Mai, *Elastic and Plastic Fracture.* Chichester, U.K.: Ellis Horwood, 1985; New York: Simon & Schuster, 1988.

3. Gordon, J. E., *The New Science of Strong Materials—Or Why You Don't Fall Through the Floor.* London: Penguin Books, 1968.

4. Gurney, C., et al., various papers in the *Proceedings of the Royal Society (London)*, e.g., 1974, vol. 340A, p. 213, which gives other references.

5. Kelly, A., *Strong Solids*. Oxford: Clarendon Press, 1964.

6. Knott, J. F., *Fundamentals of Fracture Mechanics*. London: Butterworths, 1973; New York: John Wiley.

7. Lawn, B. R., and T. R. Wilshaw, *Fracture of Brittle Solids*. Cambridge: Cambridge University Press, 1975.

8. Lawn, B. R., *Fracture of Brittle Solids*, 2nd ed. Cambridge: Cambridge University Press, 1993.

9. Polakowski, N. H., and E. J. Ripling, *Strength and Structure of Engineering Materials*. Englewood Cliffs, N.J.: Prentice Hall, 1966.

10. Tada, H., P. Paris, and G. Irwin, *The Stress Analysis of Cracks Handbook*. Hellertown, Pa.: Del Research, 1973.

## PROBLEMS

**14-1.** Give examples of materials with the following combinations of properties:
  (a) Stiff and weak.
  (b) Stiff and strong.
  (c) Flexible and strong.
  (d) Flexible and weak.

**14-2.** When a glazier scratches glass with diamond prior to cutting, is it better to cut a deep rather than a shallow groove?

**14-3.** The material of the stressed element shown in Fig. 14-26 fails by cleavage at 190 MPa and by shear at 147 MPa. Determine the fracture mode and calculate the value of $S$ when fracture occurs.

**Figure 14-26**

**14-4.** The stress concentration at the tip of a Griffith crack is approximately $\sqrt{2L/\rho}$, where $L$ is the length of the crack and $\rho$ is the crack tip radius. If fracture occurs when the "concentrated stress" is equal to the theoretical strength (i.e., $\sigma_{\text{avg}}\sqrt{2L/\rho} = \sqrt{E\gamma_s/a_0}$), we have $\sigma_{\text{avg}}$ at fracture equal to $\sqrt{E\gamma_s\rho/2La_0}$. This suggests that sharply cracked bodies (where $\rho \to 0$) have *no strength* at all! This obviously is not true. Explain.

**14-5.** (a) Tough solids (e.g., hard steel) retain considerable strength when scratched; other solids (glass and stone) do not. Explain.
  (b) Why do drinking glasses that have chips and scratches in them not break?

**14-6.** Explain why rubber is notch insensitive.

**14-7.** How can a glass cutter further weaken glass after scratching but before the act of breaking?

**14-8.** Can Griffith cracks be seen under an optical microscope?

**14-9.** What are the peculiar colored patterns that sometimes are seen on toughened windshields of cars?

**14-10.** Explain the ideas behind toughened glass.

**14-11.** Some materials that fail with no plastic extension when tested in tension are observed to exhibit some ductility when tested in compression or in pure shear. Why?

**14-12.** Coarse grain sizes in zinc and pure iron promote brittleness. Explain.

**14-13.** "Hardening without refinement of microstructure generally raises the transition temperature and promotes brittleness." Discuss this statement, illustrating your answer with reference to tempered martensite, blue brittleness, temper brittleness, hot shortness, and strain-age embrittlement.

**14-14.** An asymmetrical double-cantilever beam test piece has arms of unequal height $h_1$ and $h_2$ in contrast to the symmetrical specimen in Fig. 14-11(a). If the cut sheet has thickness $t$, determine the expression for crack propagation by equal and opposite end loads $P$.

**14-15.** Calculate the fracture toughness (in J/m$^2$) of a plate 20 mm thick with the load-extension path $ABCD$ shown in Fig. 14-27. The crack is 3.4 mm long at $A$, 3.4 mm at $B$, 8.6 mm at $C$, and 13.8 mm at $D$.

 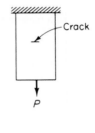

**Figure 14-27**

**14-16.** Calculate the fracture toughness of a flat plate specimen, 5 mm thick, of a polymer that yields the load–deformation data as follows:

|   | Load, N | Crack Length, mm | Deflection at Load, $\mu$m |
|---|---------|------------------|----------------------------|
| 0 | 0       | 175.0            | 0                          |
| 1 | 623     | 187.2            | 523.0                      |
| 2 | 514     | 198.3            | 562.6                      |

**14-17.** In a fracture-toughness test, the load and corresponding crosshead displacement at which cracking commenced were $P_1$ and $u_1$, respectively. After the load had fallen to $P_2$ (with displacement $u_2$), by which time the crack had propagated some distance, the test piece was unloaded. The unloading line went back to the origin. Prove that the fracture toughness $R$ is given by

$$R = \frac{\frac{1}{2}(P_1 u_2 - P_2 u_1)}{A_2 - A_1}$$

where $A_1$ is the initial crack area before cracking and $A_2$ is the final crack area after unloading (Fig. 14-28). Assume that the $R$ locus joining $P_1 u_1$ and $P_2 u_2$ is straight.

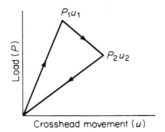

**Figure 14-28**

**14-18.** A beam of material with initial height 76.4 mm is being torn apart by equal and parallel forces $P$, as shown in Fig. 14-29. The deflection $\delta$ is measured from a fixed point as shown. From the data plotted in Fig. 14-29, calculate the fracture toughness of the solid. (Crack length $a$ is measured along the crack.)

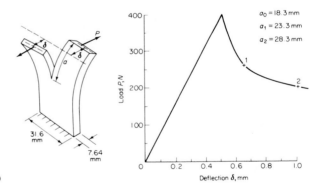

**Figure 14-29**

**14-19.** The following data for crack area, compliance, and cracking load apply to a certain structure. Determine $R$.

| Crack Area, mm$^2$ | Compliance, nm/N | Cracking Force, kN |
|---|---|---|
| 1300 | 73 | 24 |
| 1940 | 206 | 15 |
| 2580 | 447 | 12 |
| 3226 | 810 | 10 |

**14-20.** An elastic rod is embedded in a massive block as shown in Fig. 14-30. (This is reminiscent of King Arthur's sword called Excalibur.) Determine the load that will cause cracking along the interface between the rod and the block. Under what combination of properties and geometry might the rod itself break?

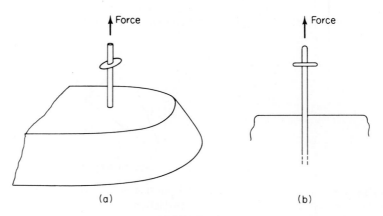

Force                                          Force

(a)                                            (b)

**Figure 14-30**

**14-21.** Select materials from App. 9 that would display notch brittleness in a comparison between plain and deeply notched round tensile bars 15 mm in diameter.

**14-22.** To overcome a failure problem, a designer decides to use thicker plates of a higher-strength steel as replacements for the broken components. Is this a good solution?

**14-23.** Rubbish often goes to grain boundaries when a metal crystallizes. What effect does this segregation have on strength properties?

**14-24.** Thin sheets are tougher than thick sheets (the tearing mode of failure being excluded). Explain.

**14-25.** Why is it difficult in principle to make strong materials?

**14-26.** Describe the likely modifications necessary in the Griffith theory to explain the mechanism of vacancy creep, (discussed in Chap. 12).

**14-27.** A V-notch Charpy specimen, shown in Fig. 14-31, has the following composition: chromium 0.95%, manganese 0.85%, carbon 0.40%, silicon 0.30%, molybdenum 0.20%, balance iron. If this specimen is heated to 850°C for 1.8 ks, oil-quenched to 20°C, reheated to 450°C for 1.8 ks, and water-quenched to 20°C, estimate the following and describe the sources of your information:

**(a)** Brinell hardness

**(b)** Fracture toughness (in $J/m^2$)

10 mm

8 mm

55 mm

10 mm

**Figure 14-31**

**14-28.** A test of an edge-cracked steel specimen 8.64 mm thick, shown in Fig. 14-32(a), gives the load–displacement curve shown when the crack grows from its initial length of 2.69 mm. In a large plate of the same material and thickness, shown in Fig. 14-32(b), calculate the length of the internal crack at which complete fracture will occur when the applied stress $\sigma = 193$ MPa.

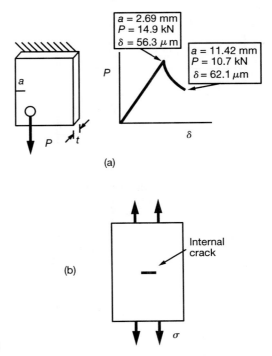

(a)

(b)

**Figure 14-32**

**14-29.** A hollow cylinder is subjected to forces as shown in Fig. 14-33(a), which cause a short, helical crack of initial length $a = 3.0$ mm to grow in a stable manner. The relationship between load $P$ and the angular displacement $\alpha$ is shown in Fig. 14-33(b). At point $B$, the crack length is 7.5 mm. Calculate the fracture toughness of this material.

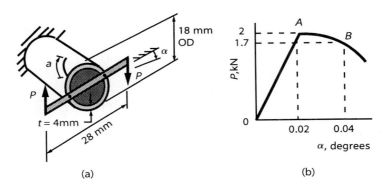

(a)

(b)

**Figure 14-33**

**14-30.** A specimen of metal Y that is 14.2 mm thick is given an ASTM standard test for $K_{Ic}$. The initial crack length is 0.525 times the specimen width. The yield strength of metal Y is 572 MPa. The load–displacement plot is shown in Fig. 14-34. Estimate $K_{Ic}$ for this metal [in MPa(m)$^{1/2}$].

Figure 14-34

**14-31.** An edge-cracked Gurney test of a specimen of metal Z that is 12.0 mm thick produced the force–displacement curve shown in Fig. 14-35(a). The edge crack was 14.3 mm long at $a$ and

(a)

Fracture Force

(b)

**Figure 14-35**

18.4 mm long at *b*. Metal Z has a yield strength of 240 MPa and an elastic modulus of 140 GPa. A very large plate of metal Z, of thickness 12.0 mm and width 2.37 m, contains a central transverse crack of length 27.2 mm [Fig. 14-35(b)]. Calculate the tensile fracture *force* (in N) that would lead to sudden fracture of this plate.

**14-32.** A standard three-point bending ASTM $K_{Ic}$ test is performed on a specimen that has the following dimensions: thickness = 4.00 mm; height = 8.00 mm; length between supports = 32.0 mm; initial crack length = 3.88 mm. The material has a yield strength of 273 MPa and an elastic modulus of 47.2 GPa, and the load-displacement curve is shown in Fig. 14-36.

(a) Calculate the critical plane strain stress-intensity factor that is obtained from these data.

(b) Is this test valid?

(c) Show why or why not.

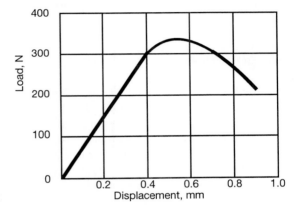

**Figure 14-36**

# 15

# *Cast Irons*

## INTRODUCTION

*Pig iron* is the first crude stage in the manufacture of iron–carbon metal alloys from iron ore. *Cast irons* are somewhat purified forms of the crude iron tapped from the blast furnace; as the name implies, they are used for making castings. In general, the alloys known as cast irons contain greater amounts of carbon than typical steels, 2 to 4% rather than less than 1%. Inspection of the iron–carbon equilibrium diagram (Fig. 8-6) shows a eutectic at 4.3% C, which has the lowest melting point (1148°C) of Fe–C alloys in that range of carbon contents; steels have melting points of about 1500°C. The iron age in prehistory began when humans were able for the first time to generate appropriate temperatures in fires to melt iron ores. Steels did not come into the picture until still higher temperatures could be generated.

Iron ores have various other elements in them, so cast irons contain significant amounts of silicon, phosphorus, sulfur, and manganese. Such additional elements modify the simple equilibrium-phase diagram in which the boundaries and temperatures are altered; thus, it is possible to predict cast-iron microstructures from Fig. 8-6 *only* for a restricted group of "pure" irons containing few if any additional elements. Furthermore, not only do impurities alter the phase diagram, but silicon in particular also encourages the formation of *free graphite*. In steels and other cast irons, carbon is combined with iron as *cementite* ($Fe_3C$), which is the stable form (either free or in pearlite) in plain-carbon steels below the eutectoid transition temperature (we ignore the minute amounts of carbon dissolved in ferritic $\alpha$-iron below the eutectoid transition temperature). In cast irons containing silicon, free graphite becomes the characteristic stable form of carbon.

Free graphite makes the iron darkish and dirty to handle; irons containing no free graphite, on the other hand, have much cementite in their microstructure, which gives a white appearance to a fracture surface. Cast irons are therefore classified broadly as *gray* or *white*.

Cast irons as made are comparatively brittle, with little reduction in area to fracture in the tensile test. Nevertheless, they can have adequate mechanical properties for some

**350**

applications, and there are plenty of uses for this cheap material (as will be discussed later); as an engineering material cast iron is second only to steel in annual tonnage produced. The extreme hardness of "chilled" cast iron was used to advantage in 1803 by A. B. Ransome in Ipswich, England, in the manufacture of plowshares, which up to that time had been made from *wrought iron*, which quickly became blunt; tremendous improvements in the speed of arable cultivation were thereby obtained. It was possible at that time to improve the ductility of white cast iron by special heat treatment, extra toughness in such *malleable* cast iron being obtained at the expense of strength. Indeed, before the advent in the 1850s of Bessemer's method of making steel on a tonnage basis, malleable cast iron and also wrought iron were the only strong ductile structural materials available. Small quantities of steel were being made at that time by the cementation and crucible processes, which were means of increasing the carbon content of wrought iron. These methods were known by the ancients, lost in medieval times, and rediscovered in the seventeenth century. Many steel artifacts from ancient times consisted of partially case-hardened wrought iron; *Damascus steel* was made by piling pieces of soft iron on pieces of high-carbon iron followed by repeated heating and hammering, which made the carbon distribution somewhat more uniform.

Nowadays there are many types of cast irons, including those alloyed with certain elements to produce particular characteristics such as heat, corrosion, and wear resistance. No other metal suitable for casting into intricate shapes possesses such a wide range of properties as cast iron.

## WHITE CAST IRONS

The microstructure of *white* irons, which have few impurities, can be interpreted more or less directly from the Fe–C equilibrium diagram (Fig. 8-6). At just below 1148°C there are two phases, $\gamma$ austenite and $\theta$ cementite, $\gamma$ having 2.11% C. For carbon contents greater than 2.11% and less than 4.30%, the microstructure is noneutectic $\gamma$ and $(\gamma + \theta)$ eutectic; for carbon contents greater than 4.30% we have $(\gamma + \theta)$ eutectic plus noneutectic $\theta$. The $(\gamma + \theta)$ eutectic is sometimes called *ledeburite* and, of course, exists only between 1148° and 727°C. As the temperature drops, $\gamma$ has a decreasing solubility for carbon (the line joining 2.11% back to 0.77%), so $\theta$ comes out of the $\gamma$, both from the free $\gamma$ and from the eutectic. Thus, the ledeburite eutectic changes composition and the free $\gamma$ throws out $\theta$ precipitate into the grain boundaries (compare the Cu–Al system, Chap. 9). Therefore, just above 727°C the phases are still $\gamma$ and $\theta$, the $\gamma$ now having 0.77% C, but the microstructure has changed from that at first solidification. For *hypoeutectic* irons (those with less than 4.30% C), the microstructure then consists of free austenite, ledeburite, and grain-boundary cementite. For *hypereutectic* irons (more than 4.30% C), we have the ledeburite eutectic plus the original free ("massive") cementite. The $(\gamma + \theta)$ eutectic at 728°C is different from that at 1147°C because of the composition of the $\gamma$. As the temperature is reduced below 727°C, all the austenite of 0.77% C becomes pearlite.

Therefore, an "ideal" hypoeutectic pure white iron would have a matrix at room temperature of pearlite and cementite (Fig. 15-1). The pearlite comes from both the free

**Figure 15-1** Hypoeutectic white cast iron, etched in 4% picral solution, showing pearlite and cementite. (Courtesy of BCIRA.)

austenite and the austenite in the $(\gamma + \theta)$ ledeburite eutectic; the cementite comes from both the $(\gamma + \theta)$ eutectic and the $\theta$ precipitate. Different pearlite spacings can be obtained by varying the cooling rate, as with steel. Martensitic structures can also be obtained by quenching.

**Example 15-1**

A white cast-iron casting of mass 15 kg is poured from a melt containing 2.7% carbon by mass. Determine the mass of eutectoid iron carbide present at room temperature.

**Solution**   From Fig. 8-6, at just less than 1148°C, the mass of noneutectic $\gamma$ is $(4.3-2.7)(15)/(4.3 - 2.11) = 10.96$ kg. At just above 727°C, $(2.11- 0.77)/(6.69-0.77) = 22.6\%$ of the $\gamma$ appears as $\theta_{ppt}$, leaving 77.4% or 8.48 kg as free $\gamma$, all of which transforms to eutectoid pearlite. Pearlite consists of $0.77/6.69 = 11.51\%$ of $\theta$. Thus, the mass of eutectoid iron carbide present at room temperature is $(0.1151)(8.48) = 0.98$ kg.

Although the foregoing description implies an equilibrium (slow) cooling rate, it would also be true only for a pure iron–carbon alloy. Even small amounts of silicon promote the dissociation of cementite $\theta$ into $\gamma$ plus graphite (above 727°C), or $\alpha$ plus graphite (below 727°C). Thus, to obtain a white-iron microstructure, the cooling rate in fact must be faster than one would think: obviously not fast enough to produce a martensite matrix but fast enough to stop any silicon from having time to catalyze the cementite dissociation to iron and graphite. Hence, the term "chilled castings" is often used, and the section size of castings is important: Because of differential cooling rates, large castings generally cannot be guaranteed to have white iron throughout, as the cooling rate in the center would be too slow.

White irons are extremely hard, with little reduction in area to fracture under tension. Depending on the pearlite spacing, the tensile strength lies between perhaps 210 and 560 MPa. The major use for white irons (which are very difficult to machine) is in wear- and abrasion-resistant applications (railroad brake shoes, pulverizing plants, shot-blasting nozzles). By appropriate cooling, a white-iron microstructure is sometimes produced on a softer (but perhaps tougher) core of gray or "mottled" iron for wear-resistant applications.

## MALLEABLE IRON

The ductility of white cast irons can be improved considerably by a reheating process whereby the iron becomes malleable. At temperatures somewhere between 800° and 900°C, the room-temperature pearlite and cementite become a $\gamma$ plus $\theta$ microstructure. If castings are left for several days in this temperature range, the $\theta$ breaks down into $\gamma$ and free graphite (G). The graphite comes out as bumpy clusters (not in smooth, round blobs, as in nodular iron described later in this chapter, nor in the sharp-ended flakes characteristic of gray irons).* On cooling the castings to room temperature, the austenite either (1) transforms on air cooling or mild quenching simply to pearlite, as we would expect, giving a *pearlitic malleable* iron with the graphite clusters sitting in a pearlite matrix, or (2) transforms on much slower furnace cooling to ferrite and additional graphite, giving *ferritic malleable* iron (Fig. 15-2).

Air cooling is too quick to allow the $\theta$ in the pearlite (which comes from $\gamma$ below 727°C) to transform, but furnace cooling is slow enough to allow this $\theta$ to dissociate (especially in the presence of some silicon), that is, P = $(\alpha + \theta) \rightarrow \alpha + (\alpha + G)$. These different microstructures are also influenced by whether an oxidizing or neutral atmosphere is used in the reheating furnace.

The fracture surface of pearlitic malleable iron tends to be shiny in comparison with ferritic malleable iron (which has more temper carbon for the same total carbon content).

**Figure 15-2** Ferritic (blackheart) malleable cast iron, etched in 5% nital solution, showing ferrite and bumpy clusters of graphite ("temper carbon"). (Courtesy of BCIRA.)

* The clusters are sometimes called *temper carbon*, as they moderate the brittleness of untreated white iron (compare with the meaning of tempered martensites, Chap. 10).

Hence, the terms *whiteheart* and *blackheart* malleable irons are applied to the pearlitic and ferritic forms.

Depending on the matrix, the tensile strength of malleable iron can be 400 to 700 MPa with 10% reduction in area to fracture. The improved ductility is attributable to the absence of cementite in the microstructure and to the fact that the round shape of the free graphite "holes" in the microstructure is relatively harmless, at least in comparison with the sharp-ended flake geometry of gray cast iron. Malleable irons are used where toughness and ductility are required for parts cast to shape, for example, in automotive gear cases, suspensions, rocker arms, wheels and axles, chain links, hand tools, trailer hitches, and boiler and manhole covers.

The malleabilizing process can be long and tedious; sometimes a week is required to heat, hold, and cool castings. Nevertheless, the effort (including the chilling precautions necessary to obtain a white-iron casting in the first place) was worth it — especially in the old days — because along with wrought iron, malleable cast iron was one of the few ductile predecessors of steel.

## GRAY CAST IRONS

Gray irons have more carbon and silicon than white irons and are associated with relatively slow cooling rates. Strictly speaking, their microstructures are not predictable from the equilibrium diagram because the large amount of silicon makes the cementite dissociate easily. Moreover, the percentage of carbon in the $\gamma$ at the eutectic and eutectoid transformations is reduced by the large amount of silicon: For example, 2% Si reduces the 2.11% maximum solubility of C in $\gamma$ to 1.5% and reduces the pearlite eutectoid carbon content to 0.7% or thereabouts instead of 0.77%. The percentage of carbon in the ledeburite eutectic is also reduced from 4.3% to 3.7% for 2% Si, and the transformations take place over a range of temperatures even under equilibrium conditions.

In as-cast gray irons cooling from the melt, the cementite that forms between 1148 and 727°C (i.e., the $\theta$ in the ledeburite eutectic and the free $\theta$ precipitated as a result of the change in solubility of C in $\gamma$ iron) is encouraged by silicon to decompose into iron (i.e., $\gamma$) and graphite. Of crucial importance, because of the effect on subsequent mechanical properties, is the fact that the graphite comes out as *sharp-ended flakes*. On cooling below 727°C, the $\gamma$ becomes pearlite, but the usual cooling rates in casting are not slow enough to allow the $\theta$ in this pearlite to dissociate and give yet more graphite, so at room temperature *pearlitic gray* cast iron results.

With much slower cooling (or by reheating pearlitic gray iron back into the $\gamma$ region followed by furnace cooling) the $\theta$ in the pearlite can also break down so that at room temperature only ferrite $\alpha$ and graphite flakes are seen, because pearlite = $(\alpha + \theta) \rightarrow \alpha + (\alpha + G)$, plus all the graphite already there from the $\theta$ broken down above 727°C — hence the term *ferritic gray* cast iron.

The graphite in both these cases, on cooling from the melt, always appears as flakes that look like black worms under a microscope (Fig. 15-3). The flakes inhibit deformation of the normally ductile ferrite and produce a very brittle material. They are essentially sharp cracks, as the graphite with which they are filled has negligible strength. The size and distribution of these flakes determine the particular strengths of gray irons, but in general terms they have practically zero reduction in area to fracture with tensile strengths ranging between 140 and 420 MPa, depending on the matrix. Gray cast irons are much stronger in compression than tension, the harmful cracks being closed up. Because of the wide range of flake sizes in different grades of gray iron and the consequent effect on the stiffness of the material, the Young's modulus can vary, in extreme

**Figure 15-3**   Flake gray cast iron, etched in 4% picral solution. (Courtesy of BCIRA.)

cases, between about 70 and 150 GPa, the smaller flake sizes of the stronger irons resulting in the higher moduli.

**Example 15-2**

Two tensile specimens of cast iron, both of initial diameter 12.8 mm, are tested to fracture. Specimen A has a maximum load of 58.0 kN and a fracture diameter of 12.4 mm. Specimen B haa a maximum load of 18.6 kN and a fracture diameter of 12.8 mm. If these are pearlitic malleable iron and ferritic gray iron, respectively, describe any differences in (a) phases present, (b) shape of phases, and (c) spacing of phases that led to the measured differences.

**Solution**   (a) The phases present in specimen A are ferrite $\alpha$ and cementite $\theta$ (combined as pearlite) and free cluster graphite; in specimen B the phases are ferrite and free flake graphite. (b) The principal difference in the shape of the phases is in the type of graphite. (c) There is likely to be no difference in the spacing of the phases. The measured differences in mechanical properties arise from the different characteristics of the free graphite, namely, sharp-ended flakes in the case of the gray iron, which promote easy fracture, and bumpy clusters of graphite in the malleable iron, which are less deleterious and permit the pearlite matrix to display some of its ductility before the weakening effect of the graphite "holes" is felt.

The strengthening mechanisms that operate in the $\alpha$ and ($\alpha + \theta$) matrices in irons are the same as those in steels, discussed in earlier chapters.

**Example 15-3**

Show that the relative strengths of different pearlitic gray cast irons, namely, 140 MPa for ASTM-20 with large flake size (flakes about 250 $\mu$m long) and 420 MPa for ASTM-60 with small flake size (about 30 $\mu$m long), follow the rules of brittle fracture (Chap. 14).

**Solution**   The mechanics of fracture says that the strengths are inversely proportional to the square root of the flake size, from which

$$\frac{\sigma_{\text{ASTM-20}}}{\sigma_{\text{ASTM-60}}} = \sqrt{\frac{3}{25}} = 0.35$$

The given strength ratio is 140/420 = 0.33, so the data seem consistent.

The relatively low strength and low ductility of gray irons make them exceedingly easy to machine. Applications include automobile engine blocks, flywheels, brake drums, piston rings, machine bases, and gears.

## DUCTILE, NODULAR, OR SPHEROIDAL GRAY IRONS

The ductility of gray irons can be improved by adding *magnesium* (particularly) and/or *cerium* to the melt, which in gray irons containing only small amounts of sulfur makes the graphite come out as round blobs (not the flakes of untreated gray irons, nor the clusters of malleable white irons) directly from the melt. There is no reheating as with malleable

**Figure 15-4**   Pearlitic nodular cast iron, etched in 4% picral solution. (Courtesy of BCIRA.)

white iron. The carbon content of such irons (variously called by all the names given in the heading above) is usually kept above 3% to ensure good castability.

With different cooling rates, *pearlitic* (Fig. 15-4) or *ferritic nodular* irons may be obtained. The absence of the sharp-ended flake form of graphite, with its stress-raising problems, improves the strength and ductility of gray irons immensely. The reduction in area to fracture in tension can be as great as 25%, and the tensile strength can range between 420 and 800 MPa. Precise values for both depend on the matrix and on any subsequent heat treatments such as quenching, tempering, annealing, or surface hardening. Typical applications are for crankshafts, heavy-duty gears, and automobile door hinges.

Magnesium is above its boiling point (1110°C) at the temperatures of liquid cast iron, so certain precautions must be taken when adding this nodule-forming element; for example, it may be introduced as a nickel–17% magnesium alloy or in a pressure-tight ladle. (Similar problems occur in making brass from its elements because zinc boils at 907°C, whereas copper does not melt until 1088°C.)

Nodular iron is a much newer material than malleable cast iron, having been developed during World War II. It is replacing malleable (white) iron for many applications, because it is produced from the melt and does not require lengthy reheating procedures.

**Example 15-4**

Draw a schematic microstructure for each of the following, clearly labeling every phase: (a) pearlitic malleable cast iron, (b) ferritic nodular cast iron, (c) white cast iron heated to 950°C for 24 h and water-quenched.

**Solution**   Figure 15-5 illustrates the phases and microstructures.

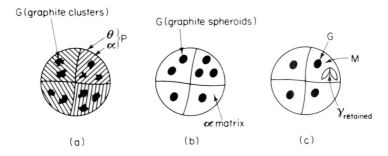

**Figure 15-5**   Solution to Example 15-4.

## INOCULATED AND ALLOY IRONS

Irons that would otherwise be white irons of low silicon and carbon content are given additions of graphitizer inoculants in the melt to cause graphite to appear in the casting in a finely dispersed flake form rather than as clusters, with a consequent increase in strength (say, to 500 MPa). For example, calcium silicide is used in the *Meehanite* range of heavy-duty irons.

Alloying elements, such as nickel, chromium, copper, and molybdenum, can be added to irons for special purposes. Nickel tends to produce a gray iron (see next section),

**Figure 15-6** "Nomag" austenitic flake cast iron, etched in 4% picral solution. (Courtesy of BCIRA.)

acting like silicon in this regard. Large amounts of nickel (up to 25%) affect the Fe–C equilibrium diagram so much that the $\gamma$ phase can exist at room temperature, giving *austenitic irons* (Fig. 15-6) which are nonmagnetic and have high electric resistance and a high coefficient of thermal expansion.

## COMPARISON AMONG DIFFERENT CAST IRONS

For engineering purposes, different types of cast irons are not normally defined in terms of chemical composition. Rather, they are specified according to mechanical properties and any other special properties required for particular purposes, with the chemistry being left to the foundry. This comes about because of the widely different compositions of the starting iron ores. Impurities other than silicon, such as sulfur, manganese, and phosphorus, can affect foundry practice, and the final properties vary markedly. Deliberate additions of alloying elements likewise can be made to produce special cast irons. A range of heat treatments can be given to cast irons to produce matrix structures varying from very soft to very hard; for example, they can be reheated into the $\gamma$ region and then heat-treated as steels are to give *martensitic* or *tempered martensitic* structures. Appendix 10 gives an indication of the range of mechanical properties usually encountered.

Figure 15-7 shows the effect of silicon in promoting the dissociation of cementite into free graphite, and thus the transition between white and gray iron microstructures. *Mottled irons* are white/gray borderline cases, and the regions of the diagram labeled "Mixture" correspond to a mixed pearlitic or ferritic microstructure. Cooling rate affects Fig. 15-7 in that as the percentage of silicon goes up, faster cooling rates are required to preserve a white-iron microstructure, and at high enough silicon contents that is not always possible.

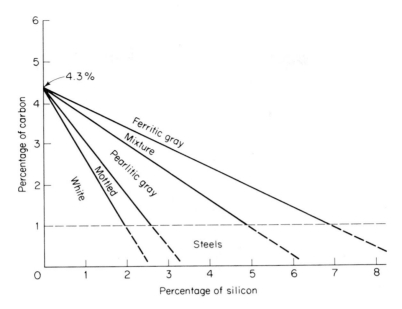

**Figure 15-7**   The effect of silicon in promoting the dissociation of cementite into free graphite. Mottled irons are borderline cases between white and gray iron.

The inherent virtues, such as good machinability, corrosion resistance, and wear resistance, possessed by gray cast iron are shared by both malleable and nodular irons. In addition, the latter have reasonable ductility, and either type can be substituted for steels in many applications. Service requirements and economics will dictate the final choice. Gray iron is lowest in cost and also the easiest to cast and obtain a sound, essentially nonporous structure. Nodular iron is similar but is subject to more shrinkage on solidification and hence requires special care in casting. Malleable iron is the most difficult to cast (as white iron), and the size of castings is more restricted (smaller) than for nodular iron. If nodular iron does not require an anneal to ensure uniformity of properties and can thus be used directly on cooling from the melt, it is much cheaper than malleable iron.

**Exercise 15-5**

Describe the influence of phosphorus in cast irons.

**Solution**   In irons made from common ores containing phosphorus, iron phosphide is probably the most important inclusion. The presence of phosphorus can offer some benefits because during casting it increases the fluidity of the iron by lowering its melting point; this feature made possible the intricate casting found in Victorian ironwork, allowing fine detail and smooth finishes. However, it tends to separate during solidification and leave a liquid network, which can subsequently freeze with markedly different contraction from the parent phases. This results in weak, porous regions, particularly in sections greater than, say, 10 mm across. In heavier-duty irons, the phosphide must be eliminated, leaving graphite as the factor controlling strength.

## WROUGHT IRON: ARMCO IRON

Wrought iron, as developed in the eighteenth and nineteenth centuries, was the result of "burning out" the excess carbon of pig iron in an attempt to produce pure iron which would be comparatively ductile. Pig iron was remelted in a puddling furnace lined with a bed of iron oxide. It took about 2 h of extremely arduous and skilled work on the part of the puddler, constantly stirring the iron with a rake or *rabble* introduced through a hole in the furnace door, before it was converted to a glowing plastic ball of wrought iron. (As the carbon content decreased, the melting point increased, and the furnace was incapable of reaching the higher temperatures.) The spongy mass of iron and slag had to be pounded or *shingled* under a heavy forging hammer to expel the cinder and to form it into a bloom suitable for conversion into bars in the rolling mill. Even so, bars rolled directly from the bloom in this way were of such poor quality as to be useless. To produce good-quality iron it was necessary to cut the bars into short lengths, bind them in bundles, forge them under the hammer, and roll them again at least once.

Rolling elongates the slag into fibers, and because they have comparatively weak interfaces, cracks become diverted along the slag fibers, producing a characteristic "woody" or fibrous fracture (Fig. 15-8). Wrought iron typically displays 25 to 35% elongation on a 50-mm gage length, with a yield strength of 200 MPa and tensile strength of 320 MPa; excessive phosphorus from the ore reduces these properties. Wrought iron has comparatively good shock resistance because of the fibrous fracture, and although rarely used nowadays, it was employed for chains, railway couplings, and crane hooks long after mild steel had become readily available. In fact, in the Aston process of manufacturing wrought iron, molten iron (already "purer" than wrought iron) from a Bessemer converter was poured into a synthetic slag to form pasty balls that could be rolled into bars.

The many thousands of miles of wrought-iron rails that were laid during the great age of railroad construction, and all the wrought-iron plates used for building bridges and ships' hulls and in boiler making in early Victorian times, were all produced by the extremely laborious process of puddling and repeated rolling. By the 1850s, puddling had become a serious bottleneck, limiting production of the basic material on which the progress of the industrial revolution depended. This led to J. Nasmyth's idea of substituting a jet of steam in place of the rabbling bar, to increase the temperature of the furnace, and was the incentive for the introduction of Kelly's, Bessemer's, and Siemens' more modern methods of making tonnage steel.

Ingot (*Armco*) iron is almost pure ferrite (say, 0.03% C in Fe). It is made by fully oxidizing metal in a basic open-hearth furnace, which is capable of coping with the melting point of pure iron (1538°C). It is thus different from wrought iron in that the slag is removed in the molten state and there are no slag fibers in the ferrite matrix. The mechanical properties are lower than those of wrought iron; for example, the yield strength is 150 MPa and the tensile strength is 280 MPa. The Brinell hardness of Armco iron is about 80 HB; note that $S_u = (3.45 \times HB)$ MPa here but that this relationship does not hold for cast irons in general.

Even purer iron can be made by electrolytic processes. The principal applications for these pure irons, such as in electric transformer cores, use their high magnetic properties.

**Figure 15-8**   Fibrous, "woody" fracture of wrought-iron nail.

## REFERENCES

1. Brick, R. M., R. B. Gordon, and A. Phillips, *Structure and Properties of Alloys*. New York: McGraw-Hill, 1965.

2. Rollason, E. C., *Metallurgy for Engineers*. London: Edward Arnold, 1956.

3. Rolt, L. T. C., *Victorian Engineering*. London: Penguin, 1970.

## PROBLEMS

**15-1.** Name the phases present in the following and describe the shape of graphite whenever it is present (all held for 2 ks):

    **(a)** Ferritic gray iron heated to 650°C.

    **(b)** Pearlitic nodular iron heated to 950°C.

    **(c)** White cast iron of 2.67% carbon at 20°C.

    **(d)** Pearlitic malleable iron at 20°C.

**15-2.** Name the phases present in the following and describe the shape of graphite whenever it is present (all held for 2 ks):

    **(a)** White cast iron of 2.37% carbon heated to 1047°C.

    **(b)** Ferritic nodular iron at 20°C.

(c) Pearlitic malleable iron heated to 950°C.

(d) Pearlitic gray iron at 200°C.

(e) 10100 steel water-quenched from 810°C to 20°C.

**15-3.** Sketch and label the shape(s) of all nonmetallic phases (including all intermetallic compounds) present for each of the following:

(a) Pearlitic gray cast iron.

(b) Ferritic malleable cast iron.

(c) Pearlitic nodular iron.

**15-4.** Give reasons, in brief, why each of the three alloys in Prob. 15-3 is manufactured.

**15-5.** A white cast-iron component with 2.40% carbon has a mass of 30 kg. Calculate the mass of noneutectoid iron carbide ($\theta$, $Fe_3C$) present at room temperature.

**15-6.** A white cast-iron specimen of mass 1.5 kg contains 2.2% C. At room temperature find

(a) The mass of eutectic iron carbide.

(b) The mass of eutectoid ferrite.

(c) The mass of noneutectic, noneutectoid iron carbide.

**15-7.** Calculate the mass of eutectic, eutectoid iron carbide present at 20°C in a 1.73-kg casting of white cast iron of 3.14% carbon.

**15-8.** (a) Calculate the mass of noneutectic eutectoid present in 7.23 kg of white cast iron of 2.67% carbon at 20°C.

(b) Calculate the mass of iron present in the total eutectoid in 7.23 kg of white cast iron of 2.67% carbon at 20°C.

**15-9.** Calculate the total mass of graphite at 20°C in a 100-kg mass of pearlitic malleable iron of 3.76% carbon.

**15-10.** Calculate the total mass of carbon in the noneutectic, noneutectoid iron carbide in a 7.38-kg bar of white cast iron of 2.69% carbon content.

**15-11.** (a) Calculate the mass of eutectic, noneutectoid iron carbide at 21°C in a white cast-iron casting of mass 1.3 kg and composition 3.4% C, 0.70% Si, 0.60% Mn, balance Fe.

(b) Calculate the mass of eutectoid, noneutectic ferrite at 21°C.

**15-12.** A ferritic ductile iron has a tensile strength of 400 MPa, whereas a pearlitic malleable iron has a strength of 670 MPa. Fully describe the heat treatments of each, show their final microstructures, label all phases, and explain why one is stronger. Which one would you expect to have the higher true strain at fracture? Why?

**15-13.** Consider the following specimens: Specimen G contains 3.5% C, 1.8% Si, 0.5% Ce, balance Fe; specimen H contains 2.1% C, 0.7% Si, balance Fe; specimen J contains 3.1% C, 1.9% Si, balance Fe. If all are air-cooled from the melt in the form of bars 10-mm in diameter, which has the greatest hardness?

**15-14.** The cast irons in two tensile tests were of identical composition. Cast iron A was an as-cast test piece 12.8 mm in diameter. Cast iron B was the same size and received the same treatment except that after casting it was reheated to 950°C for 3 ks and furnace-cooled.

| | $d_0$, mm | $P_{max}$, kN | $d_f$, mm |
|---|---|---|---|
| Cast iron A | 12.8 | 24.0 | 12.8 |
| Cast iron B | 12.8 | 17.3 | 12.8 |

(a) Explain, from a fundamental point of view, why the strengths are different.
(b) Name the phases present in each cast iron.
(c) Name the cast irons.

15-15. Describe the critical differences that would cause a designer to specify a pearlitic malleable cast iron for one application and a ferritic nodular cast iron for another application.

15-16. Although the chemical compositions of white and gray cast irons can be quite different, they are both brittle in the as-cast condition. Explain why.

15-17. Explain how the ductility of as-cast gray iron can be increased without altering the composition or reheating to the liquid state.

15-18. A relatively thick section is to be made from some type of cast iron and is to possess maximum ductility. Specify which type of cast iron you would use and explain your choice.

15-19. Typical properties of gray cast iron and fully annealed 1080 steel are given here. Sketch their two $\sigma$-$\varepsilon$ curves on log-log coordinates, marking elastic and plastic regions.
CI: $S_u = 200$ MPa; $S_y = 200$ MPa; $E = 90$ GPa; $\mu = 0.3$ $\varepsilon_f = 0.002$; % R.A. = 0%.
1080: $S_u = 650$ MPa; $S_y = 400$ MPa; $E = 210$ GPa; $\mu = 0.3$; $\varepsilon_y = 0.002$; % R.A. = 45%.

# 16

# *Ceramics and Glasses*

Ceramics are inorganic, nonmetallic solids whose microstructures in the broadest terms consist of mixtures of crystalline phases and glasses (see Fig. 4-4). The traditional family of ceramic materials incorporates clay products (china, bricks, tiles, refractories, abrasives), cement, enamels, and glasses. Many of the raw materials for these products are complex silicate and aluminum silicate minerals naturally occurring in the earth's crust. For example, ordinary clays are based on micrometer-size platelets of kaolinite crystals $[Al_2(Si_2O)(OH)_4]$; silica $(SiO_2)$ is a principal component of common glass and also of various abrasive, whiteware, and refractory formulations. Other refractories, often used as linings of metallurgical furnaces, are manufactured from nonsilicate minerals such as the chromium and magnesium ores called chromite and magnesite.

The general strength characteristic of traditional ceramics is that they are brittle—chipped chinaware and drinking glasses, cracked plaster, and so on, are not uncommon—so ceramic bodies are often only lightly stressed in use unless that loading is consistently compressive, as in masonry arch bridges or in a structural brick wall. Even so, stones such as flint, which can be broken into pieces having sharp edges, were used by early humans as tools and weapons. Despite the characteristic of brittleness, ceramic raw materials are relatively inexpensive, and a multitude of applications has been found for ceramic products. Indeed, ceramics are favored for some uses because of particular combinations of chemical, electrical, and thermal properties not found in other materials. Many new ceramics have been developed that possess special properties.

There are two principal engineering applications for ceramic materials. The first involves the traditional function of refractory solids for furnaces and linings of furnaces, whose origins are lost in antiquity. Refractory bricks are manufactured, in the main, from natural minerals. The metallurgical industries (particularly ferrous metals) are the major users of refractory materials for melting, refining and casting, and hot forming and subsequent heat treatments. As strength is not of paramount concern, such applications are not considered in detail in this book. Indeed, because they must cope with liquid slags and metals, it is their chemical behavior that is often crucial and not their strength.

The second application *does* concern strength, particularly strength at high temperatures. As ceramics are degenerate metals in essence (i.e., metal carbides, oxides, and so on), they display great thermal stability. In the development of engines of high thermodynamic efficiency and of chemical plants, the requirements of designers for materials with adequate mechanical properties at high temperatures (among which creep resistance figures strongly) could be met with ferrous alloys as long as the operating temperatures did not exceed 500°C. Among such iron-base alloys are 18 Cr and 18 Cr–8 Ni austenitic steels (see Chap. 11) and also various tungsten and molybdenum alloys. High mechanical strength can be expected only at temperatures below the recrystallization range, and because it appears impossible to raise that temperature range in iron-base alloys significantly above 800°C, recourse was had to other materials when higher temperatures were specified. Thus were developed various "superalloys" of nickel, cobalt, and chromium such as the nimonic series. However, even these alloys are eliminated for operating temperatures above 1000°C.

Nevertheless, materials are required in such applications as gas turbines, jet engines, nuclear plants, fuel burners, and high-temperature chemical plants for which a service temperature of 1000°C and higher is desirable. New types of ceramics are being developed for these applications. In addition to possessing adequate mechanical properties, these solids must clearly be resistant to oxidation, so they also find use in the chemical industry. For example, alumina ($Al_2O_3$) ceramics have high strength and high wear resistance and are good insulators. They are used in high-speed machining of metals. The carbides, borides, and nitrides of the transition elements are variously used as cutting tools, as dies, and for parts of aeroengines operating at high temperature. Traditional ceramics, although perhaps retaining their integrity at high temperatures, would not possess adequate strength (for example) in the uses mentioned here for the new ceramics.

In fact, a principal market for the new ceramics utilizes their hardness and wear resistance at ordinary temperatures rather than their high-temperature properties. For example, wire-drawing dies, seals, bearings, and thread guides in textile machinery are common applications. In the electrical industry they have wide use as insulators. Cutting tools require strength at high temperatures, and ceramics are used in high-temperature applications involving hostile environments along with high mechanical and thermal stresses. Ceramic fibers are now woven into blankets for insulation purposes and are replacing traditional firebricks on occasion.

Except for the glasses themselves, most ceramics consist of crystalline phases surrounded by glassy binders. Many of the new ceramics are purer than the traditional ones and furthermore aim to have much less glass-phase material. In some ceramics, particularly those carbides and borides used as cutting tools, the glassy phase has been replaced by a metallic binding phase. For example, "cemented" titanium carbide (TiC) can have mixtures of cobalt, chromium, and nickel as the agent binding together with the carbide which is present in powder form. Not only is some small measure of ductility conferred on the ceramic (while retaining the high strength of the hard phase), but, more importantly, the thermal shock resistance is improved. When the ceramic content in such systems is reduced, we arrive at metallic alloys strengthened by refractory particles as opposed to a refractory held together by a metal. These hybrid materials are called *cermets,* among which are aluminum dispersion-hardened with particles of alumina ($Al_2O_3$), called *SAP alloys,* and molybdenum alloys containing 1–2% alumina. Metal matrix composites (MMCs) have fiber and whisker reinforcements, as well as particulate, for example, SiC filaments in aluminum alloy matrices. Other types of new ceramics are discussed later in this chapter.

## SOLIDIFICATION AND SHAPING OF CERAMICS

In most ceramics other than the glasses themselves, the other hard crystalline phases solidify first, being surrounded by glassy binders. During solidification, there is a large volume change associated with formation of the solid phase (see Chap. 4). Moreover, the solid crystal regions often have anisotropic coefficients of expansion that are different from those of the vitreous matrix. The low and anisotropic thermal conductivity of these materials, when coupled with the brittle nature of the solid regions, gives rise to marked residual stress and cracking problems on solidification. Consequently, the melting and casting techniques well known in metal production cannot be used with ceramics. Furthermore, because of their brittleness, ceramics cannot be shaped by plastic deformation methods as is common for many metals. Manufacturing processes for ceramics usually start with the raw material ground into *powder* form. The method of shaping traditional ceramics is then to mix the powder into a liquid paste that has appropriate rheological (flow) properties to permit forming. Such shaping may be done by hand, as with potter's clay, or by machine, as in the extrusion of sewer pipe and some types of brick. The body is then dried and fired at a temperature appropriate for the ceramic, between say, 800 and 1800°C. Firing causes the powder particles to stick together, thus retaining the shape of the formed body. The mechanics of binding may be related to the melting of parts of the ceramic mixture, to diffusion reactions across particle interfaces, or to processes of *sintering* in which surface tension at high temperature consolidates the particles. Common cement reacts with water to form a cementative matrix and usually dries and hardens by chemical reaction at ambient, rather than furnace, temperatures; hydration may continue over a long period of time, the strength of concrete increasing in step.

There is a reduction in the overall volume of ceramic bodies after firing as the voids between the particles are eliminated. This shrinkage must be allowed for in the design. Voids are never completely eliminated, however, and it is inevitable that bodies made in this way from powders will be porous. For example, if powder particles of a single size are used, cubic packing considerations (Chap. 5) suggest that about 40% of the volume consists of voids between the particles. Practical powders, of course, have a range of sizes, and some small particles fill some of the holes between the large particles. Nevertheless, the body retains some porosity. As the size, shape, and distribution of pores markedly affect the properties of a material, *porosity* is as significant a microstructural feature in ceramics as the crystalline and glassy solid phases themselves (compare the size, shape, and distribution of free graphite in cast irons).

In order to improve density, *high pressures* are used in some processes of forming ceramic powders. The methods involve either *cold pressing* followed by *sintering* or the related method of *hot pressing*. Both techniques are used in "powder metallurgy," which concerns the manufacture of articles from metal powders. Cold pressing and sintering of dry ceramic powder mixtures is actually similar to the slurry and paste methods just described, except that high-capacity presses are used so that much higher green densities (i.e., densities before firing) may be achieved. In hot pressing, the powder is heated and pressed at the same time, so highly dense, strong bodies result. Because of the combination of simultaneous pressure and heat, lower temperatures can be employed for hot pressing than for sintering after cold pressing. Hot-pressing equipment is more expensive, however.

**Example 16-1**

Other things being equal, are ceramic powder compacts likely to be less dense or more dense than metal powder compacts after sintering or hot pressing?

**Solution**   Ceramic powders are harder than metal powders in general, so they do not flow and compact as much as metal powders. For the same reason the areas of contact between particles are smaller in ceramics than in metals; this, coupled with inherently lower rates of diffusion, means that sintering is not as efficient in ceramics as in metals. All in all, this process tends to produce more porous ceramic products.

Special versions of sintering and hot-pressing methods have been developed. For example, *liquid-phase* sintering at temperatures just above the melting point of the metallic binder is carried out in the manufacture of cermets in order that the metal phase may run into as many pores as possible to produce a dense body. *Reaction* sintering and hot pressing deliberately encourage solid-state transformations at the high temperature to obtain higher densities. In this way, dense silicon nitride ($Si_3N_4$) is immediately sintered by the same process by which it is produced chemically by heating silicon powder compacts in nitrogen at about 1400°C. Another process is *sinter forging* in which the article is forged into shape as sintering takes place.

The method of pressing determines the final density achievable. Friction between the powder particles and at the die walls means that some regions of a pressed shape may have poor compaction in relation to other regions. This means that some areas of a body are finally weaker than others. In dieless *isostatic* pressing (done cold before sintering or done hot), the object is placed in a flexible bag surrounded by fluid under hydrostatic pressure. Oddly shaped compacts may be readily pressed to give quite uniform density. Many ceramic materials prepared in these ways have porosities of less than a few percent.

As raw powder sizes are very small, bodies having an acceptably smooth surface finish can be produced. In fact, because the fired, sintered, or hot-pressed article is so hard and brittle, little can be done in the way of postforming other than by grinding, spark machining, or ultrasonic drilling, for example. In the case of $Si_3N_4$, mentioned earlier, it is possible first to form the silicon powder into the roughly dimensioned shape required and then to perform an intermediate nitriding at 100°C in nitrogen, which converts perhaps 5% of the body to $Si_3N_4$. The mechanical properties of partially nitrided articles are such as to allow easy machining. Fully nitrided articles are then obtained as already described.

Fully dense ceramic solids can be obtained under special conditions. These include single crystals of artificial gemstones, such as sapphire (a variety of alumina), which are used as jeweled bearings for delicate mechanisms. Pore-free alumina can be made from powder by adding small amounts of Li and MgO before sintering. The trade name Lucalox is used by General Electric for such fully dense, translucent $Al_2O_3$. *Glass-ceramics* form an important group of dense ceramics based on $Al_2O_3$–$SiO_2$. It can be arranged for these materials to be wholly glassy at first, in which state they can be readily shaped by conventional glass-forming techniques (as described in the next section). Subsequently, a nucleating agent such as $TiO_2$ or $ZrO_2$ causes controlled devitrification (i.e., controlled removal from the character and appearance of glass), so the final pore-free structure consists of perhaps 97% very small (micrometer-size) crystals and a small residual glassy phase. Glass-ceramics are essentially free of the interparticle and grain-boundary weaknesses commonly found in ceramic bodies made by the usual methods. Most particularly, these solids have an almost zero expansion coefficient, so they have important applications in thermal shock situations. *Pyroceram* is one trade name for glass-ceramics. Similar sorts of devitrified structures can be produced from metallurgical slags and can be used to make tiles.

## SOLIDIFICATION AND SHAPING OF GLASSES

The difficulties of differential phase shrinkage that occur in ceramics during solidification are not present in predominantly glassy solids. Therefore, instead of shaping first and then using high-temperature "welding," the raw materials of glasses can be fused at high temperature and then shaped at lower temperatures (after cooling and reheating if required) much in the same manner as metals. Thus, glass articles can be cast, sheet and plate glass can be made by rolling, glass tubes and glass fibers can be drawn, and, of course, glass blowing and blow molding are important techniques for necked containers.

Viscosity is the principal parameter determining the temperatures required for different forming processes. For mechanical working, the glass should not be too "runny" (otherwise, it will not retain its shape) yet must flow readily in order to give reasonably low working loads. The working range of viscosity is 1–10,000 kPa-s, which for common soda-lime glass means a temperature range between about 600 and 900°C. Such glass "melts" at, say, 1400°C. Relief of residual stresses arising from working processes can be performed at about 400°C when the viscosity is perhaps 1 TPa-s; this procedure is termed *annealing* in the glass trade, although *stress relieving* is the metallurgical term.

The mechanical behavior of hot glass can reasonably be represented by a linear viscous relation between stress and strain rate. Following Chap. 12, we may write, in terms of tensile stress $\sigma$ and tensile strain rate $\dot{\varepsilon}$,

$$\sigma \propto \dot{\varepsilon}$$

Now $\dot{\varepsilon} = \dot{L}/L = -\dot{A}/A$, where $L$ and $A$ are the current length and cross-sectional area, respectively, of a bar under tension. Then, writing $\sigma = P/A$, where $P$ is the tensile load, we have

$$\frac{P}{A} \propto \frac{-\dot{A}}{A}$$

That is, $\dot{A}$ must be constant (as $P$ is constant along the bar). Thus, whatever the initial cross-sectional profiles, all cross sections lose area at the same rate, and originally narrow regions do not neck-down faster than others. This feature of stability against necking is important in many glass-forming operations, not least of which is the production of glass rods (more than 1 mm in diameter) by pulling vertically upward from hot glass. Thick fibers (0.1 ~ 1 mm) are drawn from larger-diameter rods. Thin fibers, on the other hand, are drawn through nozzles in the base of electrically heated platinum crucibles.

An important development in manufacturing sheet glass is *Pilkington's float-glass* method, in which a layer of glass is cast on top of a bath of liquid tin in a protective atmosphere. The lower surface acquires the smooth finish of the tin and the top surface is "fire-polished" by external radiation; so that subsequent mechanical polishing, found necessary in ordinary plate-glass manufacture to improve the strength, is avoided. Surface damage, which markedly affects the strength of glass, as discussed in following sections, is thereby reduced.

## STRUCTURE OF CERAMICS AND GLASSES

As described in Chap. 4, the changes in viscosity with temperature of glassy, amorphous substances are so gradual that no clear-cut melting or freezing temperature can be identified. The glass-transition temperature is defined as that at which the viscosity passes through the (arbitrarily chosen) value of 100 TPa-s. For this reason, glass is often termed a supercooled liquid, X-ray results for the structure showing molecular arrangements normally associated with liquids.

Glass at room temperature has frozen into it a molecular arrangement appropriate to some higher temperature, called the *fictive temperature.* As a result of the rapid cooling that can occur in the production of glass fibers (which at 70 m/s can cool to room temperature from 1300°C within about 30 mm after exiting the orifice of the crucible from which they are drawn), glass may have a fictive temperature many hundreds of degrees above the glass-transition temperature $T_g$. When glass is annealed, the fictive and transformation temperatures become identical.

Common bottle and window glass is made from silica ($SiO_2$) with additions of $Na_2O$ and CaO fluxes which, among other things, reduce the temperature of fusion. Optical glass (so-called flint glass) with high refractive indices contains PbO as the active flux. When boric oxide ($B_2O_3$) is used along with the principal ingredient silica, Pyrex-type glasses with a low coefficient of expansion result. In general, in these and other glasses, the interatomic bonding is mainly covalent in nature and the amorphous array is probably an irregular three-dimensional network lattice of oxygen polyhedra akin to cross-linked network and ladder polymers (see Fig. 4-3 and Chap. 13). Under an optical microscope glasses are generally featureless.

### Example 16-2

Explain why the addition of metal oxides to silica produces softening.

**Solution**   The metal atoms effectively reduce the number of cross links in the random glass structure, so lower viscosities are attained at lower temperatures than otherwise. This behavior is analogous to plasticization in polymers, whereas the vulcanization of rubber aims to produce exactly the opposite effect (Chap. 13).

Ceramic microstructures can be determined by polishing and etching cross sections and viewing under an optical microscope with reflected light, as is done with metals. In addition, very thin ceramic sections (about 20 $\mu$m thick) can be viewed by transmitted light; this technique is popular in geological investigations.

Equilibrium-phase diagrams have been constructed for many ceramic systems; in these diagrams, mixtures of compounds are plotted instead of metallic elemental mixtures as in Chap. 8. For example, the equilibrium diagram for the MgO–CaO binary system has the general shape of that for the Ag–Cu system (Fig. 8-2), with a eutectic composed of about 66% CaO and 34% MgO at 2370°C and end regions of limiting solid solubility of 8% MgO in CaO and 17% CaO in MgO at the eutectic temperature, both decreasing to very small values at lower temperatures. Other systems have more complex diagrams to a greater or lesser extent, and some multicomponent ceramic systems have very complicated diagrams. It must be remembered that these diagrams describe the crystalline phases present and not any glassy binding phase.

**Figure 16-1**   Magnesite–chrome refractory (hard-fired); large, white grains are chrome; smaller, light-gray grains are periclase (native magnesium oxide) together with exsolved magnesia–ferrite particles; medium-gray grains are silicate particles; and black regions are pores. (Courtesy of British Ceramic Research Association Ltd.)

**Example 16-3**

In comparison with metal alloys, are there likely to be difficulties associated with the construction of phase diagrams for ceramics?

**Solution** Both cooling curve methods and quenching methods (like the construction of *TTT* curves in alloys) are used, among others. The low thermal conductivity and high viscosity of the liquid phases in many ceramic systems means that equilibrium is attained only after an exceedingly long time. Indeed, it may be said that for some systems equilibrium is never attained, and the phases produced on cooling are "metastable" phases. It can be shown on thermodynamic grounds that such phases are not the phases of lowest energy, which equilibrium demands, so it is not always clear that the observed phases represent the long-term structure. (See Example 4-1 for an example involving glass.)

A typical ceramic microstructure consists of diverse crystalline phases in a glassy matrix, perhaps together with inclusions of impurities and, most particularly, pores. The character of pores significantly affects the properties of ceramics. Some pores remain isolated, being those gaps not joined by lenses between the original powder particles. Other ceramics may have regions where the pores are interconnected. Such features affect strength, thermal conductivity, and so on. Figure 16-1 shows a polished section of magnesite–chrome refractory such as is used in steelmaking furnaces. The black regions are connected pores, which are featured prominently in this complicated microstructure. Figure 16-2 displays self-

**Figure 16-2** Self-bonded SiC, in which SiC forms on original SiC grains. White regions are excess Si. (Courtesy of British Ceramic Research Association Ltd.)

**Figure 16-3**   Pure Al$_2$O$_3$, thermally etched to show grains.  Rounded pores appear black.
(Courtesy of British Ceramic Research Association Ltd.)

bonded SiC formed by firing a mixture of carbon and SiC in an atmosphere of silicon; new silicon carbide grows on the original SiC grains. Excess silicon is shown by the white regions. Pure A1$_2$O$_3$, thermally etched to show grains, appears in Fig. 16-3; the rounded pores appear black. A much finer pore size is shown in Fig. 16-4, which is a micrograph of hot-pressed Si$_3$N$_4$.

## STRENGTHS

Extreme brittleness is the pervading characteristic of most glasses and ceramics in bulk. The room-temperature fracture toughness of typical glass is only about 20 J/m$^2$ (which with a Young's modulus of 70 GPa gives a critical stress-intensity factor $K_c$ of about 1 MPa$\sqrt{\text{m}}$). Furthermore, ordinary glass contains many surface microflaws which act as starter cracks. As described in Chap. 14, Griffith performed his classic fracture experiments on glass rods, and the microflaws in glass have come to be called Griffith cracks. They have been shown to be a few micrometers in size, with a numerical density of perhaps 10$^9$ m$^{-2}$, which is consistent with the observed tensile strength of bulk glass of about 70 MPa and toughness of about 20 J/m$^2$. Dense ceramics, whose inherent resistance to crack propagation is also comparatively small, display low practical strengths (10–100 MPa) because of similar sorts of surface flaws. In addition, porous ceramics have low engineering strengths as a result of internal voids.

**Figure 16-4**  Hot-pressed $Si_3N_4$ (MgO used as a pressing aid).  Small pores appear black. (Courtesy of British Ceramic Research Association Ltd.)

Glass in the form of thin filaments (less than 50 $\mu$m in diameter) and other ceramics such as $Al_2O_3$ and $Si_3N_4$ in fiber or whisker form usually have significantly higher tensile strengths, which can approach 4 GPa (already noted in Chap. 14). Such strengths are almost equal to the theoretical strength of approximately $E/10$ and represent in the case of glass a 50:1 increase over the strength of the bulk solid. Although a similar trend of increasing strength with decreasing size is not unknown for some metals and high polymers, the improvements are not so marked in these solids. The strength–diameter relation is explained on a *statistical* basis: There is less chance of a flaw or void being present in a small volume of material (i.e., in thin fibers rather than thick rods), so there must be a trend of higher strengths in thinner filaments and whiskers. Statistical approaches to breaking strength are discussed in a later section. The high strength of commercial glass filaments is used to advantage in glass-fiber composites (Chap. 17), and other composites exploit increases in both strength and modulus of fibers and whiskers of graphite, boron, silica, silicon nitride, and so on.

It seems clear that if the starter-crack flaws in glass and ceramic solids can be reduced in size or otherwise rendered harmless, the strength of the solids in bulk could be improved. Where do the damaging flaws come from? They arise in the main from handling, which can produce extremely high local contact stresses in rubbing, for example. Thus, freshly made glass rods (1–3 mm in diameter) produced by vertical drawing from molten glass can display the sort of strengths associated with thin filaments (0.7 ~ 4 GPa) if tested *immediately* after manufacture and if care is taken to *avoid handling* the gage section. Even so, the customary bulk strength of 70 MPa is displayed by such rods as soon as they are subjected to even the simplest handling (the scratching caused by touching with a coat sleeve on a test piece gage length, for example).

Figure 16-5 contrasts the usual strength–diameter behavior, as first observed by Griffith, with the data of F. Thomas for freshly drawn fibers and rods. Clearly, if the microcrack flaws can be eliminated, bulk glass should be as strong as thin filaments. In commercial practice, there will still be strength–size variations, but appropriate coatings *(sizes)* are used on filaments in order to protect them in transit between the glass plant and the glass-fiber manufacturing industry; extreme scatter in strengths would result if uncoated filaments were allowed to rub over one another on a winding drum.

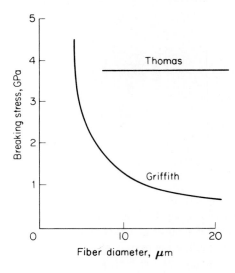

**Figure 16-5**   Strength-fiber-diameter behavior for glass filaments. Griffith results are contrasted with data on freshly drawn fibers and rods obtained by Thomas.

Sometimes it is possible to reduce the number of internal voids during or after manufacture, as in the hot isostatic pressing of ceramics. Moreover, commercial glass rods sometimes contain bubbles, and experiments show that it is possible to improve the strength to above about 1.8 GPa by floating out the bubbles by holding the bulk glass at 1400°C or so for a week prior to the rods' being made.

Methods are available for increasing the strength of surface-damaged glass. If the sharp-ended flaws in the glass can be healed up or put into a state of compressive stress, the strength should revert to the high values of undamaged glass. One method is to *acid-polish* the glass in a hydrofluoric–sulfuric acid mixture which etches away the sharp flaws. Commercial glass rods can be made to show strengths of over 1 GPa in this way; because of internal microbubbles, higher strengths are rarely achieved. Mechanical polishing of sheet glass can remove some of the worst flaws to give strengths of perhaps 100 MPa. The inverse relation between flaw size and strength (Chap. 14) is well established by experiments in which mechanically polished glass is deliberately sandblasted; the strength is reduced right down to 14 MPa. Glass and ceramics such as alumina are readily "flame-polished" and their strengths much improved thereby, provided that no chemical degradation or change occurs in the process.

**Example 16-4**

Why is pottery glazed?

**Solution** Glazes used for pottery are glasses of low thermal expansion which, because of the difference in the expansion coefficient of the base ceramic, end up in a state of residual compressive stress on cooling down. Microflaws are thereby squeezed together, and higher surface tensile stresses are required to cause fracture. In a similar manner, so-called toughened glass is rapidly quenched during manufacture from the plastic state so as to leave the free surfaces in a state of residual compressive stress.

Chemical effects play a part in determining the strengths of ceramics and glasses. Although acid polishing, as just described, smoothes out the microflaws that arise from scratching, some chemicals produce the opposite effect and can intensify the action of the flaws by enlarging them through dissolution. Here again, the fracture toughness can be altered by the environment. Glass is thus susceptible to the effects of water layers adsorbed on the surface, so atmospheric humidity affects the strength. Exposure of 120 h in 100% relative humidity reduces the tensile strength of freshly drawn rods from 3.5 GPa to 2.1 GPa; lower humidities produce smaller effects. Likewise, the phenomenon of *delayed fracture* (also wrongly called *static fatigue*) in glass, where sudden fracture at low stress occurs after a long time (see Chap. 18), is caused by small flaws slowly growing to the larger critical size necessary at that low stress to satisfy the particular fracture-mechanics formula for the test piece geometry involved.

As a result of the limited size of samples, brittleness in handling, and alignment difficulties on the testing machine, there is often not the freedom of choice of test piece geometry with ceramics and glasses as with other solids. For these reasons, three-point bend testing is popular, with the maximum outside tensile breaking stress being calculated from some beam formula. This stress is sometimes called the *modulus of rupture* of the solid. Appendix 11 gives typical room-temperature quasi-static strength values for a selection of ceramics and glasses.

The mechanical properties of concrete materials are discussed further in Chap. 17.

## THERMAL STRESSES AND THERMAL SHOCK

Materials usually expand when heated and contract when cooled. Allowance for such movements must be made in the design of engineering components. If such changes in dimensions are restrained as the temperature is changed, thermal stresses will be set up in the body. For example, a rod of length $L$ heated uniformly through a temperature interval $\Delta T$ will extend by $L\alpha \Delta T$, where $\alpha$ is the coefficient of thermal expansion. When this movement is completely resisted, it is equivalent to imposing a strain of magnitude $\varepsilon = L\alpha \Delta T/L = \alpha \Delta T$ on the rod. When the strain is within the elastic limit (as is usually the case), a compressive stress of magnitude $\sigma = E\alpha \Delta T$ is generated in the rod; if the strain is great enough to go over into the plastic region (which is uncommon), the associated stress will be given by some expression such as $\sigma_f = K\varepsilon^n$. Tensile stresses of similar magnitudes are set up when a body is fully restrained from contracting during cooling.

Brittle solids, such as ceramics and glasses, which fail at bulk elastic stresses $\sigma_f$ (given by some fracture-mechanics formula involving $K_c$ and inherent flaw size), can break under thermal stress alone when contraction is prevented. In the case of simple bars uniformly heated, fracture occurs when the temperature change from the starting condition satisfies $\sigma_f = E\alpha\,\Delta T$. The coefficients of thermal expansion of many ceramics are about (4 ~ 10) $\times 10^6/°C$; by using the bend strengths in App. 11 as appropriate indicators of the thermal failure tensile stress $\sigma_f$, it can be shown that the value of $\Delta T$ ($=\sigma_f/E\alpha$) necessary to cause fracture in many ceramics is about 100 ~ 200°C. That is, a bar in thermal and mechanical equilibrium, if fully restrained from movement, requires a ~100°C slow change in (uniform) temperature to cause fracture. In comparison, brittle metals that fail with little if any plastic flow (such as tool steels) require a greater $\Delta T$ (say, 400°C) because their $\sigma_f/E$ is larger than that for ceramics even though $\alpha$ is also greater.

In practice, components are not fully restrained. Even so, thermal stresses may develop in a body that is free to move if its temperature is not uniform. The foregoing treatment assumed the same temperature level throughout the body. However, if there are temperature gradients in the body (caused, for example, by nonuniform heating compounded by bad thermal conductivity in the body), internal stresses will be set up as different parts of the solid, which expand or contract by different amounts at the different temperature levels, are constrained in their movement by adjacent regions; the restraint is now internal, not external. The final displacement and strain field is a compromise, wherein regions that would have expanded greatly are restricted in that movement and regions that would not have expanded much are forced to suffer greater displacements than usual for the same temperature change. For example, when placed in a furnace, a piece of material at room temperature goes through a heating stage in which the outside is hotter than the inside, the magnitude of the differences in temperature depending on the conditions of heat transfer and the thermal conductivity of the solid. As the outside is restrained in its expansion by the inside, thermal stresses are set up whereby the outside is in compression and the inside goes into tension. Of course, the overall stress distribution is such that the net load and moment are zero in the free body. The thermal conductivity of metals is quite high, so the associated stresses are low and few problems are caused, particularly if the heating rate is low. In the case of ceramics and glasses, however, where the thermal conductivity is generally lower, such stresses can sometimes cause problems—not only for ovenware, and so on, but during manufacture when bodies can crack during firing.

The reverse process of cooling produces similar effects, and severe quenching can cause much more rapid changes in temperature, much more severe thermal gradients, and hence much more severe thermal stresses. Time is required to transfer the heat out of the hot, thick sections, so cooling rates are different at different locations in the body. The so-called mass effect, Jominy curves, and related problems of hardenability applying to metals are discussed in Chap. 10. Thermal gradients (which are the driving force of heat transfer) are more severe when the thermal conductivity of the solid is low, so other things being equal, the possibility of thermal cracking during quenching is more likely with ceramics and glasses than with metals. Even so, quench cracks do sometimes occur in martensite as a result of critical combinations of high thermal stress in thick sections and low $\varepsilon_f$ for martensite.

*Resistance to thermal shock* is a term used to describe the ability of a solid to withstand sudden changes in temperature. Thermal shock, as evinced by cracking of glasses into which hot liquids have been poured or by cracking of ovenware, is more of a problem with ceramics and glasses than with metals, although it must not be neglected in the design of turbine blades manufactured from refractory metals. Sometimes thermal cracking breaks a component in one go; at other times, myriad small cracks are initiated and progressively enlarge with successive thermal shocks, thereby degrading the retained strength in a cumulative fashion. For example, the starter-crack pores in some ceramics grow slowly with increasing number of shocks until the retained strength is incapable of withstanding the severity of the shock and the body breaks apart. The appearance of surface cracking on metals is called *thermal checking*; similar patterns occur in mud when ponds dry out and in varnish coatings on old furniture and paintings.

One traditional type of thermal shock test consists of successively plunging a small specimen a number of times from some high temperature into a cold fluid and then looking for surface cracks. The temperature from which the quenchings take place is increased and the process repeated until cracks are observed. The final temperature at which cracks are initiated is used to rank materials for given geometries and environments (i.e., quenching media). The type and size of specimen and the heating and quenching schedules vary from worker to worker and are arbitrary. Such tests are sometimes continued at increasing temperature levels until the already initiated cracks propagate in shock so that the specimens break completely. Then, in addition to the merit ranking based on crack initiation, some idea of the life of the part is available by adding up the number of heating and quenching cycles to failure. Insofar as the heating and quenching schedules are arbitrary, the results are quite empirical. Nevertheless, useful data can be generated for thermal cracking situations such as those encountered in interrupted machining cuts where the tool repeatedly enters and leaves the workpiece.

For a given type of specimen and given conditions of heat transfer at the interface between the initially hot body and cold-quenching medium (or initially cold body and hot medium), it is possible to perform calculations for the thermal stresses induced in a body in terms of the thermal conductivity of the body. Then, instead of the simple thermal stress of magnitude $E\alpha \, \Delta T$ produced by the same change in *uniform* temperature throughout a fully restrained body, we have

$$\sigma_f = ME\alpha \, \Delta T$$

where $M$ depends on the thermal conductivity, heat-transfer coefficient, and geometry of the body and $\Delta T$ is now understood to occur suddenly at the boundaries of the body rather than uniformly and slowly as before.

In general, $M$ varies inversely with $k$, the thermal conductivity, in a complicated fashion. Thus, the critical shock temperature difference required to cause cracking is now dependent on $k\sigma_f/E\alpha$. This parameter is sometimes known as the *spalling resistance index* and may be used to rank materials with respect to likely resistance to thermal shock. Large values of $k\sigma_f/E\alpha$ are favorable, as they mean that large thermal gradients are required before shock occurs. Materials satisfying this condition have high strength, good thermal conductivity, low stiffness, and low coefficients of thermal expansion. Appendix 11 gives values of $k\sigma_f/E\alpha$. Glasses have values less than 1 kW/m, and ceramics less than 10 kW/m. For comparison, refractory metals such as molybdenum and the nickel-based superalloys might have $k\sigma_f/E\alpha$ around 100 kW/m, and high-strength steel, even larger values.

**Example 16-5**

Describe the characteristics required of an ideal cutting tool material and relate the properties of coated, cemented solids to this ideal.

**Solution**    The final properties of engineering solids are compromises between the ideal and the attainable. Ideal tools would have: (1) high elastic moduli to minimize deflection, (2) high hardness at operating temperature to resist deformation and to provide wear resistance, (3) high fracture strength, (4) high thermal conductivity and a low thermal expansion coefficient to resist thermal shock, (5) high chemical stability to resist attack by workpiece and environment, and (6) high stability at high temperatures.

Some of these properties are mutually irreconcilable. The chemical stability and hardness of $Al_2O_3$ and TiC, especially at the temperatures encountered in cutting, make them excellent candidates as materials for cutting tools. However, they are not as strong as WC-based materials, which has limited their application. On the other hand, plain WC alloys do not have the wear resistance. Thus, TiC–WC alloys are manufactured in an attempt to combine the best features of both or to tailor the tool to the application (e.g., wear resistance is sometimes sacrificed for shock resistance in tools suitable for roughing and milling operations).

The development of disposable, indexable tool inserts has allowed even better materials, having coated layers, to be developed. The part of the tool in contact with the workpiece can be made of good-wearing material and bonded to a tough substrate with high thermal conductivity and thermal mass (which provides efficient heat removal from the cutting edge) together with high elastic modulus to provide support to the cutting edge.

Care should be taken regarding exactly what $\sigma_f$ means. It could be the stress that would initiate a crack and catastrophically break the body. However, there are situations, already mentioned, particularly with porous bodies, in which the multitude of small flaws are enlarged by a thermal shock but complete rupture does not occur. That is, *crack arrest* occurs after some propagation, and excess strain energy at crack initiation is converted to crack kinetic energy but is ultimately absorbed by the body. It can be shown that the damage, or extent of crack propagation, caused in these circumstances is inversely proportional to $ER/\sigma_f^2 = (K_c/\sigma_f)^2$, where $R$ is the dynamic fracture toughness in the shock environment and $K_c$ is the dynamic critical stress-intensity factor. This type of subcritical crack growth under shock conditions is therefore less likely to occur in solids with high toughness/strength ratios. It can also be shown that the extent of damage in bodies of a given porosity is less for a large number of small pores than for a few large pores. Appendix 11 gives values of $K_c/\sigma_f$.

The two parameters $k\sigma_f/E\alpha$ and $K_c/\sigma_f$ should correlate with the various empirical tests mentioned earlier, which establish (1) critical quenching temperatures at which surface cracks are initiated, and (2) the number of further shocks that can be withstood before complete rupture.

Subsequent room-temperature testing of ceramic bodies that have suffered thermal shock but have not been destroyed shows a degraded strength, as the fracture loads are those needed to break test pieces containing larger flaws (extent of damage) rather than flaws of the original small pore size. Figure 16-6 shows retained room-temperature bend strengths after 10 thermal cycles as a function of shock temperature interval for a type of titanium carbide cutting tool material [Fig. 16-6(a)] and a cordierite (lithium–aluminum–silicate) ceramic used in turbines [Fig. 16-6(b)]. In Fig. 16-6(a) a critical quenching temperature interval of about 400°C is shown; beyond this there is a marked

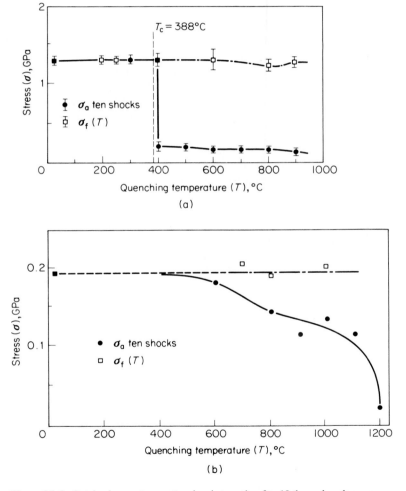

**Figure 16-6** Retained room-temperature bend strengths after 10 thermal cycles as a function of shock temperature interval for (a) a type of TiC cutting tool material, (b) a cordierite (lithium–aluminum–silicate) turbine ceramic. [From Y. W. Mai and A. G. Atkins, *J. Mat. Sci.* 10(1975), p. 1904.]

reduction in residual bend strength, which remains relatively constant thereafter. The unshocked high-temperature strength of the carbide (open points) remains constant at the room-temperature value to temperatures beyond the critical temperature. The behavior shown in Fig. 16-6(b) is different: The aftershock strength gradually decreases as the shock-temperature interval increases. The contrasting pattern is bound up in differences in the ease of crack initiation and growth at high temperatures; that is, the parameters $k\sigma_f/E\alpha$ and $K_c/\sigma_f$ change their relative rankings with temperature because $k$, $\sigma_f$, $E$, and $\alpha$ alter much more in some materials than in others.

## STATISTICAL STRENGTHS OF SOLIDS AND THE WEIBULL MODULUS

The use of materials such as ductile metals and alloys that exhibit a very narrow range of properties usually presents no significant problems in the variability of properties because the nature of the governing strengthening mechanisms controls the properties of such materials within narrow limits.   But brittle materials exhibit a *statistical variation* in strength properties.   This comes about because fracture originates at microflaws in the material, and so strength depends upon the probability of finding one or more large flaws situated normal to the maximum tensile stress. Even in a given "state of damage" (as-drawn, acid-etched, ground and polished, shot-blasted, and so on, in the case of glass), the microflaws in ceramic bodies actually vary in size and number, so it is inevitable that there will be a statistical variation in the fracture loads of test pieces (and hence the commonly quoted strengths) even when the samples are taken from the same large piece.

The designer who uses such materials requires a means to represent this variability so that he or she can design a component using material with a known probability of failure. The *Weibull modulus* was developed to respond to this need. Specifically, the Weibull modulus $m$ indicates the variability of a material's properties; a high modulus indicates little variability, and a low modulus indicates high variability. Most materials exhibit a Weibull modulus in the range of 5–25. (The use of the symbol $m$ in this context should not be confused with its use as the index of strain rate sensitivity in Chap. 12 and elsewhere.)

**Example 16-6**

If a glass rod, or other brittle material, is broken into two parts in tension, each part will apparently be stronger than the original whole. Discuss this observation.

**Solution**   The first fracture in general occurs at the first defect to fail, that is, the weakest spot, so the next defect will be less severe, and the stress needed to cause fracture will be marginally higher, assuming that the failure is due to random defects and not to homogeneous material failure. Brittle fracture usually propagates unstably from a single crack (except with very-high-energy inputs when multiple shatter cracks are produced). In general, for random effects, the worst defect is removed each time by fracture. This is a "fracture by weakest link" argument by analogy with fracture of a chain.

A strength value quoted from a given series of experiments is the arithmetic mean of all the results. Other sets of experiments on the same brittle material are unlikely to give the same result. It is not really possible to indicate an exact value for the fracture load or stress, but by using methods of statistics it is possible to quote a *probability of rupture or failure p* at a given stress $S_u$. The value $p$ is some function of $S_u$, which is a distribution curve of probability against strength. At low values of applied stress, $p \rightarrow 0$; when a body is highly stressed, $p \rightarrow 1$. A typical Weibull plot of $p$ vs. $S_u/S_{uav}$ is shown in Fig. 16-7. Alternatively, we may talk of the *probability of survival $p_s$* at the given stress $S_u$. Clearly $p + p_s = 1$. Empirically, W. Weibull in Sweden in 1939 found that the probability of survival is given by

$$p_s = \exp\left[-\left(\frac{S_u}{S_{u0}}\right)^m\right]$$ (16-1)

Here $p_s$ is the fraction of the total number $N$ of identical test pieces in a batch for which the strength *exceeds* a value $S_u$. $S_{u0}$ is a characteristic stress that is defined later. It follows that

$$p = 1 - \exp\left[-\left(\frac{S_u}{S_{u0}}\right)^m\right]$$

or

$$1 - p = \exp\left[-\left(\frac{S_u}{S_{u0}}\right)^m\right]$$

or, again,

$$\frac{1}{1-p} = \exp\left[\left(\frac{S_u}{S_{u0}}\right)^m\right]$$

where $p$ is now the fraction of the total number $N$ of identical test pieces in a batch for which the strength falls below the value $S_u$.

Note that when $S_u/S_{u0} = 1$, $p \rightarrow 1 - \exp(-1) = 0.633$. Hence the physical meaning of $S_{u0}$ is that it is the stress level for which the cumulative probability of fracture of specimens in a batch at the stress level $S_{u0}$ or lower, is 63.3%.

Note also that for $m \rightarrow \infty$, $p_s = 0$ and $p = 1$; that is, all specimens break at the same stress and there is no variability. For $m \rightarrow 0$, $p_s = 1$ and $p = 0$; the probability of fracture is the same for any value of $S_u$; that is, fracture is likely to occur at any stress level.

Taking natural logarithms,

$$\ln\left(\frac{1}{1-p}\right) = \left(\frac{S_u}{S_{u0}}\right)^m$$

Taking logs again,

$$\ln\left[\ln\left(\frac{1}{1-p}\right)\right] = m \ln\left(\frac{S_u}{S_{u0}}\right)$$ (16-2)

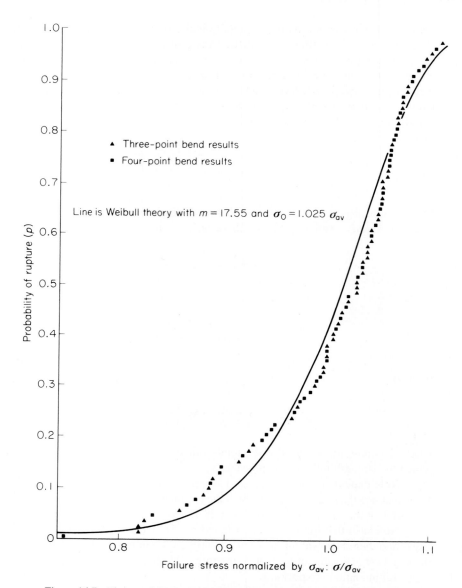

**Figure 16-7**  Three- and four-point bending failure data for reaction-bonded $Si_3N_4$. Solid line is the Weibull distribution using $m = 17.55$ and $S_{u0} = 1.025\ S_{uav}$. (Courtesy of Fulmer Research Institute, U.K.)

The Weibull modulus $m$ may be determined by plotting the double log of $1/(1 - p)$ against the log of $S_u/S_{u0}$ using data from a number of essentially identical specimens of a given material. In practice, because $S_{u0}$ is constant for the batch of samples tested, it suffices to plot $\ln\{\ln[1/(1-p)]\}$ against $\ln(S_u)$ only, if we require merely the slope $m$ of the plot and not the intercept. Each specimen is tested to fracture, and the strength $S_u$ is measured. The calculation is performed on a spreadsheet as follows:

1. Order the $S_u$ values from smallest to largest in the first column.
2. Assign the smallest value an index $i$ of 1, the next larger an $i$ of 2, and so on. List these in the second column, for a total number of tests $N$.
3. Compute

$$p = \frac{i}{N+1}$$

Enter $p$ in column 3.
4. Calculate $\ln(S_u)$ for each specimen; enter in column 4.
Calculate

$$\ln\left(\frac{1}{1-p}\right)$$

Enter this value in column 5.
6. Calculate

$$\ln\left[\ln\left(\frac{1}{1-p}\right)\right]$$

Enter this value in column 6.
7. Make a graph with $\ln\{\ln[1/(1-p)]\}$ on the vertical axis and $\ln(S_u)$ on the horizontal axis.
8. The slope of this graph $m$ is the Weibull modulus for strength.

## Example 16-7

Ten measurements of the strength of a brittle composite material, taken on nominally identical test pieces and under the same conditions of strain rate, humidity, and temperature, were as follows (in MPa): 60.00, 61.00, 59.00, 58.00, 60.50, 61.50, 58.50, 59.50, 60.00, 59.75. Determine the Weibull modulus.

**Solution**  Using the procedure described earlier, the following calculation is developed.

| Strength | $i$ | $p = i/(1 + N)$ | ln(strength) | $\ln[1/(1-p)]$ | $\ln\{\ln[1/(1-p)]\}$ |
|---|---|---|---|---|---|
| 58.00 | 1 | 0.090909091 | 4.060443011 | 0.09531018 | −2.350618656 |
| 58.50 | 2 | 0.181818182 | 4.069026754 | 0.200670695 | −1.606090045 |
| 59.00 | 3 | 0.272727273 | 4.077537444 | 0.318453731 | −1.144278086 |
| 59.50 | 4 | 0.363636364 | 4.085976313 | 0.451985124 | −0.794106012 |
| 59.75 | 5 | 0.454545455 | 4.090169191 | 0.606135804 | −0.500651220 |
| 60.00 | 6 | 0.545454545 | 4.094344562 | 0.78845736 | −0.237676951 |
| 60.00 | 7 | 0.636363636 | 4.094344562 | 1.011600912 | 0.011534137 |
| 60.50 | 8 | 0.727272727 | 4.102643365 | 1.299282984 | 0.261812562 |
| 61.00 | 9 | 0.818181818 | 4.110873864 | 1.704748092 | 0.533417353 |
| 61.50 | 10 | 0.909090909 | 4.119037175 | 2.397895273 | 0.874591383 |

A graph can be made using $\ln\{\ln[1/(1 - p)]\}$ as ordinate and $\ln(S_u)$ as abscissa, as shown in Fig. 16-8.

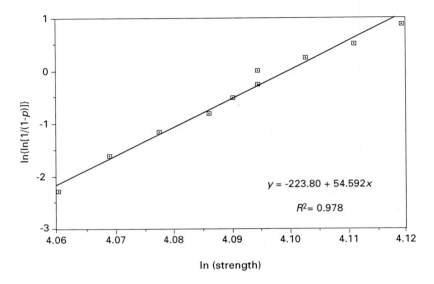

**Figure 16-8**   Plot to determine Weibull modulus $m$.

From the graph, the slope is seen to be 54.6, which is the Weibull modulus for this set of 10 specimens. Note that even if we had listed the strength in megapascals, multiplying the tabulated strength by $10^6$, the slope would be the same. Remember also that there was no need to incorporate $S_{u0}$ as explained above. It is possible to purchase special Weibull probability graph paper. If $p_s$ instead of $p$ is used, the plots fall from left to right and the slope is $-m$.

While we might expect that final fracture results from propagation of the severest crack, final fracture in a specimen where the stress varies throughout its volume might result from propagation of a smaller crack whose orientation or location relative to the stress field is appropriate. For example, a large flaw on the compressive side of a bend test-piece may very well not affect fracture, whereas a small one on the tensile side would; in this way, the breaking loads of brittle solids are likely to depend on the size of the regions under stress—particularly, perhaps, those under tensile stress.

Again, because the breaking load of a test piece reflects the load necessary for propagation of the microflaws and because, according to the mechanics of fracture, that load depends on the geometry of the test piece and the mode of deformation (Chap. 14), we should not expect the same "tensile strength" from tests in, say, simple tension and bending on specimens of different sizes. Each type of specimen is likely to have different distributions of strength about different mean values. These variations in strength are in addition to any variations to be expected on the basis of accuracy of the experiments.

Furthermore, owing to the size effect in brittle fracture (Chap. 14), where large components usually break at lower stresses than similar small components, there is a *volume effect* on the probability of rupture.  It may be shown that if $p_s$ is known for a reference body,

$$(p_s)_V = (p_s)^V$$

where $(p_s)_V$ is the probability of survival at stress level $S_u$ in a body with a volume that is $V$ times the volume of the reference body being tested in the same way.  Also, because $p = 1 - p_s$,

$$(1 - p_V) = (1 - p)^V \tag{16-3}$$

For a body of volume $V$, it is helpful to define a new parameter $B$, called the *risk of rupture*, given by

$$B = - \ln(p_s)_V$$

The risk of rupture of geometrically similar, but different size, specimens being tested in the same way depends on the volume according to

$$B = - \ln(p_s)_V = - V \ln (p_s) = - \ln(1 - p_V) = - V \ln(1 - p)$$

Using Weibull's exponential relation Eq. (16-1) for $p_s$, we have

$$B = V \left( \frac{S_u}{S_{u0}} \right)^m \tag{16-4}$$

Remember that $V$ is the volume under tensile stress only, so the parameter $B$ enables us to quantify the different risks of rupture of nominally identical test pieces depending on whether they are tested in simple tension, three- or four- point bending, torsion, and so on, where the volumes under tension are different.  For such comparisons, we must use Eq. (16-4) in its general form, that is,

$$B = \int_v \left( \frac{S_u}{S_{u0}} \right)^m dV \tag{16-5}$$

For example, a bar tested in simple tension has the same nominal tensile stress over all cross sections, but the same bar in bending has only those regions on one side of the neutral axis under tensile stress. Flaws in the compressive regions do not influence the breaking load. This means that the statistically expected average failure loads vary with the mode of deformation, as shown by the following example. Consider rectangular bars of some brittle solid which are tested (1) in simple tension and (2) in pure bending (i.e., under the action of simple moments only, without shear forces, as obtained in four-point bending). In case (1) the risk of rupture $B$ is given directly by Eq. (16-4) because the stress is independent of elemental volume position. In case (2) the tensile stress varies linearly with distance $y$ from the neutral axis according to $(y/h)S_u$, where $S_u$ is the tensile stress on the outside of the bar at fracture and $2h$ is the depth of the cross section, which also has width $w$. An elemental volume slice is $dV = Lw\,dy$, so in Eq. (16-5),

$$B = \int_0^h \left(\frac{y}{h}\right)^m \left(\frac{S_u}{S_{u0}}\right)^m Lw\,dy$$

where the limits are 0 to $h$ because we are considering only those regions suffering tensile stresses. Therefore,

$$B = \frac{(S_u/S_{u0})^m\,Lwh^{m+1}}{h^m(m+1)}$$

$$= \frac{(S_u/S_{u0})^m\,V}{2(m+1)}$$

where $V$ is the volume of the *whole* bar.

We see that the risk of rupture at stress level $S_u$ is $2(m+1)$ times as great in case (1) as in case (2). This means that the statistical mean breaking stress of nominally identical samples in pure bending will be greater, on the average, than the breaking stress in simple tension.

It may be shown that $S_{uav}$ and $B$ are related by

$$S_{uav} = \int_0^\infty \exp(-B)\,dS_u \tag{16-6}$$

For simple tension, case (1),

$$S_{uav} = \int_0^\infty \exp\left[-V\left(\frac{S_u}{S_{u0}}\right)^m\right] dS_u$$

In case (2),

$$S_{uav} = \int_0^\infty \exp\left[\frac{-(S_u/S_{u0})^m\,V}{2(m+1)}\right] dS_u$$

If the variable in the integration of case (2) is changed to $\lambda = [S_u/2(m+1)]^{1/m}$, it follows that

$$S_{uav} = [2(m+1)]^{1/m} \int_0^\infty \exp\left[-V\left(\frac{\lambda}{S_{u0}}\right)^m\right] d\lambda$$

The integral expressed in terms of $\lambda$ is the same as the integral expressed in terms of $S_u$ in case (1). Thus,

$$\frac{(S_{uav})_{bending}}{(S_{uav})_{tension}} = [2(m+1)]^{1/m}$$

As mentioned before, only when $m$ takes on very large values do the two $S_{uav}$ become similar. In other instances, if $m$ has the value 3, for example, $(2 \times 3 + 2)^{1/3} = 2$, so bending samples are "twice as strong" on the average as tension samples.

It can be shown that three-point bending, where only a small region at the center span is subjected to the greatest tensile stress, gives higher $S_{uav}$ for brittle substances that have wide scatter in their strengths (i.e., $m$ small). One reason for the popularity of modulus of rupture bending data over simple tension data is immediately apparent!

**Example 16-8**

In round bars of some solid for which $m = 3$, show that the risk of rupture in torsion is nearly five times as great as in bending.

**Solution** An elemental volume, at a distance $y$ from the neutral axis of a circular bar in bending, is given by $dV = 2L\sqrt{(r^2 - y^2)}\,dy$ (Fig. 16-9). Using $(y/r)(S_u/S_{u0})$ for the distribution of bending stress, we can readily show from Eq. (16-5) that $B = [4/15\pi]\,V(S_u/S_{u0})^3$.

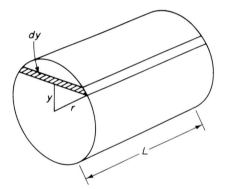

**Figure 16-9** Solution to Example 16-8, showing geometry of round bar stressed in bending.

In torsion of a circular rod, the torsional stresses $\tau$ at some radius $y$ acting on the annular volume $dV = 2\pi Ly\,dy$ is $(y/r)^3\,(\tau_u/\tau_{u0})^3$. Using Eq. (16-5), we find that the value of $B$ is $2/5\,V\,(\tau_u/\tau_{u0})^3$, where $\tau_u$ is the failure stress in shear. Pure shear is equivalent to mutually perpendicular tensile and compressive stresses of equal magnitude (Chap. 2); then if fracture is associated with a critical stress level, $S \equiv \tau$ in the foregoing. Thus, the risk of rupture in torsion is $(15\pi/4)2/5$, or nearly five times as great in torsion as in bending. The reason is clear: All the outside skin in torsion is likely to spawn a fracture, and there is thus much more chance of cracking than in bending, where only part of the surface is subject to tensile stress.

Given the laboratory data for the failure strength of a certain material, Weibull statistics can be used in design calculations to find the appropriate working stress in a component made from that material for which only a certain limited number of failures can be tolerated. For example, let us assume that three-point bending results gave $S_{u0} = 250$ MPa and $m = 15$ for some brittle solid. A component whose volume is 1000 times greater than that of the test pieces from which the $S_{u0}$ result was obtained is to be designed such that only one failure in 1000 components can be tolerated. What should the working stress be?

We have $p_V = {}^1/_{1000}$, $S_{u0} = 250$ MPa, and $V = 1000$. Consequently, in $p_V = 1 - (p_s)^V$ [Eq. (16-3)], we have from Eq. (16-1),

$$p_V = 1 - \exp\left[-\left(\frac{S_u}{S_{u0}}\right)^m V\right]$$

$$\frac{1}{1000} = 1 - \exp\left[-\left(\frac{S_u}{250}\right)^{15} 1000\right]$$

That is, $\ln 1.001 = 1000 (S_u/250)^{15}$, or $10^{-6} = (S_u/250)^{15}$, which gives $S_u = 100$ MPa. This, of course, is smaller than the stress found from the laboratory (small) test piece.

## REFERENCES

1. Bradt, R. C., D. P. H. Hasselman, and F. F. Lange, eds., *Fracture Mechanics of Ceramics.* Proceedings of a Symposium at Pennsylvania State University, 1973. New York: Plenum Press, 1974.

2. Evans, A. G., and T. G. Langdon, *Progress in Materials Science* , vol. 21, no. 314 (B. Chalmers, J. W. Christian, and T. B. Massalski, eds.). Oxford: Pergamon, 1976.

3. Gordon, J. E., *Structures—Or Why Things Don't Fall Down.* London: Penguin Books, 1978.

4. Kingery, W. D., H. K. Bowens, and D. R. Uhlmann, *Introduction to Ceramics,* 2nd ed. New York: John Wiley, 1975.

5. Pomeroy, C. D, "Concrete, An Alternative Material." Fourteenth John Player Memorial Lecture, Institution of Mechanical Engineers, London. *Proceedings,* 192 (1978), 14, p. 135.

## PROBLEMS

**16-1.** Explain why the strength of nearly all glass articles can be increased by etching off a thin surface layer with hydrofluoric acid.

**16-2.** Some glass articles can be strengthened by annealing to remove residual stresses, whereas others are strengthened by tempering, an operation that deliberately introduces residual stresses. Discuss the seeming contradiction between these two facts. (It may be of assistance to know that tempering introduces compressive residual stresses at the surface of glass.)

**16-3.** Why are $Al_2O_3$ cutting tools used with negative rakes? (The blades of woodworking planes have positive rakes; a negative-rake blade is one set with the blade trailing the direction of cut instead of attacking the direction of cut.)

**16-4.** Explain why a ceramic component subjected to a nonuniform stress field may not fail at the point of maximum nominal stress.

**16-5.** Explain why the room-temperature physical properties of glass fibers such as density, modulus, and index of refraction correspond to properties of bulk glass at higher temperatures.

**16-6.** Explain why glass rods manufactured by pulling upward from hot glass do not neck and fail.

**16-7.** Why is it said that a "silver" spoon should be put in a drinking glass before a hot liquid is poured in?

**16-8.** Distinguish between those physical properties of a solid that determine the resistance during thermal shock (1) to crack initiation and (2) to crack propagation. Using the data in App. 11, rank the various ceramic and glass solids in terms of resistance to cracking in thermal shock. Where on your ranking list do common engineering metals lie?

**16-9.** Relate the method of ranking employed in Prob. 16-8 to the two types of thermal shock behavior observed in practice, examples of which are shown in Fig. 16-6.

**16-10.** Under what circumstances is a relatively low thermal conductivity required in ceramics?

**16-11.** Show that the average strength of a series of tensile tests (which correspond to $B = 1/2$) is $(1/2)^m S_{u0}$, where the probability of rupture is related to the applied tensile stress by $(S_u/S_{u0})^m$. Show how $m$ reflects the spread of strength values.

**16-12.** A body is to be designed so that failure will not occur more than once per 10,000 components. If the working stress of the components cannot be less than 150 MPa, what stress level must be attained in the laboratory on model components whose volume is $1/100$ of the full-sized component? Your answer will depend on $m$, which is not known a priori.

**16-13.** Describe the procedure for using Weibull statistics to find the mean creep-rupture lifetime at a given experimental stress level and temperature using as an example the following test results for polyethylene pipe. Time to fracture: 207, 192, 132, 167, 224, 189, 237, 150, 198, 212 h.

Assume that the probability of survival of the $j$th specimen in ranking order of a set of $n$ is given by $[1 - (j - 0.3)/(n + 0.4)]$. How can data of this type be used to predict the expected lifetime in years of polyethylene pipe in service?

What material quality control and external variations should be taken into account? (University of Birmingham, U.K.)

**16-14.** In a sequence of fatigue tests at a fixed mean stress and cyclic stress amplitude, the number of cycles to failure is found to be 85, 117, 76, 90, 60, 104, 128 kilocycles.

Assuming that the survival probability $p_s$ of the $j$th specimen in ranking order in a set of $n$ is given by $p_s = [1 - (j - 0.3)/(n + 0.4)]$, calculate the number of stress cycles allowable for 99% probability of survival, extrapolating the graph by use of the Weibull modulus $m$ in the appropriate equation.

Why is this prediction likely to be inaccurate and how could more reliable predictions be made? (University of Birmingham, U.K.)

# 17

# Composites

In the broadest sense of the term, structural composites include any mixture of separate material components combined to give a useful material. Such a composite often possesses the best qualities of the individual constituents and sometimes displays composite properties that none of the components possess. Composite materials may be classified as follows:

1. *Fibrous composites,* having continuous or discontinuous *filaments* or *whiskers* in a matrix. Examples are wood, bone, plastics reinforced with glass fiber or carbon fiber, and metals reinforced with boron fiber or silica whiskers.
2. *Laminated composites,* having layers of various solids (such as clad metals, laminated glass, plastic-impregnated cloths and papers, and laminated fibrous composites).
3. *Particulate composites,* having particles in a matrix. Examples are concrete (stones in mortar), ABS polymers (see Chap. 13), and some rocket propellants (inorganic particles in a rubber matrix). Particulate composites having metal matrices may have either nonmetallic particles (such as ceramic particles in some metal matrix composites, MMCs) or particles of other metals (which encompasses inclusions in metals).

The discovery of very high strengths in ultrafine filaments of many solids provided the impetus for development of synthetic fibrous composite materials. Mixtures of two or more distinct materials on a very fine scale (on the order of 100 $\mu$m or less) now produce solids with a wide range of properties impossible to achieve with conventional materials. In many cases, very strong, lightweight solids result. Structures manufactured from such composites can therefore be lighter for the same working stresses or stronger for the same mass. Clearly, they find application in aeronautical and space designs, as well as in many other fields including sporting goods.

## HIGH STRENGTH

Nearly flawless crystalline solids in which dislocations cannot move may be expected to exhibit maximum strengths on the order of $E/10$ (where $E$ is the modulus of elasticity; see Chaps. 14 and 16). The only cases where strengths approaching such values have actually been observed are for very small samples of small-diameter filaments (with diameters in the range of 1 $\mu$m), usually called *whiskers*. For example, fracture strengths of iron whiskers have been measured up to 14 GPa (about $0.07E$). Measurements on whiskers of other materials likewise give high strengths: Cu, $0.02E$; Ni, $0.02E$; $Al_2O_3$, $0.05E$; SiC, $0.04E$; graphite, $0.03E$. Whiskers are discontinuous, but continuous filaments displaying respectably high strengths of materials such as glass, boron, and carbon are commercially available. For example, glass fibers can have strengths of $0.014E$; boron filaments have strengths of $0.009E$ in fibers 100 $\mu$m in diameter; and carbon fibers have strengths of $0.004E$ in fibers 8 $\mu$m in diameter. The mechanical properties of several filamentary materials are given in App. 12.

Chapters 14 and 16 discuss the role of flaws and surface quality in the strength of brittle solids and the origins of these size effects on strength, whereby small samples are likely to display values approaching the theoretical strength. Large lumps of the same solids (which in many cases cannot be manufactured anyway) display much lower strengths because there is greater likelihood of finding imperfections in the form of microcracks or dislocations in large volumes of materials. The associated statistical variation in

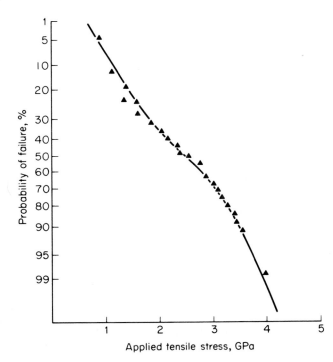

**Figure 17-1**  Cumulative strength distribution for boron-on-tungsten filaments of diameter approximately 100 $\mu$m.

strength values is also described in Chap. 16. Figure 17-1 shows a strength distribution for boron-on-tungsten filaments. When numbers of filaments with such characteristics are used in composites, there is a statistical variation in the composite strength, but it is not as far ranging as that for separate filaments because the strong filaments present help to hold up the composite structure even after some low-strength components have failed.

## MODULI

In a quest for strength, we might simply look for materials with high yield strengths. We know that the strengths of solids can often be improved by cold work, by heat treatment of some sort, or by alloying to introduce extra phases (Chaps. 2 and 9–11). However, the usefulness of ultrastrong materials is limited by their stiffnesses or elastic moduli. The reason should be clear from Fig. 17-2. If the working stress level is much increased but the modulus remains unchanged, the working strains increase in the same proportion as the strength (point $X$ to point $Y$). Translated into deflections, this means that although objects designed with strong materials can bear heavier loads (or, equivalently, require less material to take the old loads), they do so at much greater deformations. For compact items with small deflections, this may not matter, but for structures such as bridges and airplane wings, the use of structural material of low elastic modulus would lead to the unfortunate situation shown in Fig. 17-3.

If we wish to operate with these potentially ultrastrong fibrous materials at the same elastic deformations currently used with traditional engineering materials, we must look for *increased $E$* as well as increased working stress (point $X$ to point $Z$ in Fig. 17-2). Moreover, those parts of a structure in compression at high working stresses require higher moduli to prevent Euler buckling. It is possible, of course, as J. E. Gordon has pointed out, for the situation in Fig. 17-3 to be attained by materials with high $E$ if their breaking strain is great enough.

Because it has been shown that the theoretical strength of a solid is proportional to $E$, it seems that high strength and high stiffness are simultaneously satisfied in the whisker

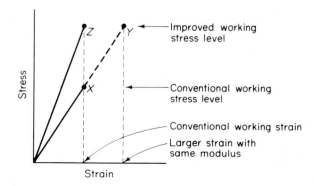

**Figure 17-2**  The need for increased stiffness as well as increased strength to keep working deflections small.

**Figure 17-3** The consequence of increased strength with no improvement in stiffness; 1.6% strain in aircraft wing spar booms (bend radius of beam = thickness/2 × strain). (After J. E. Gordon.)

and filamentary materials being discussed in this chapter. This should be contrasted with improvements in the strength of conventional materials achieved by cold working and heat treating, or other methods where $E$ is not altered.

## DENSITY

Another factor of interest in many applications is the mass of a given part. As we shall see, many strong composite materials have low mass, which enhances their attractiveness. As a measure of strength for a given mass of structure, the *specific strength* is defined here as strength divided by specific gravity. For those very strong steels with $S_u = 2$ GPa and $\rho = 7800$ kg/m$^3$ (i.e., specific gravity 7800/1000 = 7.8 because the density of water is 1000 kg/m$^3$), the specific strength is about 250 MPa. On the other hand, for graphite whiskers of strength 17 GPa and specific gravity 2.25, it is about 7.6 GPa. These figures demonstrate the potential of such high strengths: If a cable 6 mm in diameter could be made wholly of continuous carbon filaments of 1.6-GPa strength, it could support a tensile load of 45 kN and a 3-m length of the cable would weigh about 190 g. The same mass of wire rope 3 m long would support about 10 kN.

In addition to the concept of specific strength, the elastic modulus can be divided by specific gravity to give *specific elastic modulus* or *specific stiffness*. Appendix 12 lists typical values of these quantities for some common engineering materials, and App. 13 gives these values for some high-strength filamentary materials. Note the remarkable fact that the specific moduli of most common engineering solids are all about 27 GPa. Most of their specific strengths are between 20 and 200 MPa. This strongly suggests that we must look at new materials for efficient improvements in strong and stiff structures of low mass. The use of glass-filament composites represents a sort of halfway point in the transition from traditional to new materials. The specific modulus of glass filaments is the same as for traditional engineering materials, but the specific strength is perhaps 6 times that of bulk glass. The growth of the glass-fiber industry shows to what extent such composite materials can be used despite the disadvantage of large working strains. There is, in fact, an additional factor: In *bending,* specific strengths and specific moduli are more properly represented by the ratios strength/specific gravity$^2$ and modulus/specific gravity$^3$ than by the simple ratios given in App. 13 for plain tension (see Prob. 17-12). Then, because the density of plastics reinforced by glass fibers is so much lower than that of common engineering metals, the bending stiffness of unidirectional glass-fiber composites can be about 20 times the bending stiffness of the same mass of steel and 2 times that of aluminum alloy beams (see Example 17-4). Even so, Fig. 17-3 emphasizes that materials of higher specific stiffness than glass are required for many applications.

## MATRICES AND THE MAKEUP OF COMPOSITES

It is all very well to have identified high-strength, high-modulus solids, but if most of them are obtainable only in whisker or filament form, how can they be used? How can we "put together" the whiskers or filaments to make structurally useful shapes? Perhaps we can make bundles like rope or perhaps hold together the separate high-strength components in some *matrix*. The latter is the route normally chosen. By using an epoxy or polyester resin as a glue, filaments and whiskers can be made into useful structural materials. Sometimes ropes, mats, and cloths are made from filaments before these components are stacked or combined to form the final structure (a process called *laying up*), but an adhesive matrix is always necessary to hold things together. The matrix can also be a ductile metal. By analogy with reinforced concrete (which has ductile bars or meshes of steel running through a brittle stone–cement matrix) we often refer to filamentary composites as *fiber-reinforced* plastics or metals.

The matrix plays an important and complex role in composites. Apart from binding fibers together, it acts as a medium for transferring external loads on the composite to the filaments because the fibers themselves are rarely in direct contact with the external loading system. As we shall see, the interfacial bond between filament and matrix is most important. It is the load-transfer agent and, as we shall also see, is important in determining the toughness of a composite. Among other functions of the matrix are the following:

1. It can be made to align filaments in important stress directions and to permit "composite action" in tensile, compressive, and shear modes.
2. It separates fibers, thus preventing cracks from passing catastrophically from one fiber to another through the composite.
3. It protects the fibers from mechanical damage and perhaps also from environmental damage (e.g., moisture or high-temperature oxidation).
4. If ductile, it provides a mechanism for slowing down cracks.
5. If brittle, it may depend on the fibers to act as crack stoppers.

The density of most laminating resins is about 1000 $kg/m^3$. Because of the fairly large difference between this and the density of many reinforcing filaments (2000 ~ 2500 $kg/m^3$), there are important increases in the average density of fiber-reinforced composites as the number of fibers packed into the composite increases. Consequently, for structures of fixed mass made from polymer-based composites, there is an optimum number of fibers per unit volume (an optimum *volume fraction;* see next section) that gives, for example, maximum bending resistance. In metal matrix composites, on the other hand, reinforcements are lighter than the matrix. All potential metal matrices such as magnesium, aluminum, titanium, and steel have essentially the same specific stiffness, while that of a typical SiC reinforcement fiber is over four times better, so specific moduli (and specific strengths) are improved by higher volume fractions of fibers.

## VOLUME FRACTION

The amount of reinforcement in a composite is normally given in terms of its *volume fraction,* the percentage or decimal fraction of the total composite volume taken up by the filaments. Usually we wish to pack in as much reinforcement as possible, but, as in crystal packing (Chap. 4), there are geometric limits. Practical composites have 50 to 60% filamentary reinforcement.

**Example 17-1**

(a) Whiskers with specific gravity 3 are bonded with a resin of specific gravity 1.2. What percentage mass of the composite is taken up by the whiskers if their volume fraction is 0.3?
(b) If a metal matrix is used instead, with a specific gravity of 8.9 (nickel), what is the mass content of the whiskers in the composite?

**Solution**    (a) The relative mass of the whiskers is

$$\frac{3(0.3)}{3(0.3) + 1.2(0.7)} = \frac{0.9}{1.74} = 52\%$$

(b) In this case, we have

$$\frac{3(0.3)}{3(0.3) + 8.9(0.7)} = \frac{0.9}{7.13} = 13\%$$

It is seen that there are good arguments for the use of lightweight resin matrices where possible. If metal matrices are used (in order that the composite may operate at higher temperatures, for example), a penalty is paid in mass.

**Example 17-2**

What is the maximum volume fraction attainable with aligned round fibers?

**Solution**    A hexagonal cross section produces the maximum volume fraction with aligned fibers, as shown in Fig. 17-4. If there are $m$ rows of filaments with $n$ fibers in each row, the total cross-sectional area $A_f$ of the fibers is $mn\pi d^2/4$. The circumscribing matrix area $A_{total}$ is given by

$$A_{total} = \frac{md}{2}\sqrt{3}(nd)$$

$$V_f = \frac{A_f}{A_{total}} = \frac{mn\pi d^2/4}{\frac{md}{2}\sqrt{3}(nd)} = \frac{\pi}{2\sqrt{3}} = 90.7\%$$

Practical fiber-reinforced composites rarely exceed two-thirds of this value.

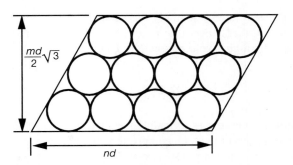

**Figure 17-4**   Solution to Exercise 17-2,
showing hexagonal array of round fibers,
with *m* rows of fibers having *n* fibers in each
row.

## COMPOSITE STRESS AND MODULUS

Consider a unidirectional array of continuous filaments embedded in some matrix.
Although the behavior of such a composite is far from being fully understood, a reasonable
approximation for both the composite modulus and strength is given by the *rule of mixtures.*

Let a tensile load $W$ be applied in the direction of the aligned filaments and assume
that all components extend equally, that is, that they have equal strains $\varepsilon_f = \varepsilon_m = \varepsilon_c$, where
the subscripts f, m, and c mean "filament," "matrix," and "composite," respectively (Fig.
17-5).

**Figure 17-5**   Load applied in the direction
of aligned filaments.

If the stresses induced in the components by this strain level are $\sigma_f$ and $\sigma_m$, then

$$W = \sigma_f A_f + \sigma_m A_m \qquad (17\text{-}1)$$

where $A_f$ and $A_m$ are the individual cross-sectional areas of the filaments and matrix.

The average composite stress is $\sigma_c = W/A_c$, where $A_c$ is the total cross-sectional area
given by $A_c = A_f + A_m$. Thus,

$$\sigma_c = \frac{\sigma_f A_f}{A_f + A_m} + \frac{\sigma_m A_m}{A_f + A_m}$$

The quantities $A_f/(A_f + A_m)$ and $A_m/(A_f + A_m)$ represent the fractions of the total cross
section taken up by the filaments and matrix. If we multiply the numerator and denominator of these area fractions by the total length of the composite, we have the volume fractions $v_f$ and $v_m$. Thus,

$$\sigma_c = v_f \sigma_f + v_m \sigma_m \qquad (17\text{-}2)$$

or, because $v_f + v_m = 1$,

$$\sigma_c = v_f\sigma_f + (1 - v_f)\sigma_m \tag{17-3}$$

This equation is the basic rule of mixtures and is clearly analogous to the lever rule for phase diagrams (Chap. 8).

Again, for the effective modulus $E_c$ of the composite, we use the fact that the stresses $\sigma_f$ and $\sigma_m$ are given by $E_f \varepsilon_f$ and $E_m \varepsilon_m$. So, in Eq. (17-1),

$$W = E_f\varepsilon_fA_f + E_m\varepsilon_mA_m \tag{17-4}$$

Now in terms of $E_c$ and $\varepsilon_c$, $W = E_c \varepsilon_c A_c$. Assuming that $\varepsilon_f = \varepsilon_m = \varepsilon_c$, we have

$$E_cA_c = E_fA_f + E_mA_m \tag{17-5}$$

which transforms as above into a volume-fraction rule of mixtures for the composite modulus:

$$E_c = v_fE_f + v_mE_m = v_fE_f + (1 - v_f)E_m \tag{17-6}$$

Graphically, Eqs. (17-3) and (17-6) are represented by Fig. 17-6.

**Figure 17-6** Meaning of Eqs. (17-3) and (17-6) for stress and stiffness in unidirectional composites.

Thus, the Young's modulus slope for a given mixture lies proportionally between those of its components, and the strength points lie similarly between those of the components.

**Example 17-3**

Boron filaments of $E_f = 380$ GPa are made into a unidirectional composite with an epoxy matrix of $E_m = 2.5$ GPa. What are the moduli parallel to the fibers (a) for $v_f = 0.05$ (i.e., 5%) and (b) for $v_f = 0.50$?

**Solution** $E_c = (1 - v_f)E_m + v_fE_f$.
(a) $E_c = (1 - 0.05)(2.5) + (0.05)(380) = 2.38 + 19 = 21.4$ GPa
(b) $E_c = (1 - 0.5)(2.5) + (0.5)(380) = 1.25 + 190 = 191$ GPa
Note that the contribution from the matrix may often be neglected.

It is clear that if the strains through the stressed composite are assumed to be the same everywhere, then the stresses in the components will be different ($\sigma = E\varepsilon$, and the $E$'s are different). We could postulate, in fact, that the stresses are equal all over and thus allow unequal strains in the components. This occurs in layered or sandwich plates loaded in compression through their thickness. We then would obtain for the composite strength and modulus "harmonic" (i.e., reciprocal) rules of mixtures:

$$\frac{1}{\sigma_c} = \frac{v_f}{\sigma_f} + \frac{1 - v_f}{\sigma_m} \tag{17-7}$$

$$\frac{1}{E_c} = \frac{v_f}{E_f} + \frac{1 - v_f}{E_m} \tag{17-8}$$

**Figure 17-7**  Different assumptions for composite action by Voigt (equal component extensions) and by Reuss (equal component stresses).

Figure 17-7 displays the two different models of composite action. Equation (17-6) was first used by W. Voigt early this century in calculating the moduli of polycrystalline solids, and Eq. (17-8) by E. Reuss a little later. In contrasting the two types of behavior (equal strains with unequal stresses or unequal strains with equal stresses), one is reminded of the mechanics of springs or resistors in parallel and in series. Real composite behavior is somewhere between the two (they are mathematical bounds to $\sigma_c$ and $E_c$) but tends toward the uniform strain case. This is because the uniform stress model suggests slippage at the interface between fiber and matrix. If the interfacial bonding between the components is good (i.e., the gluing is good), we should not expect much relative movement. The final displacement of the composite under load is a compromise movement. The parts with low $E$'s want to stretch more than those with high $E$'s but are restrained by the interfacial bonding, and vice versa.

**Example 17-4**

Show that in a composite for which $E_f \gg E_m$, the condition for maximum flexural rigidity of a beam of given mass is that the density of the matrix must be two-thirds of the overall density of the composite, that is, the mass of reinforcing fiber must be one-third of the total mass of the beam.

**Solution**   The mass $M$ of the composite beam is given by

$$M = \rho_f t_f + \rho_m t_m = \rho t$$

where $t$ is the overall thickness of a beam of given plan area, $t_f$ and $t_m$ are the equivalent thicknesses of fiber and matrix, respectively, and $\rho$ (with similar subscripts) is the density. Then,

$$\rho = v_f \rho_f + (1 - v_f)\rho_m$$

The resistance of the beam to bending is proportional to $EI$, where for a rectangular beam $I = wt^3/12$, $w$ being the planar width of the beam. For a beam of fixed mass and plan area, maximum flexural rigidity (maximum $EI$) therefore comes from maximum $Et^3$ or maximum

$$[v_f E_f + (1 - v_f )E_m]\left(\frac{M}{\rho}\right)^3 = \frac{[v_f E_f + (1 - v_f)E_m]M^3}{[v_f\rho_f + (1 - v_f)\rho_m]^3}$$

Differentiating with respect to $v_f$ and letting $E_m \approx 0$ gives the condition for a maximum as

$$v_f = \frac{\rho_m}{2(\rho_f - \rho_m)} \quad \text{or} \quad \rho_m = \frac{2\rho}{3}$$

or, equivalently, that the mass of reinforcing fiber is close to one-third of the total mass of the composite beam if $v_f$ is small. (This answer is the same as the "skin mass criterion" for maximum flexural rigidity in honeycomb sandwich laminates when the core modulus is neglected.)

**Example 17-5**

Explain why copper-coated steel wire is easier to cold-work than plain steel wire but that the maximum attainable drawing reduction of the coated wire is less than for plain steel wire.

**Solution**   The overall yield strength of copper-coated steel wire is smaller than that of steel alone. This follows from the rule of mixtures applied to the component yield strengths, that is, from

$$(\sigma_y)_{coated} = (1 - v_f)(\sigma_y)_{Cu} + v_f\,(\sigma_f)_{steel}$$

where $v_f$ refers to the volume fraction of steel. Cold working is thus easier because $(\sigma_y)_{Cu} < (\sigma_y)_{steel}$, say, 120 MPa compared with 250 MPa.

During deformation of the copper–steel composite at common uniform strain, compressive stresses are set up in the lower-strength copper coating and tensile stresses are set up in the higher-strength steel (Fig. 17-8) because the copper wants to flow at a lower stress than the steel. A simple Mohr's circle diagram illustrates the origin of these stresses during *plastic flow*. (The earlier derivations in this chapter have been put in terms of elastic deformations, but the same principles apply to plastic flow).

If the copper and steel components were being deformed individually, they would have individual Mohr's circles as in Fig. 17-8(a) and (b). When combined in a sandwich and compressed (as when passing through a wire-drawing die), the applied compressive load and stress must be the same in both, so effectively the small copper circle shifts to the left to match the steel circle applied stress [Fig. 17-8(c)]. However, this leaves an unbalanced horizontal stress (*AB*) in the weaker component. A horizontal tensile force therefore must exist in the stronger component to balance this out, so the final Mohr's circle shifts to the right [Fig. 17-8(d)]. The tensile forces generated by stresses *OD* acting on the particular area of steel balance those compressive forces generated in the cross section of copper acted on by stresses *OC*.

It is seen that if the fracture of the steel core of the wire can be described in terms of a tensile failure stress (not strictly true, but an adequate hypothesis), the presence of an additional induced tensile stress (*OD*) in the steel core produces breakages at lower total strains than in plain wire.

The presence of compressive stresses helps reinforced concrete in a similar way. Another way of describing the result is to say that the strains in the weaker part of composites are less than they would be if only the weak solid were being stressed and that in the strong component greater strains are attained for a given load in a composite than if tested alone. Use is made of this in the cold working of difficult-to-form materials such as stainless steel and titanium; their yield strengths are large, and they work-harden appreciably, but "pack-rolling" between layers of soft metals reduces the necessary working loads.

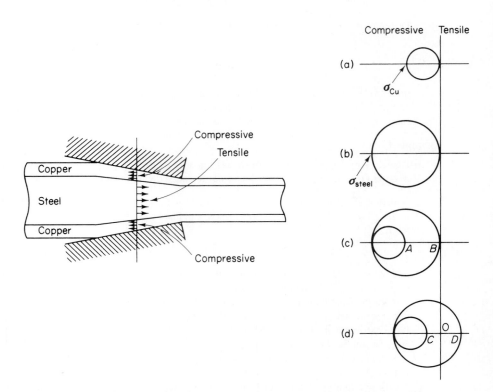

**Figure 17-8**   Solution to Example 17-5, showing stress distributions in the mechanical working of clad materials (in this case, copper-coated steel wire).

## CRITICAL TRANSFER LENGTH;
## CONTINUOUS AND DISCONTINUOUS FILAMENTS;
## ASPECT RATIOS; TENSILE STRENGTH

When a fiber-reinforced composite is loaded, it is rare for individual filaments to be acted on by the external load; we do not grab the individual filaments, which in many cases are smaller than 100 $\mu$m in diameter. Rather, the matrix acts as a medium for transferring loads to the filaments. The bond between the filament and matrix is the load-transfer agent, making the matrix and filament "move" together. The nature of the bond (chemical or mechanical) varies in different composite systems, but for convenience we can represent the interfacial bond strength by a constant shear stress $\tau$.

The existence of a bond strength builds up stresses in the filaments. We can liken the buildup of load in the filaments to the buildup of load in the rope in a tug-of-war (Fig. 17-9).

**Figure 17-9** Buildup of load and strain in a filament over a transfer length analogous to buildup of load in a tug-of-war.

The load is transferred from the matrix to the fiber only near the end regions of the fiber because as the load builds up so does the strain, and when that strain reaches the overall strain in the matrix, there is no relative slip between matrix and fiber. The adjacent matrix then becomes virtually stress-free in the region corresponding to the middle of the fiber. In this way, high-strength filaments effectively take the major portion of the external loads, as may be seen by investigating the force equilibrium on an element of length $dL$ near the end of a filament embedded in a lump of matrix [Fig. 17-10(a)].

$$\frac{\sigma_f \pi d^2}{4} + \tau \pi \, dL = (\sigma_f + d\sigma_f)\frac{\pi d^2}{4}$$

Therefore,

$$\frac{d\sigma_f}{dL} = \frac{4\tau}{d}$$

so

$$\sigma_f = \frac{4\tau L}{d} \qquad (17\text{-}9)$$

because $\sigma_f = 0$ at the end of the filament where $L$ is zero. Thus, the stress in the filament builds up linearly from the ends at the rate $\tan^{-1}(4\ \tau/d)$, and the tensile stress in the adjacent matrix decreases similarly. The stress buildup in the filament does not go on indefinitely, only over a length sufficient to transfer the particular external load into the fiber [Fig. 17-10(b)]. It is at this point that the stress in the filament becomes sufficiently high and the stress in the matrix becomes sufficiently low that the strains in the filament and in the matrix are the same. There is then no shear stress across the interface.

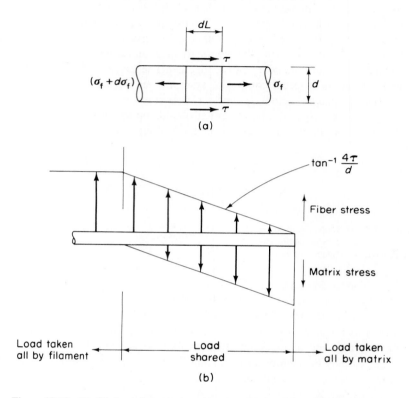

**Figure 17-10** Equilibrium of stresses in an element of the transfer length; $\sigma_f$ is the stress in the filament and $\tau$ is the interfacial stress between filament and matrix.

### Example 17-6

A single fiber of some material is embedded in a matrix subjected to a force of 40 mN. If the diameter of the filament is 10 $\mu$m and $\tau = 20$ MPa, over what length does load transfer take place?

**Solution**    When the load is fully transferred,

$$\sigma_f = \frac{4 \times 10^{-3}}{\pi(10 \times 10^{-6})^2 / 4} = 509 \text{ MPa}$$

Then, $L = (509 \times 10^6)(10 \times 10^{-6})/(4 \times 20 \times 10^6) = 64$ $\mu$m at one end of the filament. The load is transferred back to the matrix at the other end of the same filament, so $64 + 64 = 128$ $\mu$m of the filament is used up in load transfer. For the same $\tau$, thicker filaments require longer $L$s to build up to the same stress. If $d = 100\ \mu m$ in the foregoing example, $L$ would be 10 times as long, that is, $640\ \mu m$ at each end of the fiber.

There must be a limit to the load buildup, and in brittle filaments this occurs when the breaking stress of the fiber is reached. The middle length of the filament is conventionally described as being wholly subjected to this maximum stress at fracture. Obviously, it will normally break in only one spot (at some diameter variation or where the material is locally weak), so a constant fracture stress along the length is strictly unrealistic. The real stress distribution must peak at the fracture point and blend into that value elsewhere. Nevertheless, the simple distribution, with linear buildup at the ends and with the central section of the filaments capable of sustaining a stress at or near the fracture stress, is an adequate model for describing the mechanics of composites. It has already been mentioned that filaments display a statistical variation in failure strength (Fig. 17-1). Consequently, in what follows, where a value is given for the breaking strength of fibers, it must be construed as a representative average value for the family of filaments being employed.

If the average filament breaking stress is $(\sigma_f)_{max}$ the corresponding load transfer length is $(\sigma_f)_{max}(d/4\tau)$ at each end of a filament. Unless the filament length is greater than twice this value, the breaking strength is never reached [Fig. 17-11(a)]. This length is called the *critical transfer length* $L_c$ of the filament–matrix combination:

$$L_c = \frac{(\sigma_f)_{max}\, d}{2\,\tau}$$

For the general case of a long filament stressed below its breaking stress, the average working stress in filaments of length $L$ longer than $L_c$, as shown in Fig. 17-11(b), is given by

$$(\sigma_f)_{avg} = \frac{1}{L}\int_0^L \sigma_f\, dL = (\sigma_f)_{const}\left(1 - \frac{L_c}{2L}\right) \tag{17-10}$$

where $(\sigma_f)_{const}$ is the constant filament stress in the central region of a long filament. If a filament is loaded up to its breaking stress, the average stress in the fiber is $(\sigma_f)_{max}(1 - L_c/2L)$. Depending on the relative values of $L_c$ and $L$, the mean stress may be within, say, 95% of $(\sigma_f)_{max}$ (i.e., when $L_c/2L < 0.05$ or $L > 10\, L_c$) or perhaps only 50% of $(\sigma_f)_{max}$ [when $L_c/2L = 0.5$ or $L = L_c$; see Fig. 17-11(c)].

Clearly, account should be taken of these differences when the rule-of-mixtures expressions for stress [Eqs. (17-2) and (17-3)] are applied at failure. Thus, instead of

$$(\sigma_c)_{max} = (1 - v_f)(\sigma_m)_{max} + v_f(\sigma_f)_{max} \tag{17-11}$$

(a)　　　　　　　　　　　　　(b)　　　　　　　　　　　　　(c)

**Figure 17-11**　The influence of fiber length $L$ on the stress level transferred from the matrix: (a) fibers shorter than the critical transfer length $L_c$; (b) $L \gg L_c$; (c) $L = L_c$.

for breaking strength of composites (the derivation of which implicitly assumes long, continuous fibers running through the composite), it is preferable for $L \geq L_c$ to write

$$(\sigma_c)_{max} = (1 - v_f)(\sigma_m)_{max} + v_f(\sigma_f)_{max}\left(1 - \frac{L_c}{2L}\right) \qquad (17\text{-}12)$$

For $L \leq L_c$, from Fig. 17-11(a), the average stress in the filaments is less than half $(\sigma_f)_{max}$, so Eq. (17-12) becomes

$$(\sigma_c)_{max} = (1 - v_f)(\sigma_m)_{max} + v_f(\sigma_f)_{max}\left(\frac{L}{2L_c}\right)$$

The particular lengths of fibers therefore affect the value of composite breaking strengths, as the following example shows.

### Example 17-7

A composite is made up of 40% of filaments 100 $\mu$m in diameter of strength 3 GPa in a matrix of strength 80 MPa. The interfacial bond strength is 60 MPa. Determine the composite breaking strength if the filaments have average lengths of (a) 100 mm and (b) 2 mm.

### Solution

$$L_c = (\sigma_f)_{max} \frac{d}{2t} = \frac{(3 \times 10^9)(100 \times 10^{-6})}{2(60 \times 10^6)} = 2.5 \text{ mm}$$

Therefore,    (a) for $L = 100$ mm, $L_c/2L = 2.5/200 = 0.0125$ and

$\sigma_c = (1 - 0.4)(80 \times 10^6) + 0.4(3 \times 10^9)(1 - 0.0125)$

$\quad = 48 \times 10^6 + (1.2 \times 10^9)(0.9875) = 10^9(0.048 + 1.19) = 1.23$ GPa

(b) for $L = 2$ mm, $L/2L_c = 2.0/5.0 = 0.40$. So,

$\sigma_c = (1 - 0.4)(80 \times 10^6) + 0.4(3 \times 10^9)(0.40)$

$\quad = 48 \times 10^6 + (1.2 \times 10^9)(0.40) = 10^9(0.048 + 0.48) = 528$ MPa

Note the significant difference in $\sigma_c$, the reason for which lies in the values of average stress distribution in the fibers. Note that when the filaments are long, the contribution from the fibers ($\approx v_f\sigma_f$) swamps the contribution from the matrix ($\approx v_m\sigma_m$). Often in the rule-of-mixtures calculations for both $\sigma$ and $E$, we neglect the contribution from the matrix.

$L_c/2L$ can be written $(\sigma_f)_{max}(d/4\tau L)$, which shows that for a given $\tau$ the average stress in stressed filaments is close to $(\sigma_f)_{max}$ when $L/d$ is large. $L/d$ is called the *aspect ratio* of the fibers. Even if a composite is made up of discrete lengths of filament, the composite strength is the same as if they were continuous, for all practical purposes, provided that $L/d$ is large enough. As $(\sigma_f)_{max}/\tau \approx 50$ for many brittle-filament—brittle-matrix systems, large composite strengths are attainable if the aspect ratio is greater than, say, 100. For filaments 10 $\mu$m in diameter, this means that they need be only 1 mm or longer to come within 90% of the "ideal" composite strength given by the simple rule of mixtures [Eqs. (17-2) and (17-3)]. Thus, short whiskers can be used to good effect. In practice also, filaments are long and have favorable aspect ratios. Therefore, although there is a strict distinction between continuous and discontinuous filamentary composites, it is somewhat academic unless $L/d$ is small or for some reason the interfacial bond is very weak.

The simple rule-of-mixtures diagram (Fig. 17-6) is modified to take account of varying $L/d$ as shown in Fig. 17-12. A low $L/d$ prevents a large $(\sigma_c)_{max}$ from being attained both because of the low average stress in the fibers and because the maximum attainable volume fraction is restricted.

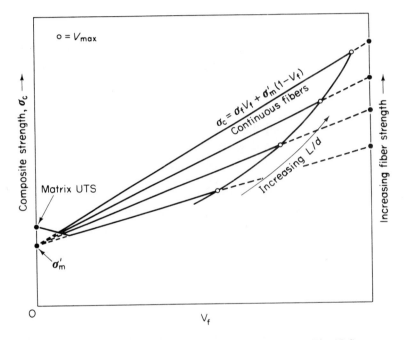

**Figure 17-12** Modifications to the simple rule-of-mixtures diagram (Fig. 17-6) to account for finite $L/d$.

Figure 17-12 also shows a modification to the simple diagram in Fig. 17-6 at the very low volume fraction end, where paradoxically it seems that the addition of fibers *reduces* the composite strength *below* that of the plain matrix. This comes about as follows: In Eq. (17-11) or (17-12), it is implicit that $(\sigma_m)_{max}$ and $(\sigma_f)_{max}$ occur at the same strain. In fact, glass filaments do tend to have roughly the same fracture strains (0.02 ~ 0.05) as many common matrices. Boron and graphite filaments have fracture strains that are smaller (less than 0.01). When the failure strain of the fiber is smaller than that of the matrix, the matrix $\sigma_m$ to be used in Eqs. (17-11) and (17-12) is not its own breaking stress $(\sigma_m)_{max}$ but rather a lower value appropriate to the failure strain of the filament. In this case, it turns out that at low filament-volume fractions, $(\sigma_c)_{max}$ from Eqs. (17-11) and (17-12) can be *less* than $(\sigma_m)_{max}$ (Fig. 17-13). Thus, until a *critical volume fraction* of filament is achieved, producing $(\sigma_c)_{max} > (\sigma_m)_{max}$, there is no gain to reinforcement.

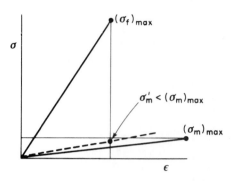

**Figure 17-13**   When the failure strain of the fiber is smaller than that of the matrix, the matrix stress is smaller than its own fracture stress. (The Voigt equal-strain model for composite behavior is assumed.)

**Example 17-8**

What is the critical volume fraction for a brittle-fiber–brittle-matrix composite made of fibers with $(\sigma_f)_{max} = 3$ GPa, $(\sigma_c)_{max} = 100$ MPa, and $E_m = 2$ GPa with $(\varepsilon_f)_{max} = 0.01$?

**Solution**   At $\varepsilon = 0.01$, $\sigma_m$ is not 100 MPa but rather is given by $(E_m)(\varepsilon_f)_{max} = (2 \times 10^9)(0.01)$ = 20 MPa. Then, if a $\sigma_c \le 100$ MPa at volume fractions less than the critical fraction, we have [using Eq. (17-11) for simplicity]

$$100 \le [(1 - v_f)20 + v_f(3000)] \text{ MPa}$$

Therefore,                                         $80 \le 2980 v_f$

So $(v_f)_{critical} = 0.0268$, or roughly 3%. Of course, practical composites have much greater $v_f$ values.

## COMPRESSION STRENGTH

When a unidirectional composite is loaded in compression parallel to the fibers, the mode of failure depends on the strength of the filament–matrix bond. If the bond is weak, the fibers and matrix detach themselves (debond) at low applied loads, the filaments buckle elastically, and the compressive strength never reaches the composite tensile strength. Even if the interfacial bond is strong, failure of the matrix in shear keeps the compressive strength lower. This is particularly true with discontinuous fiber composites, as matrix shear can occur without the necessity for elastic distortion of the filaments. Composites with high filament-volume fractions fail by *creasing in* compression, which is a shear instability (Prob. 17-22).

An approximate expression for compressive strength that applies to many, but certainly not all, systems is

$$(\sigma_c)_{max} = \frac{0.63 G_m}{1 - v_f} \qquad (17\text{-}13)$$

where $G_m$ is the matrix shear modulus. For many epoxide resins, $G_m$ is about 1 GPa, so the compressive strength of a $0.5\text{-}v_f$ composite might be about 1.2 GPa. The tensile strength of a $0.5\text{-}v_f$ boron–epoxy composite along the direction of the filaments is about $(0.5)(3.5) = 1.7$ GPa.

## ANISOTROPY IN COMPOSITES: ALIGNED AND RANDOM FIBER DISTRIBUTIONS

The strength and stiffness of composites reinforced with aligned strong fibers are highly anisotropic. The variation in tensile strength with orientation angle for two unidirectional composites is shown in Fig. 17-14. The mode of failure in an aligned continuous composite depends very much on the orientation of the fibers with respect to the applied stress. When the loading is parallel to the filaments, failure is governed by fiber tensile breakage. If the loading is shifted from the fiber direction, the matrix or the interface (whichever is weaker) will be increasingly loaded in shear until shear failure occurs even though the fiber tensile stresses are still low. When the applied stress is perpendicular to the fiber direction, failure occurs by tensile fracture of the matrix or interface, the filaments playing comparatively little part in the process. By determining the composite tensile strengths along and normal to the aligned filaments, together with the failure stress in shear parallel to the fibers, it is possible to arrive at the failure stress at any arbitrary angle using an empirical theory, similar to the Tresca condition for yielding (see Chap. 2). Failure is predicted in three distinct modes, over which one of the above three stress parameters predominates (Fig. 17-14). Failure criteria for other composites based on the Maxwell–von Mises combined-stress condition have been proposed.

## LAMINATES

Maximum use is made of the reinforcement when the composite is stressed parallel to the fibers, as with simple struts or the caps of panel stiffeners. For other applications, it is more useful to arrange the filaments in a composite in more than one direction. Thus, *laminates* are employed, which have layers with regular orientations such as the cross-ply arrangement (alternate layers at right angles, as in plywood sheets) and angle-ply schemes ($\pm\theta$ where $30° < \theta < 60°$); most applications have multiple plies with three or four angles.

The highly anisotropic character of single layers is ironed out when composites are made up from layers with different orientations. For example, Fig. 17-15 shows the variation in tensile strength with orientation for composites made with unidirectional filaments, cross-plied filaments, and angle-plied fibers. The last-mentioned composite is said to be *pseudoisotropic* in behavior, as the variation in strength with angle is not marked; note in comparison that the unidirectionally aligned system has a transverse strength of only about 8% of its strength parallel to the filaments.

Much of the difference in elastic anisotropy that determines stiffness of individual layers also becomes less marked in multilayer sheets. In consequence, well-known isotropic

**Figure 17-14** Comparison of measured strengths of composites, as a function of angle between stress axis and fiber direction, with predictions of a maximum-stress model based on strength along the filaments ($\sigma_c$), strength perpendicular to the filaments ($\sigma_t$), and shear strength parallel to the filaments ($\tau$). (Compare the Mohr's circle constructions in Chap. 2.) (a) 0.57 $v_f$ type I carbon fibers in epoxide resin; (b) 0.50 $v_f$ SiO$_2$ fibers in aluminum. [From B. Harris, *Composites*, 3 (1972), p. 152, after J. Dimmock and M. Abrahams, and P. W. Jackson and D. Cratchley, with permission.]

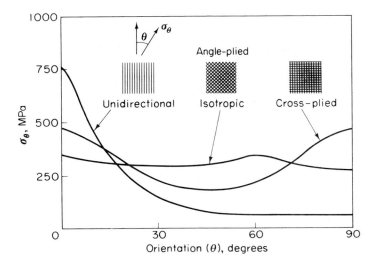

**Figure 17-15**   Variation in laminate composite tensile strength with orientation for uni-directional, cross-plied, and angle-plied laminates of glass-fiber-reinforced plastics. Angle-plied arrangement gives pseudoisotropic behavior. [After J. W. Davis and N. R. Zurkowski, 3M Corporation, with permission.]

strength-of-materials theories can often give reasonably adequate descriptions of elastic stress and strain in composites. Problems do occur, however, at the edges of laminates where interlayer and edge shear stresses are set up in the composite to counteract the rotation of one layer (individually anisotropic) relative to a differently oriented adjacent layer when subjected to external loads. Edge delamination and loss of structural integrity can result from this "scissoring" or "free-edge" problem.

If, instead of the laminae being aligned, the fibers are arranged randomly in two dimensions, the average breaking strength will be about one-third that of unidirectional composites, and if the filaments are three-dimensionally random, about one-sixth. Therefore, for the best mechanical properties, only the strongest reinforcement fibers ought to be considered for use in nonaligned composites. Of course, limits are also set on $v_f$ in random reinforcement by the difficulties of close packing.

## FRACTURE TOUGHNESS

The components of many practical filament-reinforced composites are often brittle. Consequently, although they have high fracture strengths and are stiff, such composites are usually not very tough. We have seen (Chap. 14) that for reliable use in design, a material must have an adequate resistance to crack propagation. It is essential, therefore, to ensure toughness in composites; otherwise, we shall be using expensive constituent materials to produce a composite whose potential strength we can never use because we are afraid of it breaking catastrophically if there is some local imperfection that causes a crack to start.

One peculiarity of fracture in filamentary composites is that the fracture surfaces are rarely simply planar and have bits of fiber sticking out of the matrix (Fig. 17-16). As we shall see, this phenomenon helps to produce some extra toughness in composites. The cause of the stubblelike feature is as follows: If the fibers are not continuous and a matrix crack (originating perhaps at a bubble, dust particles, or a notch in the matrix) approaches a fiber in such a way that the end of the fiber is within a length $L_c/2$ of the fracture plane, we will expect the filament end to be *pulled out* of the matrix rather than broken; this comes about simply because within $L_c/2$ of the ends of the filaments, the breaking stress is not reached [Fig. 17-10(b)]. On the other hand, continuous filaments of uniform strength should fail in the plane of the matrix crack with no pull-out; this applies also to discontinuous filaments whose ends are embedded at distances *greater* than $L_c/2$ below the main fracture plane. Now real fibers have weak points from place to place along their length and vary statistically in strength from fiber to fiber, so a fiber in the path of the crack is broken either where it crosses the plane of the crack (i.e., where the stress is greatest) or at a flaw in the fiber that is near, but not necessarily on, the plane. Thus, fracture surfaces appear "whiskery," and real composites (having continuous filaments but weak points) possess properties between the two extremes shown by those with full pull-out and no pull-out.

On statistical grounds, it is argued that the average pull-out length for random filament fracture is $L_c/4$ because the shortest is 0 and the longest $L_c/2$. This provides a

├ 20 μm ┤

**Figure 17-16**  Stubblelike fracture surface of an aligned graphite–epoxy composite. (Photograph courtesy of W. H. Durrant.)

means of approximately estimating the interfacial stress that existed before the composite broke.

**Example 17-9**

Pull-out lengths average 1 mm on a fracture surface of a filamentary composite whose fibers have $(\sigma_f)_{max}$ = 2.5 GPa and $d$ = 50 $\mu$m. Estimate the interfacial bond strength.

**Solution**   $L_c/4$ = 1 mm, so $L_c$ = 4 mm. $L_c = (\sigma_f)_{max}d/2\tau$, so

$$\tau = \frac{(\sigma_f)_{max}\, d}{2\,L_c} = \frac{(2.5 \times 10^9)(50 \times 10^{-6})}{2(4 \times 10^{-3})} = 15.6 \;\; \text{MPa}$$

When a filament fractures off the main fracture plane, debonding must take place above, below, and around the filament fracture to allow the main crack to propagate (Fig. 17-17). After the interfacial bond has been broken, there is an interfacial frictional stress $\tau'$ opposing the fibers coming out of their holes. The friction stress $\tau'$ can be as large as the interfacial stress $\tau$ before debonding (if the interfacial break occurs by slip in the matrix adjacent to the fibers), but usually it is somewhat less. Even so, work must be done pulling fibers out as the original main crack front propagates. This work contributes *significantly* to the toughness of composites.

The work of pulling out an isolated fiber can be determined as follows. In Fig. 17-17 a fiber of diameter $d$ has broken a distance $h$ below the main crack plane, where $0 < h <$ $L_c/2$. It is shown as having been pulled out a distance $x$. The total force on the cylindrical debonded area opposing pull-out at that instant is $\tau'\pi d(h - x)$. When pulled out a further distance $dx$, the work done is $\tau'\pi d(h - x)\, dx$. The total work of pull-out for the isolated fiber over the whole distance $h$ is therefore

$$\int_0^h \tau'\pi d(h - x) \; dx = \frac{\tau'\pi dh^2}{2}$$

As already explained, the shortest pull-out length is 0 and the longest is $L_c/2$; this sets limits on values taken by $h$ over a number of filaments. Thus, the average pull-out work per filament is given by

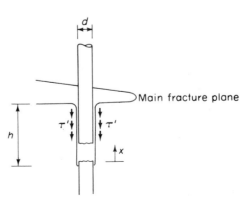

**Figure 17-17**   Debonding between filament and matrix necessary to permit main crack to propagate when fiber fractures off the principal plane of fracture.

$$\frac{1}{L_c/2}\int_0^{L_c/2} \frac{\tau' \pi dh^2}{2} \, dh = \frac{\tau' \pi dL_c^2}{24} \qquad (17\text{-}14)$$

If there are $N$ filaments in the crack advance area $A$,

$$v_f = \frac{N\pi d^2}{4A}$$

The pull-out work of these $N$ filaments divided by the advancing crack area represents the pull-out contribution $R_{\text{pull-out}}$ to the composite fracture toughness. Therefore,

$$R_{\text{pull-out}} = \frac{N(\tau' \pi dL_c^2 / 24)}{A} = \frac{v_f \tau' L_c^2}{6d} \qquad (17\text{-}15)$$

$$= \frac{v_f \tau' d(\sigma_f)_{\text{max}}^2}{24\,\tau^2} \qquad (17\text{-}16)$$

because $L_c = (\sigma_f)_{\text{max}}\, d/(2\tau)$.

**Example 17-10**

Determine the pull-out contribution to toughness in a carbon-fiber–polyester-matrix composite with the following properties: $v_f = 0.4$, $(\sigma_f)_{\text{max}} = 1.6$ GPa, $d = 8\ \mu$m, $\tau = 7$ MPa, $\tau' = 1$ MPa.

**Solution**    $R = (0.4)(8 \times 10^{-6})(1.6 \times 10^9)^2(1 \times 10^6)/24(7 \times 10^6)^2 = 6.97$ kJ/m$^2$. Notice that $L_c = (\sigma_f)_{\text{max}}\, d/(2\tau) = (1.6 \times 10^9)(8 \times 10^{-6})/2(7 \times 10^6) = 0.91$ mm, so the longest pulled-out fiber on a stubbly fracture surface would be about 0.5 mm in length, and the shortest, 0 mm.

Some other contributions to composite toughness come from the work of debonding and the creation of new surfaces, both in the main fracture plane and in the cylindrical areas around the pulled-out filaments.

Composite fracture toughness is the sum of all the various dissipative work components per unit cross-sectional area of fracture *referenced to the apparent (projected) area of the main crack*. As cylindrical debond areas and pull-out lengths are related to events off the main fracture plane, a consequence of dividing the total work of crack propagation by merely the projected area of the main matrix crack—and not the actual total area of new surfaces, which would include the cylindrical debond areas—is that a *synergism* in toughness occurs. That is, the toughness of a composite is greater than the sum of the toughnesses of the components. Thus, even though many filamentary composites are made from components that are individually brittle, the composites can have respectable resistance to crack propagation. The total toughness of typically laid-up composites with about 50% volume fraction of fibers may be 50 kJ/m$^2$. This should be contrasted with the fracture toughness of various other solids, shown in Fig. 17-18.

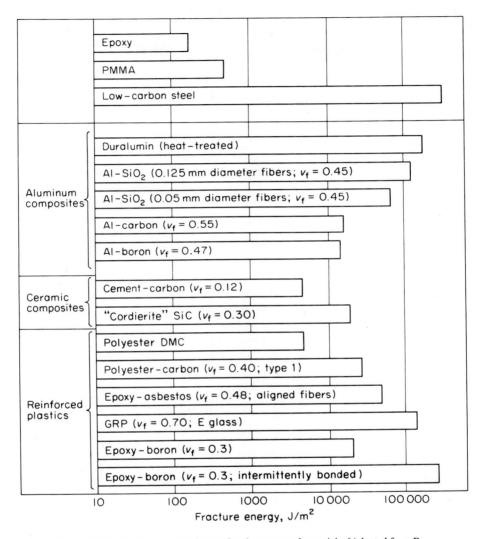

**Figure 17-18**  The fracture toughness of various types of materials. [Adapted from B. Harris, *Composites,* 3 (1972), p. 132.]

Another aspect of the cracking behavior of filamentary composites is that partially pulled-out fibers bridge the crack opening in the wake of the advancing matrix crack. The effect is enhanced the longer the pull-out lengths, as discussed in the next section. Cracked composites thus have an ability to hang together in situations where other solids of similar toughness would break apart. Note that the full contribution to toughness given by pull-out is achieved only when the filaments have pulled all the way out, at which time they no longer bridge the crack faces.

## HIGH TOUGHNESS AND HIGH STRENGTH COMBINED

Recall the rule of mixtures for composite strength [Eq. (17-12)]:

$$(\sigma_c)_{max} = (1 - v_f)(\sigma_m)_{max} + v_f(\sigma_f)_{max}\left(1 - \frac{L_c}{2L}\right) \qquad (17\text{-}12)$$

$$= (1 - v_f)(\sigma_m)_{max} + v_f(\sigma_f)_{max}\left[1 - \frac{(\sigma_f)_{max}d}{4\tau L}\right] \qquad (17\text{-}17)$$

This demonstrates that for greatest strengths the interfacial bonding must be strong, that is, $\tau$ must be large. Much research has gone into the production of strong interfaces to this end.

However, Eq. (17-16) for the pull-out contribution to toughness suggests that $\tau$ must be *small* to obtain tough composites. When $\tau$ is small, $L_c$ is long, so the fibers pull out over long lengths with a correspondingly large dissipation of toughness work. It can be shown that when the other contributions to total toughness are also considered, high-toughness composites are produced when $\tau$ is small. Thus, when a filament fractures in a composite that displays a strong interfacial bonding, a crack is formed in the matrix around the fracture, and usually this crack is energetic enough to run through the composite, breaking fibers as they are met in a "zipper" action. The work of fracture is very low.

High strength and high toughness therefore appear to be mutually exclusive. For reasons explained in Chap. 14, this is not a satisfactory situation, and full use cannot be made of the high strength because the composite is relatively brittle. Various means of getting around this impasse have been proposed.

In one method, the high tensile strength characteristic of strong interfacial filament matrix bonding can be combined with the high fracture toughness of weak interfacial bonding when the filaments are arranged to have random alternating sections of high and low shear stress (and locally low and high interfacial toughness). Such weak and strong areas can be achieved by intermittent coating of the fibers with appropriate substances (Fig. 17-19). The strong regions (especially if longer than $L_c/2$) ensure that the filament strength is picked up; weak areas randomly in the path of running cracks provide the source of toughness and pull-out work.

Extra-long pull-out lengths may also be achieved when a special mechanism of debonding at the filament–matrix interface takes place ahead of the running crack. This *Cook–Gordon mechanism* involves tensile stresses generated *parallel* to the crack, which can produce debonding and crack-tip blunting when sufficiently weak interfaces are present ahead in the path of the crack (Fig. 17-20). Such an effect is normally absent in conventionally made composites because they have continuously strong bonds. If the interfacial conditions were appropriate to allow Cook–Gordon debonding in conventional composites, the interfacial $\tau$ would probably be too low to give adequate tensile strength. Some results for intermittently bonded boron–epoxy composites are shown in Fig. 17-21. Two types of coating materials were used, one (polyurethane) that encouraged Cook–Gordon debonding and another (silicone grease) that did not. The differences in toughness with and without the

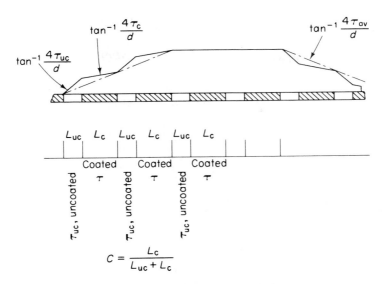

$$\tan^{-1} \frac{4\tau_c}{d} \qquad \tan^{-1} \frac{4\tau_{av}}{d}$$

$$\tan^{-1} \frac{4\tau_{uc}}{d}$$

$$C = \frac{L_c}{L_{uc} + L_c}$$

**Figure 17-19**   Intermittent bonding of filaments (alternating bands of high and low interfacial stress and low and high toughness) in order to achieve simultaneous high strength and high toughness. Subscript c means coated with low shear strength material; subscript uc means uncoated. [From A. G. Atkins, *J. Mat. Sci.*, 10 (1975), p. 819; adapted from studies of Heimbuch, Marston, Atkins, and Felbeck.]

enhanced pull-out lengths are clearly shown. As there are no significant differences in tensile strength, it suggests that a distinction should be made between the property of interfacial shear strength and the property of interfacial toughness in the coatings.

Related to the foregoing is the study by J. E. Gordon and G. Jeronimidis on wood, the toughness of which (about 9 kJ/m$^2$) is exceptionally high in relation to its other mechanical properties and to its density. The specific work of fracture is greater than the energy likely to be absorbed by fiber pull-out, so a special energy-absorbing mechanism seems to exist for

**Figure 17-20**   The Cook–Gordon debonding and crack-blunting mechanism caused by tensile stresses *parallel* to the crack, which can debond sufficiently weak interfaces in the path of the crack.

**Figure 17-21** (a) Fracture toughness and (b) tensile strengths of 0.25 $v_f$ intermittently bonded boron–epoxy composites (filament diameter 140 $\mu$m) as a function of $C$, which is the ratio of coated length to pattern-repeat distance along the fiber (repeat distances variously 19, 25, and 51 mm). Much improved toughness results from polyurethane coatings, which permit Cook–Gordon debonding. [From A. G. Atkins, *J. Mat. Sci.*, 10 (1975), p. 819; adapted from studies of Heimbuch, Marston, Atkins, and Felbeck.]

trees and presumably for other plants. The cellular microstructure of timber is like an array of drinking straws. The cellulose fibrillae of which the cells are mainly composed are disposed in a preponderantly helical manner. When pulled in tension, the cell walls buckle in such a way that longitudinal extensions of the cell up to 20% are possible. Thus, although the elastic fracture strain in wood as a whole seldom much exceeds 1%, cells close to the fracture surface typically extend 15 to 20% before breaking, absorbing much energy as they do so. Arrays of model cellulose fibers, made by winding glass and carbon filaments into hollow helices with resin, exhibit experimental fracture toughnesses up to 400 kJ/m$^2$ [cf. Fig. (17-21)].

**Example 17-11**

Thin-walled internally pressurized composite cylinders are manufactured with helically wound filaments. Show that the optimum angle of winding should be about 55° to the axis of the cylinder. Explain why, in practice, the helices are crossed.

**Solution**   In a closed thin cylinder of diameter $D$ and wall thickness $t$, internal pressure $p$ loads the system so that a hoop stress $\sigma_{hoop} = pD/2t$ and an axial stress of $\sigma_{ax} = pD/4t$ are induced (Chap. 2). Assume that the loads are taken entirely by the filaments. At fracture, the total load is $\sigma_f A_f$, where $A_f$ is the total cross-sectional area of all filaments. This load is aligned along the direction of the filaments at angle $\alpha$, say, to the axis (Fig. 17-22). The components of load in the hoop and axial directions are $(\sigma_f A_f) \sin \alpha$ and $(\sigma_f A_f) \cos \alpha$, respectively. The resolved cross-sectional areas over which these loads act are $A_f/\sin \alpha$ and $A_f/\cos \alpha$, respectively. Hence, the hoop stress is $(\sigma_f A_f) \sin \alpha /(A_f/\sin \alpha) = \sigma_f \sin^2 \alpha$ ; similarly, the axial stress is $\sigma_f \cos^2 \alpha$. The ratio of hoop stress to axial stress is thus $\sin^2 \alpha /\cos^2 \alpha = \tan^2 \alpha$ . But under a pressure $p$, these stresses are in the ratio 2:1 [$(pD/2t)/(pD/4t)$]. Optimum use of reinforcement implies equal likelihood of fracture in any direction. Hence, $\tan^2 \alpha = 2$ or $\alpha = 54°44'$. Crossed helices produce balanced strength and torsional stability and thus prevent distortion when under pressure.

# DESIGNING WITH FILAMENT-REINFORCED COMPOSITES: HYBRID COMPOSITES

Laminates of thermosetting resins (mainly polyesters and epoxies) and glass fibers, glass reinforced plastics (GRP), are extensively used as medium-cost materials for a variety of products ranging from car and boat shells to large tanks and pipes. For many structures

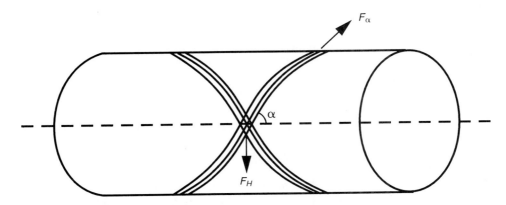

**Figure 17-22**   Helically wound filament reinforced tube. The fiber angle is $\alpha$, $F_\alpha$ is the force in a fiber, and $F_H$ is the circumferential force in the cylinder wall.

where the design parameters are deflection limits of buckling stability rather than ultimate strength, the relatively low elastic modulus of GRP is a limiting factor. Composites based on higher-modulus fibers, such as boron, carbon, and Kevlar, have greatly superior specific strengths, but their use is restricted by their high price. Even so, production aerospace structural members and some automotive components are now being manufactured (at least in part) with these stiffer filaments, particularly carbon fibers. Moreover, sporting goods such as shafts for golf clubs and frames for tennis racquets are available in carbon-fiber-reinforced plastics (CFRP).

In an attempt to bridge the cost–performance gap between GRP and, say, CFRP, small amounts of stiff fibers can be added to GRP. Such *hybrid* composites containing more than one material may be of various types and different construction, depending on the desired properties and design objectives. Stiffness and strength of laminated structures can be increased substantially by the addition of appropriate surface layers, such as carbon filaments on glass-fiber-mat-reinforced plastics.

In practical structures, the members that have to withstand compression usually make up a far higher proportion of the mass of the structure than those intended to withstand tension because of the need to prevent buckling. The geometry of cross sections becomes important. In such circumstances, optimum design for minimum mass no longer depends simply on $E/\rho$ but may be shown to depend on $\sqrt{E}/\rho$ for columnar structures or $\sqrt[3]{E}/\rho$ for panels; an $E/\rho^3$ factor arises, too, where maximum bending stiffness is needed with fixed mass (Example 17-4). It is not widely appreciated that the theoretical upper limit for composite toughness, based on fiber pull-out mechanisms, is too low for composites to compete successfully against metals in certain engineering applications. To achieve a fracture toughness on the order of hundreds of kilojoules per square meter in composites, energy must be absorbed in regions which on a molecular scale are extremely remote from the fracture plane. The tensile buckling mechanism achieves this aim, and model helically wound structures made by Gordon and Jeronimidis have works of fracture comparable to those of ductile metals. The success of certain thermoplastics such as polyethylene and nylon in replacing metals is partly due to their inherent toughness. But often they are not stiff enough; that is, they have too low a $\sqrt[3]{E}/\rho$ factor, particularly in large components. Conventional composites are stiff but not tough enough.

The Gordon–Jeronimidis helical tube solution offers the best of both worlds because high $\sqrt[3]{E}/\rho$ values can be achieved just as well by reducing $\rho$ as by increasing $E$. Holes are cheaper than carbon fibers, and the helical tube construction for synthetic composites provides holes and high work of fracture. It is a direct application to engineering of principles found in biological materials.

Appendix 13 lists representative densities and mechanical properties for a range of composite materials. Designers should remember that practical composites do not always achieve the moduli and strengths predicted by theory, or the same values obtained in the laboratory, as a result, for example, of difficulties in manufacturing large homogeneous structures (whether the fibers are aligned, in cloths, or random). Reasonable agreement with ideal behavior occurs when the loading is predominantly tensile and filaments regularly aligned (as in helically wound reinforced tubes). It is when major loadings are compressive that problems arise, as already mentioned, and then the performance of practical composites falls well below ideal behavior. Similar comments relating to tensile or compressive stress fields apply in fatigue (Chap. 18).

## METAL MATRIX COMPOSITES

Metal matrix composites (MMCs), like polymer matrix composites, may be reinforced with continuous filaments, discontinuous filaments, whiskers, or particles. They are often manufactured by the powder metallurgy route. Typical matrix materials include alloys of aluminum, magnesium, titanium, and copper, whereas typical reinforcing agents are alumina, graphite, silicon carbide, and boron carbide. The most widely available reinforcing material is silicon carbide (SiC). In continuous fiber or monofilament form ($d > 1 \ \mu m$, say), it has impressive material properties: tensile strength 3.75 GPa, Young's modulus 400 GPa, and specific gravity 3.4. The better types of graphite reinforcement are not quite as strong (2 GPa tensile strength) but are stiffer ($E = 800$ GPa) and lighter with specific gravity 2.1. Such low-density reinforcements significantly improve the specific strength and modulus of most metals, with the exception of lighter matrix materials such as aluminum alloys.

Physical properties may be predicted in the usual way using the rule of mixtures for continuous reinforcement based on the concept of shear lag where the load is transferred to the filaments from the matrix (critical transfer length, etc., Fig. 17-10). With discontinuous-reinforced MMCs, the matrix bears a significant part of the applied load, so matrix strength and toughness play an important role in the overall composite behavior. Advantage can be taken of the various strengthening mechanisms available for the matrix itself (e.g., precipitation hardening, dislocation strengthening; see Chap. 9), as well as improvements brought about by the reinforcement. That is, MMCs may be heat-treated to advantage. It was remarked in Chap. 10 that the microstructures of dual-phase steels and particulate-reinforced MMCs are alike in many ways. Because particulate reinforcements in MMCs are large in size ($1 \sim 100 \ \mu m$) compared with the size of precipitates ($<10$ nm, say) and dispersoids ($10 \sim 500$ nm, say), there is essentially no contribution toward dislocation impedance from the particle reinforcement itself. Rather the favorable properties of the reinforcement directly enhance the composite properties.

While aluminum-based MMCs are relatively inexpensive quasi-isotropic lightweight structural materials, they have the drawback of relatively low ductility as measured by tensile fracture strain, and therefore, by implication, poor formability often requiring expensive machining. For example, SAE 2124 aluminum alloy reinforced with 17% $v_f$ 3-$\mu m$ SiC particles has a tensile elongation to fracture of 12% after solution heat treatment at 530°C for 55 min followed by a water quench ($E = 99$ GPa; 0.2% offset yield stress 440 MPa; tensile strength 610 MPa). This is the manufacturer's recommended route for "highest ductility." Overaged samples (250°C for 120 min) display lower flow stresses and smaller elongations. There is, however, a significant difference in fracture behavior: The solution-treated material is brittle with hardly any necking in tension, whereas the overaged material necks considerably before fracture (even though the overall elongation is less). Paradoxically, in compression, the two types of material behave in an opposite way.

Solution heat-treated samples have lower ductilities—18% reduction in test piece height before peripheral surface cracking compared with 43% for overaged samples. The differences arise because in tension microstructural damage leading to fracture occurs by hole growth at the reinforcing particles, whereas in compression the deformation becomes localized into "shear bands" in which final fracture occurs by shear cracking [Fig. 17-23(a) and (b)].

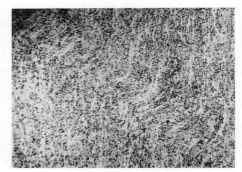

**Figure 17-23**    Aluminum–80% lithium matrix reinforced with 17% $v_f$ 3-$\mu$m SiC particles. (a) Cross section through tensile fracture surface; failure by coalescence of voids formed at reinforcing particles. (b) In compression, deformation is localized to shear bands in which fracture occurs by cracking in shear. Here the direction of loading is vertical for both micrographs. (Courtesy of Jian Jiang and Bradley Dodd.)

The improved cold compressive ductility produced by overaging suggests a new processing technique of commercial importance for aluminum-based MMCs whereby components are hot-formed to near final shape, followed by appropriate heat treatment, and then are precision cold-forged to final dimensions.

## CONCRETE MATERIALS

Conventional concrete contains Portland cement, aggregate, and water. A wide range of properties can be produced depending on the proportions of the mixture and the type of aggregate. The cement reacts with the water to form a matrix for the aggregate, the hydration sometimes continuing over many years if plenty of water is present. An adequately strong material can usually be produced in a few hours at outdoor temperatures above freezing; factory-made concrete products may be heat-cured in order to speed up production. The strength of the cementative matrix is governed mainly by porosity, which depends on the initial water/cement ratio and the extent of hydration at the time in question; consequently, strengths increase with time, and it is customary to proof-test concretes for reference purposes 28 days after manufacture. Subsequently the strength of concretes may increase by 25% a year later.

Conventional concretes have relatively low fracture toughness and hence low tensile strengths (about 3 MPa), but they do have reasonable compressive strengths (greater than 10 MPa). It is customary in civil engineering to use steel reinforcing bars to help concrete structures carry tensile loads, and if these bars are tensioned, it is possible to put the concrete into compression and effectively increase the loads that can be carried without cracks forming in the concrete. In this way, the flexural strength of tensioned reinforced concrete can be doubled over that of simple concrete.

Traditional solid reinforcing bars are comparatively large (10 ~ 50 mm in diameter), and even if steel wire is used in place of solid rods, individual strands are usually not less than a few millimeters in diameter; similarly, steel wire mesh reinforcement is not particularly small. In recent years, however, the types of filaments already described in this chapter have been successfully incorporated into concrete and cement. Fibers of steel, glass, polypropylene, and so on, can increase the tensile strength of cements to perhaps 20 MPa. An added bonus is that it becomes possible to fabricate thin sheets of fiber-reinforced cements that maintain their integrity during hardening; such techniques have long been known in the asbestos cement industry. The toughness of an asbestos cement is about 700 $J/m^2$; high-alumina cement reinforced with glass fiber can have toughnesses of about 20 $kJ/m^2$.

Concrete impregnated with polymers can also be made with compressive and tensile strengths much improved over those of conventional concrete. Polymers such as PMMA, PS, and PVC may be incorporated either as aqueous emulsions during the mixing process (when the water in the emulsion reacts with the unhydrated cement to develop strength and stiffness in the usual way) or impregnated as monomers into already hardened conventional concrete, with subsequent polymerization being effected by gamma radiation or by a thermal catalytic process. Concrete made in these ways possesses improved resistance to attack from acids.

In comparison with other engineering solids, concretes require quite small amounts of energy for their production whether calculated on the basis of energy per unit mass, per unit stiffness, per unit strength, and so on. Appendix 14 gives a broad indication of the range of values for various engineering solids. The data for different engineering materials are very revealing, particularly when the achievement of specified performance criteria using low-energy materials is of importance.

## REFERENCES

1. Clyne, T. W., and P. J. Withers, *An Introduction to Metal Matrix Composites*. Cambridge: Cambridge University Press, 1993.
2. Cottrell, A. H., *The Mechanical Properties of Matter*. New York: John Wiley, 1964.

3. Gordon, J. E., *The New Science of Strong Materials—Or Why You Don't Fall Through the Floor.* London: Penguin, 1968.

4. Jeffries, R. F., and D. J. Burrows, "Metal Matrix Composite Packaging of Microwave Circuitry." *GEC Journal of Research*, 11, 3 (1994), p. 150.

5. Jiang, J., and B. Dodd, "Workability of Aluminium-based Metal-Matrix Composites in Cold Compression." *Composites*, 26, 1 (1995), p. 62.

6. Kelly, A., *Strong Solids.* Oxford: Clarendon Press, 1964.

7. Ludema, K. C., R. M. Caddell, and A. G. Atkins, *Manufacturing Engineering.* Englewood Cliffs, N.J.: Prentice Hall, 1987.

8. National Engineering Laboratory, Glasgow, *Designing Hybrid Composite Materials.* Research Summary No. 633, 1977.

9. Pomeroy, C. D., "Concrete, An Alternative Material." Fourteenth John Player Memorial Lecture. Institution of Mechanical Engineers, London. *Proceedings,* 192, 14 (1978), p. 135.

10. Wainwright, S. A., W. D. Biggs, J. D. Currey, and J. M. Goseline, *Mechanical Design in Organisms.* London: Edward Arnold, 1976.

## PROBLEMS

**17-1.** Describe the properties of high-performance composites that make them attractive in spite of their high cost.

**17-2.** It is proposed to hang long specimens in a vertical hole 1 m in diameter in the ground, loaded only by their own mass. Find the maximum lengths that can be suspended for each of the following materials (neglect changes in gravity with the depth of the hole):

    **(a)** 1020 steel.

    **(b)** Glass-reinforced tape.

    **(c)** Boron–epoxy

    **(d)** Graphite whiskers

| Material | Average density, $Mg/m^3$ | $S_u$, MPa | $E$, GPa |
|---|---|---|---|
| 1020 steel | 7.85 | 500 | 210 |
| Glass-reinforced tape | 1.90 | 770 | 27 |
| Boron–epoxy | 2.34 | 1800 | 250 |
| Graphite | 1.65 | 20,000 | 500 |

**17-3.** Calculate the stress that can be withstood by a boron–epoxy composite having $v_f = 0.4$ at some strain level where, separately, the boron filaments require a stress of 1.4 GPa and the matrix 81 MPa.

**17-4.** Determine the critical lengths of carbon fibers 8 $\mu$m in diameter possessing $(\sigma_f)_{max} = 1.6$ GPa in a polyester matrix with $\tau = 20$ MPa. Likewise, determine $L_c$ for boron–epoxy with fibers 100 $\mu$m in diameter with $(\sigma_f)_{max} = 3$ GPa and $\tau = 70$ MPa.

**17-5.** In a tensile test of a monofilament of boron embedded in an epoxy matrix, several breaks occurred in the filament before the specimen fractured. Explain clearly how it is possible to develop a stress high enough to cause a second fracture in the broken fiber after it breaks the first time. Draw a sketch of the broken filament in order to make your explanation clear.

**17-6.** Mylar tape of 34.2 MPa strength is reinforced with continuous glass filaments to produce a composite strength of 356 MPa. If the cross section of the tape has the dimensions given in Fig. 17-24, calculate the strength of the glass filaments.

**Figure 17-24**

**17-7.** A composite specimen contains parallel graphite filament bundles 135 $\mu$m in diameter aligned in a square pattern (shown in Fig. 17-25 in cross section) in a matrix of epoxy.
**(a)** Calculate the volume fraction of filament in this specimen.
**(b)** The graphite strength is 2.93 GPa, and the epoxy stress at the fracture strain of the graphite is 83.7 MPa. Estimate the strength of the specimen.

**Figure 17-25**    157 $\mu$m

**17-8.** Graphite filaments 7.00 $\mu$m in diameter and having fracture strength 850 MPa (at a failure strain of 0.0140) are to be cast in a linear elastic epoxy matrix of strength 112 MPa (with a failure strain of 0.0670). If the filaments are to be aligned in a square matrix with the distance between filaments at least one-fourth the filament diameter (Fig. 17-26), calculate the maximum strength the composite will have.

**Figure 17-26**

**17-9.** You wish to manufacture a composite tape made of a single layer of aligned continuous glass filaments 125 $\mu$m in diameter, which have a fracture strength of 700 MPa, embedded in a matrix of resin that has a strength of 35 MPa at the same fracture strain as the glass filaments. If the tape is to be 208 $\mu$m thick and 50 mm wide and to have a strength of 200 MPa, how many strands of glass filament must it contain?

**17-10.** Tensile data for a continuous-filament glass-fiber-reinforced tape are as follows:

| Test | Fracture Stress, MPa | Orientation of Tensile Load |
|------|----------------------|------------------------------|
| A | 120 | Longitudinal to fibers |
| B | 20 | 45° to fibers |
| C | 32 | Transverse to fibers |

    **(a)** Explain the difference between the strengths in test A and test C.
    **(b)** Explain the difference between the strengths in test B and test C.

**17-11.** The world record for pole vaulting stood at 15 ft (4.6 m) in 1940. Fiberglass poles replaced bamboo poles in 1962, and by 1970 the world record stood at 18 ft (5.5 m). Other things being equal, explain how the difference in moduli between bamboo poles (about 45 GPa) and glass-fiber poles (about 40 GPa) has contributed to the improved vaulting record.

**17-12.** Show that flexural stiffness is proportional to (beam thickness)$^3$ and that consequently for a given length, width, and mass of sheet material, the specific stiffness in bending varies as $E/\rho^3$ rather than as $E/\rho$, which is used in Apps. 12 and 13 for simple tension. ($\rho$ is density.)

**17-13.** Show that in a unidirectional glass-fiber composite the maximum bending stiffness per unit mass is obtained at 25% volume fraction fibers, using $E_f/E_m = 20$ and 2.5 Mg/m$^3$ for the density of the glass filaments and 1.0 Mg/m$^3$ for the density of the laminating resin.

**17-14.** Show that in a composite for which $(\sigma_f)_{max}$ of the filaments is much greater than $(\sigma_f)_{max}$ of the matrix, the maximum bending strength is obtained when the mass of reinforcement is one-half the total mass. (This corresponds to the optimum skin mass fraction for maximum bending strengths of sandwich beams.) Explain why an optimum volume fraction exists and why in bending it is not useful to aim for the highest fiber volume fraction possible. (Your answer will be relevant to Prob. 17-13 as well.)

**17-15.** A test for fracture toughness of a cross-ply boron–epoxy composite yields the load extension curve shown in Fig. 17-27. Calculate its fracture toughness.

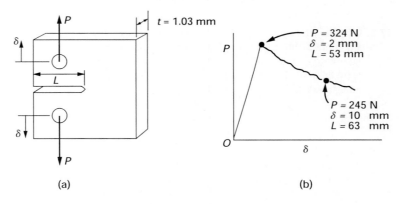

**Figure 17-27**

**17-16.** Data for a compact tension specimen of molded fiberglass of thickness 2.63 mm and crack growth increment of 6.35 mm between points 1 and 2 are given in Fig. 17-28. Calculate the fracture toughness.

**Figure 17-28**

**17-17.** Compacts of dry cement powder that are subsequently hydrated can be made with compressive strengths of over 370 MPa. Explain why they are not used commercially and suggest modifications that would permit their wider use.

**17-18.** Hardened cement paste is so alkaline that rusting of steel reinforcing material is inhibited in the absence of ingress of surface water down cracks. What is likely to happen to ordinary glass filaments if used for fiber reinforcement?

**17-19.** "Holes are enormously cheaper, both in money and in energy, than any conceivable form of high-stiffness material" (J. E. Gordon). Discuss.

**17-20.** The stress–strain characteristics of linear elastic graphite filaments 7 $\mu$m in diameter and of a linear elastic epoxy are given as follows:

| Material | Fracture Stress, GPa | Fracture Strain |
|---|---|---|
| Graphite filaments | 3.00 | 0.010 |
| Epoxy | 0.50 | 0.050 |

If you are limited to a filament volume fraction of 0.35, calculate the maximum load that can be supported by a rod 10 mm in diameter of a composite of aligned graphite and epoxy.

**17-21.** Figure 17-29 shows a crease compression failure in 40% $v_f$ carbon-fiber-reinforced epoxy. Discuss how this mode of failure arises.

**17-22.** Show that a volumetric strain of magnitude

$$\frac{\cos(\alpha - \gamma)}{\cos \alpha} - 1$$

occurs in the sheared slice in Fig. 17-29, where $\gamma$ is the shear strain in the slice and $\alpha$ is the inclination of the slice to the horizontal in the photograph. Discuss how such a volume increase may be accommodated in a fiber-reinforced composite and show that a limit is set on the axial displacement associated with this type of failure.

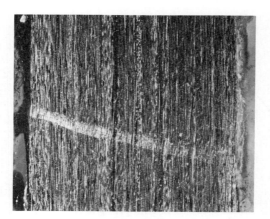

**Figure 17-29**   Specimen width is 2.5 mm.
(Courtesy of C. R. Chaplin.)

**17-23.** An open cylinder whose wall contains inextensible fibers in a crossed helical array can have shapes varying from very long and thin with a small fiber angle $\alpha$ to very short and fat with a large fiber angle (Fig. 17-22). At both extremes the volume approaches zero. Show that maximum volume occurs when $\alpha = 54°44'$.

**17-24.** Discuss the changes in cross-sectional shape of helically wound tubes as the internal pressure is altered. Hence, explain why worms can change shape easily.

**17-25.** Explain how the coefficient of thermal expansion and the thermal conductivity of MMCs may be tailored to give properties favorable for the use of such materials in electronic packaging.

# 18

# Fracture by Gradual Crack Growth: Fatigue and Stress-Corrosion Cracking

Under certain circumstances, cracks can grow gradually in a structure at average stresses below the yield stress. If this growth progresses far enough, the crack reaches a critical size that will lead to catastrophic growth (see Chap. 14) or the crack so reduces the net cross section that the tensile strength is exceeded (see Chap. 2). In either case, the structure fails suddenly and without warning. A large proportion of service fractures occur in this fashion.

The circumstances that can cause gradual growth of a crack are associated with tensile stress that may or may not be constant over a period of time, as follows: (1) *fatigue,* when the stress in a part varies cyclically with time, and (2) environmentally promoted crack growth, usually called *stress-corrosion cracking* or *hydrogen embrittlement,* which can occur in the presence of continuous tensile stresses. Fatigue seldom occurs without acting in concert with some environmental effects, except under carefully controlled laboratory conditions such as, for example, testing in a vacuum. Under service conditions, fatigue and stress-corrosion cracking often occur simultaneously.

This chapter describes the fundamental mechanisms that control these fracture modes and the conditions that promote or restrict crack growth, with the objective of assisting the engineer-designer in avoiding or minimizing gradual crack growth in structures.

## CONDITIONS FOR FATIGUE

Fatigue fracture occurs when a part is subjected to many cycles of varying stress in which at least part of each cycle is tension. The magnitude of stress required to cause fatigue is less when the number of cycles is larger. This interdependence of stress and number of cycles can be measured in a simple *rotating–bending test.* Figure 18-1 shows how a cylindrical specimen with a ground and polished cross section in its midsection is loaded in four-point bending with four equal forces $P$. The bending moment $M$ in the midsection is thus a constant $PL$ over the entire polished region, and the stress $\sigma_t$ at the bottom (tensile) side at the section of minimum diameter $d$ is thus

$$\sigma_t = \frac{32\,M}{\pi d^3} = \frac{32\,PL}{\pi d^3}$$

After the specimen has rotated one half-turn, so this same point is now on the top side, the stress becomes compressive:

$$\sigma_c = \frac{-32\,PL}{\pi d^3}$$

When this same point rotates through the positions midway between top and bottom, its stress is zero. Thus, a single point on the surface of the specimen goes through a complete sinusoidal stress cycle from tension to compression to tension with each full rotation. Although it is not necessary for the cyclic variation in stress to be either as smooth or as repetitive as this, or as symmetrical from tension to compression, this test is commonly used as a standard to measure the maximum possible fatigue life under a given stress. We will examine later in this chapter those variables for a real part that serve to reduce this life. For the present, we note that surface finish is very important in fatigue and that the standard specimens described earlier have ground and polished test sections as shown in Fig.

**Figure 18-1**   Rotating–bending test for fatigue.

18-1. In other mechanical tests with which we have dealt earlier (e.g., simple monotonic tension tests), surface condition is not as important for ductile materials.

Several identical specimens are required to establish fatigue life. For each, a value of load $P$ is selected and the specimen is rotated until it fractures. The testing machine has a cycle counter that records the number of cycles at the instant of fracture. Most tests are of relatively short duration because a machine rotating at 1800 revolutions/min (30 revolutions/s) exposes specimens to about $10^5$ cycles/h and $2.50 \times 10^6$ cycles/day. After the first specimen fractures, a new specimen is tested at a different load $P$, and so on.

Figure 18-2 is a plot of maximum stress $s$ vs. number of cycles $N$ for fatigue fracture of a steel of hardness 48 HRC and of a high-strength aluminum alloy. For this steel, stresses below about 825 MPa do not cause fracture for any number of cycles tested and presumably not for an infinite number of cycles. The stress below which fatigue does not occur in steel is called the *fatigue limit* or *endurance limit* $S_e$.

**Figure 18-2** Results of rotating–bending fatigue tests, in the form of *s–N* curves. [From ASM Committee on Analysis of Fatigue Failures, "Fatigue Failures," *Metals Handbook: Failure Analysis and Prevention,* 8th ed., vol. 10, Howard E. Boyer, ed. (Metals Park, Ohio: American Society for Metals, 1975), p. 96.]

In the absence of tests to determine the fatigue limit for a steel, the value of the mean fatigue limit $\overline{S}_e$ for a polished rotating beam specimen can be estimated from

$$\overline{S}_e = 0.5\, S_u \quad \text{for } S_u \leq 1.4 \text{ GPa}$$
$$\overline{S}_e = 700 \text{ MPa} \quad \text{for } S_u > 1.4 \text{ GPa}$$

Because the fatigue limit exhibits a wide statistical spread, these values may not provide for conservatively safe design. Actual measured values must therefore be used for design. Furthermore, this fatigue limit is reduced by numerous service factors which will be discussed later in this chapter.

Aluminum alloys do not exhibit a fatigue limit. In Fig. 18-2 the slope of the curve for the aluminum alloy indicates that for higher numbers of cycles the stress for fatigue fracture is even less than the lowest value shown in this curve. The "fatigue limit" for aluminum is often quoted as the stress for $500 \times 10^6$ cycles, but no true fatigue limit exists. The absence of a fatigue limit in aluminum alloys leads to measures such as a mandatory requirement for periodic inspection of highly stressed aluminum structures to locate fatigue cracks and replace such parts before the cracks become dangerously large. (See Chap. 14 for a discussion of the influence of crack size on fracture.)

Fatigue in engineering structures exposed to a wide and possibly random spectrum of loads depends on the variable range and absolute magnitude of the stresses produced, so exact simulation of every case is impossible. However, tests are possible with a sinusoidal variation in stress that is different from the equal tension-to-compression cycle of the rotating–bending test. All that is essential is that some portion of each cycle be tensile. For any test that varies from a maximum stress $s_{max}$ to a minimum stress $s_{min}$, the *stress ratio R* is defined as

$$R = \frac{s_{min}}{s_{max}}$$

(The symbol $R$ is also used elsewhere in this book to denote fracture toughness; because both these meanings are in common use, we use $R$ for both. The meaning should always be clear from the context.) Thus, for the rotating–bending test in Fig. 18-1, $R = -1$. A test with $R > 0$, for example, is always in tension. The number of cycles required for fracture thus depends on the stress ratio as well as the magnitude of $s_{max}$.

Other factors include the type of loading (bending, tension, or torsion, for example), the shape of the loading curve (of which sinusoidal is the simplest), the frequency of cycling, the environment, temperature, and specimen size and surface condition (notched, rough, smooth, etc.). Although it is not appropriate here to present the complete formal procedures used in designing to resist fatigue fracture (see Ref. 11 for such discussion), later in this chapter we will describe the influence of some of these variables on fatigue life. In order to understand how these factors operate, it is first necessary to present a brief discussion of the mechanisms of fatigue crack initiation and growth.

### Example 18-1

How do airlines prevent fatigue cracks in their aluminum aircraft from growing to catastrophic size?

**Solution**   Regular and very careful inspection is required to locate and remove potentially dangerous fatigue cracks. Failure to locate growing cracks has led to many aircraft disasters; in most cases these occurred in relatively new designs in which the regions susceptible to fatigue were not completely anticipated.

## FATIGUE CRACK INITIATION

Fatigue would probably not occur below the yield stress in a perfectly homogeneous material that is absolutely smooth on its external surfaces and uniformly loaded. Under elastic stresses, the material would uniformly deform elastically, and when the stress was removed the elastic deformation would be completely reversed and thus be returned to the initial configuration. No matter how many times this stress cycle is repeated, there would be no permanent change in the structure.

It is the nonuniform characteristics of real materials that lead to fatigue under cyclic stress. Consider a polycrystalline metal that contains nonmetallic impurities or inclusions, has natural dislocations that arise from solidification processes, and has an imperfectly smooth exterior surface. When a part made of this metal is stressed in the nominally elastic range, portions of many dislocations move and can contribute to very small, irreversible changes in the configuration of the part. These changes are initially much smaller than can be detected in a stress–strain test, so we consider such a load–unload cycle to be elastic, that is, "fully reversible." In fact, even the first cycle has caused some local irreversible changes.

An example of such an irreversible change under a stress ratio $R = -1$ (recall that $R = -1$ means fully reversed tension–compression) is the *Cottrell–Hull mechanism*, which we have here modified and altered for simplicity. Consider a crystal located at a free surface with a stable internal dislocation configuration such as shown in cross section in Fig. 18-3(a). The dislocations on slip plane $A$ are blocked by a barrier (e.g., dislocations on an intersecting plane) at the inner end of the plane, as are the dislocations (of opposite sign) on parallel slip plane $B$. (As this is a very special configuration, many stress cycles may be required before this kind of mechanism even begins to operate.) When a tensile stress of

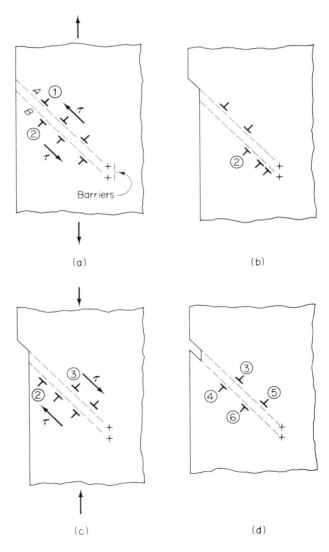

(a)

(b)

(c)

(d)

**Figure 18-3** Modified Cottrell–Hull mechanism for fatigue crack initiation.

sufficient magnitude is applied to the ends of the specimen, the local resulting shear stress $\tau$ causes dislocation 1 on slip plane $A$ to move to the free surface and to leave a step, as shown in Fig. 18-3(b). The dislocations on slip plane $B$ cannot move very far before they pile up against the barrier.

When the applied stress is compressive, as in Fig. 18-3(c), dislocation 2 on slip plane $B$ can now move freely to the surface to produce a second (and opposite) step, as in Fig. 18-3(d). The dislocations on slip plane $A$ pile up against the barrier. Thus, we have the genesis of a fatigue crack, but because its depth is roughly the Burgers vector, on the order of 300 pm, the crack and the change in shape of the specimen are still undetectable. We see here an example of a permanent change in a part previously considered elastic and fully reversible. If there are no further tension–compression cycles, the effects of the emergence of dislocations 1 and 2 in Fig. 18-3 are negligible. The infinitesimal crack formed becomes important only if it grows through many repeated stress cycles with successive emergence at the free surface of dislocations 3, 4, 5, and so on.

It is possible that dislocation 3 is slightly blocked by an impurity atom, and it is not until many cycles later that the local stresses are of sufficient magnitude to move dislocation 3, at which time the process can continue. If this process continues for enough cycles, it will eventually produce a detectable crack. If the minimum detectable crack length is, for example, 10 $\mu$m, then for this one crack in this one grain to be detected would require 10 $\mu$m/300 pm $\approx$ 30,000 cycles at the very least. (It would also require a dislocation-generation mechanism to create the large number of dislocations required.) Thus, in order to create a small initiating fatigue crack, many stress cycles are required.

This initial crack formation during fatigue is called *stage I fatigue*. The fracture surface, which is the new surface created by the slip on two parallel planes shown in Fig. 18-3(d), is therefore the plane of the slip process and is very flat when confined to a single grain. Stage I fatigue usually extends 2–5 grains from the origin, which requires slip in several adjacent grains. We have already seen (Chap. 5) that slip in fcc aluminum occurs on {111} planes in a <110> direction, so we would expect to find stage I fatigue fracture surfaces in aluminum to be {111} planes. Indeed they are, but stage I fracture surfaces in aluminum have also been observed to be {110} and {100}; these observations are not inconsistent with our simple stage I model, as the slip direction <110> also lies in the {110} and {100}. Thus, fatigue initiation processes are more complex than simple slip processes. One example of the appearance of a stage I fatigue fracture surface is shown in Fig. 18-4.

The foregoing mechanism considers only one of many possible ways that dislocations can lead to fatigue cracking. A foreign impurity particle, a surface imperfection, or a plastically deforming grain surrounded by elastic grains can act to cause local stress concentrations that will in similar ways lead to creation and growth of small fatigue cracks. The more severe the initial imperfection, the more rapidly the fatigue cracks are formed. Because of the difficulty of observing the initiation and growth of extremely small cracks in the earliest stages of fatigue and probably also because of the large number of variables involved, the dominant processes at this stage are not yet fully understood.

**Figure 18-4**  Stage I fatigue fracture surface in aluminum alloy 2024-T3 (UNS A92024). Initiation at free surface, pointer at 0, with {111} of each grain as fracture plane. Stage II begins at lower right. [From ASM Committee on Fractography by Electron Microscopy, "Interpretation of Scanning-Electron-Microscope Fractographs," *Metals Handbook: Fractography and Atlas of Fractographs,* 8th ed, vol. 9, Howard E. Boyer, ed. (Metals Park, Ohio: American Society for Metals, 1974), p. 68.]

## FATIGUE CRACK GROWTH

Once a fatigue crack is created, its growth to finite size is controlled by progressively larger scale effects than operated in its earliest stages. It is in this realm that the importance of a tensile component to the stress cycle becomes apparent. This period of crack growth is called *stage II fatigue.*

Because metals and alloys exhibit a wide range of properties, no one mechanism completely describes the events of stage II fatigue, but the essential elements are included in the discussion that follows, for a moderately ductile, medium-strength alloy. Figure 18-5(a) shows the condition at the end of stage I, where four grains have generated slip-induced cracks; each crack passes through a different grain and thus has a different orientation and produces a zigzag path. At this point, the initial crack has become long enough to produce on the next tensile cycle a stress concentration at the crack tip that is high enough to cause local yielding, over the shaded region shown as circular here. The local strain hardening that results then causes this region suddenly to crack transversely under the tensile stress, as in Fig. 18-5(b). The growth of this crack is limited to the width of the damaged region during this cycle, so the crack progresses only across the width of the shaded region. In the context of Chap. 14, here denoting $K_c$ as the initial stress-intensity factor required for crack growth and $K$ as the stress-intensity factor resulting from the applied load, we can say that in the shaded region in Fig. 18-5(a), $K \geq K_c$, so the crack grows, but it stops when it reaches the undamaged region of higher $K'_c$ where $K < K'_c$. It is likely then that when the transverse crack runs out of the strain-damaged region, it will revert for a short distance to the shear mode and create a new, small strain-hardened region (shaded) and then stop.

Under either tensile stress reduction or stress reversal to compression, the crack partially or completely closes up [Fig. 18-5(c)]. Because the region in the vicinity of the new tip has been locally strain-hardened, it may not close up entirely during the compression portion of the cycle. The process is then repeated on the next application of a tensile stress [Fig. 18-5(d)].

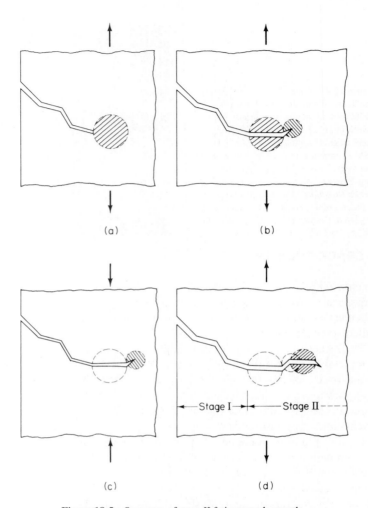

**Figure 18-5**    Sequence of stage II fatigue crack growth.

Stage II fracture is thus very different in operation from stage I fracture and produces a very different fracture appearance. The terminus of crack advance for each cycle in stage II is usually a slight change in the direction of the crack, as shown in Fig. 18-5(d), when the crack extends into the undamaged region. This change in crack contour extends along the full length of the crack, perpendicular to the plane of the diagram, and appears as a narrow band or *striation* on the fracture surface. Figure 18-6 shows the appearance of such striations when the fluctuating stress amplitude is constant; as expected, the advance of the crack is approximately the same distance for each cycle. Figure 18-7 exhibits the typical variation in striation width that results from the variable stress amplitude of a part that fatigued in service. Additional examples of service-fatigue fractures are given here and in Chap. 19.

**Figure 18-6** Fatigue striations in TEM fractograph of Type 302 stainless steel (UNS S30200) at 427°C, $R = 0.2$, $\sigma_{max} = 124$ MPa, which failed after 1.6 million cycles. The nearly vertical dark line at left is a *tear ridge* which connects adjacent patches of striations. Propagation direction is downward. [From John A. Fellows, "Transmission-Electron-Microscope (TEM) Fractographs from Replicas," *Metals Handbook: Fractography and Atlas of Fractographs,* 8th ed., vol. 9, Howard E. Boyer, ed. (Metals Park, Ohio: American Society for Metals, 1974), p. 333.]

**Figure 18-7** Stage II service fracture in a 0.06% C steel pin in reversed bending. This TEM micrograph exhibits several features frequently found in service fractures. The striations are irregularly spaced as a result of varying stress amplitude. Postfracture squashing of the surface, resulting from the compression portion of the stress cycle, can be seen in several locations. The squashing also has partially closed up some surface cracks, which causes the replica to be torn when removed, seen in the micrograph as large, feathery, dark blotches. Note the tear ridge running from upper center to lower right. (See Example 18-2 for a discussion of its fracture growth direction.)

**Example 18-2**

    In which direction is the crack growing in Fig. 18-7?

    **Solution**   Toward the lower right. A discontinuity of some sort, possibly a grain boundary, runs from the upper left to the lower right and acts to retard the growth of the crack. This tear ridge leads to a backward curving, a kind of *drag effect*, of the striations in the vicinity of the discontinuity. This effect, if it occurs, can also be useful in discriminating between postfracture damage marks, or scrapes, and true striations because scrapes (as well as many striations) appear virtually straight and parallel at this high magnification.

    Service fatigue fractures are seldom as clear and orderly as in Fig. 18-6 for at least two reasons: (1) Most service fractures experience stress cycles of varying amplitude, so the resultant stage II striations are irregularly spaced, and (2) most metals do not exhibit striations as smooth and regular as is often found in aluminum alloys. In fact, different metals exhibit a wide range of characteristics of stage II striations. Figure 18-8 shows the appearance of fatigue in a medium-carbon steel.

**Figure 18-8**   SEM micrograph of fatigue striations in a 0.54% C steel I-beam on a crane subjected to repeated impact loads. The dotted marker is 7.5 $\mu$m long.

    Under torsion failure, fatigue striations in steel can be different than in tension. Figure 18-9 shows such striations in steel.

    Fatigue in a magnesium alloy may have an appearance closer to that seen in aluminum, as shown in Fig. 18-10.

    One of the more unusual-appearing fatigue pattern occurs in nodular iron, which consists of small spheroids of graphite surrounded by ferrite and pearlite (see Chap. 15). Ordinary ductile fracture occurs with the crack passing between the graphite spheroids and the surrounding matrix, while fatigue cuts through the graphite spheroids. Figure 18-11 shows the appearance of both modes.

**Figure 18-9** SEM micrograph of fatigue striations from combined tension and torsion in a rigging swivel. The outer surface of the shaft is to the upper right. Note that because this patch of striations is buried in a kind of valley and actually undercuts the wall to the left, these lines could not have resulted from postfracture damage. The dotted marker is 38 $\mu$m long.

Die-cast aluminum can exhibit unique fatigue characteristics. Figure 18-12 shows fatigue fracture in a die-cast transaxle housing.

Prosthetic devices such as total hip replacements, bone plates, and dental implants are made of (1) stainless steel, (2) cobalt–chrome–molybdenum, or (3) titanium–aluminum–vanadium. In most prosthetic applications, fatigue can be a serious problem. Figure

**Figure 18-10** Field emission microscope micrograph of fatigue fracture in a die-cast Mg–Al–Zn alloy motorcycle backing plate. The striations lie in the 10:00–4:00 direction, while the nearly vertical pattern of wavy lines are tear ridges which connect the adjacent planes on which fatigue occurs. Dotted marker is 3.8 $\mu$m.

**Figure 18-11**  SEM micrograph of fatigue fracture (right and center) and ductile fracture (along left edge) of a McPherson strut wheel knuckle of nodular iron. Note that most of the graphite spheroids at the right are cut by the fatigue fracture but at the left are bypassed by the ductile fracture. Note also that the patches of fatigue striations are not precisely parallel to each other. Dotted marker is 380 $\mu$m.

18-13 shows the characteristics of fatigue in austenitic stainless steel, and Fig. 18-14 shows fatigue in cobalt–chrome–molybdenum. Each exhibits unique characteristics.

Service-fatigue fracture often occurs in parts that are not constantly in use, so a stage II fatigue crack may remain at a fixed length for some time. Ordinary rust concentrated at the fracture tip that can occur during these periods leads to a feature clearly visible to the unaided eye called *progression marks*, *clamshell marks*, or *beach marks* (because of their

**Figure 18-12**  FEM micrograph of fatigue fracture in a die-cast transaxle housing of UNS A03800 aluminum alloy. Contrast the irregular intermittent patches visible here with the smooth striations of stainless steel in Fig. 18-6. Dotted marker is 15 $\mu$m.

**Figure 18-13**   SEM micrograph of fatigue fracture in the bone plate of a compression
hip screw of austenitic stainless steel, UNS 31603 (316L), which is 17 Cr–12 Ni–2.5 Mo
and <0.03 C. Striations in the form of irregular ridges are commonly seen in this alloy.
Marker bar is 10 $\mu$m.

**Figure 18-14**   SEM micrograph of fatigue
striations in a cobalt-base dental implant
with 28 Cr–6 Mo. The triangular wedged
pattern across striations seen in this alloy is
one of the most unusual forms of fatigue.

similarity to the marks left in the sand at the ocean shore by the receding tide). Figure 18-
15 shows classic beach marks. A marked change in maximum stress range extending for a
period of time can also cause beach marks, representing the line of demarcation between
regions of high- and low-stress stage II fracture. Progression marks are not a positive indication of fatigue but only of gradual and interrupted crack growth, as progression marks

can also result from interrupted stress-corrosion cracking. This will be discussed later in this chapter.

## FINAL FRACTURE

The last stage of fracture, stage III, is not fatigue at all. When the stage II fatigue crack becomes long enough so that $K = K_c$ under the stress amplitude applied, where $K_c$ is the critical stress-intensity factor of the undamaged metal, the part fractures completely and suddenly. Alternatively, for ductile metals, when the stage II fatigue crack becomes so large that the remaining cross section of metal can fail by tensile overload, the part fails suddenly by a ductile mode. Thus, in either case, the final stage III fracture occurs without prior warning. The sudden fracture of the tractor axle in Fig. 18-15 thus occurred without prior warning and left the rough fracture surface visible in the lower portion of the photograph.

**Figure 18-15**   Fatigue fracture in a tractor axle, showing progression marks over upper two-thirds of fracture surface. The crack initiated at the approximately 11:30 position (where 12:00 is straight up). Crack growth direction is down in the photograph, culminating in sudden fracture over the rough oval-shaped region at the bottom. (Axle courtesy of Richard B. Felbeck. Photograph courtesy of W. H. Durrant.)

It is quite common to find incomplete stage III fracture interspersed with stage II fracture, as in Fig. 18-16. This can occur when insufficient energy is present in the stressed structure to drive the stage III fracture completely through the part. Figure 18-16 shows the propensity of the stage II fatigue fracture to favor growth at the free surfaces of the part (the left and right edges of the figure) and thus to tend to straighten the contour of the crack tip, while the stage III fracture favors growth of the central, internal region of the crack and thus tends to further bend the contour of the crack tip.

**Figure 18-16** Alternate fatigue (light, narrow bands) and ductile fracture in forged 7075-T6 aluminum (UNS A97075). This is a transmitted light photograph of a plastic replica. [From ASM Committee on Use of Fractography for Failure Analysis, "Use of Fractography for Failure Analysis," *Metals Handbook: Fractography and Atlas of Fractographs,* 8th ed., vol. 9, Howard E. Boyer, ed. (Metals Park, Ohio: American Society for Metals, 1974), p. 120.]

## FACTORS INFLUENCING FATIGUE

For a fatigue crack to initiate and grow, the following are essential:

1. The stress must vary.
2. At least a portion of each stress cycle must be tensile.
3. The crack must initiate at or very near a free surface or a large discontinuity.

It is obvious that a semi-infinity of possible combinations of stress–time history can lead to fatigue in a given part and that the introduction of additional variables even further complicates the problem of predicting fatigue fracture. These variables have been simplified for design purposes, and some of the references (Refs. 10 and 11) cited at the end of this chapter describe the procedures used in design. We will therefore discuss here the practical service factors that can alter fatigue life from the point of view of how these factors influence the initiation and propagation of fatigue cracks.

The *stress level* has already been discussed in terms of the *s–N* curve (Fig. 18-2), for fully reversed tension–compression. Another way to demonstrate sensitivity of fatigue to stress is to show the rate of growth in fatigue of preexisting cracks of length *a*. Figure 18-17 shows the effect of stress range $\Delta\sigma = \sigma_{max} - \sigma_{min}$ on the rate of growth of a crack in identical specimens, presumably for the same stress ratio *R*. Thus, any increase in $\Delta\sigma$ will increase crack growth rate and decrease fatigue life. Stress concentrations resulting from details such as notches, fillets, holes, keyways, splines, gear teeth, and screw threads all lead to higher fatigue crack growth rates. When these factors are unavoidable, the designer must accordingly take account of their weakening effect.

Note in Fig. 18-17 that the slope of each curve at any point is *da/dN* and that this value increases as *N* increases, which means that as the crack grows longer it grows at a faster rate. This should be expected because for a given $\Delta\sigma$ a longer crack leads to a higher

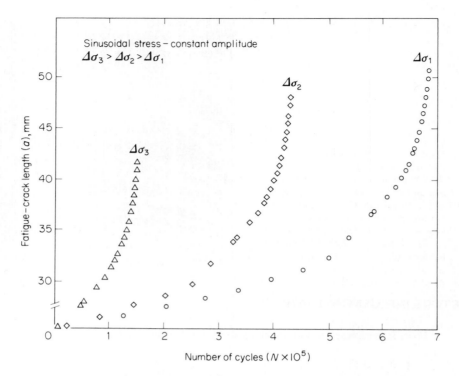

**Figure 18-17**   Effect of cyclic stress range on crack growth. [From Stanley T. Rolfe and John M. Barsom, *Fracture and Fatigue Control in Structures: Applications of Fracture Mechanics*, ©1977, p. 233. Reprinted with permission of Prentice Hall, Inc. (Englewood Cliffs, N. J.).]

value of the change $\Delta K$ in applied stress-intensity factor. Since the applied average stress $\sigma$ determines the applied stress-intensity factor $K$, via the appropriate fracture-mechanics formula for the crack geometry and type of loading (see Chap. 14), changes in stress intensity at the crack tip ($\Delta K$) are determined by the changes in remote stress ($\Delta\sigma$) in the same formula. The rate of crack growth $da/dN$ is thus a function of $\Delta K$. Figure 18-18 (discussed in the next paragraph) is an example of this form of presenting crack growth data: The rate of growth of the fatigue crack of length $a$ for each cycle $N$, $da/dN$, is the ordinate, and the range of stress-intensity factor $\Delta K$ that results from the cyclic variation in $\Delta\sigma$ is the abscissa.

The stress ratio $R$ influences fatigue life in a predictable manner. If the stress range $\Delta\sigma$ is held constant, increasing $R$ means that the maximum tensile stress will be increased, while the minimum stress becomes more tensile. Figure 18-18 shows this effect in a titanium alloy. For a given $\Delta K$, say, 20 MPa$\sqrt{\text{m}}$, $da/dN$ for $R = 0.50$ is roughly 5 times greater than $da/dN$ for $R = 0.05$. Another way of expressing this relationship is to say that when $R$ is increased, the magnitude of permissible $\Delta K$ and $\Delta\sigma$ for infinite life decreases; thus, for higher mean stress $\sigma_m$, where

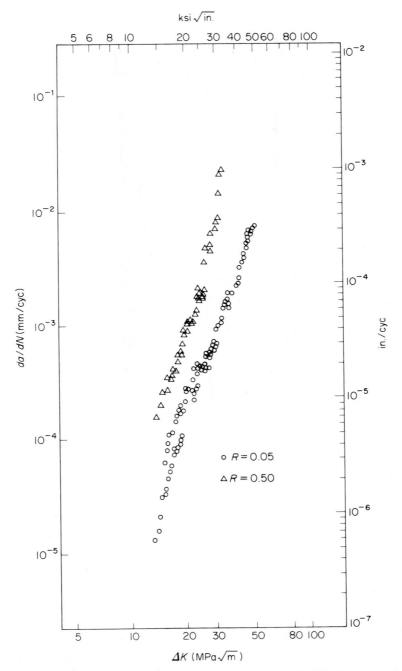

**Figure 18-18** Effect of load ratio $R$ on crack propagation rate in Ti–8 Al–1 Mo–1 V in an argon environment. [From Richard W. Hertzberg, *Deformation and Fracture Mechanics of Engineering Materials* (New York: John Wiley & Sons, 1976), p. 509, Fig. 13.32.]

$$\sigma_m = \frac{\sigma_{max} + \sigma_{min}}{2}$$

the allowable $\Delta\sigma = \sigma_{max} - \sigma_{min}$ must be reduced if fatigue is to be avoided. The designer uses this relationship as the *modified Goodman line*, shown in Fig. 18-19. As the mean stress $\sigma_m$ increases, the stress range for high-cycle fatigue fracture decreases. Note that $\sigma_m$ may exceed the yield stress $S_y$, so a typical design calculation needs to be further limited to hold $\sigma_{max} < S_y$ so as to avoid yield failure.

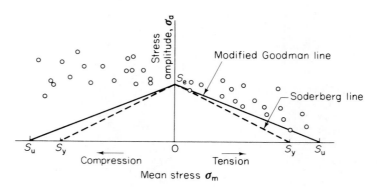

**Figure 18-19**  Typical high-cycle fatigue failures (circles) over a range of stress values. Note that these failures all occur at values of stress amplitude $\sigma_a$ that exceed the limits set by the modified Goodman line. The Soderberg line, which is similar to the modified Goodman line except that it runs to the yield stress, is shown for comparison. [From Joseph E. Shigley, *Mechanical Engineering Design*, 3rd ed. (New York: McGraw-Hill, 1977), p. 202, Fig. 5-24b.]

### Example 18-3

If the form of the graph of $da/dN$ vs. $\Delta K$ is known (e.g., Fig. 18-18), can we calculate the number of cycles that a broken part was subjected to in its lifetime?

**Solution**   We can estimate its life, but we cannot precisely determine the number of crack-growth cycles it experienced except by actually counting all the stage II striations (see Fig. 18-7) across the fracture. Note that because the $da/dN$ data can record only the growth of the crack after it is long enough to be detected, the number of cycles required for initiation of the crack and its rate of growth prior to reaching detectable length are unknown. Thus, we cannot integrate the crack growth from the condition $a = 0$. The best we can do is make some estimate of early crack-growth conditions from tests and predict the remaining life of the part once a crack of observable length has formed. Techniques for estimating fatigue life are discussed in more detail in Chap. 19.

The total number of service-stress cycles in many steel structures is much less than $10^6$, so the designer can design for higher stress than the fatigue limit $S_e$ (see Fig. 18-2). The smaller the number of cycles, the higher the allowable stress. When the local maximum tensile stresses in a perceptible region around the crack tip are above the yield stress, fatigue usually occurs within less than about $10^4$ or $10^5$ cycles; this behavior is called *low-cycle fatigue.* A simple and extreme example is the fracture of a low-carbon steel nail after about 100 cycles of plastic reversed bending. In this case, the successive striations that define each step of crack growth are visible at relatively low magnification, at least in the latter stages of fracture when each striation is rather wide, as in Fig. 18-20. Design procedures have been established (see Ref. 11) for estimating the service stresses that parts can tolerate under low-cycle fatigue loading, usually in the range $10^3$–$10^6$ cycles.

**Figure 18-20** Fracture surface of a common low-carbon steel nail of diameter 2.5 mm fractured in low-cycle fatigue by reversed bending in 222 cycles. Fatigue fractures initiated at the lower right, where the tensile bending stress was higher, and at the upper left, with the final fracture at the region of broad striations in the upper left center of the SEM micrograph.

*Surface finish,* and surface conditions in general, are critical to fatigue. A "smooth" surface is never really without imperfections of some kind to provide a value for surface crack length $a$, no matter how small. Thus, the smoother a surface is, the smaller the value of $a$ is and the greater the value of $\Delta\sigma$ must be in order to reach a given value of $\Delta K$. Alternatively, if the value of $\Delta\sigma$ remains constant, a smoother surface experiences a reduced $\Delta K$ and consequently will sustain more cycles to fracture than will a rougher surface. Figure 18-21 shows the effect of increasing initial crack length on the number of cycles required to reach a given (critical) crack length.

To achieve the longest fatigue life for a given material therefore requires the smoothest possible surface. Grinding and polishing provide the best commercially attainable surface quality in metals, and such treatment is usually required for parts subjected to severe fatigue conditions. Axles and shafts subjected to bending stresses and high numbers of cycles are therefore normally ground and polished. A surface finish of poorer quality, such as an ordinary machined surface, an as-cast or as-forged surface, a rusty or corroded surface, or, even worse, a gouged or undercut surface, will substantially reduce fatigue life below the $S_e$ obtained in the ground and polished rotating–bending test.

**Example 18-4**

Which is more sensitive to surface damage, a polished steel surface or an as-forged steel surface?

**Solution** The polished surface will suffer greater *reduction* in fatigue life than the as-forged surface as a result of surface damage. The as-forged surface already has numerous crack-initi-

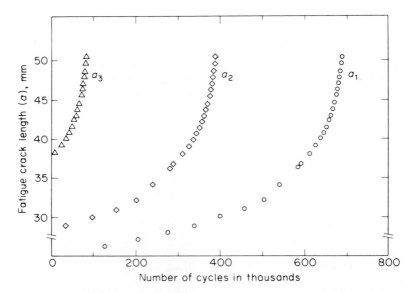

**Figure 18-21**   Effect of initial crack length on crack growth; $a_3$, $a_2$, and $a_1$ are initial crack lengths; $a_3 > a_2 > a_1$ [From Stanley T. Rolfe and John M. Barsom, *Fracture and Fatigue Control in Structures: Applications of Fracture Mechanics*, © 1977, p. 234. Reprinted with permission of Prentice Hall, Inc. (Englewood Cliffs, N. J.).]

ation regions because of its roughness, so it is less sensitive to further surface damage. The polished specimen, because it is relatively free of surface flaws at the outset, will experience a substantial loss in fatigue life when its surface is damaged.

Figure 18-22 shows fatigue fractures that initiated at an improperly machined undercut between two teeth of an automotive steering gear. This has the effect of producing a notch in a region that is already under higher stresses because of the shape of the teeth; under such circumstances, fatigue life can be markedly reduced. A polished part must be carefully protected from damage before and during its service life if it is to fulfill the designer's expectations. Small, nonmetallic inclusions located at or very close to a free surface can likewise act as initiating cracks. Large, macroscopic, nonmetallic inclusions, seams, cracks, holes, and voids located inside a part can act as a free surface and allow initiation of fatigue cracks if the local stress is sufficiently high.

It is possible to improve fatigue life through surface hardening. The usual methods are carburizing, nitriding, and shot-peening. Carburizing is usually applied to steels of low (about 0.20%) carbon content, which are selected to give good toughness. The part is heated in an atmosphere containing carbon so as to increase the carbon content by diffusion into a very thin surface layer to, say, 0.60 to 0.80% C; then it is quenched to transform this outer layer to martensite, which is then stress-relieved. Nitriding similarly introduces enough nitrogen into the surface to produce hard nitrides. In both cases, this very hard layer inhibits the slip processes that lead to early crack initiation (Fig. 18-3) and thus contributes substantially to increased fatigue life far out of proportion to its small fraction of

(a)

**Figure 18-22** (a) Teeth of a fractured automotive steering gear, with fractured parts held together. Fatigue fracture initiated along an improperly machined slot that produced an undercut between the first two teeth on the right. (Note that the undercut appears even worse at the left side of the slot between the middle two teeth.) (b) Fracture surface of the left (larger) portion of the gear shown in (a). The three white markers indicate possible fatigue initiation sources; the left marker shows a partial crack that is not visible here but is located exactly along the undercut.

(b)

the volume of the whole part. At the same time, the part retains its toughness because it consists almost completely of a relatively soft low-carbon steel.

Shot-peening not only hardens a thin layer at the surface but when properly done also leads to compressive *residual stresses* at the surface. Under service loads, the net tensile stresses in this surface thus become more compressive than they would otherwise be, with a consequent increase in fatigue life.

The environment surrounding a part during fatigue can profoundly affect the rate of crack growth. Because this is a special case involving mechanisms different from those discussed so far, corrosion fatigue will be discussed at the end of this chapter after stress-corrosion cracking.

## STRESS-CORROSION CRACKING

Sometimes a crack can propagate slowly when a part is under a constant tensile stress with no cyclic stresses applied. This acutely hazardous mode of fracture is called *stress-corrosion cracking.* In contrast to ordinary corrosion (such as rust), which leads to a more-or-less uniform and gradual loss of metal at a free surface exposed to a corrosive environment,

stress-corrosion cracking leads to crack growth in the presence of an appropriate environment (see later) that occurs most rapidly where tensile stresses are highest. Furthermore,
the cracks that result are perpendicular to the tensile stress. Thus, stress-corrosion cracking
leads to the worst possible situation: the growth of cracks in regions of highest tensile
stress and oriented in the most severely weakening manner.

The essential features of stress-corrosion cracking can be demonstrated in a simple
experiment. A cast flat bar of polystyrene, 3.2 mm thick and 13 mm wide, is forced between
fixed blocks so as to cause it to bend to a predetermined radius of curvature $r$ [Fig. 18-
23(a)]. To a first approximation, the tensile stress on the top center face of the bar is related

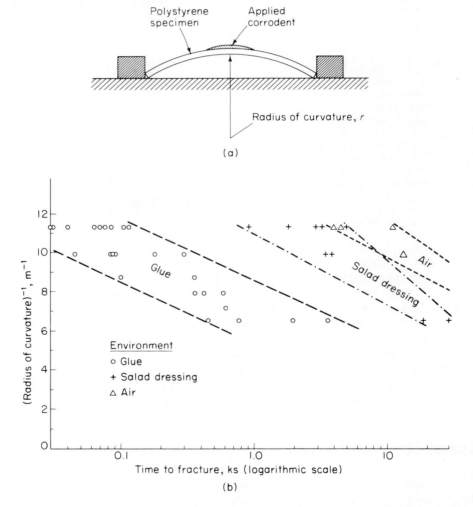

**Figure 18-23**   Experimental stress-corrosion cracking in polystyrene: (a) test setup; (b)
results.

to $1/r$. (Of course, because relaxation occurs in a viscoelastic material such as polystyrene, as described in Chap. 12, the stress drops throughout the experiment. But the behavior of all specimens is similar, so the results will still be comparable.) As soon as the specimen is in place, a stress-corroding environment suitable for polystyrene is applied to the top center surface of the specimen, which is in tension. Acetone is such a substance; in fact, pure acetone is so effective as to be useless for test purposes because a drop of acetone on the top surface leads to almost instantaneous fracture. We have found that acetone-based glue and ordinary salad dressing provide suitably slow stress-corrosion cracking in these tests. The substance is placed on the top center surface, and this is time zero. A series of parallel flaws then grow normal to the top surface in the region of the corrodent until the longest flaw provides, in conjunction with the stress, the condition $K = K_c$ (see Chap. 14) and the part breaks. This fracture occurs suddenly and without warning, in much the same manner as service fractures. Figure 18-23(b) shows that the rate of crack growth is sensitive to the corrodent and is slower for lower stress. The fact that fracture also occurs (very slowly) in air, in the absence of any deliberately applied stress-corrosion cracking environment, suggests that air itself, or probably the moisture in the air, also acts as a mild corrodent.

This same effect of moisture sometimes occurs in glass and is often called (improperly) "static fatigue." As no cycling of stress is involved, the phenomenon is better called *delayed fracture*: A crack in a bottle grows gradually under the combined effects of residual tensile stress present because of faulty heat treatment and any applied tensile stress (such as from internal pressure) until after an elapsed time of perhaps years, the crack reaches critical dimensions and the bottle explodes. This behavior is another example of stress-corrosion cracking, and anyone who has ever lost an unopened bottle of expensive Scotch to this phenomenon (as one of the authors did) will never forget the event.

Many metals and alloys of commercial importance are susceptible to stress-corrosion cracking in certain special environments. Fortunately, the number of corrodents for each alloy is small, otherwise metals would be of limited usefulness. For example, austenitic stainless steels (see Chap. 11) are generally sensitive to stress-corrosion cracking in chloride environments. Figure 18-24 shows the rather wide range of cracking sensitivity of austenitic stainless steels of different compositions in boiling magnesium chloride. Most alloys exhibit some sensitivity to stress-corrosion cracking in certain environments. One example is the so-called season cracking of moderately cold-worked 70 Cu–30 Zn brass in the presence of ammonia; the tensile stress is provided by the residual stress that remains after cold working, but when the brass is severely cold-worked, the residual stress is reduced and the brass becomes less susceptible to stress-corrosion cracking.

The *threshold stress* for stress-corrosion cracking is that tensile stress below which stress-corrosion cracking does not occur, even for long times. The corresponding *critical threshold stress-intensity factor for stress-corrosion cracking* $K_{scc}$ thus represents the stress intensity below which cracking does not occur. The shape of the curves in Fig. 18-24 is somewhat similar to that of the *s–N* curves in Fig. 18-2, where the fatigue limit is analogous to the threshold stress in Fig. 18-24. Typical values of threshold stresses required for the onset of stress-corrosion cracking are from 10 to 70% of the yield stress. Stress-corrosion cracking is thus a potentially serious problem because fracture can occur under static stress well below the yield stress.

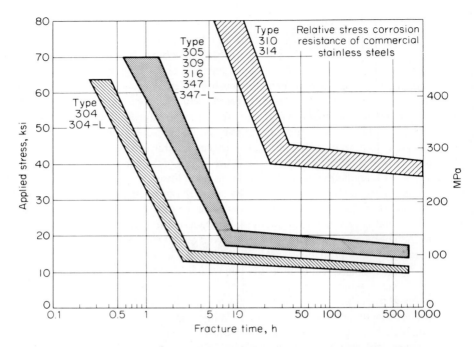

**Figure 18-24**  Stress-corrosion cracking of austenitic stainless steels in boiling magnesium chloride. See App. 7 for composition of these stainless steels. [From Denhard, E., *Corrosion,* 16, 7 (1960), p. 131.]

The mechanisms that lead to stress-corrosion cracking are associated with changes in the bulk solid in the region of the tip of a crack. (Note the similarity to the mechanisms of fatigue, where cyclic stress also alters the properties of the bulk solid in the region of the crack tip.) If we assume the presence of surface flaws, cracks, or irregularities, which are present in all solids, the corrodent, either liquid or gaseous, diffuses to the tip of the flaw, where tensile stresses are highest [Fig. 18-25(a)]. The actual mechanism by which the corrodent causes cracking may be either (1) physical dissolution of the interatomic bonds in the region of highest tensile stress, or (2) a local reduction in energy required to create new fracture surface, with a consequent local reduction in the $K_c$ required for cracking. In either case the effect is the same, and a short length of new crack is formed at the tip of the prior flaw [Fig. 18-25(b)]. The increment of new crack is short because only a very limited amount of the bulk solid has been altered by the corrodent, which has only limited access. Thus, we see that the crack grows gradually and continuously [Fig. 18-25(c)], not by short jumps as in fatigue. Figure 18-26 shows the early branched cracks that develop in 304 stainless steel (UNS S30400) exposed to boiling magnesium chloride, for which data were given in Fig. 18-24. Such branching of cracks is characteristic of many stress-corrosion cracks. The rate of crack growth is limited by (1) the rate of diffusion of fresh corrodent to the new tip of the

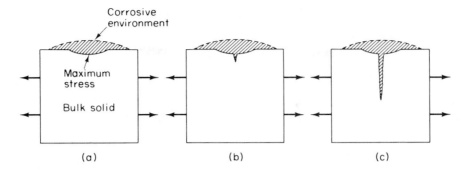

**Figure 18-25** Macroscale mechanism of stress-corrosion cracking: (a) initial condition; (b) after short crack growth; (c) after more crack growth.

**Figure 18-26** Stress-corrosion cracks at the root of a notch in 304 stainless steel exposed to boiling 42% magnesium chloride solution. [From H. L. Logan, *The Stress Corrosion of Metals* (New York: John Wiley & Sons, 1966), p. 120, Fig. 4.9.]

crack, (2) the effectiveness of the corrodent in producing cracking, and (3) the local stress level. This process continues until the crack reaches critical dimensions and the entire part fails.

The gross rate of crack growth can vary as the environmental stress-corrosion cracking conditions vary in concentration and temperature and as the average stress varies. Thus, the fracture surface may exhibit *progression marks* or *beach marks* that resemble those of a fatigue fracture. Figure 18-27 shows the appearance of a stress-corrosion crack (the small, concentric crescents) and the final sudden fracture (the balance of the fracture surface) in a high-strength steel. Thus, beach marks indicate only that a crack has grown

**Figure 18-27**   Progression marks from two origins of stress-corrosion cracking, at top. Steel of hardness 50 HRC under static load in distilled water. [From ASM Committee on Fractography by Light Microscopy, "Interpretation of Light-Microscope Fractographs," *Metals Handbook: Fractography and Atlas of Fractographs,* 8th ed., vol. 9, Howard E. Boyer, ed. (Metals Park, Ohio: American Society for Metals, 1974), p. 42.]

gradually over time. Furthermore, the absence of beach marks does not indicate that the fracture is not the result of fatigue or stress-corrosion cracking.

On a microscopic scale, however, the presence of striations is positive evidence of fatigue, never of stress-corrosion cracking alone. The fracture mode of stress-corrosion cracking can be either intergranular or transgranular, so discriminating between stress-corrosion cracking and single-load fracture is often difficult. However, when a local region of intergranular cracking leads to gross fracture by a quasi-cleavage mode, this evidence points strongly to stress-corrosion cracking as the cause of the first crack in cases where sustained tensile stress and a potential stress-corrosion environment are present.

Since each alloy is susceptible to its own very limited set of environments, a full description of all possible hazardous combinations is beyond the scope of this book. Some of the references listed at the end of this chapter give more detailed information. Appendix 15 lists just a few of these known combinations. Among the common metals that are relatively more susceptible to stress-corrosion cracking are low-carbon steel, high-strength steels, austenitic stainless steels, high-strength aluminum alloys, and brasses and other copper alloys. Because the effective concentration of the corrodent can be quite low, it is essential that the service environment be carefully studied by the designer. The designer has a responsibility either to avoid using a material susceptible to the anticipated service environment or to protect the material from exposure to the environment. Figure 18-28 shows an intergranular fracture that resulted from exposure of an aluminum ladder, probably to road salt, in the presence of local residual stresses. In aluminum alloys, the only explanation for intergranular fracture of parts that have not been heated to temperatures close to melting is stress-corrosion cracking.

### Example 18-5

What should a designer do if the only choice is to use a material susceptible to stress-corrosion cracking, and without adequate protection?

**Figure 18-28**   SEM micrograph of stress-corrosion cracking in a heat-treated ladder rail of aluminum alloy UNS A96005-T6, which contains 0.75 Si and 0.50 Mg. Fracture initiated at holes punched for bracket attachment, which provided residual stress; the environment was probably road salt, as the ladder had been transported for some time in the back of a pickup truck in Michigan. This failure (and another similar one) is particularly interesting because the 6000 series of aluminum alloys are generally listed in the handbooks as being free from stress-corrosion cracking. Marker is 120 $\mu$m.

**Solution**   If the manufacturing process can be altered so that no residual tensile stress exists, and if the service conditions do not involve sustained residual stress, then stress-corrosion cracking cannot occur, even in an unfavorable environment.

If sustained tensile stress is unavoidable and after careful testing the designer has determined that the part can operate for a reasonable time before serious cracks develop, the user should be warned to inspect the part for cracks at prescribed intervals in the same way that aluminum alloy structures are inspected for fatigue cracks. Such a procedure should be undertaken only after thorough study of all the variables in the service environment.

## HYDROGEN EMBRITTLEMENT

Atomic hydrogen can diffuse through solid metal because its effective atomic radius is relatively small. Once inside the bulk metal, the hydrogen can combine in the form of gaseous hydrogen at internal inclusions, voids, and discontinuities. These small bubbles can achieve a pressure as high as 1.3 GPa for maximum hydrogen concentration and are thus able to produce small local fractures. This mode of failure is called *molecular hydrogen cracking*. Under tensile stress as low as 40% of yield stress sustained for a period of time, these fractures may join up and eventually lead to sudden complete fracture of the part in the same fashion as stress-corrosion cracking. Furthermore, there is evidence that hydrogen acts to reduce the surface energy at the tip of a crack under tensile stress, and to this extent hydrogen is just another environment that can lead to stress-corrosion cracking. For this reason, we can often treat hydrogen embrittlement as a special case of stress-corrosion cracking. But because of the problem of the irreversible small cracks produced by gaseous hydrogen, the subject of hydrogen embrittlement warrants a separate discussion.

A second mechanism by which hydrogen can damage a polycrystalline metal involves the diffusion of hydrogen to grain boundaries where it can cause cracking when a relatively small tensile stress is applied. Thus, *hydrogen damage* at grain boundaries can lead to *hydrogen stress cracking* when stressed.

Hydrogen embrittlement causes trouble in many high-strength metal alloys, chiefly in high-strength steels such as AISI-SAE 4340 (UNS G43400) quenched and tempered to above 1.4 GPa and high-carbon plain-carbon steels quenched and tempered to above 1.4 GPa. The hydrogen can come from any of several sources during manufacture and service: pickling, electroplating, stray currents, and service environments. Hydrogen can be charged into a part when the part is the cathode in an aqueous conducting solution formed by the dissociation of water to $H^+$ and $O^-$, where the $H^+$ picks up an electron from the cathode and is free to diffuse into the cathode as atomic hydrogen. Figure 18-29 shows an intergranular fracture surface located near a fracture origin that resulted from exposure of a 0.62% C steel rivet under residual stresses to a dilute solution of hydrochloric acid with an applied voltage of 6 V for 1.2 ks. If a system itself constitutes an electrolytic cell, hydrogen embrittlement can occur even without an applied voltage; for the example in Fig. 18-29, the voltage was applied in order to accelerate and control the process.

If the part is not under tensile stress during and after the charging and no local fractures have occurred, the hydrogen can usually be driven off by heating the part to 150°C to 200°C for several hours. If the part is not later exposed again to hydrogen, there will be no damage. If the part has been plated, the plating will hinder the desorption of hydrogen, so longer baking times (e.g., 190°C for 3–4 h) are employed, with even longer times for cadmium plating.

If the part is under tensile stress (from applied service loads or from residual stresses) after being charged with hydrogen and the hydrogen concentration and time are

**Figure 18-29**    Intergranular fracture from hydrogen embrittlement in high-carbon steel.

sufficient to cause local cracking, then the part has been permanently damaged. No amount of baking will eliminate the cracking.

Hydrogen embrittlement can be mitigated as follows:

1. Use a lower-strength alloy. For example, steels of strengths below 700 MPa are generally resistant to hydrogen embrittlement, steels from 700 to 1000 MPa are slightly susceptible, and steels over 1000 MPa are very susceptible. Within each group of alloys, certain classes have been found to have high resistance to stress-corrosion cracking. Reference 4 provides detailed rankings of alloys according to their susceptibility to stress-corrosion cracking.
2. Avoid treatments and service conditions that can lead to hydrogen absorption.
3. Bake to remove hydrogen as described here.
4. Reduce residual stresses.
5. Reduce severity of notches under tensile stress.

Martensitic and precipitation-hardening stainless steels can experience hydrogen embrittlement, especially in their higher-strength forms (see Chap. 11). Baking can eliminate hydrogen picked up during manufacture, but only isolation from hydrogen-producing environments can protect these alloys from cracking in service. In the absence of other practical protection, these stainless steels must be heat-treated to substantially reduced strength levels to provide protection from hydrogen damage.

Fractographic analysis of hydrogen embrittlement is substantially the same as for stress-corrosion cracking. Although it is known that the initial cracking of quenched and tempered steel is always intergranular and on prior austenite grain boundaries for hydrogen embrittlement, and that hydrogen damage may produce some other subtle differences in the fracture, there are no unambiguous fractographic differences that permit us to separate general hydrogen-embrittlement fracture from stress-corrosion cracking. Thus, hydrogen embrittlement should be treated as a special case of stress-corrosion cracking.

## CORROSION FATIGUE

In nearly every instance, fatigue in service occurs in conjunction with the possibility of corrosion and/or stress-corrosion cracking; this is called *corrosion fatigue*. In its simplest form, ordinary corrosion can lead to surface pitting that will substantially reduce the fatigue limit of, for example, a polished part. In addition, the presence of either residual tensile stresses or the tensile component of the fatigue stress cycle (especially for tests with stress ratio $R > 0$) can lead to simultaneous crack growth by fatigue and by stress-corrosion cracking. Figure 18-30 shows the marked increase in crack growth rate *da/dN* that results when a 17-4 PH stainless steel (UNS S17400) is tested in sea water. Thus, in addition to the basic mechanisms of fatigue and stress-corrosion cracking, there exists the additional possibility of interaction between the two fracture modes. For example, it is possible as a result of local strain hardening at the crack tip (see Fig. 18-5) that during periods of rest

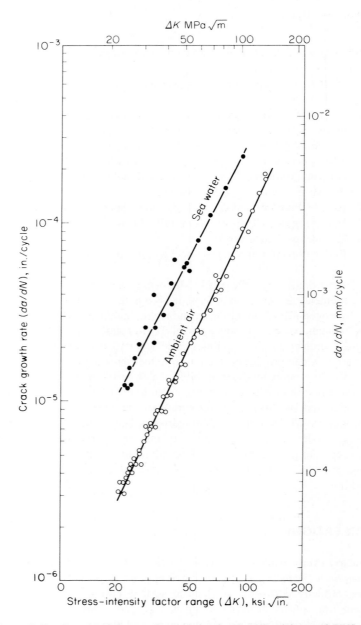

**Figure 18-30** Corrosion-fatigue crack growth data for 17-4 PH stainless steel (UNS S17400). [From T. W. Crooker, F. D. Bogar, and W. R. Cares, "Effects of Flowing Natural Seawater and Electrochemical Potential on Fatigue-Crack Growth in Several High-Strength Marine Alloys," *Corrosion-Fatigue Technology,* ASTM STP 642, H. L. Craig, Jr., T. W. Crooker, and D. W. Hoeppner, eds. (Philadelphia: American Society for Testing and Materials, 1978), pp. 189–201. Copyright American Society for Testing and Materials. Adapted with permission.]

from fatigue cycling (or when the service cyclic stresses are low) stress-corrosion cracking can lead to more rapid crack growth than would occur in the absence of fatigue damage.

We may thus expect some of the characteristics of both fatigue and stress-corrosion cracking to be associated with corrosion-fatigue fracture. Corrosion pits are often seen on the fracture surface, presumably forming after local fracture occurs. In fact, the entire fracture surface may be coated with corrosion products that are difficult to remove; these non-conducting products appear white in a scanning electron microscope because they build up an electron charge. Figure 18-31 shows such a service fracture, in which the harmful environment may have been calcium chloride (road salt) and water. Other characteristics of

**Figure 18-31**   SEM micrograph of corrosion-fatigue service fracture of a case carburized AISI 8650 (UNS H86500) steel track pin from a battle tank that had traveled 2521 mi (4056 km). Intergranular fracture and surface corrosion appear to dominate and thus mask any possible fatigue striations.

corrosion-fatigue fracture may include beach marks visible to the naked eye, striations visible in an electron microscope, and secondary cracking parallel to the main crack in the initiation region. Corrosion fatigue exhibits no single unique characteristic that sets it apart from other fracture modes.

In contrast to stress-corrosion cracking, corrosion fatigue does not require a narrow range of specific environments. Any effective corrodent may contribute to corrosion-fatigue fracture, and higher concentrations are generally more effective. These observations suggest the possibility of a fairly simple mechanism for corrosion fatigue, as already mentioned, in which crack growth by corrosive attack in the region of the crack tip is accelerated by the high local plastic strain in this region that results from fatigue cycling. At the same time, this picture may be complicated somewhat by the fact that the corrodent may remove the hard surface film from the metal, allow dislocations to escape more easily, and thus promote growth of the fatigue crack by a shear mode of the sort described by Fig. 18-3. The reduction in fatigue life by such a mode is shown in Fig. 18-32 for an Al–Zn–Mg alloy, where the length of initial crack growth by shear is greater under the corrosive condition.

**Figure 18-32**    Reduction of fatigue strength in an Al–Zn–Mg alloy tested in 3% NaCl solution. [From P. J. E. Forsyth, *The Physical Basis of Metal Fatigue* (New York: American Elsevier, 1969), p. 116, Fig. 6.13.]

# REFERENCES

1. American Society for Metals, *Metals Handbook,* 8th ed., vol. 10, *Failure Analysis and Prevention.* Metals Park, Ohio: American Society for Metals, 1975.

2. Atkins, A. G., and Y.-W. Mai, *Elastic and Plastic Fracture*, 2nd ed. Chichester, U.K.: Ellis Horwood, New York: Simon & Schuster, 1988.

3. Beachem, Cedric D., ed., *Hydrogen Damage.* Metals Park, Ohio: American Society for Metals, 1977.

4. Brown, B. F., *Stress Corrosion Cracking Control Measures.* Washington, D.C.: U.S. Dept. of Commerce, National Bureau of Standards, Monograph 156, June 1977.

5. Brown, B. F., ed., *Stress Corrosion Cracking in High Strength Steels and in Titanium and Aluminum Alloys.* Washington, D.C.: Naval Research Laboratory, U.S. Govt. Printing Office, 1972.

6. Craig, H. L., Jr., T. W. Crooker, and D. W. Hoeppner, eds., *Corrosion-Fatigue Technology,* ASTM STP-642. Philadelphia: American Society for Testing and Materials, 1978.

7. Forsyth, P. J. E., *The Physical Basis of Metal Fatigue.* New York: American Elsevier, 1969.

8. Logan, Hugh L., *The Stress Corrosion of Metals.* New York: John Wiley, 1966.

9.  Mann, J. Y., *Fatigue of Materials*. London and New York: Melbourne University Press and Cambridge University Press, 1967.

10. Rolfe, Stanley T., and John M. Barsom, *Fracture and Fatigue Control in Structures*. Englewood Cliffs, N. J.: Prentice Hall, 1977.

11. Shigley, Joseph E., *Mechanical Engineering Design,* 3rd ed. New York: McGraw-Hill, 1977.

12. Suresh, Subra, *Fatigue of Materials.* New York: Cambridge University Press, 1991.

## PROBLEMS

**18-1.** Why is it not possible to estimate the prior number of fatigue cycles that have been applied to a broken part by dividing the observed width of the fatigue-fracture surface by the observed width of a single striation?

**18-2.** Why might fatigue striations be difficult to observe in a brittle metal?

**18-3.** Why might fatigue striations be difficult to observe in a very ductile metal?

**18-4.** What fractographic evidence can indicate whether a fracture is solely the result of fatigue or if stress-corrosion cracking has occurred?

**18-5.** What are the minimum essential requirements for stress-corrosion cracking to occur in a part in service?

**18-6.** Why do most parts in service with alternating loads not survive as long as would be indicated by the results of a standard rotating–bending test?

**18-7.** What are the differences between stage I and stage II fatigue crack growth?

**18-8.** What do beach marks or progression marks indicate?

**18-9.** What characteristics or properties of a fabricated metal part can be altered (or specified differently by the designer) in order to increase the fatigue life of the part?

**18-10.** What measures can be taken to minimize the effects of stress-corrosion cracking?

**18-11.** How can hydrogen embrittlement be avoided?

**18-12.** What is the difference between fatigue striations and progression marks (also called beach marks or clamshell marks)?

**18-13.** What does acetone-based glue do to polystyrene under tensile stress? Under compressive stress? Explain why in each case.

**18-14.** An 85-mm-diameter tractor axle that failed by fatigue has the appearance of Fig. 18-15. We have been asked to estimate the number of stress cycles that this part experienced (recognizing the inherent inaccuracies of such a calculation), based on the following assumptions and simplifications:

(a) Stage I fatigue required 65% of the total number of cycles.

(b) Each stress cycle during stage II fatigue produced one striation.

(c) The average striation spacing at the beginning of stage II was 80 nm.

(d) The average striation spacing grew linearly across the stage II region and was 750 nm at the end of stage II.

(e) Stage III is 28 mm wide in the direction of crack growth. Estimate the total number of stress cycles for this shaft.

# 19

# Failure Analysis

When an engineering structure fails to perform its intended function, a *failure analysis* should be made. The impact of such an analysis can range from the potential economic crippling of a nation's economy down to searching for a replacement for a broken pencil lead or rubber band, which hardly warrants analysis. When an engine pylon of a DC-10 jumbo jet fractured on takeoff from O'Hare Airport in Chicago in mid-1979, with the loss of 273 lives, the resulting grounding of all DC-10s for many weeks while a failure analysis was performed led to severe problems for airlines and passengers alike. Even if no lives are lost and no damage occurs as the result of a failure, the design engineer and the user have a responsibility to determine the cause of the failure so that similar events can be avoided in the future. And when injury, death, and property damage occur, the question of responsibility for the failure is raised because insurance and other contractual obligations are usually involved. In cases of dispute, these matters are often taken to courts of law for ultimate resolution under statutes of *product liability,* which are designed to aid in establishing responsibility.

This chapter describes the process of analysis of failures that involve deformation and fracture of metals. Excluded from discussion here are such matters as design deficiencies, the necessity of adequate guarding devices to protect users of equipment, and other engineering failures such as inadequate flow rates or heat-transfer rates. The job of the *failure analyst* is to determine the original, proximal cause of a failure. In this task the analyst has two big advantages over the designer, for he or she knows (1) that a failure did occur and (2) where it occurred. As we shall see, metallurgical failure analysis may also reveal the mode of failure. For example, the DC-10 designers stated during the investigation following the crash that they had never considered the possibility of the mode of failure that actually occurred (fatigue fracture of the pylon). The designer has a much broader and in some ways more difficult job than the failure analyst, for he or she must consider absolutely all possible modes of failure and then design a structure to avoid all of them.

It is our aim here to describe the procedures of failure analysis so that the reader will be able, if necessary, to determine the reasons for a failure. Because this subject is still rather new as a formal area of inquiry (although it has, of course, been carried on for thousands of years), there is limited published literature on the subject (see the references at the end of the chapter). We shall therefore draw on our own experience for examples to illustrate the range of solutions that one might employ in the course of analyzing failures.

## EXAMINATION OF PARTS

**Visual inspection.** Usually the first information available to a metallurgical failure analyst is that obtained from broken parts. (Some failures may involve only plastic deformation and no fracture, but even so we shall include them when we speak of "broken parts" or "failed parts.") Sometimes a brief visual examination may reveal the fracture *mode* and the *proximal cause* of the fracture very easily. However, an investigator must be careful not to become firmly attached to any one failure theory until after the investigation is completed. *Working hypotheses* for the failure should be established, with the purpose of the investigation being eventually to exclude all hypotheses except one, if possible. In many cases the initial hypothesis stands up throughout the entire investigation. Figure 19-1 shows two photographs of a void 16 mm in diameter that initiated fracture in the neck of a cast-iron gooseneck used to charge molten zinc alloy under pressure into a die-casting mold. Preliminary examination showed that the surface of the void was markedly different from the fracture surface, and this hypothesis ultimately proved to explain the fracture.

**Low-power optical inspection.** Once the part is visually examined, closer inspection with a low-power magnifying glass of about 3× magnification will reveal details of areas of interest. Not only should the fracture surface be carefully examined, but the entire part should be searched for cracks, scrapes, gouges, upset areas, machining marks, and other evidence of forces applied to the part. A 3× glass is often a more useful device in failure analysis than an expensive electron microscope. Figure 19-2 shows an indentation in the lip of a ball-joint socket of the steering linkage of an automobile that was involved in a crash. The essential question in such a situation is: Did this indentation occur before the accident, or is it a result of the accident? Calculations (see Example 19-5 later and Ref. 7) showed that forces sufficiently large (in this case, 29.5 kN) to lead to such an indentation could only have arisen during a crash, indicating that this portion of the steering mechanism was intact at the instant of the crash. Obviously, then, a fracture found elsewhere in this portion of the steering mechanism could not have been responsible for the crash, which, in the absence of other fractures led in this case to the conclusion that the car crashed because of driver error.

**Example 19-1**

How can we be certain that the indentation shown by the arrow in Fig. 19-2 came from the ball-joint stud and not from some other source?

**Solution**   We cannot be absolutely certain, but we can be reasonably sure. If the indentation dimensions and curvature match the portion of the ball-joint stud that contacts the indentation when the ball-joint stud is rocked into contacting position, then it is reasonably certain that

(a)

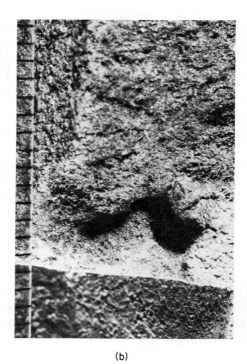

**Figure 19-1** Casting void in the fracture surface of a cast-iron gooseneck. (a) A portion of the tip of the fractured gooseneck with void (arrow) near the region of maximum circumferential stress. (b) Closeup of the void.

(b)

this caused the indentation. Such reasoning is usually sufficient for both engineering analysis and courtroom testimony.

Indentations and scratch marks can indicate prior points of contact and thus reveal something of the history of a part. Figure 19-3 shows marks on the inside of the base of an automobile bumper jack that tipped over the only time it was used. These marks indicate

**Figure 19-2**   Indentation in the lip of a
ball-joint socket.

that the jack column was improperly inserted into the base, vertically rather than at an
angle as it was designed to be. Such accidental misuse could have been prevented in this
case by a simple guide installed in the base to prevent incorrect insertion of the column.

**Figure 19-3**   Rub marks (arrow) indicating
incorrect seating of column in the base of an
automobile bumper jack.

Sometimes the shape, orientation, and location of an indentation, in addition to its
size, exclude some possible explanations for a failure. Figure 19-4 shows an elliptical
indentation of maximum dimension 3.4 mm on the cylindrical surface of the shaft of a
McPherson strut that was found fractured following an accident. The defendant claimed
that the accident was caused by driver error, that the indentation was produced during the
accident rollover when this portion of the shaft made contact with the square-edged upper
surface of the bump cap in which the shaft slides, and that the resultant bending moment
caused the fracture at a point well below this dent. Three observations regarding this indentation show that this scenario is impossible:

- *Shape.* The shape of the indentation could not have resulted from contact with a
  square-edged part but only with a cylindrical surface.
- *Orientation.* The major axis of the elliptical indentation is oriented at approximately
  84° to the axis of the shaft. This may not be obvious at first examination. However,

**Figure 19-4** Elliptical indentation located approximately crosswise on the cylindrical surface of a McPherson strut shaft. The major axis of the ellipse is approximately 84° to the axis of the shaft, which is vertical in the photograph; "up" in the photograph also corresponds approximately to "up" on the vehicle.

the coils of the spring surrounding the shaft are oriented at 84° to the axis of the shaft and are thus parallel to the major axis of the indentation during partial compression. Thus, it is clear that this indentation occurred from contact with the spring during the accident (because the forces were large), following the fracture and while the spring was partially compressed, as the strut could not physically have made contact with the spring until after the strut was fractured.

- *Location.* The indentation is located on the strut at a point that during normal service (i.e., prior to fracture) is never closer than 10 mm to the bump cap that was alleged to have caused the indentation.

The *absence* of indentations and scratches is sometimes as important as their presence. A tractor being towed, shortly after a steering system overhaul, suddenly veered off the road and severely injured the driver. The only possible explanation was either (1) that the driver was careless and was going too fast, smashed the vehicle, and sheared the (missing) pin that connected the upper end of the steering shaft to the lower end through a collar, or (2) that the critical pin fell out and caused loss of steering. Because the steering system was operating just prior to the accident, the pin could not have been missing fol-

**Figure 19-5** Undamaged hole that held pin connecting upper and lower portions of tractor steering column.

lowing the overhaul. Figure 19-5 shows the completely undamaged hole for the pin in the lower end of the steering column; calculations showed that the pin could not have sheared off without also indenting the inside of the hole in the shaft. The repair mechanic admitted on his second deposition that he may have neglected to install a retaining clip on the pin, thus allowing the close-fitting pin to fall out after a short period of service.

**Optical microscope inspection.** A binocular optical microscope of low power (30–70×) provides the next level of magnification useful for fractography. Damage that may have occurred to the fracture surface for other reasons after the time of fracture is usually visible in such a microscope. Many instances of such *postfracture damage* result from complex events that occur during and after a collision or crash, so the observed damage often cannot be related to contact with a particular part. But when mating fracture surfaces are forced into contact after the initial fracture is complete, the postfracture damage may extend over large fractions of the fracture area because the surfaces match so well. Furthermore, one would expect to find the damaged regions visible on both portions of the mating surfaces, although not necessarily in perfectly symmetric locations when relative motion of the fractures is possible. Such a damaged region can be seen on a portion of the fracture surface of a punch press drive pin in Fig. 19-6(a). This end of the pin broke off as

(a)

**Figure 19-6** Punch press drive pin. (a) Small fractured end of pin that caused malfunction, showing extensive postfracture damage. (b) Larger matching portion of pin, showing fatigue fracture in the lower 20 to 30% of the fracture surface and postfracture damage in concave region (arrow) of final cleavage fracture. (One or more small pieces between these two parts were lost through secondary fractures, as these surfaces do not match over their entire area.)

(b)

a result of fatigue arising from a poorly machined fillet radius, was mashed at least once against the pin, got caught in the small gap between the driving flywheel and driven wheel of the punch press, caused a "double trip," and injured the operator. The concave fracture surface of the larger part of the pin, shown in Fig. 19-6(b), could have been damaged only by the matching convex fracture surface in Fig. 19-6(a).

Prior indentation damage can be the cause of a fracture. The fracture surface of a fixation rod used inside a fractured upper leg bone is shown in Fig. 19-7. This rod had been installed in a woman's broken femur following an automobile accident to hold the bones in place during healing. After 11.5 months (30 Ms) the rod broke, the result of fatigue failure. At the fatigue initiation site there is a surface indentation (arrow); a similar indentation exists on the matching fracture surface, indicating that the two indentations are two sides of a single dent that existed before the fracture. Because of the sensitivity of fatigue to poor surface condition (Chap. 18), this dent led to the initiation of fatigue at that point.

**Figure 19-7**  Fatigue fracture of an intramedullary rod that initiated at a prior surface indentation (arrow).

**Example 19-2**

Why does a dent promote fatigue?

**Solution**  To repeat the arguments given in Chap. 18, a surface dent represents a region of stress concentration where local slip can induce cracks before general slip occurs. Furthermore, the plastic deformation that has occurred under the dent restricts slip in this region and thus leads to earlier fracture than in an undamaged region. Good surface quality is thus essential to long fatigue life.

**TEM, SEM, and FEM.**   The transmission electron microscope (TEM), the scanning electron microscope (SEM), and the field emission microscope (FEM) can usually reveal the fracture mode more clearly and explicitly than optical microscopy. In the magnification range of 25–10,000×, most of the important fractographic discriminations are possible.

Because of the possibility that high magnification of surface contours may be required to determine the fracture mode, the fracture surface should be kept in the condition that it was immediately following fracture, if at all possible. The matching fracture surfaces should not be brought together to see how well they fit; this matter can be established with the surfaces 5 mm apart if there is any question. During the initial examination, the fracture surfaces should not be touched with the fingers or any tools, or cleaned or wire-brushed, and the surfaces should be kept free of moisture. A thin layer of light rust-preventive oil is effective for protection of ferrous metals if they must be stored before TEM, SEM, or FEM examination.

The TEM was described briefly in Chap. 5 for use in the examination of thin foils of metal to observe dislocations that become visible because of Bragg diffraction of the electron beam. When used for fractography, the TEM sends an electron beam of 20–100 kV through a very small (diameter about 2 mm) *replica* of the fracture surface to be examined. This polymer replica of the fracture surface is first shadowed at an angle with a dense metal, often chromium, to accent the hills and valleys of the surface contours. Then a thin layer of carbon, which is relatively transparent to electrons, is coated over the entire surface to provide some strength, and the plastic is then dissolved away so that it will not be vaporized in the electron beam. The entire replica is usually supported at the outset by a very fine wire grid of about 125-$\mu$m spacing, so when the replica is viewed in the TEM the fracture region is visible between the lines of the wire grid. The fracture contours are revealed because the electron beam is blocked off in those regions where the chromium has been deposited.

One replica 2 mm in diameter can require several hours of study at high magnification for a complete examination. For example, at 10,000× the apparent area to be studied through the fluorescent viewing screen, which is about 100 mm wide, would be about 20 m in diameter if viewed at one time. However, it is seldom necessary to examine the entire replica because in many instances a number of traverses indicate the characteristic fracture modes present. But when searching for isolated patches of fatigue striations or intergranular fracture surface, more thorough examination may be required.

**Example 19-3**

How does the investigator know where on the fracture surface to take replicas?

**Solution**    This decision is based primarily on the observations from earlier optical examinations. From experience it is possible to select regions that may exhibit fatigue, stress-corrosion cracking, or brittle cleavage, for example. If a preliminary study of replicas is inconclusive, more replicas may be required.

The SEM allows direct examination of any electrically conducting surface that can fit into the specimen chamber (usually less than approximately 100 mm in diameter). The image is formed by the secondary electrons re-emitted at an angle from the surface that is being bombarded with a narrow electron beam. This incident beam is moved over a line-by-line raster (where the electron beam sweeps along a horizontal line, then along a second horizontal line just under the first, and so on, until it has covered a selected rectangular area, and then it starts over again at the top line) covering the very narrow width of the region being examined, and the subsequent enlarged image is synchronized with the raster on a television viewing screen. Since the texture and orientation of each point on a surface control the intensity of the secondary electron emission from that point, this procedure provides an image of the fracture surface. The viewer must become knowledgable in interpreting images, particularly regarding intensity, because secondary electrons do not behave the same as reflected light: Nearly vertical surfaces emit *more* electrons, and flat, horizontal surfaces emit *fewer* electrons. In addition, the tops of ridges are bright lines because these are favored regions for electron emission.

The TEM and the SEM each have advantages in fractography. Obviously, the SEM cannot be used in many instances when the fracture surface cannot be cut because many fractured parts are more than 100 mm in diameter or are too long to fit into an ordinary specimen chamber. If such destruction of the part is not permitted, then the SEM usually cannot be used. Since the TEM replica does not alter the fracture surface except by stripping off some of the corrosion and other contaminants clinging to the surface, any number of TEM replicas can be made without damage to the surface. Furthermore, the replicas provide a permanent and easily stored record of the surface. A lifetime of TEM replicas can be stored by a failure analyst in a volume of 0.1 L. When the SEM can be used, its greatest assets are the following: (1) At lower powers, in the range 20–100×, the SEM provides much greater depth of focus than is possible with optical microscopy; (2) at higher powers, up to perhaps 5000×, the SEM reveals the contours of rough fracture surfaces, holes, tears, cracks, and fissures with great accuracy; and (3) specimen preparation for the SEM is very simple, consisting only of cleaning the sample and fastening it to a mount with conducting adhesive.

The TEM cannot be used for very rough fracture surfaces, as the plastic of the replica that flows into cracks either tears when it is stripped or forms a smooth surface across rough fractures and thus does not accurately represent the actual surface. However, for relatively smooth, subtle contours, the TEM is very powerful because the shadowing of the replica accentuates small contour differences, which the SEM cannot do. Because of the inherently greater complexity of the SEM, it does not provide resolution as good as the TEM at higher magnifications, for example, above 10,000×.

The FEM provides some of the advantages of both the TEM and the SEM. The raster viewing system of the FEM is the same as for the SEM. However, using field emission as a source of electrons allows use of a much finer filament which is not heated. For a given accelerating voltage, the FEM provides a much sharper electron source with consequent improvement in resolution. A large (20 × 20 mm) replica of the fracture surface is taken and then coated with a thin (10-nm) layer of gold and mounted on a thin aluminum base. The gold surface of the replica permits secondary electron emission in the FEM in the same way that the actual fracture surface acts in the SEM. Although similar plastic replicas of fracture surfaces can be used in the SEM, in order to obtain good resolution voltages on the order of 20 kV are required; but 20-kV electrons would vaporize a plastic replica and thus cannot be used. If the voltage in the SEM is reduced so as to avoid damaging the replica, the resolution becomes so bad that magnifications to the range 5000–10,000× cannot be used. But the higher inherent resolution of the FEM permits use of 2–5 kV accelerating voltages on plastic replicas up to magnifications of at least 10,000×. Even these lower voltages can leave marks on the replica if sufficient care is not taken. Thus, the FEM allows the examination of replicas of a fracture surface without the necessity of cutting the fracture to make it small enough to fit into the chamber. The quality of FEM micrographs of fracture replicas is virtually as good as that of SEM micrographs of the actual fracture, except that deep cracks and holes cannot be replicated.

**Quasi-cleavage fracture.**    Numerous examples of fracture-mode discrimination by TEM or SEM examination are illustrated in Refs. 2, 5, 6, and 9. The topography of *quasi-cleavage fracture* in nodular iron can be seen in Fig. 19-8. Local cleavage planes, which are (100) in iron, join together with adjacent planes that are slightly out of plane, and the connecting cliffs are called *tear ridges*, though they are really cliffs. These small cliffs join together with other small cliffs to produce progressively larger cliffs; this can be seen in the upper right of Fig. 19-8. The resulting pattern resembles the tributaries of a river; these are called *river markings*, and downstream in the river corresponds to the local direction of propagation of the fracture.

**Figure 19-8**    SEM micrograph of quasi-cleavage fracture in nodular iron, resulting from local stress concentration. Propagation here is generally from right to left. The usual fracture mode in ductile iron is shown on the left side of Fig. 18-11. Dotted marker is 10 μm.

Quasi-cleavage fracture in steel exhibits the same appearance as in irons. Figure 19-9 shows quasi cleavage in a 0.47% C steel pedestal of a truck crane which resulted from the sudden stage III propagation of a fatigue crack.

Even in aluminum, which is fcc and does not exhibit cleavage fracture in the usual sense, brittle fracture occurs with an appearance very similar to quasi cleavage in steel and iron. Figure 19-10 is an FEM micrograph of high-speed stage III fracture propagation from a fatigue crack in a die-cast aluminum transaxle housing.

**Ductile fracture.**    The typical fracture appearance of ductile fracture is that of *microvoids,* where small voids form and grow by slip processes until the walls between the voids tear apart in the final stages of fracture. The fracture surface appears as a series of

**Figure 19-9** FEM micrograph of quasi-cleavage fracture in 0.47% C steel. River marks are clearly visible. Note that the use of replication for FEM examination has hardly altered the appearance of the fracture surface; a few wisps of replica material can often be seen, such as the torn region at the upper right. Dotted marker is 136 $\mu$m.

**Figure 19-10** FEM micrograph of brittle fracture in die-cast aluminum UNS A03800, resulting from rapid propagation of a fatigue fracture. This is part of the same fracture shown in Fig. 18-12. Dotted marker is 30 $\mu$m.

craters of a wide range of possible sizes. Figure 19-11 is an SEM micrograph of ductile fracture in the core of a case-hardened automobile axle of 0.51% C; this fracture mode is what the designer expects from the core of a case-hardened part because it provides substantial energy absorption before complete fracture.

Replicas for use in the FEM do not show ductile microvoid fracture as clearly as they do other modes because the fracture is so rough. Figure 19-12 shows ductile

**Figure 19-11**  SEM micrograph of ductile microvoid fracture of the core of an automobile axle. Dotted marker is 15 $\mu$m.

**Figure 19-12**  FEM micrograph of ductile microvoid fracture in a die-cast aluminum. The very flat regions result when the replicating material bridges over a hole too deep to replicate. Regions that appear here as mounds represent microvoids, as the replica is a mirror image of the fracture. Dotted marker is 100 $\mu$m.

microvoid fracture in the same die-cast aluminum alloy transaxle housing as shown in Figs. 18-12 and 19-10.

**Fatigue fracture and stress-corrosion cracking.**  Fatigue can appear either with or without striations (see Chap. 18), and the striations can be regular, irregular, more-or-less straight, very wavy, or even discontinuous and interspersed with cleavage or

**Figure 19-13**  FEM micrograph of intergranular fracture in the hardened case of a 0.40% C steel shaft that had been Cr-plated. The fracture mode could have resulted from either improper heat treatment or hydrogen embrittlement. Dotted marker is 20 $\mu$m.

microvoid fracture. Stress-corrosion cracking and hydrogen embrittlement may appear as intergranular fracture, where the environmental damage has caused the fracture to proceed along grain boundaries, and these appear as clear facets at distinct angles to each other. An ongoing problem with case-hardened steel shafts involves the grain-boundary damage that results from improper heat treatment, with consequent intergranular fracture at stresses lower than the nominal strength of the part. Figure 19-13 exhibits the fracture of a 0.40% C shaft of a McPherson strut that had been induction-hardened to produce a case depth of ~1 mm. Such fracture under normal service conditions either can lead to complete fracture when the crack becomes large enough or can produce intermittent fatigue fracture under cyclic loading conditions.

Figure 19-14 exhibits intergranular fracture mixed with fine ductile microvoid fracture, a behavior that is more favorable than the intergranular fracture in Fig. 19-13 but is still to be avoided. A well-manufactured hardened-steel part should fracture by fine microvoids or quasi cleavage, not by intergranular fracture, for it has then achieved the highest possible fracture stress of which the material is capable. The part shown in Fig. 19-14 is a steel sector shaft of a steering gear that had been induction-hardened to 63 HRC.

But stress-corrosion cracking and fatigue may also occur with no such characteristic features, and the TEM, SEM, or FEM examination may not provide unambiguous discrimination of the fracture mode. Thus, analysis by TEM, SEM, or FEM should be considered just one more technique available to the failure analyst that may provide additional information to be used in conjunction with information obtained through other means.

An example of the discrimination made possible by the TEM, SEM, or FEM is shown in Fig. 19-15. A pin [manufactured from AISI 4320 (UNS G43200) and case-hardened to 60 HRC] of one link in the lifting chain of a forklift truck exhibits very faint beach marks at the initiation site indicated by the arrow in Fig. 19-15(a). But one would have expected the

**Figure 19-14**   SEM micrograph of intergranular fracture mixed with ductile microvoid fracture in a case-hardened steel of 63 HRC. Dotted marker is 30 $\mu$m.

designer to have considered fatigue as the chief fracture mode against which to design. Figure 19-15(b) shows the intergranular fracture surface observed in the TEM at the initiation site; in the absence of fatigue striations (which are often not observed in fatigue of high-hardness steel), it is not clear whether fatigue was significant, but hydrogen embrittlement and stress-corrosion cracking are certainly potential fracture modes. We would not have known this without the TEM. Sometimes unexpected modes are found with a TEM. Figure 19-16(a) shows the fracture surface of one column of a stepladder 3 m high; this fracture might be expected to be the characteristic microvoid fracture of a high-strength aluminum alloy (UNS A96005). However, a routine TEM examination showed extensive fatigue striations [Fig. 19-16(b)] throughout the short flange region on the right in Fig. 19-16(a). None of the experts involved in this case had ever seen fatigue fracture in an aluminum ladder before, but subsequent stress analysis showed how it could occur. The evidence held up through the trial; when true striations are present, fatigue has definitely occurred.

**Example 19-4**

How do we know that the pattern in Fig. 19-16(b) represents true fatigue striations?

**Solution**   This TEM micrograph has several features that strongly indicate fatigue. (1) The striations are not perfectly straight; sometimes postfracture rubbing of a fracture surface leaves very straight parallel lines, and this possibility is thus excluded here. (2) The striations are unevenly spaced, and this is characteristic of service fatigue resulting from a range of stress levels. (3) Bending of the striations around the barrier in the center of the micrograph, the drag effect, indicates a crack-growth direction toward the top of the micrograph; this direction agrees with the apparent crack-propagation direction at the macroscopic level.

The only published explanation to date for intergranular fracture of aluminum alloys is stress-corrosion cracking. So when we observe intergranular fracture in aluminum, as

(a)

(b)

**Figure 19-15** Chain pin of forklift truck. (a) Initiation site at bottom center (bar is 1 mm long). (b) Intergranular fracture at initiation site (bar is 5 $\mu$m long).

(a)

(b)

**Figure 19-16**   Fracture surface of a column of an aluminum alloy (UNS A96005) ladder.
(a) Full cross section of column, with flat, featureless fracture in short flange region at
right, changing to ductile fracture and tearing in long flange region at left. (b) TEM
micrograph of short flange region fracture, showing fatigue striations; most of the short
flange fracture region exhibited similar striations (bar is 5 $\mu$m long).

discussed in Chap. 18, we can unambiguously attribute it to stress-corrosion cracking.
Unfortunately, we cannot be so certain of our assessment in the case of other metals, for
sometimes stress-corrosion cracking can occur by a transgranular mode. If we recognize
that slip systems, as well as grain boundaries, can be anodic to the surrounding bulk crystals, then stress-corrosion cracking can take place on slip planes if enough dislocations
occur on one plane. This phenomenon can occur for metals with low *stacking fault energy*,

which means metals in which cross slip is difficult; stainless steels are in this category, and we see transgranular fracture described as one of their modes for stress-corrosion cracking. But transgranular fracture is much more difficult to identify by study of the fracture surface alone; only identification of the crystallographic orientation of the fracture surface can indicate transgranular stress-corrosion cracking, and such analysis is seldom performed.

Many published TEM, SEM, and FEM fractographs are of fracture surfaces made under controlled laboratory conditions where the fracture is carefully protected from the elements. The fracture analyst may be discouraged when the quality of his or her fractographs resembles, for example, Fig. 18-31, the result of weeks or months of exposure to rain, mud, and road salt. Nevertheless, under close inspection the essential features of the fracture mode can usually be determined and an accurate analysis made.

Sometimes it becomes important to determine the chemical composition of a portion of a fracture surface. Such was the case in the fracture of the tungsten carbide cutting tip of a wood-cutting circular saw blade. The plaintiff's expert claimed that a contaminant visible (optically) on the fracture surface was brazing alloy that had entered a crack present in the tungsten carbide tip at the time it was brazed to the steel blade. An attachment for the SEM provides a means for measuring the characteristic energy of the secondary X-rays emitted by each element on or near a surface bombarded with an electron beam. By measuring the strength of the secondary beam at each energy level, it is possible to determine semiquantitatively the relative abundance of each element present on the surface. In the case of the saw blade tip, none of the elements of the brazing alloy was found to be present in the surface contaminant in question. In fact, some of the contaminant was vaporized under the highly concentrated beam required for local analysis, so it was concluded that chiefly nonmetallic material, not brazing alloy, had collected on the surface.

There are numerous other techniques for analysis of surfaces, but they are generally slow and very expensive. In most cases, the failure analyst can obtain all the high-magnification information needed from a TEM, SEM, or FEM. For information that must be limited to the outer few interatomic layers, the Auger scanning electron microscope is essential because it operates at electron voltage levels of a few hundred volts so the electron beam penetrates very little into the surface. Thus, any secondary X-rays emitted must arise from the surface layers; this prevents the overwhelming fraction of elements in the base metal from masking the composition of the surface layer, as occurs when using a conventional high-voltage SEM for secondary X-ray analysis.

**Photographs.**    Before any changes or cuts are made in failed parts, detailed photographs should be taken. Probably not 1 photograph in 20 will prove essential, but at the outset it is impossible to know which ones will be useful. For future reference, the analyst should include in some photographs a scale of length. A detailed log of all photographs taken should also be kept, in which exact orientations of all parts are noted.

Informed failure analysts must take their own pictures because no commercial photographer, no matter how skilled, can discriminate the important from the trivial. Figure 19-17(a) is an overall view of one half of a failed cam 1.35 m in diameter from a 71-MN forging press, and Fig. 19-17(b) shows 6 of the 12 separate fatigue fractures that precipi-

(a)

**Figure 19-17**  Failed forging press cam. (a)
One half of cam, with fractured end at left.
(b) Six fatigue fracture sites along periphery
of fracture; the largest fatigue fracture, at
left, is about 50 mm wide.

(b)

tated this failure. Without the second photograph, the analysis would have been incomplete. Likewise, Fig. 19-18(a) shows an interesting but irrelevant fracture of the end of the tank of a railroad tank car; this fracture resulted from a collision caused by brittle fracture of the main support frame [Fig. 19-18(b)] attributable to a poor design detail and very low temperature.

If the failure analyst is fortunate enough to be able to visit the site of a failure, observations and photographs taken there can be of great help in understanding the reasons for a fracture. For example, the general condition of the track of a railroad siding (Fig. 19-19) provided valuable support to an allegation of poor installation and/or maintenance that caused fracture of the track, derailment, and the death of a bystander.

**Dimensions.**   The physical dimensions of key parts involved in a failure should be measured and recorded. Because at this stage the investigator usually does not know the cause of the failure, much more information must be recorded than will probably be

(a)

(b)

**Figure 19-18**  Railroad tank car. (a) The most conspicuous fracture, the result of a collision that followed the original fracture. (b) Brittle fracture of the main support frame that occurred at very low air temperature. Arrow indicates fracture origin.

needed. If a subsequent stress analysis (see later) becomes necessary, the critical dimensions of the part can be very important. Later, detailed drawings of the part can be obtained and compared with the observed dimensions. Such a comparison will show whether the part is worn, bent, or damaged.

Figure 19-20(a) shows the apparently normal left rear axle flange of an automobile that lost its left rear wheel while traveling on a highway at 30 m/s, rolled over, and caused

(a)

**Figure 19-19**   Railroad siding. (a) Wavy track. (b) Old, partially fractured track reinstalled at siding. The spalled side of the track, at top right of photograph, is here installed so as to be on the outside of the track, the side that is not worn by wheel flanges.

(b)

two deaths and five severe injuries. Careful measurements showed this outer surface to have a convexity of about 0.8 mm across its face, while the manufacturing drawings required that it be at least absolutely flat or have a maximum *concavity* of about 0.1 mm across its face. Such a subtle discrepancy in manufacturing tolerance led to improper contact between axle flange, brake drum, and wheel, so the wheel lug bolts rather than friction with the face of the wheel supported the radial forces between the wheel and axle flange. The resulting fatigue fracture of the region surrounding the lug bolts allowed the wheel to come off over the lug nuts [Fig. 19-20(b)] and caused the accident. An example of the widespread fatigue striations found in the cracks in the wheel is shown in Fig. 19-20(c). Without careful measurement of the axle flange we would not have known that the observed fatigue fractures originated from a manufacturing error.

Sometimes careful observation of the configuration of a failed part can be very revealing. A car drove into a canal and the driver was drowned, but an autopsy showed that he had a high concentration of carbon monoxide in his blood. The cast-iron exhaust manifold was found to be fractured, and it was at first uncertain whether (1) the impact and/or thermal shock of the water caused it to crack or (2) it was previously cracked and had

(a)

(b)

**Figure 19-20**   Automobile wheel failure.
(a) Outer face of left rear axle flange, which
had been improperly manufactured with a
surface that is slightly convex rather than
slightly concave. (b) Inner face of the left
rear wheel, which failed in fatigue from
transfer of the radial wheel forces by the lug
bolts instead of by friction with the wheel
face. Note the small cracks at the periphery
of the holes shaped by the hexagonal lug
nuts. (c) Fatigue striations typical of many
such regions observed during TEM examination of the fracture surfaces of these
cracks. Scale bar is 1 $\mu$m long.

leaked the exhaust gas that asphyxiated the driver. Figure 19-21 shows one portion of the
mismatching fracture surfaces of the exhaust manifold pressed tightly together (such a
practice, even though it violates the stricture against bringing fracture surfaces together, is
sometimes necessary), indicating that some wear had occurred after the fracture and prior
to the accident. This observation, coupled with the absence of any other possible source of
carbon monoxide in this case, demonstrated that the cracked manifold and leaky cowling
seal that allowed the gas to enter the heater duct to the passenger compartment were
responsible for the accident.

Measurement of the dimensions of indentations present on parts can be used to estimate the magnitude of the force required to produce the indentation (Ref. 7). Although this
knowledge alone seldom explains the reasons for the fracture, it often limits the possible
time of occurrence of the indentation, as discussed later and illustrated by Fig. 19-2.

**Figure 19-20** (cont.)                                    (c)

**Figure 19-21** Fractured sections of a cast-iron exhaust manifold pressed tightly together. Prior wear at several points, such as at arrow, show that the fracture had occurred earlier and had caused the asphyxiation of the driver.

## HARDNESS TESTS

Of all the mechanical tests, hardness measurement is the fastest and the least destructive (see Chap. 2). The hardness of all but the most brittle steels (see Fig. 2-25) and of many nonferrous metals can be used to determine the approximate strength. Appendix 4 gives approximate conversions between the more common hardness scales and an estimate of strength. For lower-hardness alloys, such as aluminum, the hardness (usually Rockwell E) can be used in conjunction with a chemical analysis (see later discussion) to determine the probable heat treatment. From this knowledge, the yield and ultimate tensile strength can be obtained from handbooks such as Ref. 3.

The wide variety of hardness scales allows the failure analyst to select the one that measures the hardness accurately and with the least damage to the part. For this reason, the 3000-kg$_f$ (29.4-kN) Brinell test is less used in failure analysis except for some cast irons where point-to-point variation in hardness may require the averaging effect of the relatively large Brinell indentation. For steels, Rockwell C (for hard steels) and Rockwell B (for soft steels) are fairly standard. Often a complex part cannot be properly supported to withstand the 100 kg$_f$ (981 N) or 150 kg$_f$ (1.47 kN) of the Rockwell B or C test, respectively. The Rockwell A scale, using 60 kg$_f$ (588 N), provides an alternative. Even better, the superficial Rockwell scales, which require a separate testing machine, provide hardness measurements using forces of 15, 30, and 45 kg$_f$ (147, 294, and 441 N) and either a diamond brale (the N scales) or a $^1/_{16}$-in. (1.6-mm)-diameter ball (the T scales). The superficial scales must be used with caution, however, because they measure hardness in a very thin layer of the surface and over a very small region. If the hardness of the surface is different from the underlying bulk hardness, then either the surface must be ground off or a different hardness test must be used. For very thin parts, the superficial test is often the only choice available.

The simplest application of the result of a hardness test is for comparison with the hardness specified by the manufacturer. A more specific use of hardness in failure analysis is in utilizing the derived strength values in a stress analysis of the structure. When a brass safety latch of an 8.9-kN steel hook failed to prevent a large steel panel from falling off the hook, comparison of the hardnesses and yield strengths of the brass latch with the cold-worked steel latch for which the unit was originally designed showed that the steel latch would have been about 2.5 times as strong as the substitute brass and probably would have protected against the failure. (It was never clear why the manufacturer substituted an expensive metal such as brass for low-carbon steel in a noncorrosive application.)

A rare use of hardness is to infer the temperature–time history of a part that has been through a fire. If a heat-treated part is reheated, the hardness can change, and this information may be useful in inferring the conditions in a fire. Such a case is discussed under operational tests in a later section of this chapter.

## MUTUAL INDENTATION HARDNESS ANALYSIS

Following a failure, parts often exhibit indented regions where they were in contact with other parts. It is possible to estimate the magnitude of the forces required to produce observed indentations (Ref. 7). Once the forces are known, it is often possible to establish unequivocally whether they arose prior to or at the instant of the failure. Such knowledge can thus be of immense aid to failure analysis.

As described in Chap. 2, hardness $H$ in general is the force $F$ required to produce an indentation divided by the projected area $A$ of the indentation:

$$H = \frac{F}{A} \tag{19-1}$$

In the Brinell test, for example, the numerical value of HB is the force $F$ (in $kg_f$, unfortunately) divided by the actual area of the spherically indented surface (in $mm^2$); for present purposes the actual area is considered to be roughly equal to the projected area $A$.

For a real work-hardening material, the hardness of an indented (softer) flat surface is related to the yield stress $Y_{soft}$ at the completion of a conventional ball indentation (such as Brinell):

$$H = H_{soft} = cY_{soft} \tag{19-2}$$

where $c$ is a function of geometry; $c = 2.8$ in the Brinell test, where the indenting ball is always much harder than the material being tested. For other very hard indenters of different geometries, and also when the two contacting bodies have hardnesses comparable with one another, $c$ changes. Experiments show that the geometric constant $c$ is a function of the shape and relative hardnesses of the two indenting surfaces, up to the point, for example, where a ball is about 2.5 times the hardness of a flat; beyond this there is no further change. These results are given in Fig. 19-22 for a number of representative geometries of contact-

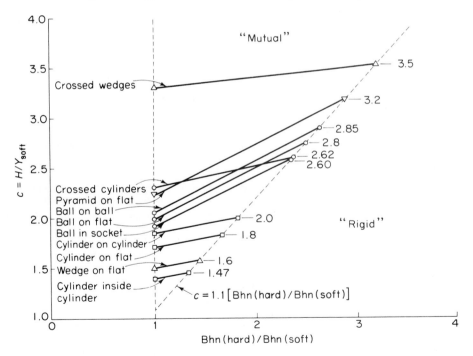

**Figure 19-22**   Geometric factors $[c = (H/Y)_{soft}]$ for many geometries, for use in mutual indentation hardness analysis. [From A. G. Atkins and D. K. Felbeck, "Applying Mutual Indentation Hardness Phenomena to Service Failures," *Met. Engr. Quart.,* 14, 2 (May 1974), p. 55. © American Society for Metals. Reprinted with permission. Also *Chartered Mech. Eng.* 21, 6 (June 1974), p. 78.]

ing surfaces such as ball on flat (hardness test), wedge on flat, and crossed cylinders. The hardnesses of both soft and hard contacting surfaces must be known or estimated.

Combining Eqs. (19-2) and (19-1) gives, for the general case of the force $F$ for any indenter against a softer surface,

$$F = H_{\text{soft}} A_{\text{soft}}$$

$$F = c Y_{\text{soft}} A_{\text{soft}} \tag{19-3}$$

Since the softer material usually is permanently indented more, the value $c = (H/Y)_{\text{soft}}$ from Fig. 19-22 can be used in Eqs. (19-2) and (19-3) to determine the force $F$ that caused an indentation in the soft surface of known Brinell hardness HB. The yield strength $Y_{\text{soft}}$ is known from HB:

$$Y_{\text{soft}} = \frac{\text{HB}}{2.8}$$

This leads to

$$F = c\left(\frac{\text{HB}}{2.8}\right) A_{\text{soft}} \tag{19-4}$$

But because the numerical units of hardness $H$, as measured on the Brinell scale HB, are $\text{kg}_f/\text{mm}^2$, for a force $F$ in newtons and an area $A_{\text{soft}}$ in $\text{mm}^2$,

$$F[\text{N}] = c A_{\text{soft}}[\text{mm}^2] \frac{H\,[\text{HB, kg}_f/\text{mm}^2]}{2.8} \frac{9.807\ \text{N}}{\text{kg}_f}$$

$$F = 3.5\, c\, A\, H \tag{19-5}$$

$F$ [N] is the force required to cause an indentation of projected area $A$ [mm²] in a material of hardness $H$ [HB, kg$_f$/mm²] under geometric and relative hardness conditions that give the constant $c$.

### Example 19-5

If the indentation in Fig. 19-2 has an area of 34.8 mm², the hardness of the indented ball-joint socket is measured as 85 HRB and the hardness of the stud is estimated to be about 300 HB, show how the required indentation force of 29.5 kN mentioned earlier was derived.

**Solution** From App. 4, 85 HRB is equivalent to 165 HB (3000 kg$_f$). Thus, HB(hard)/HB(soft) = 300/165 = 1.82. From Fig. 19-22, for "cylinder inside cylinder," $c$ has reached the constant value of 1.47 [for all HB(hard)/HB(soft) above 1.34]. Then, from Eq. (19-5), $F = (3.50)(1.47)(165)(34.8) = 29.5$ kN. Note that the hardness of the stud has no effect on the answer if it is greater than 1.34(165 HB) = 221 HB because the stud is not indented.

## METALLOGRAPHIC EXAMINATION

In some cases, if the analyst is permitted to cut out a portion of the part at a location removed from the fracture site, a routine metallographic examination can be performed. A thin sliver of the part cut out is mounted in plastic, polished, and directly examined with a metallographic microscope in the unetched condition (for porosity, cracks, seams, etc., and for graphite size and distribution in cast irons) and/or in the etched condition to determine the microstructure. Seldom does such examination alone lead to an explanation for a failure, but it does supplement and support other tests such as the measurement of hardness and observations of the fracture surface.

In cases not involving lawsuits, it is often desirable to cut out a sample containing one cut perpendicular to the fracture surface. (Such destruction of evidence is almost never allowed in a lawsuit.) Such a specimen, mounted and polished, can provide information that is unavailable from surface fractography concerning the possible multiple-branching paths of secondary cracks, for example, as in Fig. 18-26, or whether the main crack is intergranular or transgranular.

## SPECTROGRAPHIC ANALYSIS

After the sliver has been removed for metallographic examination, a small sample (about 5–10 g; for steel this is a cube about 10 mm on each side) removed from the part can be quantitatively analyzed for chemical composition. (We often make an additional hardness test on this sample at this stage before it is analyzed because it is of convenient size and can easily be ground off. Because the sample will be consumed in the analysis, no further damage is done to the parent part.) Most failure analysts do not maintain the necessary specialized and expensive equipment and staff needed for spectrographic analysis, so a commercial laboratory is generally used.

Knowing the chemical composition of a failed part usually provides no new information and is therefore just a routine check against specifications. However, discrepancies occur often enough to warrant the test. For example, in a case of fatigue failure of a pin in a paper-shearing machine that led to injury of the operator, the pin was analyzed and found to be a plain-carbon steel of 0.06% C, whereas the specification was for 0.45% C. This discrepancy explained the fatigue fracture, in conjunction with a stress analysis of the conditions in the fractured region. In this case, a replacement pin of the same dimensions and part number as the failed pin, ordered from the manufacturer's distributor, had a composition of 0.17% C, still not close to the 0.45% C in the specification. It was clear that the manufacturer was not maintaining close control over the steel used for this pin; one may wonder how many other failures resulted from other pins that did not meet specification.

A spectrographic analysis can sometimes save a lot of investigative time. A stainless-steel boat-hoist strap, plastic-coated to prevent damage when lifting boats, fractured and dropped a 12-m-long boat on its owner and killed him. Early in the study, a sample of the strap was sent out for chemical analysis in the expectation that the grade of stainless steel used might have had some special problems because the fracture appeared rusted. When the analysis showed that it was 0.18% C plain-carbon steel, the case was over.

## ADDITIONAL INFORMATION

Although examination of the parts usually provides the chief activity of the failure analyst, full use should be made of all other available information that is pertinent to the failure.

**Witnesses.**  People who were present at the time of a failure are usually questioned about what they saw. Their reports, in the form of statements or interviews, or as formal depositions, often provide useful supplementary information to the analyst. However, all such information should be considered to be of secondary value as compared with the hard facts obtained from examination of the parts. People's memories can be unreliable, especially at a time of crisis, and often much time has passed before they can be questioned in detail regarding their observations.

The statements of completely disinterested people should be carefully examined because they should be free of apparent bias. Even then, deductions from physical observations should be given priority. But associates of an injured person often act so at to protect the injured person as well as themselves and their reputations. In most states in the United States, an employer cannot be sued by an injured employee because the employer is automatically obligated to provide workmen's compensation insurance to pay any injured employee. But the employer may still be motivated to provide a public image of doing no wrong, so statements, test results, and other information provided by the employer should be weighed as if coming from a possibly biased source. Injured parties to a lawsuit or employees of a defendant, even if qualified as experts, who stand to gain from a successful outcome of the suit, may or may not provide all the facts in the course of an inquiry; their potential gain from their side's winning the lawsuit may lead them to less than complete disclosure. These cautions to the independent failure analyst apply no matter who the client is, for plaintiff and defendant alike have been known to withhold facts from their own experts in their zeal to win a case. The expert should therefore listen cautiously to all he or she is told and carefully weigh it against what is seen and measured.

**Example 19-6**

What should a failure analyst do on becoming convinced that a client's product liability case is without merit?

**Solution**   He or she should tell the client as quickly as possible. Society is served better by engineers who accept reality than by those who distort it to suit their own ends.

**Police, fire department, and safety inspectors' reports.**  In many instances such formal reports are available to the analyst. Their value usually lies in the indisputable, hard facts of the investigation: date, time, physical locations of parts, injured people, photographs, and compass orientation. Conclusions as to cause should be considered with caution. Most police officers, except those with special training, simply are not qualified to say with confidence that an accident resulted from the broken steering component observed after the accident. And most firefighters are not very good at determining the origin of a fire; expert fire investigators are specially trained for this purpose.

**Other experts.**  Factual reports of qualified accident investigators can be very useful if they are narrowly limited to on-site observations and avoid reaching conclusions beyond their expertise. For studies involving areas outside his own field, the failure analyst should not hesitate to recommend that other experts be brought into a case: experts in, for example, accident reconstruction, hydraulics, structures, dynamics, or specialists in the specific equipment involved, such as automobiles, cranes, or aircraft. We work regularly with people in such areas and are thus able to complete an analysis much more quickly than without them. An expert who tries to deal in a very broad technical area becomes so general that he does not serve his clients well. We once participated in a case in which the expert for the other side testified that he was an expert in automobile design, manufacturing processes, materials failure analysis, accident reconstruction, nodular iron, mechanics and dynamics, environmental factors, and human factors!

**Drawings and specifications.** As already mentioned, comparison of the dimensions, hardness, and chemical composition of a part with the specified values may be of great assistance in establishing the cause of failure. Even if the part meets all specifications, such a comparison should be made in order to eliminate these deficiencies as possible causes of failure.

**Example 19-7**

If a part is found not to meet its design specifications, such as in dimensions or hardness, is that sufficient evidence to identify the cause of a failure?

**Solution**  No. A *causal relationship* must be established between any design deficiency and a resulting failure before a failure can be thus explained. A substandard windshield wiper blade on an automobile does not explain an alleged steering linkage failure. Much litigation time is wasted in such allegations, in what might be termed "smokescreen" failure analysis, when an expert attempts to hide a nonexistent causal relationship behind a cloud of irrelevancies.

The design capacities of a structure should be determined from the manufacturer. Use of a device beyond its capabilities has been the cause of many failures. Figure 19-23 shows the fatigue fracture of the yoke of a universal joint of a large-capacity end loader. (An earlier investigator had already butchered the evidence by cutting out the center of the fractured area.) While the machine was driving down a public highway, the yoke fractured and the drive shaft flew out, severely injuring a bystander. The history of this machine indicated that after manufacture it had been adapted for heavy-duty use in a landfill operation, which involved the addition of heavy protective underplating armor and installation of a heavy-duty torque converter that increased the torque transmitted by this joint. Because small increases in stress above the fatigue limit can drastically reduce the life of a

**Figure 19-23**  Yoke of universal joint of an end loader (having a section cut out earlier), showing two fatigue fracture origins (arrows).

part, this failure was attributed to the excessive loads on the part which were not considered in the original design.

Sometimes a second-year engineering student could have determined the reasons for a failure. In one such case, a sling made of wire rope 9.5 mm in diameter failed while supporting a weighing hopper being installed inside a grain elevator; one man fell 10 m to his death. The safe rated capacity of the wire rope alone (for a factor of safety of 5) was 10.8 kN, and this was confirmed by subsequent tensile testing of the wire rope. The total known gravity force exerted by the hopper, the two men standing on it, and the hoisting block was 32.7 kN. Thus, the load was clearly well above the safe rated load, but this does not necessarily explain the failure. The important point was that the strength of the wire rope when used as a sling around a hook (as in this accident) was reduced to 60% of the rated strength, or 33.1 kN. This is virtually identical to the actual load at failure, 32.7 kN. It was never established whether the men involved in setting up this installation knew either the mass of the hopper or the fracture load of the wire rope.

Failure of a wheel as a result of convexity of the axle flange was discussed earlier. Another example of the profound influence of small dimensional changes is the fracture of the stud of a ball joint connecting two links in the steering system of a large farm combine, as shown in Fig. 19-24(a). The slight mismatch between the taper of the shank and the taper of the plate to which it was attached [Fig. 19-24(b)] led to an unexpected bending moment at the bottom end of the taper, with consequent fatigue fracture [Fig. 19-24(c)] that caused the combine to run off the road and be seriously damaged. Such discrepancy from correct dimensions could arise either from manufacturing errors or from wear.

**Example 19-8**

In Fig. 19-24(c), how is it possible to discriminate between fatigue and, for example, excessive tensile stress from overtorquing the nut on the shank?

**Solution**    (1) The simplest discrimination results from the macroscopic appearance of the fracture. Fracture progression lines are clearly visible in Fig. 19-24(c). (2) This pattern is typical of bending fatigue (Ref. 5), not overload tension. (3) Measurements on the parts showed them to correspond to Fig. 19-24(b) rather than Fig. 19-24(a), as was intended by the designer. (4) Overload fracture from overtorquing usually occurs at the time of overtorquing or not at all, as subsequent relaxation in service of the torsional moment on the bolt (shank) lowers the maximum principal tensile stress from its initial value.

The best efforts of the designer may be worthless if the part is not manufactured as designed. Figure 19-25(a) shows a correct weld of two legs to the vertical column of a steel chair. Figure 19-25(b) shows that if this weld is incomplete (in this case the legs were not adequately welded to the column), the chair can fail under normal service loads. It caused a 120-kg man to fall and be injured.

If the design permits excessive wear or accidental abuse to occur during normal service by a reasonable and careful operator, failure can occur. A motor crane [Fig. 19-26(a)] had its wire rope fail while lifting a very light load of two men in a bucket, injuring them both. Lack of guides for the wire rope as originally installed had allowed the rope when slack to be caught on the lower corner of the boom. Figure 19-26(b) shows the wear at this point that had occurred prior to the accident. This location is out of the view of the operator. This led to either wear or damage so severe that the wire rope could fail under a very

**Figure 19-24**   Ball-joint shank of steering linkage of farm combine. (a) As designed, shank taper fits taper of supporting plate. (b) As measured after the failure, shank has +0.04 mm possible lateral motion, with consequent high stress at fracture site (the dimensions are exaggerated in the diagram). (c) Fatigue fracture; scale bar is 2 mm long.

light load. The problem was solved after the accident by the addition of guards to prevent the misguiding.

**Applicable standards.**   No part has ever failed in service solely because it did not meet the requirements of standards written to serve as guidelines for safe practice. But the failure analyst should consult standards, especially if a failed component was pur-

(a)

(b)

**Figure 19-25** Improperly welded joint between vertical column of a steel chair (horizontal in photos) and the legs. (a) Correct weld joining two legs and column. (b) Incorrect weld, joining only two legs, did not penetrate to column (arrow).

chased to comply with a particular standard. Standards for metals, materials, dimensions, tests, and safe practice have been established by such organizations as American Society of Mechanical Engineers (ASME), Society of Automotive Engineers (SAE), American Society for Testing and Materials (ASTM), American Iron and Steel Institute (AISI), and American National Standards Institute (ANSI), to name but a few. Superimposed on these are state and federal governmental requirements such as by the Occupational Safety and Health Administration (OSHA).

For the failure analyst, these standards often give a good view of industry practice, and they are frequently very useful in describing functions and designs of parts. But standards are certainly not absolute because each serves a different purpose. For example, the people who manufacture wire rope want their wire rope to perform as well as possible in service, so they recommend that the ratio of the diameter of the sheave (pulley) to the diameter of the wire rope be a relatively large (e.g., for 6 × 25 FW*) 36:1, with a minimum of 24:1. (Larger sheave diameter causes less bending and consequently leads to longer life

---

*This is a wire rope made of 6 strands, each strand made of 25 wires [19 larger wires plus 6 much smaller filler wires (FW)].

(a)

**Figure 19-26** Wire rope failure because of misguiding. (a) Right side of motor crane with drum and wire rope at upper left. (b) Lower left end of boom (boom extends upward to the left in photo) worn from wire rope that came out of U-shaped trough of boom (left arrow). Guard to prevent such misguiding (right arrow) was added after accident.

(b)

for wire rope.) The SAE, on the other hand, wants hoisting equipment to be kept to moderate size and cost, so for the same wire rope it recommends a minimum 12:1 ratio of sheave diameter to wire rope diameter.

The failure analyst must carefully examine any deviations from applicable standards to determine their possible influence on the failure in question. Even if the part was incorrectly made or did not meet some standard, if this error or deviation did not cause the failure, then it is irrelevant. Efforts to suggest incompetence through elaborate presentation of such irrelevant information may help win cases by obfuscation before juries, but they waste a lot of time and do not further the cause of impartial failure analysis.

## MISAPPLICATION

Sometimes a part or assembly is simply used in a manner for which it was never intended. Figure 19-27 shows 7 of about 100 gear teeth (made of UNS G51400 steel) that failed by fatigue in three separate but identical 233-kW screw pump installations designed to pump purified water from a sewage treatment plant up a 5-m head in order to dump it into an adjacent river.

**Figure 19-27**   Fatigue fractures of teeth of an overloaded reduction gear.

When identical failures occur in separate installations, one suspects similar causes. In this case, the gear reduction system specified by the designer was not rated at a high enough capacity for this application, and the consequent excessive stress led to the fatigue failures.

When an adjustable aluminum alloy staging 5 m long used for lowering workers from the roof of a building failed, it dropped two men 25 m to the ground, killing one and injuring the other. Investigation showed that the structural frame of the staging was not overloaded, but it had been overloaded at an earlier time, damaged, and repaired by riveting. The person who authorized the repair of course had not performed a stress analysis on the structure; a subsequent stress analysis revealed that, although the riveted joint appeared to be as strong in tension as the continuous beam it replaced, the shift of the tensile force carried by the original continuous strut to a shear force in the replacement machine screws that substituted for the strut led to a reduction in load-carrying capacity by a factor of 4. Thus, a misapplication consisting of repair by an unqualified person led to injury and death.

**Example 19-9**

How could the manufacturer of a device of such sophisticated design as the aluminum alloy staging (see above) prevent such an accident?

**Solution**   Since the purpose of the design and material in this application is to produce a very light but strong staging, any change in the construction of the part can easily lead to excessive stress. In such a case, the simplest way to avoid improper repair is to place conspicuous and permanent warnings on the structure instructing the user to scrap it if it is damaged and stating that it cannot be safely repaired.

# ANALYSIS

Many failures can be fully explained without further analysis, as we have seen, for reasons such as improper strength, dimensions, or chemical composition, or obvious abuse or overload. But in some instances an analysis of the conditions that existed at the time of the failure can lead to an order-of-magnitude answer that may either exclude a potential explanation for the failure or include it as possible. We mentioned in the preceding section a case where stress analysis of a repaired staging led to full explanation for a failure. In this section, we show how fracture analysis and dynamics can be used together to evaluate the validity of a proposed explanation of an accident.

An unusual case involved a *question* of fracture, that is, whether a fracture had occurred at all. A man fell while skiing and flew into a large tree, so injuring himself as to be paralyzed from the chest down. Witnesses and friends said he was an excellent skier and observed that his tracks during the turn just prior to the accident intersected a small tree of about 20-mm diameter and apparently cut it off. The fracture surface of the tree appeared fresh and was flush with the surface of the snow. In the litigation that followed (about 4 years later), the plaintiff skier claimed that he had struck the stump of this small tree and severed the top off it, leading to his fall and injury. He could not prove the existence of the stump except by the statements of some of the witnesses and one photograph that showed the tree stump as little more than a dot in the snow, for the area of the accident site had been completely altered by a bulldozer during the spring following his accident. The owner of the ski area had removed not only the stump but even the large tree that the plaintiff had struck. The defendant ski slope owner claimed there was never any stump, that the plaintiff "caught an edge" of his ski in the snow, flipped over, and struck the large tree.

As the plaintiff had struck the large tree with the back of his head after completing about three-quarters of a forward revolution in the air, the truth of this matter hinged on the dynamics of the situation: (1) Could the change in angular momentum from fracturing a tree lead to the three-quarter turn, and/or (2) could catching an edge in the snow lead to the three-quarter turn? Distances, velocities, and elevations were all known, so all that was needed were estimates of stresses for fracturing green wood (handbook values) and for "fracturing" Michigan snow (determined by experiment). In brief, question (1) turned out to be well within the range of possibility, while question (2) was impossible, by a factor of at least 4. The defendant settled while the jury was deliberating but before they had reached a decision.

A question regarding whether one possible explanation for fracture was valid was likewise settled by a simple dynamic analysis. A horizontal platform suspended near the top and inside a 200-m-high chimney tipped and dropped three men to their deaths. One possibility proposed was that the crane operator had dropped a loaded bucket on the platform and caused one of the platform's welded support lugs to fracture. A simple analysis of the forces that would result from such an elastic impact showed that if the weld had been sound, the bucket could have been dropped from 4 m above the platform and would not have produced forces that exceeded the strength of the weld. Because there was no possibility that the bucket could have been dropped from such a height, this possibility was completely excluded.

## OPERATIONAL TESTS

Sometimes a test that to a limited extent simulates the conditions of failure can help in understanding what is possible and what is impossible. Such a test is seldom used to simulate fully the circumstances of a failure because there are usually unknown variables in an actual failure that can make a difference in test results.

Even a simple test can be very informative. Two boys were jogging down a hall in a high school, and when they tried to turn right to enter a doorway, one of them skidded, he said, past the doorway and 5.2 m further to the end of the hall, crashed through a wire-reinforced window in an exit door and severely cut his arm. He sued the school for having a floor that was too slippery. These facts were subject to testing, which involved measurement of the coefficient of friction between the actual floor and the kind of shoes worn by the plaintiff, using a student of about the same mass. The resulting friction force was then used to calculate the initial speed (8.2 m/s) that would be required for the plaintiff to slide the 5.2-m distance from the doorway and hit the glass door; this speed approaches the Olympic record for the 400-m dash, 9.1 m/s, which would not be considered jogging. Furthermore, it would have been nearly impossible for a person to remain upright during such a slide because the coefficient of friction was 0.35 and the resultant drag force from the floor would have caused him to rotate forward. From such a simple test it was concluded that the plaintiff and his friend were just running too fast and did not slow down in time to avoid hitting the door; sliding on the floor for 5.2 m did not occur and thus had no influence on the accident.

**Example 19-10**

Why did the wire-reinforced window break? If a completely unbreakable window had been in place, the plaintiff would not have been cut.

**Solution**   The wire-reinforced window met the building safety code (see the earlier discussion of standards) in effect at the time of construction, and it is capable of withstanding a higher force than a nonreinforced window. Because no window can be completely unbreakable or completely "safe," we must compromise between cost and safety, trying to achieve the greatest increase in safety for the least cost. Other solutions, such as reducing window size, changing its location, or completely eliminating it, should also be considered by the designer.

Simple laboratory experiments can often test a failure hypothesis. To establish why a blade fractured in a power lawn mower, we simply ran the unit with a new blade. Speed measurement revealed that the speed governor did not function, and the excessive speed had thus contributed to the fracture. Another simple test was made in a case involving an injury sustained after a car fell off an air-activated automobile jack and the injured mechanic claimed that the gripping portion of the jack had slipped off the bumper because the jack was poorly designed. To test this claim, a car of the same model and year was lifted several times with the jack under the most unstable conditions possible, and the car never fell. This led to the conclusion that some other factor, not revealed in the plaintiff's testimony (see the earlier section on witnesses), had caused the jack to fall.

Knowledge of the effect of heat treatment on microstructure can define limits on the possible temperature–time history of a heat-treated part that has been through a fire. In one instance, a plaintiff claimed that a fire on his premises was caused by the absence of a synthetic rubber disk in a valve seat that allowed a gas regulator (pressure reducer) to pass gas of too high a pressure to a gas-fired space heater, leading to the fire. He had reached this conclusion because after the fire the disk was missing from the aluminum alloy disk holder to which it ought to have been glued when manufactured. The aluminum alloy was 2011-T3 (UNS A92011), specified to be quenched and aged at room temperature, which gave 98 HRE (121 HB) hardness as heat-treated; production samples of this part were tested and found to be about $98 \pm 3$ HRE. The hardness of the disk holder in question was 48 HRE, so clearly it had been through a fire. Although the exact duration of the fire could not be established, an assumption of 1 h at temperature appeared to be reasonable. New aluminum disk holders were therefore exposed to 1 h at temperatures ranging from 316°C to 538°C (it begins to melt at about 538°C), and then half were air-cooled and half were furnace-cooled. A second assumption was that the actual cooling rate fell somewhere between these two extremes. The lowest exposure temperature that gave a final hardness of 48 HRE was 360°C, and the highest was 538°C. Because the synthetic rubber of the missing disk degrades into ashes at about 480°C, these tests showed that it was consistent to have the final hardness of the disk holder 48 HRE and at the same time to have achieved high enough temperatures to completely destroy all evidence of the synthetic rubber disk. As is often the case, these test results did not *prove* that the rubber disk was in place before the fire but only demonstrated the consistency of that hypothesis.

Thus, such tests are used in conjunction with other evidence and observations to identify the cause of a failure.

**Example 19-11**

How do we know that the disk holder was not improperly heat-treated by the manufacturer to an initial hardness of 48 HRE?

**Solution** We do not, and no test can answer this question. A completed failure analysis seldom excludes all other explanations, but a rational person tends to accept the most probable explanation. If we are to accept the premise in this question, we will have to agree that two manufacturing errors occurred, in sequence: incorrect heat treatment of the disk holder and omission of the rubber disk. Such a situation seems much less probable than the idea that the disk was heated to 480–538°C in the fire.

Sometimes a full-scale field test under carefully controlled conditions can provide useful data. A young man operating a very small end loader [Fig. 19-28(a)] broke his neck while unloading compacted silage from a trench silo in the winter when some of the silage fell from a height of about 5 m. He claimed that the silage fell on the bucket of the end loader and caused the vehicle to tip suddenly forward. Either the resulting upward acceleration of the back end of the vehicle or the deceleration when the back end of the vehicle hit the ground broke his neck in compression, he claimed, and he sued the manufacturer for not having designed the shock-absorbing system of the seat so as to prevent such an injury. (He did not sue for lack of a guard cage around the driver because he had declined this option at the time of purchase.) We made some preliminary calculations of the angular acceleration of the vehicle that would result from the maximum possible amount of silage (95 kg) falling on the bucket from a 5-m height. The maximum acceleration of a rigid seat, obtained from this calculation, was insufficient to break his neck. But the calculation involved many simplifying assumptions, notably an estimate of the dynamic behavior of the tires under such an impact.

We therefore recommended a full-scale test, complete with an instrumented dummy having mechanical properties similar to those of a human body. An enormous mass of silage (5000 kg), far greater than the bucket could hold, was dropped on the bucket of the small end loader from a 5-m height [Fig. 19-28(b)], and it barely lifted the rear wheels off the ground. The resulting measured maximum acceleration (7.8 m/s$^2$ or 0.8 $g$'s) of the dummy was the deceleration that occurred when the vehicle returned to the ground and was well below that required to break a person's neck. The other extreme possibility was the deceleration from the vehicle's falling on its rear wheels after a full forward tip. Figure 19-28(c) shows the unit tipped fully forward, held by a jack, at a 42° angle. When the jack was suddenly pulled out, the maximum vertical deceleration of the seat was 27.5 m/s$^2$ (2.8 $g$'s) and of the dummy 39.2 m/s$^2$ (4.0 $g$'s) when the wheels struck the ground. Such deceleration could indeed cause injury, but on the basis of the first test there was no evidence that dropping any amount of silage on the bucket could cause the vehicle to tip forward 42°. It was concluded that the plaintiff's neck had been broken when some silage fell directly on his head; in winter moist silage freezes and is dug out in clumps, and it is altogether possible that a large clump fell directly on the plaintiff. There was no other consistent explanation possible after the test results were analyzed.

(a)

(b)

**Figure 19-28** Test of acceleration of
dummy on small end loader. (a) Prior to
silage drop; note instrumentation wires. (b)
After silage drop. (c) Prior to reverse drop of
end loader to ground.

(c)

## ASSESSMENT

Most of the failures described in this chapter were explained rather clearly at some point in the investigation by simple measurements or tests. The formal procedure of evaluation is thus usually so direct that we are hardly aware of the process. But, in fact, we must consider all tenable hypotheses for the failure at the outset of the investigation and then eliminate each in turn as possibilities following completion of the tests and analyses. If we have done everything correctly, if all the facts are known accurately, and we have considered all the possibilities, we should be left with only one hypothesis at the end of the study. In this way, we are often able to establish whether damage caused an accident or was the result of an accident, as has been discussed. But usually we lack some of the facts, and sometimes we are even given incorrect, misleading information, so the failure analyst must work through a study rather like a detective trying to solve a crime.

Sometimes a complete understanding of a failure comes rather late. In a fracture of a wire rope of an overhead hoist during a traverse of the entire rail-supported upper crane housing, operated by a person standing on the floor using a drop pedestal control, the block and hook of the hoist, *carrying no load,* dropped on the operator-plaintiff's head and caused severe injury. This was a real puzzle, as the wire rope could safely support many times the mass of the block and hook. The entire wire rope involved, which was about 50 m long, was examined but yielded no explanation. During the trial, the plaintiff's expert described the physical surroundings in the building housing the hoist and showed photographs of several large floor-mounted presses in this area. When the expert launched into an irrelevant discussion about the ratio of sheave diameter to wire rope diameter being too small, it was clear that this certainly had no bearing on this failure. Something had to have applied a large force to break the wire rope, but what? That evening the entire wire rope was examined again, stretched out in a parking lot. This led to evidence that had been overlooked during the earlier examination: a set of several sharp kinks about 10 m from the upper end of the wire rope that had been noticed before but forgotten because they did not seem relevant at the time. Just prior to the failure, the height of the hoist and the position of the block were such as to place this kinked region of the wire rope where it could have caught on the heavy press bolted to the floor. So instead of lifting just the block and hook, the plaintiff was unknowingly attempting to lift this large bolted-down machine. When the wire rope broke at a rather sharp edge of the sheave housing of the block, it dropped the block and hook. If the witnesses to the accident were aware that the wire rope had caught on the bolted-down machine, they never mentioned it.

## REFERENCES

1. American Society for Metals, *Case Histories in Failure Analysis.* Metals Park, Ohio: American Society for Metals, 1979.

2. American Society for Metals, *Metals Handbook,* 8th ed., vol. 10, *Failure Analysis and Prevention.* Metals Park, Ohio: American Society for Metals, 1975.

3. American Society for Metals, *Metals Handbook,* 9th ed., vol. 2, *Properties and Selection: Nonferrous Alloys and Pure Metals.* Metals Park, Ohio: American Society for Metals, 1979.

4. American Society for Metals, *Metals Handbook,* 9th ed., vol. 9, *Metallography and Microstructures.* Metals Park, Ohio: American Society for Metals, 1985.

5. American Society for Metals, *Metals Handbook,* 9th ed., vol. 11, *Failure Analysis and Prevention.* Metals Park, Ohio: American Society for Metals, 1986.

6. American Society for Metals, *Metals Handbook,* 9th ed., vol. 12, *Fractography.* Metals Park, Ohio: American Society for Metals, 1987.

7. Atkins, A. G., and D. K. Felbeck, "Applying Mutual Indentation Hardness Phenomena to Service Failures." American Society for Metals, *Source Book in Failure Analysis.* Metals Park, Ohio: American Society for Metals, 1974, p. 364.

8. Bannantine, Julie A., Jess J. Comer, and James L. Handrock, *Fundamentals of Metal Fatigue Analysis.* Englewood Cliffs, N.J.: Prentice Hall, 1990.

9. Engel, Lothar, and Hermann Klingele, *An Atlas of Metal Damage.* Englewood Cliffs, N.J.: Prentice Hall, 1981. Techniques of failure analysis using the SEM are described; also included is an excellent collection of SEM micrographs covering a wide range of fracture conditions.

10. Guy, Henry, *Fractography and Microfractography.* Philadelphia: Heydon, 1979.

11. Hertzberg, Richard W., *Deformation and Fracture Mechanics of Engineering Materials.* New York: John Wiley, 1976.

12. Rolfe, Stanley, T., and John M. Barsom, *Fracture and Fatigue Control in Structures.* Englewood Cliffs, N.J.: Prentice Hall, 1977.

13. Shank, M. E., *A Critical Survey of Brittle Failure in Carbon Plate Steel Structures Other Than Ships.* Ship Structure Committee Report, Serial No. SSC-65. Washington, D.C.: Ship Structure Committee, National Academy of Sciences, 1953. (Also reprinted as Welding Research Council Bulletin No. 17.)

14. Smith, Charles O., *The Science of Engineering Materials,* 2nd ed. Englewood Cliffs, N.J.: Prentice Hall, 1977.

## PROBLEMS

**19-1.** Describe all the possible fracture mode(s) that can produce each of the following characteristics:
   **(a)** Cleavage fracture.
   **(b)** Beach marks.
   **(c)** Intergranular fracture.
   **(d)** Striations.
   **(e)** Microvoids.

**19-2.** Name four kinds of fracture appearance that might be observed through examination of fracture surfaces with a magnifying glass. Tell what inferences can be drawn from each.

**19-3.** Why is an optical microscope not used to observe and identify fracture surface features such as intergranular fracture, microvoids, and striations?

**19-4.** In what ways is the transmission electron microscope better than the scanning electron microscope? In what ways is the SEM better than the TEM? What can the FEM do that neither the TEM or the SEM can do?

**19-5.** What information can be deduced from a plastically indented region observed after an accident?

**19-6.** What purposes are served by chemical analysis of a failed part?

**19-7.** What does a hardness test reveal to the failure analyst?

**19-8.** Under what circumstances should a metallographic examination be made following a failure?

**19-9.** If a full-scale test is being considered as an aid to analyzing a failure, what cautions should be observed before the test is performed if it is to have significance for the actual failure?

**19-10.** Name two observable fracture characteristics which when found in combination constitute a *unique* identification of fracture mode for each of the following. (Note that the combination of the two constitutes a positive identification of the fracture mode, although either of the characteristics may also be observed in other fracture modes.)

   **(a)** Fatigue in aluminum alloy
   **(b)** Hydrogen damage in high-strength AISI-SAE 4340 steel
   **(c)** Ductile fracture of AISI-SAE 1020 steel
   **(d)** Stress-corrosion cracking of high-strength aluminum alloy.

**19-11.** **(a)** Without emphasis on the derivation of the theory, explain the main features of engineering significance that occur when a sphere is loaded against a flat surface. In particular, emphasize the changes in the character of the conditions that occur as the load is increased.
   **(b)** An engineering joint, represented in simplified form in Fig. 19-29, is part of a mechanism that has broken at another place. It has been argued that the failure was caused by excessive loads. Examination shows that the socket has been indented with a circular area of indentation of approximately 4.1 mm diameter; the hemispherical surface has a similar, if somewhat shallower, indentation. Both parts are of steel, the socket has a hardness of 150 HB, and the spherical component has a hardness of 225 HB.

$R_1 = 20.0$ mm
$R_2 = 40.0$ mm

**Figure 19-29**   Courtesy of Dr. J. F. Archard.

Make an estimate of the load to which the component has been subjected. Discuss the way in which the spherical surface might have been indented, despite the fact that it has a higher hardness than the socket; thus, suggest an approximate value for the hardness required to avoid any indentation when in contact with the socket. (Courtesy of J. F. Archard.)

**19-12.** Following an accident, a tie rod (part A) of an automobile steering system was found to have been indented as shown in Fig. 19-30 by another shaft, call it part B. The projected area of the indentation is 17 mm². The tie rod is an air-cooled AISI-SAE 1030 steel. Estimate the magnitude of the force between the two parts that caused the indentation if the shaft (part B) is later found to exhibit no indentation.

**19-13.** A ball peen hammer strikes a thick plate of air-cooled 0.20% C plain-carbon steel and produces an indentation 3.67 mm in diameter. Estimate the maximum force applied by the hammer.

**Figure 19-30**

**19-14.** The linkage shown in Fig. 19-31 is subjected to tensile forces $F$ of 25 kN. Parts A and B and the nut D are made of AISI 1045 steel oil-quenched from 825°C to 20°C. The machine screw C, of diameter 9.65 mm, is made of AISI 4140 steel oil-quenched from 825°C to 20°C and tempered at 200°C for 2 ks.

Estimate the dimensions of the indentation on the inner wall of the hole of part B that would result from the tensile force $F$. Show all calculations and cite all sources of data that you use.

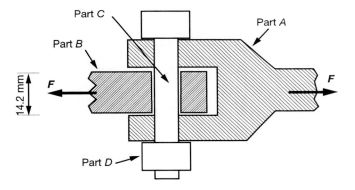

**Figure 19-31**

**19-15.** Two solid cylinders, each of 12.0 mm diameter and 30.0 mm length, have hardnesses of 217 HB (cylinder A) and 330 HB (cylinder B). These cylinders are forced down against a very hard, flat anvil as shown in Fig. 19-32.

If the force $F$ is 20.0 kN, show in a drawing the shape and dimensions of any deformation of cylinder A (the lower cylinder) that would be visible after the load is removed and the parts separated.

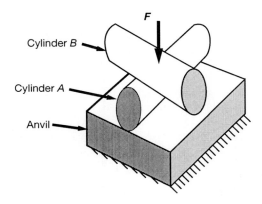

**Figure 19-32**

# Appendices

**APPENDIX 1.** PREFIXES FOR USE
WITH SI UNITS

| Multiplication factor | Prefix | SI symbol |
|:---:|:---|:---:|
| $10^{18}$ | exa | E |
| $10^{15}$ | peta | P |
| $10^{12}$ | tera | T |
| $10^{9}$ | giga | G |
| $10^{6}$ | mega | M |
| $10^{3}$ | kilo | k |
| $10^{-3}$ | milli | m |
| $10^{-6}$ | micro | $\mu$ |
| $10^{-9}$ | nano | n |
| $10^{-12}$ | pico | p |
| $10^{-15}$ | femto | f |
| $10^{-18}$ | atto | a |

**APPENDIX 2**.  FACTORS FOR CONVERSION TO SI UNITS

Conversion factors in the following table are given as a number between 1 and 10, followed by the letter E (for exponent), a plus or minus symbol, and two digits that indicate the power of 10 by which the number must be multiplied to obtain the value in the appropriate SI unit. For example, to convert from pound-force/inch$^2$ (psi) to pascal (N/m$^2$), the factor listed is 6.894 757 E + 03. Thus, 1 psi is 6.894 757 kPa, and 1000 psi is 6.894 757 MPa.

| To convert from | To | Multiply by |
|---|---|---|
| *Area* | | |
| foot$^2$ | m$^2$ | 9.290 304 E – 02 |
| inch$^2$ | m$^2$ | 6.451 600 E – 04 |
| *Energy and Work* | | |
| Btu (Int. Table) | J | 1.055 056 E + 03 |
| calorie (Int. Table) | J | 4.186 800 E + 00 |
| erg | J | 1.000 000 E – 07 |
| foot-pound-force | J | 1.355 818 E + 00 |
| watt-second | J | 1.000 000 E + 00 |
| *Energy/Area (Toughness)* | | |
| erg/cm$^2$ | J/m$^2$ | 1.000 000 E – 03 |
| foot-pound-force/inch$^2$ | J/m$^2$ | 2.101 522 E + 03 |
| inch-pound-force/inch$^2$ | J/m$^2$ | 1.751 268 E + 02 |
| *Force* | | |
| dyne | N | 1.000 000 E – 05 |
| kilogram-force | N | 9.806 650 E + 00 |
| pound-force | N | 4.448 222 E + 00 |
| poundal | N | 1.382 550 E – 01 |
| *Length* | | |
| angstrom | m | 1.000 000 E – 10 |
| foot | m | 3.048 000 E – 01 |
| inch | m | 2.540 000 E – 02 |
| mile (U.S. nautical) | m | 1.852 000 E + 03 |
| mile (U.S. statute) | m | 1.609 344 E + 03 |
| *Mass* | | |
| grain | kg | 6.479 891 E – 05 |
| pound-mass | kg | 4.535 924 E – 01 |
| slug | kg | 1.459 390 E + 01 |
| *Mass/Volume (Density)* | | |
| g/cm$^3$ | kg/m$^3$ | 1.000 000 E + 03 |
| pound-mass/inch$^3$ | kg/m$^3$ | 2.767 990 E + 04 |
| slug/foot$^3$ | kg/m$^3$ | 5.153 788 E + 02 |
| *Power* | | |
| Btu (Int Table)/hour | W | 2.930 711 E – 01 |

**APPENDIX 2.**   (cont.)

| To convert from | To | Multiply by |
|---|---|---|
| foot-pound-force/s | W | 1.355 818 E + 00 |
| horsepower | W | 7.456 999 E + 02 |

| | Pressure or Stress | |
|---|---|---|
| atm (760 torr) | Pa | 1.013 250 E + 05 |
| bar | Pa | 1.000 000 E + 05 |
| cm of mercury (0°C) | Pa | 1.333 22 E + 03 |
| cm of water (4°C) | Pa | 9.806 38 E + 01 |
| dyne/cm$^2$ | Pa | 1.000 000 E − 01 |
| kg-force/cm$^2$ | Pa | 9.806 650 E + 04 |
| kg-force/mm$^2$ | Pa | 9.806 650 E + 06 |
| newton/m$^2$ | Pa | 1.000 000 E + 00 |
| pound-force/inch$^2$ (psi) | Pa | 6.894 757 E + 03 |
| torr (mm Hg, 0°C) | Pa | 1.333 22 E + 02 |

| | Temperature | |
|---|---|---|
| degree Celsius (°C) | K | $t_K = t_C + 273.15$ |
| degree Fahrenheit (°F) | K | $t_K = \dfrac{t_F + 459.67}{1.8}$ |

| | Time | |
|---|---|---|
| day (mean solar) (d) | s | 8.640 000 E + 04 |
| hour (mean solar) (h) | s | 3.600 000 E + 03 |
| year (365-day) (y) | s | 3.153 600 E + 07 |
| year (tropical) | s | 3.155 693 E + 07 |

| | Velocity | |
|---|---|---|
| foot/minute | m/s | 5.080 000 E − 03 |
| km/hour (km/h) | m/s | 2.777 778 E − 01 |
| mile/hour (U.S. statute) | m/s | 4.470 400 E − 01 |

| | Viscosity | |
|---|---|---|
| centipoise | Pa-s | 1.000 000 E − 03 |
| poise | Pa-s | 1.000 000 E − 01 |
| pound/force-s/ft$^2$ | Pa-s | 4.788 026 E + 01 |

| | Volume | |
|---|---|---|
| barrel (oil, 42 U.S. gal) | m$^3$ | 1.589 873 E − 01 |
| fluid ounce (U.S.) | m$^3$ | 2.957 353 E − 05 |
| foot$^3$ | m$^3$ | 2.831 685 E − 02 |
| gallon (Imperial liquid) | m$^3$ | 4.546 122 E − 03 |
| gallon (U.S. liquid) | m$^3$ | 3.785 412 E − 03 |
| inch$^3$ | m$^3$ | 1.638 706 E − 05 |
| liter (L) | m$^3$ | 1.000 000 E − 03 |

| | Miscellaneous Special Conversions | |
|---|---|---|
| ksi $\sqrt{\text{in}}$. | MPa $\sqrt{m}$ | 1.098 843 E + 00 |
| mile/gallon (U.S. liquid) | km/L | 4.251 437 E − 01 |

**APPENDIX 3.** PLASTIC STRESS–STRAIN CONSTANTS FOR METALS

| | | | $\sigma = K(\varepsilon_{CW} + \varepsilon)^n$ | |
|---|---|---|---|---|
| UNS | Metal | Condition | $K$, MPa | $n$ |
| A91100 | 1100 Al | Annealed | 182 | 0.20 |
| A92024 | 2024 Al | T-4 | 700 | 0.15 |
| C11000 | Cu | Annealed 1 h at 540°C | 550 | 0.55 |
| C11000 | Cu | Annealed 1 h at 820°C | 480 | 0.48 |
| C26000 | 70–30 brass | Annealed 1 h at 540°C | 770 | 0.56 |
| C33100 | 70–30 leaded brass | Annealed 1 h at 680°C | 735 | 0.50 |
| — | Armco iron | Normalized | 560 | 0.32 |
| G10200 | 1020 steel | As hot-rolled | 800 | 0.22 |
| G10450 | 1045 steel | As hot-rolled | 980 | 0.14 |
| H43400 | 4340 steel | As hot-rolled | 1500 | 0.10 |
| G52986 | 52100 steel | Annealed 1 h at 820°C | 1470 | 0.07 |
| | | Spheroidized | 1150 | 0.18 |
| S30400 | 18–8 stainless steel | Annealed 1 h at 870°C | 1500 | 0.51 |

Note: $\varepsilon_{CW}$ is the prior true prestrain from cold work.

**APPENDIX 4.** APPROXIMATE HARDNESS CONVERSIONS FOR METALS

A.  Hardness conversions: hardened steel and hard alloys*

| Rockwell scale | | | | | Dph, 10 $kg_f$ | HB, 3000 $kg_f$ | Tensile strength | |
|---|---|---|---|---|---|---|---|---|
| C | A | 15N | 30N | 45N | | | $10^3$ psi (approx.) | MPa (approx.) |
| 80 | 92.0 | 96.5 | 92.0 | 87.0 | 1865 | — | — | — |
| 79 | 91.5 | — | 91.5 | 86.5 | 1787 | — | — | — |
| 78 | 91.0 | 96.0 | 91.0 | 85.5 | 1710 | — | — | — |
| 77 | 90.5 | — | 90.5 | 84.5 | 1633 | — | — | — |
| 76 | 90.0 | 95.5 | 90.0 | 83.5 | 1556 | — | — | — |
| 75 | 89.5 | — | 89.0 | 82.5 | 1478 | — | — | — |
| 74 | 89.0 | 95.0 | 88.5 | 81.5 | 1400 | — | — | — |
| 73 | 88.5 | — | 88.0 | 80.5 | 1323 | — | — | — |
| 72 | 88.0 | 94.5 | 87.0 | 79.5 | 1245 | — | — | — |
| 71 | 87.0 | — | 86.5 | 78.5 | 1160 | — | — | — |
| 70 | 86.5 | 94.0 | 86.0 | 77.5 | 1076 | — | — | — |
| 69 | 86.0 | 93.5 | 85.0 | 76.5 | 1004 | — | — | — |
| 68 | 85.5 | — | 84.5 | 75.5 | 942 | — | — | — |
| 67 | 85.0 | 93.0 | 83.5 | 74.5 | 894 | — | — | — |
| 66 | 84.5 | 92.5 | 83.0 | 73.0 | 854 | — | — | — |
| 65 | 84.0 | 92.0 | 82.0 | 72.0 | 820 | — | — | — |
| 64 | 83.5 | — | 81.0 | 71.0 | 789 | — | — | — |
| 63 | 83.0 | 91.5 | 80.0 | 70.0 | 763 | — | — | — |
| 62 | 82.5 | 91.0 | 79.0 | 69.0 | 739 | — | — | — |
| 61 | 81.5 | 90.5 | 78.5 | 67.5 | 716 | — | — | — |
| 60 | 81.0 | 90.0 | 77.5 | 66.5 | 695 | 614 | — | — |

**APPENDIX 4A.** (cont.)

| | | Rockwell scale | | | Dph, 10 $kg_f$ | HB, 3000 $kg_f$ | Tensile strength | |
|---|---|---|---|---|---|---|---|---|
| C | A | 15N | 30N | 45N | | | $10^3$ psi (approx.) | MPa (approx.) |
| 59 | 80.5 | 89.5 | 76.5 | 65.5 | 675 | 600 | — | — |
| 58 | 80.0 | — | 75.5 | 64.0 | 655 | 587 | — | — |
| 57 | 79.5 | 89.0 | 75.0 | 63.0 | 636 | 573 | — | — |
| 56 | 79.0 | 88.5 | 74.0 | 62.0 | 617 | 560 | — | — |
| 55 | 78.5 | 88.0 | 73.0 | 61.0 | 598 | 547 | 301 | 2080 |
| 54 | 78.0 | 87.5 | 72.0 | 59.5 | 580 | 534 | 291 | 2010 |
| 53 | 77.5 | 87.0 | 71.0 | 58.5 | 562 | 522 | 282 | 1940 |
| 52 | 77.0 | 86.5 | 70.5 | 57.5 | 545 | 509 | 273 | 1880 |
| 51 | 76.5 | 86.0 | 69.5 | 56.0 | 528 | 496 | 264 | 1820 |
| 50 | 76.0 | 85.5 | 68.5 | 55.0 | 513 | 484 | 255 | 1760 |
| 49 | 75.5 | 85.0 | 67.5 | 54.0 | 498 | 472 | 246 | 1700 |
| 48 | 74.5 | 84.5 | 66.5 | 52.5 | 485 | 460 | 237 | 1630 |
| 47 | 74.0 | 84.0 | 66.0 | 51.5 | 471 | 448 | 229 | 1580 |
| 46 | 73.5 | 83.5 | 65.0 | 50.0 | 458 | 437 | 221 | 1520 |
| 45 | 73.0 | 83.0 | 64.0 | 49.0 | 446 | 426 | 214 | 1480 |
| 44 | 72.5 | 82.5 | 63.0 | 48.0 | 435 | 415 | 207 | 1430 |
| 43 | 72.0 | 82.0 | 62.0 | 46.5 | 424 | 404 | 200 | 1380 |
| 42 | 71.5 | 81.5 | 61.5 | 45.5 | 413 | 393 | 194 | 1340 |
| 41 | 71.0 | 81.0 | 60.5 | 44.5 | 403 | 382 | 188 | 1300 |
| 40 | 70.5 | 80.5 | 59.5 | 43.0 | 393 | 372 | 182 | 1250 |
| 39 | 70.0 | 80.0 | 58.5 | 42.0 | 383 | 362 | 177 | 1220 |
| 38 | 69.5 | 79.5 | 57.5 | 41.0 | 373 | 352 | 171 | 1180 |
| 37 | 69.0 | 79.0 | 56.5 | 39.5 | 363 | 342 | 166 | 1140 |
| 36 | 68.5 | 78.5 | 56.0 | 38.5 | 353 | 332 | 162 | 1120 |
| 35 | 68.0 | 78.0 | 55.0 | 37.0 | 343 | 322 | 157 | 1080 |
| 34 | 67.5 | 77.0 | 54.0 | 36.0 | 334 | 313 | 153 | 1050 |
| 33 | 67.0 | 76.5 | 53.0 | 35.0 | 325 | 305 | 148 | 1020 |
| 32 | 66.5 | 76.0 | 52.0 | 33.5 | 317 | 297 | 144 | 993 |
| 31 | 66.0 | 75.5 | 51.5 | 32.5 | 309 | 290 | 140 | 965 |
| 30 | 65.5 | 75.0 | 50.5 | 31.5 | 301 | 283 | 136 | 938 |
| 29 | 65.0 | 74.5 | 49.5 | 30.0 | 293 | 276 | 132 | 910 |
| 28 | 64.5 | 74.0 | 48.5 | 29.0 | 285 | 270 | 129 | 889 |
| 27 | 64.0 | 73.5 | 47.5 | 28.0 | 278 | 265 | 126 | 869 |
| 26 | 63.5 | 72.5 | 47.0 | 26.5 | 271 | 260 | 123 | 848 |
| 25 | 63.0 | 72.0 | 46.0 | 25.5 | 264 | 255 | 120 | 827 |
| 24 | 62.5 | 71.5 | 45.0 | 24.0 | 257 | 250 | 117 | 807 |
| 23 | 62.0 | 71.0 | 44.0 | 23.0 | 251 | 245 | 115 | 793 |
| 22 | 61.5 | 70.5 | 43.0 | 22.0 | 246 | 240 | 112 | 772 |
| 21 | 61.0 | 70.0 | 42.5 | 20.5 | 241 | 235 | 110 | 758 |
| 20 | 60.5 | 69.5 | 41.5 | 19.5 | 236 | 230 | 108 | 745 |

*From Vincent E. Lysaght and Anthony DeBellis, *Hardness Testing Handbook* (American Chain & Cable Co.). Reprinted with permission of Page-Wilson Corporation, which holds the copyright.

B. Hardness conversions: soft steel, gray and malleable cast iron, and most nonferrous metals*

| | Rockwell scale | | | | | HB, 500 kg$_f$ (10-mm ball) | HB, 3000 kg$_f$ and Dph, 10 kg$_f$ | Tensile strength | |
|---|---|---|---|---|---|---|---|---|---|
| | | | | | | | | 10$^3$ psi (approx.) | MPa (approx.) |
| B | 15T | 30T | 45T | E | A | | | | |
| 100 | 93.0 | 82.0 | 72.0 | — | 61.5 | 201 | 240 | 116 | 800 |
| 99 | 92.5 | 81.5 | 71.0 | — | 61.0 | 195 | 234 | 112 | 772 |
| 98 | — | 81.0 | 70.0 | — | 60.0 | 189 | 228 | 109 | 752 |
| 97 | 92.0 | 80.5 | 69.0 | — | 59.5 | 184 | 222 | 106 | 731 |
| 96 | — | 80.0 | 68.0 | — | 59.0 | 179 | 216 | 103 | 710 |
| 95 | 91.5 | 79.0 | 67.0 | — | 58.0 | 175 | 210 | 101 | 696 |
| 94 | — | 78.5 | 66.0 | — | 57.5 | 171 | 205 | 98 | 676 |
| 93 | 91.0 | 78.0 | 65.5 | — | 57.0 | 167 | 200 | 96 | 662 |
| 92 | 90.5 | 77.5 | 64.5 | — | 56.5 | 163 | 195 | 93 | 641 |
| 91 | — | 77.0 | 63.5 | — | 56.0 | 160 | 190 | 91 | 627 |
| 90 | 90.0 | 76.0 | 62.5 | — | 55.5 | 157 | 185 | 89 | 614 |
| 89 | 89.5 | 75.5 | 61.5 | — | 55.0 | 154 | 180 | 87 | 600 |
| 88 | — | 75.0 | 60.5 | — | 54.0 | 151 | 176 | 85 | 586 |
| 87 | 89.0 | 74.5 | 59.5 | — | 53.5 | 148 | 172 | 83 | 572 |
| 86 | 88.5 | 74.0 | 58.5 | — | 53.0 | 145 | 169 | 81 | 558 |
| 85 | — | 73.5 | 58.0 | — | 52.5 | 142 | 165 | 80 | 552 |
| 84 | 88.0 | 73.0 | 57.0 | — | 52.0 | 140 | 162 | 78 | 538 |
| 83 | 87.5 | 72.0 | 56.0 | — | 51.0 | 137 | 159 | 77 | 531 |
| 82 | — | 71.5 | 55.0 | — | 50.5 | 135 | 156 | 75 | 517 |
| 81 | 87.0 | 71.0 | 54.0 | — | 50.0 | 133 | 153 | 74 | 510 |
| 80 | 86.5 | 70.0 | 53.0 | — | 49.5 | 130 | 150 | 72 | 496 |
| 79 | — | 69.5 | 52.0 | — | 49.0 | 128 | 147 | — | — |
| 78 | 86.0 | 69.0 | 51.0 | — | 48.5 | 126 | 144 | — | — |
| 77 | 85.5 | 68.0 | 50.0 | — | 48.0 | 124 | 141 | — | — |
| 76 | — | 67.5 | 49.0 | — | 47.0 | 122 | 139 | — | — |
| 75 | 85.0 | 67.0 | 48.5 | — | 46.5 | 120 | 137 | — | — |
| 74 | — | 66.0 | 47.5 | — | 46.0 | 118 | 135 | — | — |
| 73 | 84.5 | 65.5 | 46.5 | — | 45.5 | 116 | 132 | — | — |
| 72 | 84.0 | 65.0 | 45.5 | — | 45.0 | 114 | 130 | — | — |
| 71 | — | 64.0 | 44.5 | 100 | 44.5 | 112 | 127 | — | — |
| 70 | 83.5 | 63.5 | 43.5 | 99.5 | 44.0 | 110 | 125 | — | — |
| 69 | 83.0 | 62.5 | 42.5 | 99.0 | 43.5 | 109 | 123 | — | — |
| 68 | — | 62.0 | 41.5 | 98.0 | 43.0 | 107 | 121 | — | — |
| 67 | 82.5 | 61.5 | 40.5 | 97.5 | 42.5 | 106 | 119 | — | — |
| 66 | 82.0 | 60.5 | 39.5 | 97.0 | 42.0 | 104 | 117 | — | — |
| 65 | — | 60.0 | 38.5 | 96.0 | — | 102 | 116 | — | — |
| 64 | 81.5 | 59.5 | 37.5 | 95.5 | 41.5 | 101 | 114 | — | — |
| 63 | 81.0 | 58.5 | 36.5 | 95.0 | 41.0 | 99 | 112 | — | — |
| 62 | — | 58.0 | 35.5 | 94.5 | 40.5 | 98 | 110 | — | — |
| 61 | 80.5 | 57.0 | 34.5 | 93.5 | 40.0 | 96 | 108 | — | — |
| 60 | — | 56.5 | 33.5 | 93.0 | 39.5 | 95 | 107 | — | — |
| 59 | 80.0 | 56.0 | 32.0 | 92.5 | 39.0 | 94 | 106 | — | — |
| 58 | 79.5 | 55.0 | 31.0 | 92.0 | 38.5 | 92 | 104 | — | — |
| 57 | — | 54.5 | 30.0 | 91.0 | 38.0 | 91 | 103 | — | — |

**APPENDIX 4B.**   (cont.)

| | | Rockwell scale | | | | HB, 500 kg$_f$ (10-mm ball) | HB, 3000 kg$_f$ and Dph, 10 kg$_f$ | Tensile strength | |
|---|---|---|---|---|---|---|---|---|---|
| | | | | | | | | 10$^3$ psi (approx.) | MPa (approx.) |
| B | 15T | 30T | 45T | E | A | | | | |
| 56 | 79.0 | 54.0 | 29.0 | 90.5 | — | 90 | 101 | — | — |
| 55 | 78.5 | 53.0 | 28.0 | 90.0 | 37.5 | 89 | 100 | — | — |
| 54 | — | 52.5 | 27.0 | 89.5 | 37.0 | 87 | — | — | — |
| 53 | 78.0 | 51.5 | 26.0 | 89.0 | 36.5 | 86 | — | — | — |
| 52 | 77.5 | 51.0 | 25.0 | 88.0 | 36.0 | 85 | — | — | — |
| 51 | — | 50.5 | 24.0 | 87.5 | 35.5 | 84 | — | — | — |
| 50 | 77.0 | 49.5 | 23.0 | 87.0 | 35.0 | 83 | — | — | — |
| 49 | 76.5 | 49.0 | 22.0 | 86.5 | — | 82 | — | — | — |
| 48 | — | 48.5 | 20.5 | 85.5 | 34.5 | 81 | — | — | — |
| 47 | 76.0 | 47.5 | 19.5 | 85.0 | 34.0 | 80 | — | — | — |
| 46 | 75.5 | 47.0 | 18.5 | 84.5 | 33.5 | — | — | — | — |
| 45 | — | 46.0 | 17.5 | 84.0 | 33.0 | 79 | — | — | — |
| 44 | 75.0 | 45.5 | 16.5 | 83.5 | 32.5 | 78 | — | — | — |
| 43 | 74.5 | 45.0 | 15.5 | 82.5 | 32.0 | 77 | — | — | — |
| 42 | — | 44.0 | 14.5 | 82.0 | 31.5 | 76 | — | — | — |
| 41 | 74.0 | 43.5 | 13.5 | 81.5 | 31.0 | 75 | — | — | — |
| 40 | 73.5 | 43.0 | 12.5 | 81.0 | — | — | — | — | — |
| 39 | — | 42.0 | 11.0 | 80.0 | 30.5 | 74 | — | — | — |
| 38 | 73.0 | 41.5 | 10.0 | 79.5 | 30.0 | 73 | — | — | — |
| 37 | 72.5 | 40.5 | 9.0 | 79.0 | 29.5 | 72 | — | — | — |
| 36 | — | 40.0 | 8.0 | 78.5 | 29.0 | — | — | — | — |
| 35 | 72.0 | 39.5 | 7.0 | 78.0 | 28.5 | 71 | — | — | — |
| 34 | 71.5 | 38.5 | 6.0 | 77.0 | 28.0 | 70 | — | — | — |
| 33 | — | 38.0 | 5.0 | 76.5 | — | 69 | — | — | — |
| 32 | 71.0 | 37.5 | 4.0 | 76.0 | 27.5 | — | — | — | — |
| 31 | — | 36.5 | 3.0 | 75.5 | 27.0 | 68 | — | — | — |
| 30 | 70.5 | 36.0 | 2.0 | 75.0 | 26.5 | 67 | — | — | — |
| 29 | 70.0 | 35.5 | 1.0 | 74.0 | 26.0 | — | — | — | — |
| 28 | — | 34.5 | — | 73.5 | 25.5 | 66 | — | — | — |
| 27 | 69.5 | 34.0 | — | 73.0 | 25.0 | — | — | — | — |
| 26 | 69.0 | 33.0 | — | 72.5 | 24.5 | 65 | — | — | — |
| 25 | — | 32.5 | — | 72.0 | — | 64 | — | — | — |
| 24 | 68.5 | 32.0 | — | 71.0 | 24.0 | — | — | — | — |
| 23 | 68.0 | 31.0 | — | 70.5 | 23.5 | 63 | — | — | — |
| 22 | — | 30.5 | — | 70.0 | 23.0 | — | — | — | — |
| 21 | 67.5 | 29.5 | — | 69.5 | 22.5 | 62 | — | — | — |
| 20 | — | 29.0 | — | 68.5 | 22.0 | — | — | — | — |
| 19 | 67.0 | 28.5 | — | 68.0 | 21.5 | 61 | — | — | — |
| 18 | 66.5 | 27.5 | — | 67.5 | — | — | — | — | — |
| 17 | — | 27.0 | — | 67.0 | 21.0 | 60 | — | — | — |
| 16 | 66.0 | 26.0 | — | 66.5 | 20.5 | — | — | — | — |
| 15 | 65.5 | 25.5 | — | 65.5 | 20.0 | 59 | — | — | — |
| 14 | — | 25.0 | — | 65.0 | — | — | — | — | — |
| 13 | 65.0 | 24.0 | — | 64.5 | — | 58 | — | — | — |

**APPENDIX 4B**.   (cont.)

| B | 15T | 30T | 45T | E | A | HB, 500 kg$_f$ (10-mm ball) | HB, 3000 kg$_f$ and Dph, 10 kg$_f$ | 10$^3$ psi (approx.) | MPa (approx.) |
|---|---|---|---|---|---|---|---|---|---|
| | Rockwell scale | | | | | | | Tensile strength | |
| 12 | 64.5 | 23.5 | — | 64.0 | — | — | — | — | — |
| 11 | — | 23.0 | — | 63.5 | — | — | — | — | — |
| 10 | 64.0 | 22.0 | — | 62.5 | — | 57 | — | — | — |
| 9 | — | 21.5 | — | 62.0 | — | — | — | — | — |
| 8 | 63.5 | 20.5 | — | 61.5 | — | — | — | — | — |
| 7 | 63.0 | 20.0 | — | 61.0 | — | 56 | — | — | — |
| 6 | — | 19.5 | — | 60.5 | — | — | — | — | — |
| 5 | 62.5 | 18.5 | — | 60.0 | — | 55 | — | — | — |
| 4 | 62.0 | 18.0 | — | 59.0 | — | — | — | — | — |
| 3 | — | 17.0 | — | 58.5 | — | — | — | — | — |
| 2 | 61.5 | 16.5 | — | 58.0 | — | 54 | — | — | — |
| 1 | 61.0 | 16.0 | — | 57.5 | — | — | — | — | — |
| 0 | — | 15.0 | — | 57.0 | — | 53 | — | — | — |

*From Vincent E. Lysaght and Anthony DeBellis, *Hardness Testing Handbook* (American Chain & Cable Co.). Reprinted with permission of Page-Wilson Corporation, which holds the copyright.

C.   Hardness conversions: steels*

| Brinell 10-mm steel ball Load 3000 kg$_f$ | | Vickers diamond pyramid hardness, kg$_f$/mm$^2$ | Rockwell | |
|---|---|---|---|---|
| Diameter, mm | Hardness, kg$_f$/mm$^2$ | | C 150-kg$_f$ load 120° diamond cone | B 100-kg$_f$ load $^1/_{16}$-in. steel ball |
| 2.20 | 780[a] | 1,150 | 70 | — |
| .25 | 745[a] | 1,050 | 68 | — |
| .30 | 712[a] | 960 | 66 | — |
| .35 | 682[a] | 885 | 64 | — |
| .40 | 653[a] | 820 | 62 | — |
| .45 | 627[a] | 765 | 60 | — |
| .50 | 601[a] | 717 | 58 | — |
| .55 | 578[a] | 675 | 57 | — |
| .60 | 555[a] | 633 | 55 | ~ 120 |
| .65 | 534[a] | 598 | 53 | — |
| .70 | 514[a] | 567 | 52 | — |
| .75 | 495[a] | 540 | 50 | — |
| .80 | 477[a] | 515 | 49 | — |
| .85 | 461[a] | 494 | 47 | — |
| .90 | 444 | 472 | 46 | — |
| .95 | 420 | 454 | 45 | ~ 115 |

**APPENDIX 4C**.   (cont.)

| Brinell 10-mm steel ball Load 3000 kg$_f$ | | Vickers diamond pyramid hardness, kg$_f$/mm$^2$ | Rockwell | |
|---|---|---|---|---|
| Diameter, mm | Hardness, kg$_f$/mm$^2$ | | C<br>150-kg$_f$ load<br>120° diamond cone | B<br>100-kg$_f$ load<br>$^1/_{16}$-in. steel ball |
| 3.00 | 415 | 437 | 44 | — |
| .05 | 401 | 420 | 42 | — |
| .10 | 388 | 404 | 41 | — |
| .15 | 376 | 389 | 40 | — |
| .20 | 363 | 375 | 38 | ~ 110 |
| .25 | 352 | 363 | 37 | — |
| .30 | 341 | 350 | 36 | — |
| .35 | 331 | 339 | 35 | — |
| .40 | 321 | 327 | 34 | — |
| .45 | 311 | 316 | 33 | — |
| .50 | 302 | 305 | 32 | — |
| .55 | 293 | 296 | 31 | — |
| .60 | 285 | 287 | 30 | 105 |
| .65 | 277 | 279 | 29 | 104 |
| .70 | 269 | 270 | 28 | 104 |
| .75 | 262 | 263 | 26 | 103 |
| .80 | 255 | 256 | 25 | 102 |
| .85 | 248 | 248 | 24 | 102 |
| .90 | 241 | 241 | 23 | 100 |
| .95 | 235 | 235 | 22 | 99 |
| 4.00 | 229 | 229 | 21 | 98 |
| .05 | 223 | 223 | 20 | 97 |
| .10 | 217 | 217 | 18 | 96 |
| .15 | 212 | 212 | 17 | 96 |
| .20 | 207 | 207 | 16 | 95 |
| .25 | 201 | 201 | 15 | 94 |
| .30 | 197 | 197 | 13 | 93 |
| .35 | 192 | 192 | 12 | 92 |
| .40 | 187 | 187 | 10 | 91 |
| .45 | 183 | 183 | 9 | 90 |
| .50 | 179 | 179 | 8 | 89 |
| .55 | 174 | 174 | 7 | 88 |
| .60 | 170 | 170 | 6 | 87 |
| .65 | 167 | 167 | 4 | 86 |
| .70 | 163 | 163 | 3 | 85 |
| .75 | 159 | 159 | 2 | 84 |
| .80 | 156 | 156 | 1 | 83 |
| .85 | 152 | 152 | — | 82 |
| .90 | 149 | 149 | — | 81 |
| .95 | 146 | 146 | — | 80 |

**APPENDIX 4C**. (cont.)

| Brinell 10-mm steel ball Load 3000 kg$_f$ | | Vickers diamond pyramid hardness, kg$_f$/mm$^2$ | Rockwell | |
|---|---|---|---|---|
| Diameter, mm | Hardness, kg$_f$/mm$^2$ | | C 150-kg$_f$ load 120° diamond cone | B 100-kg$_f$ load $^1/_{16}$-in. steel ball |
| 5.00 | 143 | 143 | — | 79 |
| .05 | 140 | 140 | — | 78 |
| .10 | 137 | 137 | — | 77 |
| .15 | 134 | 134 | — | 76 |
| .20 | 131 | 131 | — | 74 |
| .25 | 128 | 128 | — | 73 |
| .30 | 126 | 126 | — | 72 |
| .35 | 123 | 123 | — | 71 |
| .40 | 121 | 121 | — | 70 |
| .45 | 118 | 118 | — | 69 |
| .50 | 116 | 116 | — | 68 |
| .55 | 114 | 114 | — | 67 |
| .60 | 111 | 111 | — | 66 |
| .65 | 109 | 109 | — | 65 |
| .70 | 107 | 107 | — | 64 |
| .75 | 105 | 105 | — | 62 |
| .80 | 103 | 103 | — | 61 |
| .85 | 101 | 101 | — | 60 |
| .90 | 99 | 99 | — | 59 |
| .95 | 97 | 97 | — | 57 |
| 6.00 | 96 | 96 | — | 56 |

*From D. Tabor, *The Hardness of Metals*. (Oxford: Clarendon Press, 1951).

[a]With the standard steel ball, Brinell hardness values above about 400 are not considered reliable.

HARDNESS CONVERSION: The values given here are approximate only and apply only to steels of uniform chemical composition and uniform heat treatment. The values are less reliable for nonferrous materials and are not recommended for case-hardened steel.

**APPENDIX 5.**  MECHANICAL PROPERTY RANGES FOR COMMON ENGINEERING
MATERIALS AT ~ 20°C

| Material | Density, kg/m$^3$ | Elastic modulus, GPa | Strength, MPa | Critical stress intensity factor, MPa $\sqrt{m}$ | Nom. fracture strain — | Maximum strength/ sp. gr., MPa |
|---|---|---|---|---|---|---|
| Ductile steel | 7850–7870 | 200 | 350–800 | 170 | 0.2–0.5 | 100 |
| Cast iron | 6950–7700 | 55–140 | 140–650 | 5–20 | 0.0–0.2 | 90 |
| Cast Al alloys | 2570–2950 | 65–72 | 130–300 | — | 0.01–0.14 | 100 |
| Wrought Al alloys | 2620–2820 | 68–72 | 55–650 | 15–100 | 0.05–0.30 | 200 |
| Polymers | 900–2200 | 0.1–21 | 5–190 | 0.5–3 | 0.0–8.0 | 100 |
| Low-alloy steel | 7800–7900 | 200 | 1000–2000 | 15–170 | 0.05–0.25 | 250 |
| Glasses, ceramics | 2200–4000 | 40–140 | 10–140 | 0.2–5 | 0.0 | 50 |
| Stainless steels | 7600–8000 | 200 | 300–1400 | 55–120 | 0.10–0.30 | 200 |
| Copper alloys | 8200–8900 | 100–117 | 300–1400 | 10–100 | 0.02–0.65 | 170 |
| Titanium alloys | 4400–4800 | 116 | 400–1600 | 50–140 | 0.05–0.30 | 350 |
| Mg alloys | 1740–1840 | 45 | 80–370 | — | 0.01–0.25 | 200 |
| Moldable glass fiber resin | 1700–2100 | 11–17 | 55–440 | 20–50 | 0.003–0.015 | 200 |
| Graphite–epoxy | 1500 | 200 | 1000 | 45–>120 | 0–0.02 | 650 |
| Tool steels: Hardness to 68 HRC; stable up to 600°C; strength to 2000 MPa | | | | 30–100 | | |
| Sintered carbides: Hardness at 20°C, 86 to 93 HRA (69 to well above 70 HRC equivalent); to 85 HRA max. at 760°C | | | | 3–21 | | |
| Superalloys: Strength is approximately 140 MPa at 650°C, 50 MPa at 750°C, 25 MPa at 900°C | | | | — | | |

**APPENDIX 6.**  PROPERTIES OF PURE METALS AT ~ 20°C

| Metal | Symbol | Structure | Lattice const., pm | | Atomic rad., pm | Density, kg/m³ | Melting point, | | Elastic modulus, GPa | Shear modulus, GPa | Symbol |
|---|---|---|---|---|---|---|---|---|---|---|---|
| | | | a | c | | | °C | °F | | | |
| Aluminum | Al | fcc | 405 | — | 143 | 2 700 | 660 | 1220 | 62 | 24 | Al |
| Chromium | Cr | bcc | 288 | — | 125 | 7 190 | 1875 | 3407 | 248 | 95 | Cr |
| Copper | Cu | fcc | 362 | — | 128 | 8 960 | 1083 | 1981 | 110 | 42 | Cu |
| Gold | Au | fcc | 408 | — | 144 | 19 300 | 1063 | 1945 | 80 | 31 | Au |
| Iron | Fe | bcc | 287 | — | 124 | 7 870 | 1538 | 2800 | 196 | 76 | Fe |
| Lead | Pb | fcc | 495 | — | 175 | 11 400 | 327 | 621 | 14 | 5 | Pb |
| Magnesium | Mg | hcp | 321 | 521 | 160 | 1 740 | 650 | 1202 | 44 | 17 | Mg |
| Molybdenum | Mo | bcc | 315 | — | 136 | 10 200 | 2610 | 4730 | 324 | 125 | Mo |
| Nickel | Ni | fcc | 352 | — | 125 | 8 900 | 1453 | 2647 | 207 | 80 | Ni |
| Platinum | Pt | fcc | 393 | — | 139 | 21 400 | 1769 | 3217 | 73 | 28 | Pt |
| Silver | Ag | fcc | 409 | — | 144 | 10 500 | 961 | 1761 | 76 | 29 | Ag |
| Tin | Sn | bct | 583 | 318 | — | 7 300 | 232 | 449 | 43 | 17 | Sn |
| Titanium | Ti | hcp | 295 | 468 | — | 4 510 | 1668 | 3035 | 116 | 45 | Ti |
| Tungsten | W | bcc | 316 | — | 137 | 19 300 | 3410 | 6170 | 345 | 133 | W |
| Vanadium | V | bcc | 304 | — | 132 | 6 100 | 1900 | 3450 | 131 | 50 | V |
| Zinc | Zn | hcp | 266 | 495 | 133 | 7 130 | 420 | 787 | — | — | Zn |

**APPENDIX 7.** COMPOSITIONS OF SOME STAINLESS STEELS*

| Type number | C | Mn Max. | P Max. | S Max. | Si Max. | Cr | Ni | Mo | UNS |
|---|---|---|---|---|---|---|---|---|---|
| 17-4 PH | 0.07 Max. | 1.00 | 0.040 | 0.030 | 1.00 | 15.50/ 17.50 | 3.00/ 5.00 | | S17400 |
| | | | | Nb 0.15/0.45, Cu 3.00/5.00 | | | | | |
| 17-7 PH | 0.09 Max. | 1.00 | 0.040 | 0.040 | 1.00 | 16.00/ 18.00 | 6.50/ 7.75 | | S17700 |
| | | | | Al 0.75/1.50 | | | | | |
| 301 | 0.15 Max. | 2.00 | 0.045 | 0.030 | 1.00 | 16.00/ 18.00 | 6.00/ 8.00 | | S30100 |
| 302 | 0.15 Max. | 2.00 | 0.045 | 0.030 | 1.00 | 17.00/ 19.00 | 8.00/ 10.00 | | S30200 |
| 303 | 0.15 Max. | 2.00 | 0.20 | 0.15 Min. | 1.00 | 17.00/ 19.00 | 8.00/ 10.00 | 0.60[a] Max. | S30300 |
| 304 | 0.08 Max. | 2.00 | 0.045 | 0.030 | 1.00 | 18.00/ 20.00 | 8.00/ 10.50 | | S30400 |
| 304L | 0.030 Max. | 2.00 | 0.045 | 0.030 | 1.00 | 18.00/ 20.00 | 8.00/ 12.00 | | S30403 |
| 305 | 0.08 Max. | 2.00 | 0.045 | 0.030 | 1.00 | 19.00/ 21.00 | 10.50/ 13.00 | | S30500 |
| 309 | 0.20 Max. | 2.00 | 0.045 | 0.030 | 1.00 | 22.00/ 24.00 | 12.00/ 15.00 | | S30900 |
| 309S | 0.08 Max. | 2.00 | 0.045 | 0.030 | 1.00 | 22.00/ 24.00 | 12.00/ 15.00 | | S30908 |
| 310 | 0.25 Max. | 2.00 | 0.045 | 0.030 | 1.50 | 24.00/ 26.00 | 19.00/ 22.00 | | S31000 |
| 310S | 0.08 Max. | 2.00 | 0.045 | 0.030 | 1.50 | 24.00/ 26.00 | 19.00/ 22.00 | | S31008 |
| 314 | 0.25 Max. | 2.00 | 0.045 | 0.030 | 1.50/ 3.00 | 23.00/ 32.00 | 19.00/ 22.00 | | S31400 |
| 316 | 0.08 Max. | 2.00 | 0.045 | 0.030 | 1.00 | 16.00/ 18.00 | 10.00/ 14.00 | 2.00/ 3.00 | S31600 |
| 316L | 0.030 Max. | 2.00 | 0.045 | 0.030 | 1.00 | 16.00/ 18.00 | 10.00/ 14.00 | 2.00/ 3.00 | S31603 |
| 317 | 0.08 Max. | 2.00 | 0.045 | 0.030 | 1.00 | 18.00/ 20.00 | 11.00/ 15.00 | 3.00/ 4.00 | S31700 |
| 321 | 0.08 Max. | 2.00 | 0.045 | 0.030 | 1.00 | 17.00/ 19.00 | 9.00/ 12.00 | | S32100 |
| | | | | Ti 5 × C Min. | | | | | |
| 347 | 0.08 Max. | 2.00 | 0.045 | 0.030 | 1.00 | 17.00/ 19.00 | 9.00/ 13.00 | | S34700 |
| | | | | Nb 10 × C Min. | | | | | |
| 348 | 0.08 Max. | 2.00 | 0.045 | 0.030 | 1.00 | 17.00/ 19.00 | 9.00/ 13.00 | | S34800 |
| | | | Nb 10 × C Min., Co 0.20 Max, Ta 0.10 Max. | | | | | | |

**APPENDIX 7.** (cont.)

| Type number | C | Mn Max. | P Max. | S Max. | Si Max. | Cr | Ni | Mo | UNS |
|---|---|---|---|---|---|---|---|---|---|
| AM-350 | 0.07/ 0.11 | 0.50/ 1.25 | 0.040 | 0.030 | 0.50 | 16.00/ 17.00 | 4.00/ 5.00 | 2.50/ 3.25 | S35000 |
| | | | | N 0.07/0.13 | | | | | |
| 403 | 0.15 Max. | 1.00 | 0.040 | 0.030 | 0.50 | 11.50/ 13.00 | | | S40300 |
| 405 | 0.08 Max. | 1.00 | 0.040 | 0.030 | 1.00 | 11.50/ 14.50 | | | S40500 |
| | | | | Al 0.10/0.30 | | | | | |
| 410 | 0.15 Max. | 1.00 | 0.040 | 0.030 | 1.00 | 11.50/ 13.50 | | | S41000 |
| 416 | 0.15 Max. | 1.25 | 0.060 | 0.15 Min. | 1.00 | 12.00/ 14.00 | | 0.60[a] Max. | S41600 |
| 416 Se | 0.15 Max. | 1.25 | 0.060 | 0.060 | 1.00 | 12.00/ 14.00 | | | S41623 |
| | | | | Se 0.15 Min. | | | | | |
| 420 | Over 0.15 | 1.00 | 0.040 | 0.030 | 1.00 | 12.00/ 14.00 | | | S42000 |
| 430 | 0.12 Max. | 1.00 | 0.040 | 0.030 | 1.00 | 16.00/ 18.00 | | | S43000 |
| 430F | 0.12 Max. | 1.25 | 0.060 | 0.15 Min. | 1.00 | 16.00/ 18.00 | | 0.60[a] Max. | S43020 |
| 430F Se | 0.12 Max. | 1.25 | 0.060 | 0.060 | 1.00 | 16.00/ 18.00 | | | S43023 |
| | | | | Se 0.15 Min. | | | | | |
| 431 | 0.20 Max. | 1.00 | 0.040 | 0.030 | 1.00 | 15.00/ 17.00 | 1.25/ 2.50 | | S43100 |
| 440A | 0.60/ 0.75 | 1.00 | 0.040 | 0.030 | 1.00 | 16.00/ 18.00 | | 0.75 Max. | S44002 |
| 440B | 0.75/ 0.95 | 1.00 | 0.040 | 0.030 | 1.00 | 16.00/ 18.00 | | 0.75 Max. | S44003 |
| 440C | 0.95/ 1.20 | 1.00 | 0.040 | 0.030 | 1.00 | 16.00/ 18.00 | | 0.75 Max. | S44004 |
| 442 | 0.20 Max. | 1.00 | 0.040 | 0.030 | 1.00 | 18.00/ 23.00 | | | S44200 |
| 446 | 0.20 Max. | 1.50 | 0.040 | 0.030 | 1.00 | 23.00/ 27.00 | | | S44600 |
| | | | | N 0.25 Max. | | | | | |

*Society of Automotive Engineers, Inc., *Unified Numbering System for Metals and Alloys*, 2nd ed., Warrendale, Pa: Society of Automotive Engineers, 1977. © Society of Automotive Engineers, Warrendale, Pa. Reprinted with permission.
[a]Optional.

**APPENDIX 8.**   REPRESENTATIVE MECHANICAL PROPERTIES OF POLYMERS

| Thermoplastics | $T_g$, °C | $E$, GPa | $\sigma_y$, MPa | Izod impact J/m of notch | Rockwell Hardness | Acronym or trade name |
|---|---|---|---|---|---|---|
| Acrylonitrile–butadiene–styrene | 80 | 2.5 | 50 | 500 | 70 HRR | ⎰ABS, |
| Glass-filled | | 7.0 | 110 | 200 | | ⎱Cyclolac |
| Polyamides | 60 | 2.8 | 70 | 53 | 119 HRR | ⎰Nylon 6, |
| 30% glass-filled | | 7.0 | 150 | 145 | 121 HRR | ⎱Durethan |
| Polycarbonate | 150 | 2.5 | 68 | 800 | 70 HRM | PC, Lexan |
| 30% glass-filled | | 6.0 | 90 | >1000 | 86 HRM | |
| Polyethylene | | | | | | |
| Low-density | −20 | 0.18 | 11 | — | 43 HRD | PE |
| High-density | | 0.56 | 30 | 300 | 65 HRD | |
| Polyethylene terephthalate | 67 | 8 | 135 | — | — | ⎧PETP, Dacron Terylene fibers; Melinex film |
| Polymethylmethacrylate | 100 | 2.8 | 70 | 35 | 99 HRM | PMMA, Perspex, Plexiglas |
| Poly-4-Methylpentene-1 | 55 | 1.4 | 28 | 350 | 70 HRL | TPX |
| Polyoxymethylene | −13 | 3.6 | 70 | 74 | 94 HRM | POM, Acetal Delrin, Hostaform |
| Polyphenyleneoxide | 180 | 2.4 | 62 | 210 | 90 HRR | PPO, Noryl |
| Glass-filled | | 8.4 | 120 | 60 | | |
| Polyphenylenesulphide | 150 | 3.3 | 75 | 20 | 123 HRR | PPS, Ryton |
| 40% glass-filled | | 7.8 | 150 | 20 | 123 HRR | |
| Polypropylene | 0 | 1.3 | 35 | 70 | 90 HRR | PP |
| Glass-filled | | 7.0 | 55 | 100 | 100 HRR | |
| Polystyrene | 100 | 2.7 | 50 | 200 | 70 HRM | PS |
| Polysulfone | 200 | 2.5 | 70 | 700 | 120 HRR | |
| 30% glass-filled | | 8.3 | 117 | 960 | 84 HRM | |
| Polytetrafluoroethylene | 120 | 0.4 | 25 | 160 | 52 HRD | PTFE, Teflon |
| Polyvinylchloride | 80 | 2.0 | 25 | 100 | 110 HRR | PVC |

| Thermosets | $T_g$, °C | $E$, GPa | $\sigma_y$, MPa |
|---|---|---|---|
| Alkyds (glass-filled) | (200) | 15 | 45 |
| Epoxides | (150–250) | 3 | 60 |
| Epoxides (glass-filled) | (150–250) | 21 | 140 |
| Melamines (glass-filled) | (200) | 17 | 50 |
| Phenolics (glass-filled) | (200–300) | 18 | 80 |
| Polybutadienes (glass-filled) | (200) | 11 | 65 |
| Polyesters (glass-filled) | (200) | 14 | 70 |
| Polyimides (glass-filled) | (350) | 21 | 190 |
| Silicones (glass-filled) | (300) | 8 | 40 |
| Ureas | (80) | 7 | 60 |
| Urethanes (solid) | (100) | 7 | 70 |

**APPENDIX 9.** TYPICAL VALUES OF QUASI-STATIC TOUGHNESS AND YIELD STRESS

| Material type | $E$, GPa | $\sigma_y$, MPa | $R$, kJ/m$^2$ | $K_c$, MPa$\sqrt{m}$ | $K_c/\sigma_y$, $\sqrt{m}$ |
|---|---|---|---|---|---|
| Glasses | 70 | 138 | 0.01 | 1 | 0.01 |
| Ceramics | 300 | 280 | 0.1 | 5 | 0.02 |
| PMMA | 2.8 | 70 | 0.5 | 1 | 0.02 |
| Cast iron | 100 | 300 | 1 | 10 | 0.03 |
| Boron–fiber epoxy composites | 170 | 2000 | 30 | 70 | 0.04 |
| Tough polymers (Polycarbonate) | 2.5 | 63 | 4 | 3 | 0.05 |
| High-strength aluminum-base alloys | 70 | 600 | 28 | 44 | 0.07 |
| High-strength steels (4340) | 210 | 1400 | 140 | 170 | 0.12 |
| High-strength titanium-base alloys | 110 | 1100 | 170 | 136 | 0.12 |
| Maraging steel | 210 | 960 | 150 | 180 | 0.20 |
| Low-carbon steels | 210 | 280 | 140 | 170 | 0.60 |

**APPENDIX 10.** TYPICAL COMPOSITIONS AND MECHANICAL PROPERTIES OF CAST IRONS

| Category | Class | Typical composition,% | | | Tensile strength, MPa | HB | Percentage elongation at fracture; 50 mm gage length |
|---|---|---|---|---|---|---|---|
| | | Carbon | Silicon | Manganese | | | |
| Gray | ASTM-20 large flakes | 3.20 | 2.30 | 0.65 | 140 | 155 | ~0 |
| | ASTM-60 small flakes | 2.65 | 1.35 | 0.65 | 430 | 230 | ~0 |
| Ductile (nodular) | 60-45-10 | 3.2–4.1 plus | 1.8–2.8 0.1% Mg + Ce | 0.80 | 420 | 150 | >10% |
| | 120-90-02 | | same range | | 840 | 260 | >2% |
| White | Cupola | 3.5 | 0.7 | 0.6 | 315 | 400 | ~0 |
| Malleable | 32510 (ferritic) | 2.5 | 1.2 | 0.4 | 385 | 140 | 18% |
| | 80002 (pearlitic) | 2.3 | 1.2 | 0.8 | 700 | 250 | 5% |

Young's modulus $E$ is about 170 GPa for nodular and malleable irons. For gray irons $E$ varies inversely with flake size (compliance of body); for example, for ASTM-20 class (250-$\mu$m flakes), $E = 80$ GPa, and for ASTM-60 (30-$\mu$m flakes), $E = 140$ GPa. See Example 15-3.

**APPENDIX 11.** REPRESENTATIVE ROOM-TEMPERATURE MECHANICAL PROPERTIES OF CERAMICS AND GLASSES

| | Melting point, °C | Tensile or bend strength ($\sigma_f$), MPa | Young's modulus ($E$), GPa | Fracture toughness ($R$), J/m² | Coefficient of thermal expansion ($\alpha$), °C$^{-1}$ ×10$^{-6}$ | Thermal conductivity ($k$), W/mK | Spalling index ($k\sigma_f/E\alpha$), kW/m | Toughness/strength ratio ($K_c/\sigma_f$) = $\sqrt{ER}/\sigma_f$, $\sqrt{m}$ |
|---|---|---|---|---|---|---|---|---|
| Typical porous insulating firebrick | ? | ~15 | 20 | ? 30 | 7 | 1 | 0.1 | 0.05 |
| Alumina (Al$_2$O$_3$) | 2050 | 300 (700 whisker) | 380 | 80 | 8 | 38 | 3.8 | 5.5 |
| Magnesia (MgO) | 2850 | 100 | 315 | 70 | 14.8 | 44 | 1 | 0.5 |
| Silicon carbide (SiC) | 2300 (decomposes) | 200 | 350 | 50 | 4.5 | 209 | 26 | 0.02 |
| Silicon nitride (Si$_3$N$_4$) | 1900 (sublimes) | 600 (1400 whisker) | 320 | 80 | 2.5 | 8 | 6 | 0.01 |
| Titanium carbide (TiC), Ni/Mo binder | 3250 | 200 | 350 | 400 | 7 | 17 | 1.4 | 0.6 |
| Tungsten carbide (WC), Co binder | 2620 | 350 | 700 | 1000 | 1 | 34 | 1.4 | 0.8 |
| Common bulk glass | Fuses at 1600 | 70 (up to 4000 filaments) | 70 | 20 | 10 | 1 | 0.1 | 0.2 |
| Pyroceram glass ceramic | Fuses at 1600 | 190 | 126 | ? 40 | 10 | 3 | 0.5 | 0.1 |
| Concrete, reinforced | ? | 20 | 35 | 30 | 7 | 0.5 | 0.1 | 0.1 |
| Carbon graphite | 3600 (sublimes) | 30 | 6 | 100 | 3 | ~100 | 170 | 0.3 |

NOTE: There is much more variation in properties of these types of solid than metals, variations occurring with production methods. Properties also vary with temperature and rate, for example, $k$ for Al$_2$O$_3$ is 4 W/mK at 1000°C and is 13 W/mK at 1000°C for SiC.

**APPENDIX 12.** DENSITY AND MECHANICAL PROPERTIES OF BULK
MATERIALS AND FILAMENTS

|                        | $\rho$, kg/m$^3$ | $S_u$, MPa | $S_u$/sp. gr., MPa | $E$, GPa | $E$/sp. gr., GPa |
|------------------------|------------------|------------|---------------------|----------|-------------------|
| **Bulk Material**      |                  |            |                     |          |                   |
| Low-carbon steel       | 7850             | 400        | 51                  | 210      | 27                |
| High-tensile steel     | 7850             | 2000       | 250                 | 210      | 27                |
| Aluminum alloy         | 2700             | 550        | 200                 | 74       | 27                |
| Magnesium alloy        | 1800             | 350        | 190                 | 45       | 25                |
| Nickel                 | 8860             | 490        | 55                  | 210      | 24                |
| Copper                 | 8940             | 240        | 27                  | 120      | 13                |
| Titanium alloy         | 4500             | 1000       | 220                 | 120      | 27                |
| Wood                   | 600              | 60         | 100                 | 14       | 23                |
| Glass                  | 2500             | 60         | 24                  | 70       | 28                |
| **Filaments**          |                  |            |                     |          |                   |
| Glass                  | 2500             | 1000       | 400                 | 70       | 28                |
| Iron                   | 7850             | 14000      | 1800                | 210      | 27                |
| Nickel                 | 8860             | 3800       | 430                 | 210      | 24                |
| Copper                 | 8940             | 2900       | 320                 | 120      | 13                |
| Alumina ($Al_2O_3$)    | 4000             | 18700      | 4700                | 390      | 98                |
| Silicon carbide        | 3200             | 24100      | 7500                | 560      | 175               |
| Boron                  | 2500             | 3450       | 1400                | 380      | 152               |
| Graphite               | 2250             | 21000      | 9300                | 750      | 334               |
| Carbon                 | 2250             | 1600       | 710                 | 360      | 160               |
| Kevlar 49              | 1440             | 2700       | 1900                | 130      | 90                |
| Nylon                  | 1100             | 400        | 360                 | 4        | 3.6               |
| Polypropylene          | 900              | 900        | 1000                | 8        | 9                 |

**APPENDIX 13.** DENSITY AND MECHANICAL PROPERTIES OF COMPOSITES

| Material                 | $\rho$, kg/m$^3$ | $S_u$, MPa | $S_u$/sp. gr., MPa | $E$, GPa | $E$/sp. gr., GPa |
|--------------------------|------------------|------------|---------------------|----------|-------------------|
| Boron–5505 epoxy         | 1990             | 1590       | 800                 | 207      | 104               |
| HTS graphite–5208 epoxy  | 1550             | 1480       | 950                 | 172      | 111               |
| Kevlar 49–resin          | 1380             | 1380       | 1000                | 76       | 55                |
| E glass–1002 epoxy       | 1800             | 1100       | 610                 | 39       | 22                |
| Boron–6061Al             | 2600             | 1490       | 570                 | 214      | 82                |
| T50 graphite–2011 Al     | 2580             | 566        | 220                 | 160      | 62                |

**APPENDIX 14.** ENERGY CONSUMPTION IN MANUFACTURE OF VARIOUS ENGINEERING SOLIDS

| | Energy/mass, GJ/1000 kg | Energy/mass-stiffness, $m^3$/1000 kg | Energy/mass-strength, $m^3$/1000 kg |
|---|---|---|---|
| Aluminum | 200 | 2.8 | 870 |
| Cast iron | 45 | 0.3 | 300 |
| Low-carbon steel | 50 | 0.24 | 240 |
| Glass | 20 | 0.3 | 333 |
| Brick | 6 | 0.1 | 30 |
| Concrete | 2 | 0.05 | 80 |
| Wood | 1 | 0.07 | 10 |
| Polyethylene | 45 | 90 | 1500 |
| Carbon-fiber-reinforced composite | 4000 | 27 | 6000 |

**APPENDIX 15.**   SOME COMMON KNOWN STRESS-CORROSION CRACKING
(AND HYDROGEN-EMBRITTLEMENT) ENVIRONMENT-MATERIAL COMBINATIONS

| | Low-C steel | High-strength steel | Austenitic stainless steel | Al + alloys | Cu + alloys | Ni + alloys | Ti + alloys |
|---|---|---|---|---|---|---|---|
| Ammonia | X | | | | X | | |
| Ammonium chloride | | | X | | | | |
| Ammonium hydroxide | | | | | X | | |
| Calcium chloride | | | X | | | | |
| Carbon dioxide or monoxide (wet) | X | X | | | | | |
| Carbon tetrachloride | X | X | X | | | | |
| Chlorides (wet) | | X | X | X | | | X |
| Chlorine | | X | X | | | | |
| Chromic acid | X | | | | X | X | |
| Ethanol | | | | | | | X |
| Ferric chloride | X | | | | X | | |
| Hydrogen | | X | X | | | | |
| Hydrogen sulfide (wet) | | | X | X | | | |
| Magnesium chloride | X | | X | | X | X | |
| Manganese chloride | | | X | | | | |
| Mercuric chloride | | | X | X | X | X | |
| Mercury | | | X | X | X | X | |
| Methanol | | | | | | | X |
| Nitric acid | X | | | | X | | |
| Oxygen | | | X | | X | | |
| Potassium chloride | | | X | | X | | |
| Potassium hydroxide | X | | X | | | X | |
| Sodium carbonate | | | X | | | | |

**APPENDIX 15.**  (cont.)

| | Low-C steel | High-strength steel | Austenitic stainless steel | Al + alloys | Cu + alloys | Ni + alloys | Ti + alloys |
|---|---|---|---|---|---|---|---|
| Sodium hydroxide | X | | X | | | X | |
| Steam | | | | X | | X | |
| Sulfides (wet) | | X | | | | | |
| Sulfur dioxide | | | | | X | | |
| Toluene | | | | | X | | |

NOTE: Not all alloys within a group are necessarily susceptible; see references for specific alloys. [ASM v. 10; *Corrosion Data Survey* (Metals Section), 5th ed. (Houston, Tex.: National Association of Corrosion Engineers). © National Association of Corrosion Engineers. Reprinted with permission.]

# Index